Pattern Recognition and Image Processing

Pattern Recognition and Image Processing

Editor:

Ivan Stanimirović, Ph.D.

ArclerPress

www.arclerpress.com

Pattern Recognition and Image Processing

Editor: Ivan Stanimirović, Ph.D.

708 3rd Avenue, 6th Floor
New York
NY 10017
United States of America
www.arclerpress.com

ISBN: 978-1-68094-447-1

Arcler Press LLC publishes wide variety of books and eBooks. For more information about Arcler Press and its products, visit our website at www.arclerpress.com

About the Editor

Ivan Stanimirović, Ph.D.

Ivan Stanimirović gained his PhD from University of Niš, Serbia in 2013. His work spans from multi-objective optimization methods to applications of generalized matrix inverses in areas such as image processing and computer graphics and visualisations. He is currently working as an Assistant professor at Faculty of Sciences and Mathematics at University of Niš on computing generalized matrix inverses and its applications.

Contents

List of Contributors

Christian Hennig
Department of Statistical Science, University College London, Gower Street, London WC1E 6BT, United Kingdom

Gustaf Kylberg
Vironova AB, Gävlegatan 22, SE-11330 Stockholm, Sweden

Ida-Maria Sintorn
Vironova AB, Gävlegatan 22, SE-11330 Stockholm, Sweden
Uppsala University, Department of Information Technology, Lägerhyddsvägen 20, SE-75105 Uppsala, Sweden

Sebastian Hegenbart
Department of Computer Sciences, University of Salzburg, Jakob-Haringer Strasse 2, 5020 Salzburg, Austria

Andreas Uhl
Department of Computer Sciences, University of Salzburg, Jakob-Haringer Strasse 2, 5020 Salzburg, Austria

Gang Liu
College of Applied Mathematics, Chengdu University of Information Technology

Ting-Zhu
College of Applied Mathematics, Chengdu University of Information Technology
Research Center for Image and Vision Computing/School of Mathematical Sciences, University of Electronic Science and Technology of China

Huang
Research Center for Image and Vision Computing/School of Mathematical Sciences, University of Electronic Science and Technology of China

Xiao-Guang
School of Science, Huaihai Institute of Technology

Lv and Jun Liu
Research Center for Image and Vision Computing/School of Mathematical Sciences, University of Electronic Science and Technology of China

David Pickup
School of Computer Science and Informatics, Cardiff University, Queen's Buildings, 5 The Parade, Roath, Cardiff CF24 3AA, United Kingdom

Xianfang Sun
School of Computer Science and Informatics, Cardiff University, Queen's Buildings, 5 The Parade, Roath, Cardiff CF24 3AA, United Kingdom

Paul L. Rosin
School of Computer Science and Informatics, Cardiff University, Queen's Buildings, 5 The Parade, Roath, Cardiff CF24 3AA, United Kingdom

Ralph R. Martin
School of Computer Science and Informatics, Cardiff University, Queen's Buildings, 5 The Parade, Roath, Cardiff CF24 3AA, United Kingdom

Jose Rivera-Rubio
Imperial College London, South Kensington Campus, London SW7 2AZ, United Kingdom

Ioannis Alexiou
Imperial College London, South Kensington Campus, London SW7 2AZ, United Kingdom

Anil A. Bharath
Imperial College London, South Kensington Campus, London SW7 2AZ, United Kingdom

Siyamalan Manivannan
CVIP, School of Computing, University of Dundee, UK

Wenqi Li
CVIP, School of Computing, University of Dundee, UK

Shazia Akbar
CVIP, School of Computing, University of Dundee, UK

Ruixuan Wang
CVIP, School of Computing, University of Dundee, UK

Jianguo Zhang
CVIP, School of Computing, University of Dundee, UK

Stephen J. McKenna
CVIP, School of Computing, University of Dundee, UK

Michael J.V. Leach
Heriot-Watt University, Edinburgh Campus, Edinburgh, Scotland, United Kingdom

Ed.P. Sparks
Heriot-Watt University, Edinburgh Campus, Edinburgh, Scotland, United Kingdom

Neil M. Robertson
Heriot-Watt University, Edinburgh Campus, Edinburgh, Scotland, United Kingdom

Heinz Hofbauer
University of Salzburg, Department of Computer Sciences, Austria

Andreas Uhl
University of Salzburg, Department of Computer Sciences, Austria

Zhi Jin
Department of Electrical and Electronic Engineering, Xi'an Jiaotong-Liverpool University, Ren Ai Road 111, 215123 Suzhou, People's Republic of China
Department of Electrical Engineering and Electronics, University of Liverpool, Merseyside

Tammam Tillo
Department of Electrical and Electronic Engineering, Xi'an Jiaotong-Liverpool University, Ren Ai Road 111, 215123 Suzhou, People's Republic of China

Jimin Xiao
Department of Electrical and Electronic Engineering, Xi'an Jiaotong-Liverpool University, Ren Ai Road 111, 215123 Suzhou, People's Republic of China

Yao Zhao
BX, Liverpool, UK

Peter Wild
University of Reading, School of Systems Engineering, Whiteknights, Reading RG6 6AY, United Kingdom

Petru Radu
University of Reading, School of Systems Engineering, Whiteknights, Reading RG6 6AY, United Kingdom

Lulu Chen
University of Reading, School of Systems Engineering, Whiteknights, Reading RG6 6AY, United Kingdom

James Ferryman
University of Reading, School of Systems Engineering, Whiteknights, Reading RG6 6AY, United Kingdom

Ying-Nong Chen
Department of Computer Science and Information Engineering, National Central University

Chi-Hung Chuang
Department of Applied Informatics, Fo Guang University

Hsin-Min Lee
Department of Computer Science and Information Engineering, National Central University

Chih-Chang Yu
Digital Education Institute, Institute for Information Industry

Kuo-Chin Fan
Department of Computer Science and Information Engineering, National Central University

Mariusz Oszust
Department of Computer and Control EngineeringRzeszow University of Technology

Krzysztof Bartyzel
Department of Computer Image AnalysisJohn Paul II Catholic University of Lublin

Bogdan Smolka
Faculty of Automatic Control, Electronics and Computer ScienceSilesian University of Technology

Damian Kusnik
Faculty of Automatic Control, Electronics and Computer ScienceSilesian University of Technology

A. Tonazzini
Istituto di Scienza e Tecnologie dell'Informazione-CNR, Via G. Moruzzi, 1, 56124 Pisa, Italy

P. Savino
Istituto di Scienza e Tecnologie dell'Informazione-CNR, Via G. Moruzzi, 1, 56124 Pisa, Italy
e-mail: pasquale.savino@isti.cnr.it

E. Salerno
Istituto di Scienza e Tecnologie dell'Informazione-CNR, Via G. Moruzzi, 1, 56124 Pisa, Italy
e-mail: emanuele.salerno@isti.cnr.it

Preface

Pattern recognition and image processing are rapidly growing technologies within engineering and computer science. Algorithms for pattern recognition aim to observe the environment, learn to distinguish patterns of interest from their background and make pattern classes. A series of methods are developed to group individual patterns into specific classes according to their common properties, resulting in different pattern classes. Because of emerging applications which are computationally demanding, a wide specter of algorithms is developed to deal with problems such as data mining, classification of multimedia data, and obtaining biometric features like face and fingerprint recognition. Contrast enhancement, labeling of connected components, segmentation and feature detection methods are created based on the existing signal processing algorithms.

Clustering is a widely used concept in pattern recognition and image processing. It becomes scientific not through uniqueness but through transparent and open communication. Various desirable characteristics of clusterings and various approaches to define a context-dependent truth are listed, and it is discussed what impact these ideas can have on the comparison and the choice of clustering methods in practical applications. This subject is discussed in detail in the first section of this book. The following six chapters of the book present methods for solving problems of pattern and texture recognition. The remaining content of this book focuses on the advances of specific methods and algorithms in the field of image processing. The specific advances discussed in this book include the use of dynamic detection and rejection of liveness-recognition, a system to recognize patterns at cellular and specimen levels, specific scale-adapted features and various techniques to encode visual structure. Advances in solving TV deblurring and denoising problems and the suppression of mixed Gaussian and impulsive noise in color images are also investigated.

Dynamic detection and rejection of liveness-recognition pair outliers for spoofed samples in true multi-modal configuration with its inherent challenge of normalization is investigated. Bootstrap aggregating (bagging) classifiers for fingerprint spoof-detection algorithm are presented. Experiments on the latest face video databases and fingerprint spoofing database illustrate the efficiency of proposed techniques.

A system to recognize patterns at cellular and specimen levels, in images of HEp-2 cells is developed. Ensembles of SVMs were trained to classify cells into six classes based on sparse encoding of texture features with cell pyramids, capturing spatial, multi-scale structure. A similar approach was used to classify specimens into seven classes. Detailed descriptions and extensive experiments with various features and encoding methods are provided.

Specific scale-adapted features are computed in reference to the estimated scale of an image, based on the distribution of scale normalized Laplacian responses in a scale-space representation. Intrinsic-scale-adaption is performed to compute features, independent of the intrinsic texture scale, leading to a significantly increased discriminative power for a large amount of texture classes. In a final step, the rotation- and scale-invariant features are combined in a multi-resolution representation, which improves the classification accuracy in texture classification scenarios with scaling and rotation significantly.

A linear time complexity method for computing a canonical form is suggested. This method is using Euclidean distances between pairs of a small subset of vertices. This approach has comparable retrieval accuracy but lower time complexity than using global geodesic distances, allowing it to be used on higher resolution meshes, or for more meshes to be considered within a time budget.

Vision is one of the most important of the senses, and humans use it extensively during navigation. Different types of image and video frame descriptors that could be used to determine distinctive visual landmarks for localizing a person based on what is seen by a camera that they carry are computed. For each type of descriptor, different techniques to encode visual structure and to search between journeys to estimate a user's position are also tested. The techniques include single-frame descriptors, those using sequences of frames, and both color and achromatic descriptors. The results suggest that appearance-based information could be an additional source of navigational data indoors. This offers a complementary approach to methods based on simultaneous localization and mapping (SLAM) algorithms.

A method of utilising social context and scene context to improve behaviour analysis is developed. It is found that in a crowded scene the application of Mutual Information based social context permits the ability to prevent self-justifying groups and propagate anomalies in a social network, granting a greater anomaly detection capability. Scene context uniformly improves the detection of anomalies in both datasets.

Visual security metrics are deterministic measures with the (claimed) ability to assess whether an encryption method for visual data does achieve its defined goal. These metrics are usually developed together with a particular encryption method in order to provide an evaluation of said method based on its visual output. Therefore, a methodology for assessing the performance of security metrics based on common media encryption scenarios is considered.

New inexact explicit shrinkage formulas for one class of these relevant problems are introduced, whose regularization terms have translation invariant overlapping groups. These results are applied to TV deblurring and denoising problems with overlapping group sparsity. Thus, alternating direction method of multipliers to iteratively solve them is used. It is shown that the choice of interpolation method when rotating textures greatly influences the recognition capability. Lanczos 3 and B-spline interpolation are comparable to rotating the textures prior to image acquisition, whereas the recognition capability is significantly and increasingly lower for the frequently used third order cubic, linear and nearest neighbour interpolation. It is also shown that including generated rotations of the texture samples in the training data improves the classification accuracies. For many of the descriptors, this strategy compensates for the shortcomings of the poorer interpolation methods to such a degree that the choice of interpolation method only has a minor impact.

To enable an appropriate and fair comparison, a new texture dataset is introduced which contains hardware and interpolated rotations of 25 texture classes. An optimisation approach to the design of a binary descriptor is proposed, in which the detected keypoint is described using several, scale-dependent patches. Each such patch is divided into disjoint blocks of pixels, and then, binary tests between blocks' intensities, as well as their gradients, are used to obtain the binary string. Since the number of image patches and their relative sizes influence the descriptor creation pipeline, a simulated annealing algorithm is used to determine them, optimising recall and precision of keypoint matching. The simulated annealing is also used for dimensionality reduction in long binary strings.

A new filter was created by improving the standard Kuwahara filter. It allows more efficient noise reduction without blurring the edges and image preparation for segmentation and further analyses operations. One of the biggest and most common restrictions encountered in filter algorithms is the need for a declarative definition of the filter window size or the number of iterations that an operation should be repeated.

A novel technique designed for the suppression of mixed Gaussian and impulsive noise in color images is proposed. The new denoising scheme is based on a weighted averaging of pixels contained in a filtering block. The main novelty of the proposed solution lies in the new definition of the similarity between the samples of the processing block and a small window centered at the block's central pixel. Instead of direct comparison of pixels, a measure based on the similarity between a given pixel and the samples from the neighborhood of the central pixel is utilized. This measure is defined as the sum of distances in a given color space, between a pixel of the block and a certain number of most similar samples from the filtering window. The main advantage of the proposed scheme is that the new similarity measure is not influenced by the outliers injected into the image by the impulsive noise and the averaging process ensures the effectiveness of the new filter in the reduction of Gaussian noise.

Editor
Ivan Stanimirović, Ph.D.

CHAPTER 1

WHAT ARE THE TRUE CLUSTERS?

Christian Hennig

Department of Statistical Science, University College London, Gower Street, London WC1E 6BT, United Kingdom

ABSTRACT

Constructivist philosophy and Hasok Chang's active scientific realism are used to argue that the idea of "truth" in cluster analysis depends on the context and the clustering aims. Different characteristics of clusterings are required in different situations. Researchers should be explicit about on what requirements and what idea of "true clusters" their research is based, because clustering becomes scientific not through uniqueness but through transparent and open communication. The idea of "natural kinds" is a human construct, but it highlights the human experience that the reality outside the observer's control seems to make certain distinctions between categories inevitable. Various desirable characteristics of clusterings and various approaches to define a context-dependent truth are listed, and I discuss what impact these ideas can have on the comparison of clustering methods, and the choice of a clustering methods and related decisions in practice.

INTRODUCTION

Cluster analysis is about finding groups in a set of objects. Cluster analysis is used in many different areas with many different aims (see Section 3 for examples). Researchers who apply cluster analysis in practice often want to know whether the clusters that they find are truly meaningful in the sense that they correspond to a real underlying grouping. Researchers in the field of cluster analysis are interested in whether and which methods are better at finding the true clusters correctly. In most cluster analysis literature, however, explanations of what "true" or "real" clusters are, are rather hand-waving. It is widely acknowledged that there is no agreed definition of what a cluster is, and in the majority of papers in which new cluster analysis methods are proposed, the authors do not give a general and formal definition of what the "true clusters" are that their method is supposed to find.

The aim of this paper is to offer a philosophically informed attitude toward the problem of choosing, assessing and interpreting cluster analysis methods and clusterings. Section 2gives an overview of thoughts in philosophy and cognitive science regarding clustering and categorization. Afterward the paper turns to considerations and implications that are directly related to the theory and practice of data-based cluster analysis.

The groups that cluster analysis sets out to find are characterized by data that can take various forms such as values of variables, dissimilarities or weighted edges in a graph. The groups may form a partition of the object set, but they may also be overlapping or non-exhaustive. Group memberships

Citation: Christian Hennig, What are the true clusters?, Pattern Recognition Letters, Volume 64, 15 October 2015, Pages 53-62, ISSN 0167-8655, http://dx.doi.org/10.1016/j.patrec.2015.04.009.

may be crisp or fuzzy. Some of the discussion here was written with crisp partitions in mind, some apply to Euclidean space or a given dissimilarity measure, but most thoughts are more general.

There is a good reason why there is no generally accepted unique definition of true clusters. In different applications, cluster analysis is used with different aims, and the researchers have different ideas of what should make the objects belong together that are in the same cluster. The term "cluster" does not mean the same to all researchers in all situations. This is acknowledged in general overviews and books about cluster analysis, but seems to be ignored by many authors of specialist work who try to convince readers that a certain method is best for finding the "true/natural/real" clusters. Even where it is acknowledged, this often takes the form of a "general health warning", and consequences regarding the selection and comparison of methods and the interpretation of results are rarely spelled out. Is it possible to escape the alternative to either make the hardly justifiable assumption that there is a unique "true/natural/real" clustering against which the quality of cluster analysis methods can be objectively assessed, or to think that cluster analysis is somehow arbitrary and "more of an art than a science" [1]?

My perspective is that of a statistician with expertise in cluster analysis and a strong interest in the philosophical background of statistics and data analysis. A key idea of this paper is that, given that it depends on the context and clustering aim what a "good" clustering is, researchers need to characterize what kind of clusters are required for a given real clustering problem, and what kind of clusters the different clustering methods are good at finding, or in other words, what problem-specific "truth" researchers are interested in. Similar ideas have recently been discussed in [2] and [1]. The present paper can be seen as contributing to the research program sketched in those papers, but also as enrichening their perspective by adding further philosophical and statistical considerations.

In Section 2 I will sketch the philosophical basis of the present paper, which complements constructivism with Hasok Chang's pluralist active scientific realism, and I will discuss the concepts of "natural kinds" and "categorization". Section 3 lists and discusses various context-dependent clustering aims. Section 4 is about how "true" clusters could be defined in statistical or data analytic terms so that they can be used for comparing and assessing different clustering methods. Section 5 discusses some practical consequences, particularly regarding choice and comparison of cluster analysis methods, and rationales for certain methodological decisions such as dimension reduction.

PHILOSOPHICAL BACKGROUND

Constructivism and Science

In the present paper I focus on the question what clusters are "true" and/or "real". Truth and reality, and to what extent they can be observed, are controversial issues in philosophy. My starting point in this respect is my constructivist philosophy of mathematical modeling as outlined in [3], which is connected to radical constructivism [4]and social constructionism [5]. Radical constructivism is based on the idea that the perception and world-view of human beings can be interpreted as a construction by the body and the brain of the individual, which is seen as a self-organizing system. Social constructionism focuses on the construction of a common world-view of social systems by means of communication. "Construction" refers to the activity of the body, the brain, and communicative activity within social systems, setting up perceptions and world-views. Construction is largely unconscious or semi-conscious, and is not arbitrary but subject to constraints. It is not claimed that individuals or social systems are free to construct any arbitrary perception or world-view. Experience tells us that perception is rather severely constrained and shaped by what we perceive to be a reality outside of ourselves.

I distinguish observer-independent reality, personal reality and social reality. The observer-independent reality is only accessible to humans by observation, which means that there is no way to make sure which of its features are really observer-independent, but it is usually perceived as the source of constraints for personal and social constructs. The perceptions of individuals, together with their thoughts and feelings, make up their personal reality. Part of most personal and social realities is the belief that much personal perception represents or reflects the observer-independent reality. This belief is normally based on the experience of consistency between different sensory perceptions, at different times and from different positions, and on the confirmation of the existence of many of the perceived items by communication with others. It is therefore the result of active accommodation of perceptions.

Social reality is made up by communication between individuals. It is carried by social systems, which may overlap and may partly lack clear borderlines, although some social systems such as formal mathematics are rather clearly delimited. Personal and social realities influence each other. According to the point of view taken here, science is a social attempt to construct a consensual and stable view of the world, which can be shared by everyone and is open to criticism and scrutiny in free exchange. In this sense, science aims at a view that is as independent as possible of the individual observer, and is therefore connected to a traditional realist view, according to which science aims at finding out the truth about observer-independent reality. But constructivists are pessimistic regarding an observer-independent access to reality, and assess the success of science based on stability, agreement and pragmatic use instead of referring to objective truth. A scientific world-view with which constructivists can agree needs to acknowledge the existence and legitimacy of diverse personal and social realities and is therefore inherently pluralist. A tension between a drive for unification and general agreement and a necessity to allow space for diverse realities in order to allow for criticism and creative progress is an essential implication of the scientific idea. Central tools of science are mathematics, which aims at setting up and exploring concepts that are clear and well defined independently of the different personal and social points of view and at statements about which absolute agreement is possible, and measurement, which unifies observations of reality in a way that they can be processed by mathematical means.

Constructivism is often accused of denying the existence of the observer-independent reality altogether by calling it "a construct", but actually, being as stable and ubiquitous a construct as the observer-independent reality seems to be in most personal and social realities, it is as real as anything can get in constructivism.

Active Scientific Realism

Although constructivism is often interpreted as anti-realist, I complement my constructivist view here by the "active scientific realism" introduced by Hasok Chang [6]. In the abstract of his Chapter 4, Chang writes: *"I take reality as whatever is not subject to ones will, and knowledge as an ability to act without being frustrated by resistance from reality. This perspective allows an optimistic rendition of the pessimistic induction, which celebrates the fact that we can be successful in science without even knowing the truth. The standard realist argument from success to truth is shown to be ill-defined and flawed. I also reconsider what it means for science to be "mature", and identify humility rather than hubris as the proper basis of maturity. The active realist ideal is not truth or certainty, but a continual and pluralistic pursuit of knowledge."* Chang's use of the term "reality" refers to what is vital for the success of the scientific idea, namely to confront scientific work continually with the observed realities that individuals and social systems experience as outside their control. In agreement with my constructivist view, active scientific realism values a plurality of perspectives. The term "truth" is constructivist used in both Chang [6] and the constructivist literature as a relative concept "internal to systems of practice". For example, within the mathematical formal system, "truth" is a rather unproblematic concept due to the clear rules by which it can be ensured, whereas the truth-value of the statement "the German Democratic Republic was a democracy" depends on which characteristics of a political system are taken as essential for being a democracy, which differs between social systems.

The emphasis of the strong role of communication and language is an aspect that constructivism adds to active scientific realism. In this respect I follow Fleck [7], a pioneer work regarding the role of communication and social systems ("thought collectives") for scientific knowledge. Fleck showed how scientific facts are shaped by the specific way how collectives of scientists conceptualize their field.

Natural Kinds

"Natural kinds" in philosophy refer to the idea that there are some "naturally" separated classes in observer-independent reality, which, for traditional realists, correspond to "true clusters". For example, biological species and chemical elements are considered as candidates for being natural kinds [8]. There is much controversy about what constitutes natural kinds (e.g., common properties, behaving homogeneously according to natural laws). The concept runs counter to the constructivist view that what is perceived as "kinds" is constructed by human activity and language and depends on the conditions of observation and practice of living of the observers. For such reasons, for example Goodman [9] rejected the term "natural" for kinds. Hacking [10] argued that "natural kinds" should refer to kinds that are connected to human activity and utility, which allows for non-uniform and more pluralist kinds. According to him, the concept links a nominalist inclination with a traditional realist view of "nature". He also suggested that many classes that can be seen as natural in some sense are not "natural kinds", and that this term may be reserved for a few very special kinds.

I agree with Goodman that the term "natural" is not helpful, at least if it is used in order to suggest that some categorizations have a special authority by matching observer-independent reality. What is valuable about the concept of "natural kinds" is that it describes a human experience that certain categorizations seem impossible to escape when confronted with Chang's "reality outside our control". Such an experience always has to be framed by the make-up of the personal and social realities that are involved, it may change, and controversy persists even about central candidates for natural kinds such as biological species [11] and chemical elements [6]. Still, it highlights that when following an active scientific realist agenda, phenomena should not be lumped arbitrarily into classes, but that scientific observation should be used to guide classification in a stable way that should aim at general agreement; by which I mean agreement about the legitimacy and use of the classification as opposed to its uniqueness.

Categorization

From the constructivist point of view, although we experience "reality outside our control", the categorization of its phenomena is a constructive human activity, and any idea of "true" or "really meaningful" categories is located in personal and social reality. In order to understand such an idea it therefore seems promising to look at work in cognitive science about human categorization. Van Mechelen et al. [12] review cognitive theories of categorization with a view to connecting them to inductive data analysis including clustering. Although no explicitly pluralist position is taken in that book, the various presented theories seem to apply to different kinds of categories used by human beings in different circumstances. Many of these theories correspond to formal approaches to cluster analysis, for example that categorization can be based on defining features, prototypes and exemplars, or family resemblance (similarity). From a constructivist perspective, von Foerster [13] saw "objects" in human perception as eigenvalues (fixed points) of recursive coordinations of actions, which has a reflection in self-organizing clustering algorithms. Because of the exchange between cognitive science and artificial intelligence research, this should not be surprising. However, formal and algorithmic views of categories have strong limitations, and it has been pointed out that in order to understand human categorization, context such as the conditions of the human body, a metaphorical or theoretical framework in which a category is embedded [14], Chapter 7 of [12] and the ever-changing social and communicative environment [5] need to be taken into account.

Another line of research concerns intuitive clustering by humans of two dimensional point clouds, regardless of the meaning of the points, see [15] and [16], with mixed results in the sense that there are predominant strategies such as looking for high density areas and for shapes of similar kinds ("model fitting"), but there is also considerable variation, although Lewis et al. [17] argue that humans and particularly experts are more consistent in assessing clusterings than existing cluster validation indexes.

Overall, categorization seems to work in rather pluralist and context-dependent ways, as is also acknowledged in more recent publications on categorization [18] and [19]. It may be controversial to what extent cluster analysis methods are meant to reflect human categorization. One could argue that "true clusters" should have a more scientific and well-defined character than the concepts that humans normally use. Furthermore, clustering often aims at finding categories that are thought of as determined by unobserved features, which differs from forming categories from what is observed. The theories discussed in this section are relevant in artificial intelligence applications where the aim is to simulate human categorization, and they can also inspire methodological ideas in clustering, but their potential to define "true clusters" as targets for data analysis is limited.

CLUSTERING AIMS AND CLUSTER CONCEPTS

A list of Aims of Clustering

That there is no generally accepted definition of a cluster is not surprising, given the many different aims for which clusterings are used. Here are some examples:

- delimitation of species of plants or animals in biology,
- medical classification of diseases,
- discovery and segmentation of settlements and periods in archeology,
- image segmentation and object recognition,
- social stratification,
- market segmentation,efficient organization of data bases for search queries.
- There are also quite general tasks for which clustering is applied in many subject areas:
- exploratory data analysis looking for "interesting patterns" without prescribing any specific interpretation, potentially creating new research questions and hypotheses,
- information reduction and structuring of sets of entities from any subject area for simplification, effective communication, or effective access/action such as complexity reduction for further data analysis, or classification systems,
- investigating the correspondence of a clustering in specific data with other groupings or characteristics, either hypothesized or derived from other data.
- Depending on the application, it may differ a lot what is meant by a "cluster", and cluster definition and methodology have to be adapted to the specific aim of clustering in the application of interest.

Realist and Constructive Aims of Clustering

A key distinction can be made between "realist" and "constructive" aims of clustering. Realist aims concern the discovery of some meaningful real structure (referring to what is experienced as "reality outside our control", see Section 2). Constructive aims refer to the researchers' intention to split up the data into clusters for pragmatic reasons, regardless of whether there is some essential real difference between the resulting groups. The connection between "realist" and "constructive" clustering aims and realist and constructivist philosophy is not straightforward. Nothing stops a realist from being interested in data compression and from therefore having a constructive clustering aim. On the other hand, a constructivist can legitimately be interested in realist clustering aims, although she would maintain that the idea of clusters that are real and meaningful in the observer-independent reality is a personal and social construct.

The distinction between realist and constructive clustering aims is not clear cut. As follows from Section 2, researchers with realist clustering aims should not hope that the data alone reveal real structure; constructive impact of the researchers is needed to decide what counts as real.

The key issue in realist clustering is how the real structure the researchers are interested in is connected to the available data. This requires subject matter knowledge and decisions by the researchers. "Real structure" is often understood as the existence of an unobserved categorical variable, the values of which define the "true" clusters. But neither can it be taken for granted that the categories of such a variable are the only existing ones that could qualify as "real clusters", nor do such categories necessarily correspond to data analytic clusters. For example, male/female is a meaningful categorization of human beings, but there may not be a significant difference between men and women regarding the results of a certain attitude survey, let alone separated clusters corresponding to sex. Usually the objects represented in a dataset can be partitioned into real categories in many ways. Also, different cluster analysis methods will produce different clusterings, which may correspond to patterns seen as "real" in potentially different ways. This means that in order to decide about appropriate cluster analysis methodology, researchers need to think about what data analytic characteristics the clusters they are aiming at are supposed to have. I call this the "cluster concept" of interest in a study.

The real patterns of interest may be more or less closely connected to the available data. For example, in biological species delimitation, the concept of a species is often defined in terms of interbreeding (there is some controversy, see [11]). But interbreeding patterns are not usually available as data. Species are nowadays usually delimited by use of genetic data, but in the past, and occasionally in the present in exploratory analyses, species were seen as the source of a grouping in phenotype data. In any case, the researchers need an idea about how true distinctions between species are connected to patterns in the data. Regarding genetic data, knowledge needs to be used about what kind of similarity arises from persistent genetic exchange inside a species, and what kind of separation arises between distinct species. There may be subgroups of individuals in a species between which there is little actual interbreeding (potential interbreeding suffices for forming a species), e.g., geographically separated groups, and consequently not as much genetic similarity as one would naively expect. Furthermore there are various levels of classification in biology, such as families and genii above and subspecies below the level of species, so that data analytic clusters may be found at several levels, and the researchers may need to specify more precisely how much similarity within and separation between clusters is required for species.

Such knowledge needs to be reflected in choice of the cluster analysis method. E.g., species may be very heterogeneous regarding geographical distribution and size, and therefore a clustering method that penalizes large within-cluster distances too heavily such as *k*-means or complete linkage is inappropriate.

In some cases, the data are more directly connected to the cluster definition. In species delimitation, there may be interbreeding data, in which case researchers can specify the requirements of a clustering more directly. This may imply graph theoretic clustering methods and a specification of how much connectedness is required within clusters, although such decisions can often not be made precise because of missing information arising from sampling of individuals, missing data, etc. On the other hand, the connection between the cluster definition and the data may be less close, as in the case of phenotype data used for delimiting species, in which case some speculation is needed in order to decide what kind of clustering method may produce something useful.

In many situations different groupings can be interpreted as real, depending on the focus of the researchers. E.g., social classes can be defined in various ways. Marx made ownership of means of production the major defining characteristic of different classes, but social classes can also be defined by looking at patterns of contact, or occupation, or education, or wealth, or by a mixture of these [20]. In this case, a major issue for data clustering is the selection of the appropriate variables and measurements, which implicitly defines what kinds of social classes can be found.

The example of social stratification illustrates that there is a gradual transition rather than a clear cut between realist and constructive clustering aims. According to some views (such as the Marxist one) social classes are an essential and real characteristic of society, but according to other views, in many societies there is no clear delimitation between supposedly "real" social classes, despite the existence of real inequality. Social classes can then still be used as a convenient tool for structuring the inequality.

Regarding constructive clustering aims, it is obvious that researchers need to decide about the desired "cluster concept", i.e., about the characteristics that their clusters should have. This needs to be connected to the practical use that is intended to be made of the clusters.

Where the primary clustering aim is constructive, realist clustering may still be of interest. If indeed some real grouping structure is manifest in the data, many constructive aims will be served well by having this structure reflected in the clustering. E.g., market segmentation may be useful regardless of whether there are really meaningfully separated groups in the data, but it is relevant to find them if they exist.

Desirable Characteristics of Clusters

Here is a list of potential characteristics of clusters that may be desired, and that can be checked using the available data. Several of these are related with the "formal categorization principles" listed in Section 14.2.2.1 of [12].

- Within-cluster dissimilarities should be small.
- Between-cluster dissimilarities should be large.
- Clusters should be fitted well by certain homogeneous probability models such as the Gaussian or a uniform distribution on a convex set, or by linear, time series or spatial process models.
- Members of a cluster should be well represented by its centroid.
- The dissimilarity matrix of the data should be well represented by the clustering (i.e., by the ultrametric induced by a dendrogram, or by defining a binary metric "in same cluster/in different clusters").
- Clusters should be stable.
- Clusters should correspond to connected areas in data space with high density.
- The areas in data space corresponding to clusters should have certain characteristics (such as being convex or linear).
- It should be possible to characterize the clusters using a small number of variables.
- Clusters should correspond well to an externally given partition or values of one or more variables that were not used for computing the clustering.
- Features should be approximately independent within clusters.
- All clusters should have roughly the same size.
- The number of clusters should be low.

When trying to measure these characteristics, they have to be made more precise, and in some cases it matters a lot how exactly they are defined. Take no. 1, for example. This may mean that all within-cluster dissimilarities should be small (i.e., their maximum, as required by complete linkage clustering), or their average, or a high quantile of them. These requirements may look similar at first sight but are very different, e.g., regarding the integration of outliers in clusters. Having large between-cluster dissimilarities may emphasize gaps by looking at the smallest dissimilarities between two clusters, or it may rather mean that the cluster centroids are well distributed in data space. As another example, stability can refer to sampling other data from the same population (this may play a privileged role in hypothesis driven repeated experiments aiming at reproducible results, which is often identified with the scientific method; see [21] for some results and critical remarks), to adding "noise", or to comparing results from different clustering algorithms.

Some of these characteristics conflict with others in some datasets. E.g., connected areas with high density may include very large distances, and may have shapes that are undesired in specific applications (e.g., non-convex). Representing objects by centroids well may require some clusters with little or no gap between them. Stability is often easier to achieve with few clusters; but more clusters may be required in situations where clusters need to be very homogeneous.

Deciding about such characteristics is the key to linking the clustering aim to an appropriate clustering method. E.g., if a database of images should be clustered so that users can be shown a single image to represent a cluster, centroid representation is most important. Useful market segments need to be addressed by non-statisticians and should therefore normally be represented by few variables, on which dissimilarities between members should be low. Section 5 outlines how the listed characteristics can help with the selection of a clustering method in practice.

The idea of listing potentially desirable characteristics of clusterings for helping with the selection of clustering methods is central also to [2], but the axiomatic characteristics listed there are strikingly different from the present list. As necessary for the theoretical analysis, the characteristics in [2] are formal. One reason for the differences may be that the aim of the authors was to prove general theorems, and therefore they went for characteristics that make such theorems possible. Ackerman et al. [22] and [23]investigated cluster analysis approaches with respect to further formal characteristics, which are related to some of the characteristics listed above. Ultimately, characteristics need to be formalized to be used in practical analyses, in which case at least some of them (distance to centroids, quality of representation of the data and fit by probability models) also serve to measure information loss through clustering. Similar considerations can be found in [1], which are closer to the present approach, but somewhat less detailed. Ultimately, the characteristics listed here need to be formalized, too, to be used in practical analyses.

DEFINITIONS OF TRUE CLUSTERS

There is no agreed definition of what true clusters are in reality, but mathematical formalism allows to give a clear definition (a mathematical model) of true clusters based on mathematical objects. In different situations, different kinds of clusters are of interest, and a mathematical definition of true clusters cannot be unique. It is necessarily idealized and abstract, and discrepancies between such a definition and the more complex and informal ideas that researchers have about reality should not be suppressed just (see[3]).

Still, an explicit formal definition of true clusters has important benefits. It communicates the cluster concept in a specific setup in a clear way, and it provides a transparent framework for comparing methods. It may also stimulate the development of new methodology. In the literature on clustering methods, clear definitions of the specific clustering problem to be solved are often missing, probably because authors feel that such definitions could not properly cover the clustering problem in general. But this means that a chance is missed to clarify the understanding of what kind of problem a method is good or not so good for.

For every formal definition there need to be arguments why it formalizes a reasonable cluster concept researchers could be interested in, so it needs to be related to desirable characteristics of clusters. Definitions of true clusters can be based on the data, which are measurements that therefore "live" in the system of mathematical formalism. This is only appropriate if what makes a certain subset of the data a true cluster according to the researchers can indeed be defined from the data alone. For realist clustering aims, true clusters need to be defined based on a certain truth "behind" the data. There are two possibilities for doing this. Firstly, one could assume that in the "mathematical world" there is true clustering information for all observations, which is available in principle but not used by the clustering method. Secondly, one could assume that the data are generated by a true probability model, and then define the truth in terms of this model.

Definitions Based on the Data Alone

Let $\mathbf{x}_1, \ldots, \mathbf{x}n$ be n observations in Rp. k-means clustering is defined by choosing $k$cluster mean vectors $\mathbf{a}_1, \ldots, \mathbf{a}k$ and a cluster assignment function γ: $\{1, \ldots, n\} \mapsto \{1, \ldots, k\}$ so that $\sum_{i=1}^{n} \|\mathbf{x}_i - \mathbf{a}_{\gamma(i)}\|^2$ is minimized. The solution of this problem could be called "the true clustering".

Is this appropriate? It could be, namely if the real aim is to find a clustering with k clusters in which all observations are represented optimally (in the sense of averaging the squared Euclidean distance) by the centroid of the cluster to which they are assigned. On the other hand, if in the situation of interest clusters should rather correspond to high-density regions, clusters defined as "true" by k-means can be inappropriate, see Fig. 2 for an example. Note also that for defining true clusters according to the k-means criterion, $k$has to be assumed to be known.

Is such a definition helpful? If the k-means objective function is used to define the true clusters, obviously k-means clustering is the best clustering method, and this may look tautological, although it is still of interest to investigate to what extent different algorithms are successful for minimizing the objective function.

In principle, if the objective function that defines a clustering method corresponds exactly to the loss function of the practical problem for which a clustering is required, there is no point to look for other clustering methods. The same holds for methods that are not defined by optimizing an objective function but, e.g., are stable states reached by an algorithm, as long as this is for solving a practical problem properly formalized by the algorithm. In this sense, most clustering methods implicitly define their own truth. A practical implication is that the definition of a clustering method often gives strong information about what kind of clustering problem the method is good for.

However, in most clustering applications the aims of clustering do not directly translate into a specific cluster analysis method, be it through matching the practical "loss" with the method's objective function or otherwise. In general, the choice of the practical "loss" and therefore the objective function or more generally the clustering principle needs to be supported by validation techniques and background information.

In some other situations it is possible to define a clustering problem based on the data alone without corresponding directly to any available clustering method. An example for this is the optimal approximation of the distance matrix of the data by an ultrametric induced by a dendrogram produced by a hierarchical clustering method. Another approach would be the definition of an aim-dependent cluster quality index as a weighted mean of appropriately scaled statistics measuring cluster characteristics as listed inSection 3.3 (in [1] there is a related discussion of measuring and optimizing "usefulness" of clusters). In an implicit manner, internal cluster validation indexes [24] such as the average silhouette width attempt to aggregate desirable features of clusterings, and "true clusters" could be defined by optimizing them, although such criteria are usually designed with the aim of defining a too general notion of cluster quality, which does not take into account the differences between clustering aims in practice.

If "truth/quality" is defined in such a way, one could try to optimize the cluster quality index directly. This is often not computationally feasible, and also in some cases desirable characteristics need to be combined in other ways than just averaging them (for example, one may be interested in constrained optima of objective functions, putting an upper bound on within-cluster distances). So there is still a place for clustering methods that do not directly optimize a quality index. Also, clustering applications in which the idea of truth refers to the observed data alone are probably a small minority; particularly it implies that the data cover all objects of interest and are not only a sample from which the researchers want to generalize.

Some other work explores notions of "clusterability" of data [25]. Ackerman and Ben-David [26] reveal that there are several reasonable notions that contradict each other in many situations.

Definitions Based on External Information

In comparisons of cluster analysis methods in the literature, authors often use datasets for which there is a given "true classification". Often these are standard examples for supervised classification such as Fisher's famous Iris dataset in which there are measurements on 150 Iris plants from three different subspecies. Clustering methods can generate clusterings ignoring the true classification to which they then can be compared.

This is an artificial situation. In reality cluster analysis is applied to find clusters that are not yet known. The appeal of this approach is that realistic datasets can be used and that it is usually easy to argue that the true given classes are meaningful. But often measuring the performance of clustering methods on datasets with given true classes is not very informative. How informative it is depends on to what extent the true classes in such cases are good models for the true clusters the researcher wants to find in a new dataset with unknown truth. This is hardly ever discussed. Usually, it is not investigated to what extent the true given classes have the desired characteristics of clusters in the situation of interest. There is no guarantee that true classes from supervised classification problems qualify as "data analytic clusters" (in the sense of the previous subsection), and it may not be reasonable to expect a good clustering method to find them. Furthermore, there is no guarantee that the given true classes are the only categorical variable that qualifies for defining true classes; there could be further (unobserved) variables defining alternative true classes.

Although such real datasets with given true classes can contribute to the comparison of clustering methods, the approach seems to be overused in the literature, and where it is used, more care is required for exploring what can be learned for other datasets without known classes from the "success" of certain methods to recover known true classes.

The same applies to the presentation of datasets for which authors refer to some "truth" without a formal definition, just appealing to the reader's (usually Euclidean) intuition. E.g., data distributed on a ball about the origin together with data distributed around a much wider circle about the origin with a hole in the middle that separates it from the central ball are often presented as an illustration that "*k*-means does not work", not reproducing the clustering the authors declare to be true by fiat. This clustering is based on separation, but the biggest distances in the dataset occur within a cluster, namely the wider circle, so this qualifies as "true cluster" in some respects but not others. Euclidean intuition is irrelevant in a large number of clustering problems (e.g., with categorical variables or non-Euclidean dissimilarities) and should not be overrated as reliable indicator of "truth" in Euclidean setups either. Again, such data can be used in a constructive way for evaluating clustering methods, but reference to the specific characteristics of the given true clustering needs to be made.

External information can also be used in other ways to define cluster quality (and therefore implicitly the "true clusters" by optimizing quality). In applications where clustering is used instrumentally for some other aims of data analysis, for example for data compression in order to predict an external variable, different clusterings can be compared according to quality measures related to the final aim, e.g., prediction quality.

Definitions Based on Probability Models

Assuming that data are generated from probability models is the standard technique for defining true underlying but unobserved clusters. It can then be investigated by (asymptotic) theory or systematic simulation whether cluster analysis methods find such clusters. There are various approaches to define true clusters based on probability models. Most straightforward are mixture models of the form $f(\mathbf{x}) = \sum_{j=1}^{k} \pi_j f_{\theta_j}(\mathbf{x})$,where data $\mathbf{x}$ are assumed to be i.i.d. generated from a distribution with density f with is a mixture of parametric densities $f_{\theta j}$. This models that $\mathbf{x}$ is generated from mixture component $f_{\theta j}$ with probability ϖ_j, and data can be simulated by simulating the true component memberships first. The usual interpretation is that the true clusters correspond to the mixture components. Clusterings computed from the data $\mathbf{x}_1, \ldots, \mathbf{x}n$can be compared to the true component memberships for simulated data.

Although such a definition gives researchers a much clearer idea of the involved cluster concept than using a given true class for real data, there are several issues with this approach.

Firstly, the family of mixtures of distributions of the form f needs to be identifiable, i.e., no two sets of parameters $\{(\varpi_1, {}_1), \ldots, (\varpi_k, {}_k)\}$ should generate the same probability measure. This is fulfilled for most popular mixture models including Gaussian mixtures. If mixtures are considered in full generality of the concept, however, identifiability cannot be taken for granted. Uniform distributions on connected sets can be pieced together from uniform distributions on subsets in different ways. Gaussian mixtures can be written down as mixtures of truncated Gaussians, which are no longer identifiable. This indicates that parametric families that generate identifiable mixtures are chosen rather for technical reasons than because they would be particularly qualified for representing a clustering "truth" in reality.

Secondly, identifying clusters with mixture components may intuitively not be justified. The parametric family needs to be chosen in such a way that the f can indeed be interpreted as "cluster shaped", as prototypical models for clusters of interest. But two parameters ${}_1$ and ${}_2$ may be so close to each other that the mixture of distributions $\pi_1 f_{\theta_1} + \pi_2 f_{\theta_2}$ may be unimodal, and may look so homogeneous that it would be inappropriate to split it up into two clusters in a real application. Fig. 1 shows a density contour of a Gaussian mixture with five components but only four modes, two of which are not separated by a deep density valley. Fig. 2 shows some data generated from this mixture. It strongly depends on the application whether it is appropriate to interpret this distribution as generating five clusters. Note that there are very large distances within some of the mixture components, and it is hard to argue that the points from component 3 "belong together". One may wonder whether mixtures of homogeneous distributions such as the Gaussian should be interpreted as single clusters if their mixture is homogeneous enough, which allows for more flexible cluster shapes, but violates identifiability and requires the researcher to define under what conditions mixture components should be merged [27].

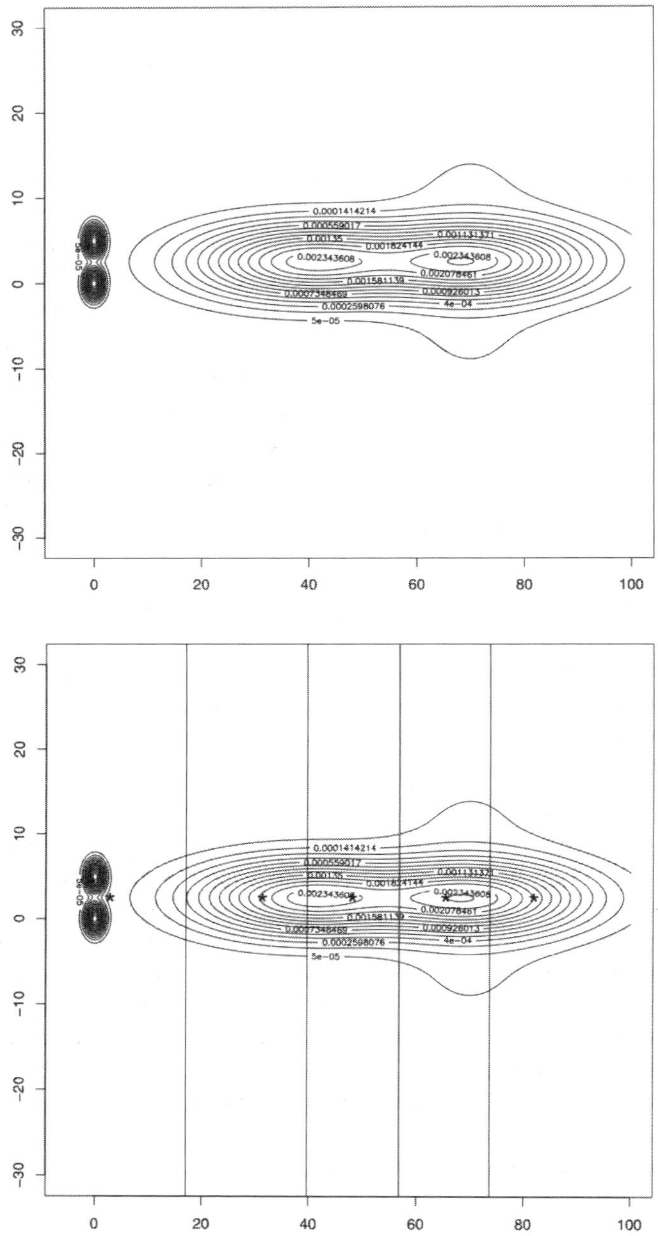

Figure. 1: Density contour of a mixture of five Gaussian distributions (mean vectors are (0, 0), (0, 5), (40, 2.5), (70, 2.5); there are two components centered at (70, 2.5) with different covariance matrices). Below: optimal 5-means partition and mean vectors (asterisks).

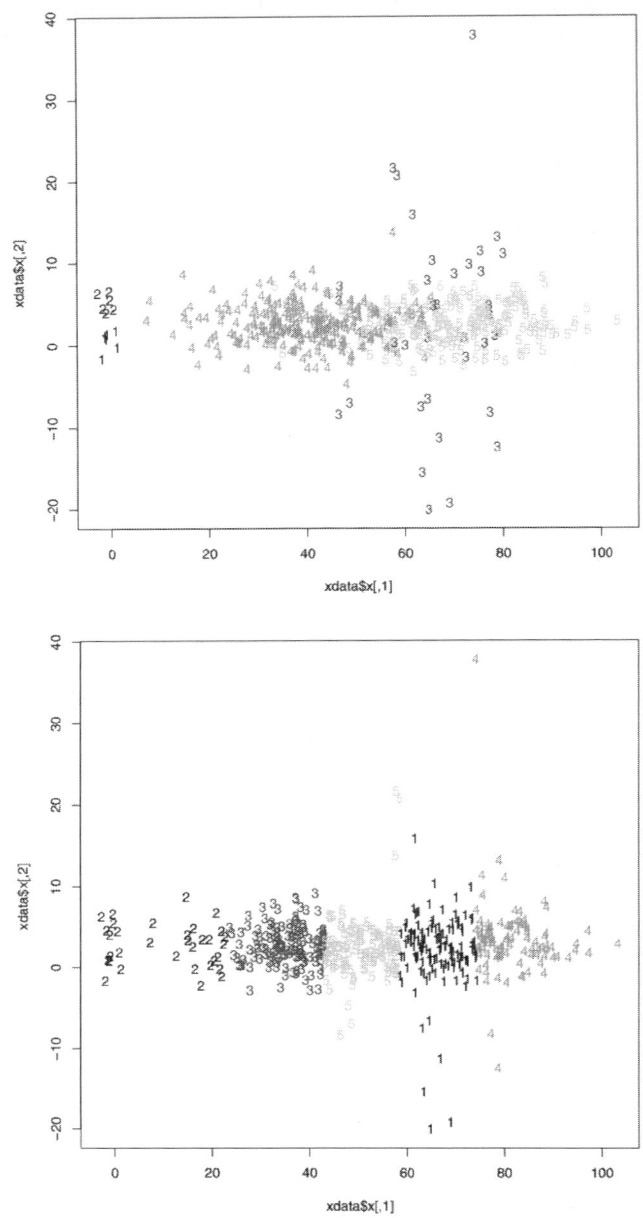

Figure. 2: Data generated from model in Fig. 1, above: mixture components from which observations were generated, below: 5-means clustering.

Thirdly, statisticians do not believe that parametric probability models hold precisely in reality, but true clusters as mixture components are only well defined if the mixture model holds precisely. This problem is worse for mixture models than elsewhere in parametric statistics, because if data come from a distribution with a density g that is slightly different than $f = \sum_{j=1}^{k} \pi_j f_{\theta_j}$ with a certain k, g can (under weak assumptions) be approximated arbitrarily well by a mixture f^+ of distributions of the form f with $k^+ > k$mixture components, which means that g can be approximated by a distribution with more and potentially quite different true clusters, despite being so close to f that it would require a very large dataset to tell f and g apart.

Despite such problems, defining true clusters as mixture components at least communicates a clear idea of a "cluster prototype model", and allows tests whether clustering methods recover the true clusters in such mixtures. Such tests can be expected to favor clustering methods that are based on parameter estimators (e.g., maximum likelihood, ML). A more comprehensive evaluation needs to consider models that are approximately but not precisely equal to such mixtures, and cases in which the interpretation of single mixture components as clusters breaks down, e.g., because mixtures of several components are homogeneous in some sense.

Alternatively, true clusters could be defined as high density level sets or attraction areas of density modes of distributions. This requires only the weaker nonparametric assumption that a density exists. Although this is more general than the mixture approach and allows for more flexible cluster shapes (which may or may not be desired), it does not solve all the problems connected to the mixture approach. For every distribution *P* with a density and *k* modes there are distributions without an existing density and distributions with an arbitrarily higher number of density modes that are so similar to *P* that they cannot be distinguished by an arbitrarily large amount of data [28]. As the mixture model approach, the density-based approach does not generalize to a full neighborhood of *P*.

A third approach is to define true clusters through statistical functionals of distributions. This allows, for example, to generalize the definition of *k*-means to distributions *P*, defining true underlying (unobserved) *k*-means-type clusters, by defining $\mathbf{a}_1, \ldots, \mathbf{a}k$ and $\gamma: \mathrm{R}p \mapsto \{1,\ldots,k\}$ as minimizers of $\int \|\mathbf{x} - \mathbf{a}_{\gamma(\mathbf{x})}\|^2 dP(\mathbf{x})$. For some other clustering methods (including ML estimation for mixtures) corresponding notions of truth can be defined in similar ways; see Section 4.1 for comments on adapting the cluster definition to a certain method. The formalization using probability models allows the investigation of the asymptotic properties of the methods. E.g., Pollard [29] proved the consistency of*k*-means applied to data as estimator for the *k*-means functional. Such functionals can in principle be defined for any distribution; a density is not required, but in case of the *k*-means functional existence of second moments is necessary. The *k*-means functional can still vanish or change rapidly in the neighborhood of any distribution *P*. Davies [30]argued (for linear regression) that statisticians should be interested in estimating globally defined and continuous functionals of distributions, because only such functionals cannot change arbitrarily in the neighborhood of a distribution. The clustering problem, though, is inherently discontinuous in borderline situations where a cluster splits, where the number of clusters changes or is misspecified (as far as I know, all currently existing functional-type definitions of true clusters require the number of clusters to be fixed).

These different approaches to define the truth illustrate that the clustering problem does not boil down to estimating the underlying distribution. Genuinely different true clusterings can be defined for the same distribution. The distribution showed in Fig. 1 is a mixture of five Gaussian components, has four density modes and (with appropriate level set cutoff is) three high density level-sets. The right side shows the true 5-means-type functional partition of the distribution. This may look counter-intuitive, and it is important to argue that any definition of true clusters based on a distribution formalizes a clustering that has certain desirable characteristics. But in the specific case that researchers want to find cluster centroids so that observations can be represented optimally by the centroids in the *k*-means sense, even such a counter-intuitive partition can be seen as "true".

Limitations of Formal Definitions

All the definitions listed above have shortcomings. Definitions based on the data alone do not reflect the idea of an unobservable underlying truth and of generalization of results to entities that were not observed. An external true clustering is usually not available in reality. Using it for assessment of clustering quality where it exists may not help much to clarify the characteristics of the clustering methods. Known "true" classes in datasets where they exist may deviate systematically from unknown classes of interest in real clustering problems. Definitions based on probability models suffer from instability. Sometimes a researcher may have a loss function in mind that formalizes the practical problem, but often this involves an unobservable truth and cannot be directly computed on the data alone, in which case it relies on model assumptions and the comments inSection 4.3 apply.

In any case, researchers may have a more complex informal idea of a cluster in mind than what can be captured by a formal definition. The definitions of true clusters should be taken as helpful constructs that support clarification and transparent comparison of methods, but they should not be taken as the ultimate clustering truth. Researchers may also complement formal definitions by less formal descriptions of more general cluster shapes they are interested in, for example "our method should find elliptical clusters with light tails that can reasonably be approximated by Gaussian distributions but are separated well enough that there is a density valley (depth to be defined) between them". Methods can then be compared by distributions that fit this description. Despite all the shortcomings, it would be a strong progress for scientific communication to accompany the introduction of new clustering methods regularly with an explicit definition of the clustering problem.

IMPLICATIONS FOR CLUSTER ANALYSIS RESEARCH AND PRACTICE

Choice of a Clustering Method in Practice

If researchers want to find true or real clusters, they have to specify what kind of truth they are interested in and what should constitute a "real" cluster. An appropriate clustering method can be found by connecting the characteristics of the clustering method to what is desired according to the researchers' cluster concept. Some methods optimize certain characteristics directly (such as *k*-means for representing cluster members by centroids), and in further cases experience and research suggest typical behavior (*k*-means tends to produce clusters of roughly equal size and spherical shape, whereas methods looking for high-density areas may produce clusters of very variable size and shape). Other characteristics such as stability are not involved in the definition of most clustering methods, but can be used to validate clusterings and to compare clusterings from different methods by use of resampling techniques [31]. Realist clustering aims can often be related to desirable characteristics that can be computed from the data. A more direct approach to method choice for realist clustering aims is possible if the researchers can specify a probability model and a formal definition of truth for the problem under study. Methods with good statistical properties for estimating this truth qualify for being chosen, preferably if they can still do a good job if the model assumptions are slightly violated. Even realist clustering is a constructive act in the sense that the researchers need to construct their concept of "real/true" clusters, and in the interest of scientific communication it is desirable to make this explicit.

The task of choosing a clustering method is made harder by the fact that in many applications more than one of the listed characteristics is relevant. Clusterings may be used for several purposes, and desired characteristics may not be well defined, e.g., in exploratory data analysis, or in cases where the connection between the interpretation of the clusters and the data is rather loose.

The specification of a cluster concept that captures a researcher's informal idea of true clusters is a hard problem, too. Often researchers only find out that their initial specification was not appropriate if they see what clustering this yields from their data. I have come across such situations often in advisory work. E.g., researchers may realize that the used methodology needs to enforce the connection of their clustering to an external variable to which their clustering should be related, but which they did not specify initially because they believed that this would happen automatically. Or they realize that small clusters are useless for them only after finding out that their initially preferred method produces such small clusters in their data. This illustrates the value of active scientific realism as complement to constructivism (and the value of cluster validation); the researcher's constructs are required, but the researchers should be open to change them responding to input from the reality outside their control.

Comparison of Clustering Methods

Although in reality the choice of a clustering method needs to depend on the context and the clustering aim, research comparing clustering methods independently of specific applications is useful because it adds to the understanding of the characteristics of the clustering methods. However, as mentioned in Section 4.1 already, in most published comparisons of clustering methods the authors seem to be far too keen to produce simple rankings of methods without providing any insight regarding what can be learned about the suitability of different methods for different clustering aims. I have hardly seen any study in which different clusterings of the same data or of data from the same probability model have been treated as legitimate and were used to tell the implicit cluster concepts of different models apart ([22], [23] and [27] are examples where this is done). Characteristics such as those listed in Section 3.3 could be used to evaluate what clustering methods do best according to various different characteristics datasets without given truth, and they could also be used to characterize the true classes in situations where these classes are given, which could help to understand more precisely what can be learned from the performance in these cases. Mixture models with a range of true parameters and component distributions are occasionally used in comparative studies in a slightly more pluralist way with the result that different methods "win" different mixtures, although usually without questioning the idea that there is only one true clustering for any fixed choice of mixture parameters. Looking at various fixed sets of parameters and distributions is more informative for understanding the methods in detail than aggregating simulations with randomly chosen parameters, as some authors seem to prefer, probably because this approach can generate a single ranking of methods out of many different models.

Context-Driven vs. Data-Driven Decision Making

There are a number of other decisions that have to be made when carrying out a cluster analysis, such as standardization and transformation of variables, definition of a dissimilarity measure etc. Similar considerations as before apply regarding the idea that there is a single "best" way of doing this, and their dependence on the context and the clustering aim. A number of these decisions is discussed in [20].

Here is an exemplary remark regarding variable selection and dimension reduction. Many methods are currently advertised for performing this task automatically. Often they are motivated by their performance in probability models with a few truly informative and some further homogeneous "noise" variables (often following a Gaussian or uniform distribution). These models capture the idea that indeed some variables are relevant for clustering and some others are not, abstracted from the meaning of these variables. But in real applications, in which the variables have a meaning that is of substantial importance for the clustering task, choosing different variables changes the meaning of the resulting clustering. E.g., in a dataset of students with marks on a number of courses and some standard socio-demographic information, one may be interested for different reasons in clusterings of the marks from science courses, those from humanities courses, all courses combined, the socio-demographic information, or all information combined. It cannot be decided by automatic techniques in which of these clusterings the researchers should be interested, and whether certain variables "do not cluster" and whether they then should not be involved in the computation of the clustering of interest depends on the context and the clustering aims.

Regarding the choice of a dissimilarity measure, consider again the example of data on a central ball and data on a separated ring around it. In Section 4.2 it was mentioned that 2-means (based on Euclidean data) partitions such a dataset in a way different from ball vs. ring. Assuming that ball vs. ring is the correct partition, one could argue that one should use a different, data driven, dissimilarity (e.g., a path-based distance) for such data. But if both the Euclidean distance and the use of 2-means have a context-driven justification, it is more appropriate to question the intuitive assumption about what the correct partition is.

CONCLUSIONS

It seems to me that a misguided desire for uniqueness and context-independent objectivity makes many researchers reluctant to specify desired characteristics and to choose a clustering method accordingly, because they hope that there is a universally optimal method that will just produce "natural" clusters. Probably for such reasons there is currently only very little research investigating the characteristics of methods in terms of the various cluster characteristics that could be of interest in different applications of clustering. Also probably many researchers are worried about the fact that too strong subjective impact could bias analyses and conclusions and could violate the principles of science because it will yield results that clearly depend on the observer, see Section 2.1.

As pointed out before, there is a tension between the scientific goal of general agreement and the acknowledgment of individual differences and the unavoidable impact of the individual's point of view. Indeed it is important that individual decisions and their rationale are made transparent, and that they are made in such a way that the "reality outside our control" still can deliver its message. E.g., variables should be chosen, because they are relevant for the research question of interest, and not because they produce a specific clustering that the researcher wants to promote for some reason. There are a number of reasons to make decisions in a data dependent manner, particularly if the initial analysis of the data reveals that the researchers did not properly formalize their aims (see Section 5.1), in which case a confirmation on new data (or left out validation data) without making data dependent decisions will normally be required to convince the audience that the results are meaningful.

The philosophical perspective presented here tries to explain how cluster analysis can at the same time be strongly dependent on contexts, aims and decisions of the researcher, but also scientific, transparent and clear regarding its underlying concepts and aims, and open to impact from Chang's reality outside our control.

I think that the general philosophical considerations apply to much wider areas of statistics and data analysis; in cluster analysis the plurality of definitions, approaches and ideas of truth is particularly striking and better visible than elsewhere, but believing in a unique "natural" truth has problematic implications elsewhere as well.

REFERENCES

1. U. von Luxburg, R. Williamson, I. Guyon, Clustering: Science or art? JMLR: Workshop Conf. Proceed. 27 (2012) 65–79.
2. M. Ackerman, S. Ben-David, D. Loker, Towards property-based classification of clustering paradigms, in: Advances in Neural Information Processing Systems (NIPS), 2010, pp. 10–18.
3. C. Hennig, Mathematical models and reality—A constructivist perspective, Found. Sci. 15 (2010) 29–49.
4. E. von Glasersfeld, Radical Constructivism: A Way of Knowing and Learning, Falmer Press, London, 1995.
5. J. Shotter, Conversational Realities, Sage, London, 1993.
6. H. Chang, Is Water H2O? Evidence, Realism and Pluralism, Springer, Dordrecht, 2012.
7. L. Fleck, The Genesis and Development of a Scientific Fact, University of Chicago Press, Chicago, 1979.
8. A. Bird, E. Tobin, Natural kinds, in: E.N. Zalta (Ed.), The Stanford Encyclopedia of Philosophy, winter 2012 ed., The Metaphysics Research Lab, Stanford University, Stanford, CA, 2012.
9. N. Goodman, Ways of Worldmaking, Hackett, Indianapolis, 1978.
10. I. Hacking, A tradition of natural kinds, Philos. Stud. 61 (1991) 109–126.
11. B. Hausdorf, Progress toward a general species concept, Evolution 65 (2011) 923– 931.
12. I. Van Mechelen, J. Hampton, R.S. Michalski, P. Theuns, Categories and Concepts—Theoretical Views and Inductive Data Analysis, Academic Press, London, 1993.
13. H. von Foerster, Observing Systems, Intersystems Publications, Seaside, CA, 1981.
14. G. Lakoff, Women, Fire, and Dangerous Things, University of Chicago Press, Chicago, 1987.
15. J. Santos, J.M. de Sa, Human clustering on bi-dimensional data: An assessment, Technical Report No. 1, INEB Instituto de Engenharia Biomedica, Porto, Portugal, 2005.
16. J. Lewis, A psychophysical investigation of clustering, in: N.A. Taatgen, H. van Rijn (Eds.), Proceedings of the 31th Annual Conference of the Cognitive Science Society, Austin, Texas, 2009, pp. 315–320.
17. J. Lewis, M. Ackerman, V. de Sa, Human cluster evaluation and formal quality measures, in: Proceedings of the 34th Annual Conference of the Cognitive Science Society, Austin, Texas, 2012, pp. 1870–1875.
18. F.G. Ashby, W.T. Maddox, Human category learning, Ann. Rev. Psychol. 56 (2005) 149–178.
19. L.J. Rips, E.E. Smith, D.L. Medin, Concepts and categories: Memory, meaning, and metaphysics, in: K.J. Holyoak, R.G. Morrison (Eds.), The Oxford Handbook of Thinking and Reasoning, Oxford University Press, 2012, pp. 177–209.
20. C. Hennig, T.F. Liao, Comparing latent class and dissimilarity based clustering for mixed type variables with application to social stratification (with discussion), J. Roy. Stat. Soc. Ser. C 62 (2013) 309–369.
21. S. Ben David, U. von Luxburg, D. Pal, A sober look at clustering stability, in: G. Lugosi, H.U. Simon (Eds.), Learning Theory (Proceedings of the 19th Annual Conference on Learning Theory), Springer, Berlin, 2006, pp. 5–19.
22. M. Ackerman, S. Ben-David, S. Branzei, D. Loker, Weighted clustering, in: Proceedings of the 26th AAAI Conference on Artificial Intelligence, Palo Alto, California, 2012, pp. 858–863.
23. M. Ackerman, S. Ben-David, S. Sabato, D. Loker, Clustering oligarchies, in: Proceedings of the 16th AISTATS International Conference on Artificial Intelligence and Statistics, 2013, pp. 66–74.
24. H. Xiong, Z. Li, Clustering validation measures, in: C.C. Aggarwal, C.K. Reddy (Eds.), Data Clustering: Algorithms and Applications, CRC Press, Boca Raton (FL), 2014, pp. 571–606.
25. M.-F. Balcan, A. Blum, S. Vempal, A discriminative framework for clustering via similarity functions, in: Proceedings of the 40th STOC Symposium on Theory of Computing, Association for Computing Machinery (ACM), New York, 2008, pp. 671–680.
26. M. Ackerman, S. Ben-David, Clusterability: A theoretical study, J. Mach. Learn. Res. 5 (2009) 1–8.
27. C. Hennig, Methods for merging gaussian mixture components, Adv. Data Anal. Classif. 4 (2010) 3–34.
28. D.L. Donoho, One-sided inference about functionals of a density, Annal. Stat. 16 (1988) 1390–1420.

29. D. Pollard, Strong consistency of k-means clustering, Annal. Stat. 9 (1981) 135– 140.

30. L. Davies, Aspects of robust linear regression, Annal. Stat. 21 (1993) 1843–1899.

31. R. Tibshirani, G. Walther, Cluster validation by prediction strength, J. Comput. Graph. Stat. 14 (2005) 511–528.

CHAPTER 2 ON THE INFLUENCE OF INTERPOLATION METHOD ON ROTATION INVARIANCE IN TEXTURE RECOGNITION

Gustaf Kylberg[1] and Ida-Maria Sintorn[1,2]

[1]Vironova AB, Gävlegatan 22, SE-11330 Stockholm, Sweden

[2]Uppsala University, Department of Information Technology, Lägerhyddsvägen 20, SE-75105 Uppsala, Sweden

ABSTRACT

In this paper, rotation invariance and the influence of rotation interpolation methods on texture recognition using several local binary patterns (LBP) variants are investigated.

We show that the choice of interpolation method when rotating textures greatly influences the recognition capability. Lanczos 3 and B-spline interpolation are comparable to rotating the textures prior to image acquisition, whereas the recognition capability is significantly and increasingly lower for the frequently used third order cubic, linear and nearest neighbour interpolation. We also show that including generated rotations of the texture samples in the training data improves the classification accuracies. For many of the descriptors, this strategy compensates for the shortcomings of the poorer interpolation methods to such a degree that the choice of interpolation method only has a minor impact.

To enable an appropriate and fair comparison, a new texture dataset is introduced which contains hardware and interpolated rotations of 25 texture classes. Two new LBP variants are also presented, combining the advantages of local ternary patterns and Fourier features for rotation invariance.

INTRODUCTION

In many computer vision and image analysis applications, the texture of an object is an important property that can be utilized for classification or segmentation procedures. However, texture analysis in digital images is not a trivial task and numerous texture descriptors have been proposed. In some applications, e.g. face recognition as in [1], the orientation of the object is known, while in many other applications the orientation of an object may be arbitrary and hence also the texture. In the latter case, the texture can be rotated to a main orientation or principal direction, see e.g. [2] where the Radon transform is used to accomplish this, or alternatively, a texture descriptor invariant to rotation can be used. A third way to achieve rotation invariance is to add rotated versions of the textures to the training data. This technique of adding a priori information to achieve invariance towards something through adding virtual training samples is explored in [3].

Rotation invariant texture descriptors have been widely studied, as reviewed in [4]. Texture descriptors invariant to viewpoint (adding even more degrees of freedom) have also been studied in for example [5]. However, the problem of arbitrary viewpoints is not addressed in this paper.

Citation: Gustaf Kylberg and Ida-Maria Sintorn, On the influence of interpolation method on rotation invariance in texture recognition, DOI: 10.1186/s13640-016-0117-6.

Rotation invariance for a texture descriptor can be achieved locally (at each pixel position) or globally (for the region/patch investigated). In [6], clusters of filter bank responses, denoted textons, are computed and the histogram of occurring textons are used as features. They propose a filter bank, denoted MR8, for which global rotation invariance is achieved by using the maximum response over six different orientations. A possible advantage with globally rotation invariant descriptors is that such a descriptor can retain the distribution of local orientations while this information will be lost in a local rotational invariant descriptor. Note however, that in the case of MR8, this information is lost when only keeping the maximum response over orientations. In [7], a globally rotation invariant descriptor retaining distributions over orientations is introduced based on Fourier transformed responses from Gabor filter banks. In [8], a local rotation invariant descriptor is introduced based on the local binary pattern (LBP) descriptor and [9, 10] proposed globally rotation invariant descriptors based on Fourier transformed of the LBP.

The LBP descriptor has over the past decades resulted in a whole family of texture descriptors. For this study on rotation invariance, we have selected the classic LBP descriptor together with seven extensions, including two different approaches to rotation invariance. The descriptors are the following: (i) the classic LBP descriptor [8], (ii)LBP^{ri}—the approach to local rotation invariance [8], (iii) LBP^{DFT}—an approach to rotation invariance where globally rotation invariant Fourier features are extracted [10], (iv) $ILBP$ (improved LBP) which is a more noise robust extension of the LBP [11], (v) $ILBP^{ri}$ using the approach to local rotation invariance, (vi) $ILBP^{DFT}$ using the approach to global rotation invariance, (vii) LTP^{DFT} (local ternary patterns) where three rather than two states are considered in the local neighbourhood ([12]) and using the approach to global rotation invariance, (viii) $ILTP^{DFT}$ combining ILBP, LTP and the approach to global rotation invariance. By studying these descriptors, we can get a baseline performance from the classic LBP and see how two well used extensions to the LBP perform. In addition, we can see how invariant the two different approaches to rotation invariance are.

In this paper, we investigate and compare the following: local rotation invariance, global rotation invariance, including rotations in the training data, and the effect of the different interpolation methods when rotating textures, in the setting of retaining discriminant texture information. In order to do this, we introduce a new texture dataset, the Kylberg Sintorn Rotation Dataset. The dataset includes images of hardware-rotated textures as well as texture images rotated by the interpolation kernels: nearest neighbour, linear, third order cubic, cubic B-spline and Lanczos 3. The dataset has 25 classes of different types of textured surfaces and is publicly available [13]. The images in the dataset are acquired in raw format avoiding compression artefacts.

THE KYLBERG SINTORN ROTATION DATASET

There are, to our knowledge, two texture datasets available which contain images of rotated textures; the Outex dataset [14] and the Mondial Marmi dataset [10].

The Mondial Marmi dataset contains images of slabs of Italian marble. The textured surfaces are rotated prior to image acquisition to allow for studying rotation invariance in texture analysis. The imaging is done with a compact camera storing the images in JPEG format with notable compression artefacts.

The Outex dataset is a large dataset with 320 texture classes imaged at different resolutions, orientations and illuminations. Unfortunately, there are periodic stripe-like artefacts in the images. In addition, for a given class, orientation, resolution and scale, the image data to generate samples from is rather limited.

Due to, for our purposes, undesirable artefacts in these rotation texture datasets (periodic stripes in Outex and JPEG compression in Mondial Marmi), a new dataset with rotations of textures was acquired. The Kylberg Sintorn Rotation Dataset is a generic texture dataset with similar types of textured surfaces as in the Outex dataset but with the same general dataset structure (many samples of each texture-rotation combination) as the Mondial Marmi dataset. Furthermore, the acquisition setup used in the Kylberg Sintorn Rotation Dataset avoids the aforementioned limitations and artefacts.

The new dataset contains 25 texture classes, mainly consisting of fabrics and arranged textured surfaces using small articles. Figure 1 shows an example patch from each class. The images were acquired using a Canon EOS 550D DSLR camera with a Sigma 17–70-mm zoom lens. The camera was mounted above the textured surfaces, as shown in Fig. 2. Fluorescent lights were placed on two sides of the camera, just above the lens opening. Focus and exposure settings were manually set once for each texture class. The 5184×3456 images were acquired as lossless compressed raw files (CR2). The raw files were corrected for lens distortion, chromatic aberration and vignetting formed by the Sigma lens. The corrections were performed according to the settings in the "Adobe (SIGMA 17-70 mm F2.8-4 DC Macro OS HSM, Canon)" lens profile in Adobe Photoshop CS5 and then saved as lossless PNG files. The images were next converted to grey scale as 0.2989 R+0.5870 G+0.1140 B, where R, G and B are the red, green and blue intensities, respectively. The selected imaging conditions led to spatially over-sampled images, and to get the textures in a more suitable scale for the relatively small local neighbourhoods used in many texture descriptors, the images were sub-sampled to half the original size (2 592×1 728 pixels) using a Lanczos 3 kernel. By this, the influence of sensor noise was also reduced.

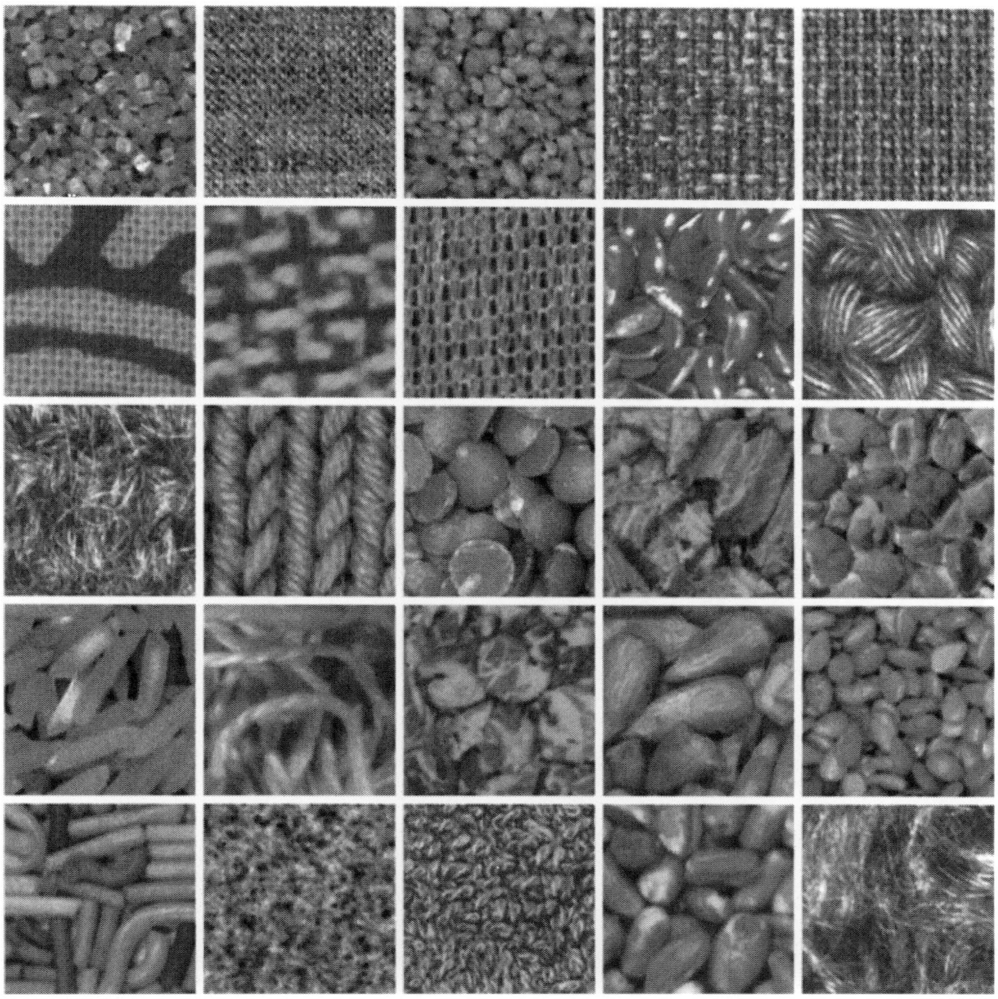

Figure. 1: Examples from the 25 texture classes in the Kylberg Sintorn Rotation Dataset. Row 1: cane sugar, canvas, couscous, fabric 1, fabric 2. Row 2: fabric 3, fabric 4, fabric 5, flax seeds, knitwear 1. Row 3: knitwear 2, knitwear 3, lentils, oatmeal, pearl sugar. Row 4: rice, rug, rye flakes, seeds 1, seeds 2. Row 5: sprinkles, floor tile, towel, wheat, wool fur

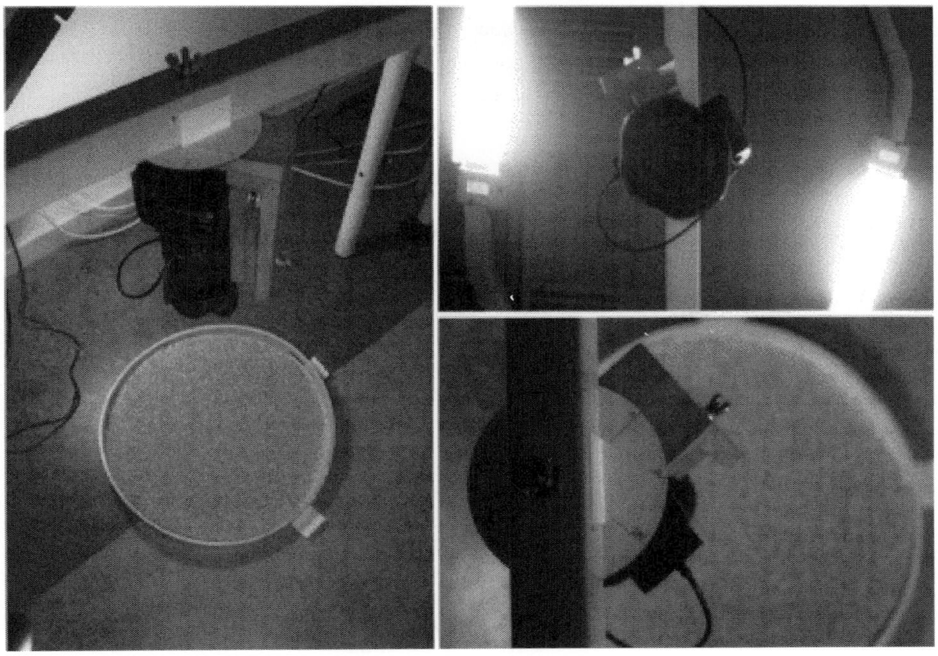

Figure. 2: The texture dataset image acquisition setup. From *left* to *right*: overview, view from the camera towards the texture surface beneath, view from the texture surface towards the camera and lights

Hardware rotation

The acquisition setup allows for rotation of the camera around the central axis of the camera lens. By rotating the camera rather than the textured surface, the same lighting conditions are kept throughout the image acquisition. For each texture class, one image is acquired for each orientation. The textures are imaged in the nine orientations $\in\{0°,40°,80°,\ldots,320°\}$ chosen not to be even multiples of 90° for which the choice of interpolation methods would not make a difference.

Rotation by interpolation

Five interpolation methods were used: nearest neighbour, linear, third order cubic, cubic B-spline and Lanczos. The images from the zero orientation were used to interpolate the eight other orientations using each of the interpolation approaches. For the interpolation methods based on convolution, Fig. 3 shows 1-D versions of the kernels in spatial as well as in the Fourier domain. The ideal sinc function is also shown for reference. The 1-D definitions of the interpolation methods are as follows:

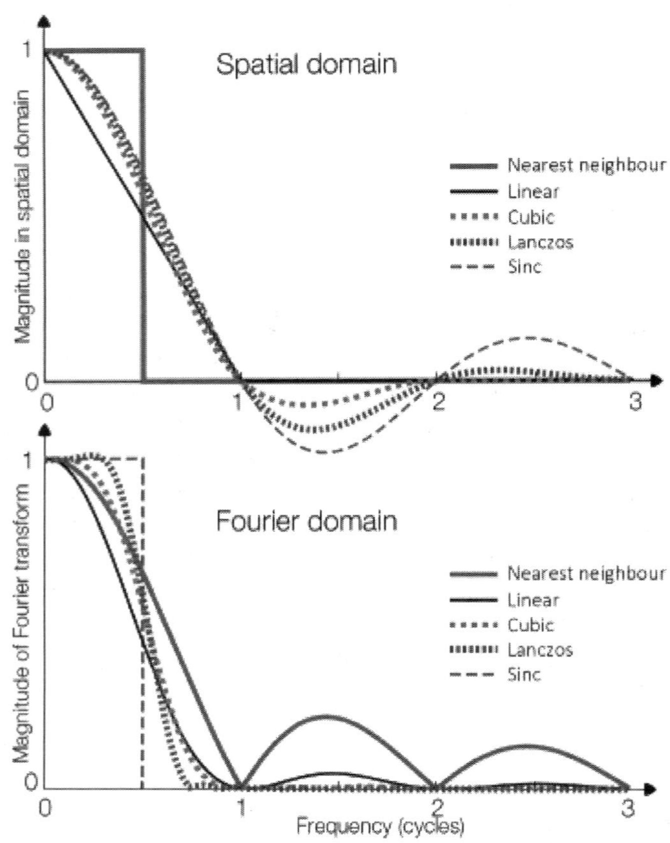

Figure. 3: Interpolation kernels. 1-D plots of the interpolation kernels and the sinc function in the spatial domain (*top*) and Fourier domain (*bottom*)

Nearest neighbour

$$k_{\text{nearest}}(x) = \begin{cases} 1, & |x| \in [0, 0.5] \\ 0, & \text{otherwise} \end{cases}$$

Linear

Assuming a linear transition between pixel intensities gives

$$k_{\text{linear}}(x) = \begin{cases} 1 - |x|, & |x| \in [0,1] \\ 0, & \text{otherwise} \end{cases}.$$

Cubic

The third order cubic interpolation kernel is a third order polynomial defined as in [15]:

$$k_{\text{cubic}}(x) = \begin{cases} \frac{3}{2}|x|^3 - \frac{5}{2}|x|^2 + 1, & |x| \in [0,1] \\ -\frac{1}{2}|x|^3 + \frac{5}{2}^2 - 4|x| + 2, & |x| \in (1,2] \\ 0 & \text{otherwise} \end{cases}$$

Lanczos

The Lanczos kernel is introduced in [16] and is a windowed approximation to the sinc function:

$$k_{\text{Lanczos}}(x) = \begin{cases} \text{sinc}(x)\text{sinc}(x/a), & x \in (-a, a) \\ 0 & \text{otherwise} \end{cases},$$

where

$$\text{sinc}(x) = \frac{\sin(x)}{x},$$

and $a \in \mathbb{N}$ sets the number of lobes to include of the sinc-function. In this paper, a=3 is used, resulting in the Lanczos kernel shown in Fig. 3. Interpolation of 2-D signals using Lanczos kernels was introduced in [17].

Spline interpolation

In contrast to the interpolation methods based on convolution, spline interpolation fits piecewise polynomials between the data points. The interpolated values are then the values the polynomial assumes at the new sample positions. The cubic B-spline interpolation used here, introduced in [18], fits a third order polynomial of the following form:

$$k_{\text{Bspline}}(x) = \begin{cases} \frac{1}{2}|x|^3 - |x|^2 + \frac{4}{6}, & |x| \in [0,1] \\ -\frac{1}{6}|x|^3 + |x|^2 - 2|x| + \frac{8}{6}, & |x| \in (1,2] \\ 0 & \text{otherwise} \end{cases}$$

The pieces fit smoothly together, forming a continuous function, going through the original data points.

MATLAB R2012b and the toolbox DIPimage [19] were used for all the interpolations.

Texture Sample Generation

All the images, rotated by hardware and software, are divided into smaller texture samples. The region of an image that is common among the different orientations is a disk centred in the image with the diameter that equals the height of the images (1728 pixels). The largest square within this disk (with the side $\lfloor 1728/\sqrt{2} \rfloor = 1221$) is divided into 100 sub-squares with a size of 122×122 pixels (since $\lfloor 1222/10 \rfloor = 122$). The partitioning scheme is illustrated in Fig. 4. Each texture sample is intensity normalized to have a mean value of 127 and a standard deviation of 40. The mean is set to be centred in the interval [0,255]. The standard deviation is set empirically so that intensity values in the dataset generally do not get mapped to integers ∉[0,255] while retaining most of the dynamic range of the dataset.

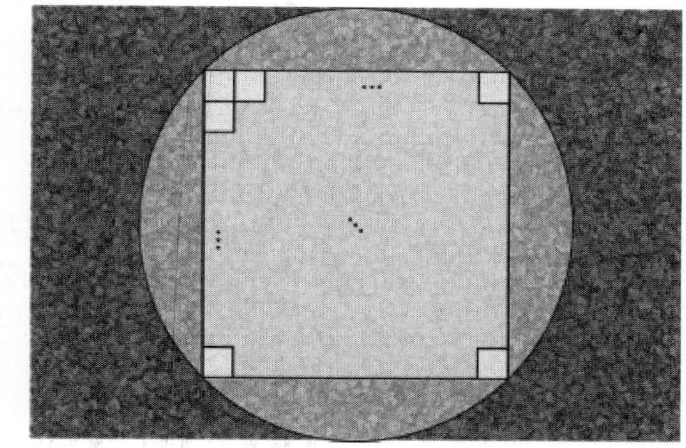

Figure. 4: Partitioning scheme. Illustration of the partitioning scheme used to extract texture samples

LBP-BASED DESCRIPTORS

In the original definition of LBP [20], the eight connected neighbours of a centre pixel are considered. The neighbours are thresholded by the intensity of the centre pixel and placed in a clockwise order producing an 8-bit binary code. The feature vector consists of the occurrences (histogram) of the different binary codes in a region/patch. The neighbourhood has been generalized to N samples on a radius R from the centre pixel in [8], here denoted $LBP_{N,R}$. The local binary code for the pixel position (x_c, y_c) with grey value g_c is defined as:

$$LBP_{N,R}(x,y) = \sum_{p=0}^{N-1} s(g_p - g_c)2^p, \tag{1}$$

where

$$s(x) = \begin{cases} 1, & x \geq 0 \\ 0, & \text{otherwise} \end{cases}. \tag{2}$$

If a point p does not coincide with a pixel centre, linear interpolation is used to compute the grey value g_p. Finally, the histogram of occurring binary codes in a region is the feature vector of this region.

One way of making LBP more robust to noise is to threshold the value of a point, g_p with the mean value of the neighbourhood (including the centre pixel), g_{mean}, rather than with the value of the centre pixel g_c. The centre pixel is also thresholded with the mean value and included in the binary code. This descriptor is called improved local binary patterns (ILBP) and was introduced in [11]. It is defined as

$$ILBP_{N,R}(x,y) = \sum_{p=0}^{N-1} s(g_p - g_{\text{mean}})2^p + s(g_c - g_{\text{mean}})2^N, \tag{3}$$

where

$$g_{\text{mean}} = \frac{1}{N+1}\left(\sum_{p=0}^{N-1} g_p + g_c\right), \tag{4}$$

and the function s is defined as in Eq. 2. Note that p_c is part of the binary code making it N+1 bits long.

In [21], local ternary pattern (LTP) descriptor is proposed. The difference between neighbouring values g_p and the centre pixel value g_c are encoded with three values using a threshold value t:

$$s_3(g_p, g_c, t) = \begin{cases} 1, & g_p \geq g_c + t \\ 0, & g_c - t \leq g_p < g_c + t \\ -1, & \text{otherwise} \end{cases}. \tag{5}$$

In our implementation, as described above, the interval coded with zero is half-bound while in [21] it is open. Instead of using a code with base 3 to encode the three states in Eq. 5, LTP uses two binary codes representing the positive and the negative components of the ternary code, i.e., two binary codes coding for the two states {−1, 1}. These binary codes are collected in two separate histograms, and as a last step, the histograms are concatenated to form the LTP feature vector [21].

In analogy with the extension of LBP to ILBP, where the neighbourhood mean value (g_{mean}) is used as the local threshold and the centre pixel (g_c) is included in the code, LTP can be extended to ILTP. This was done in [22]. For ILTP, the same scheme of using two concatenated binary codes as the final feature vector as for LTP is employed [22].

Rotation Invariance

There are different approaches to making the classical LBP descriptor rotation invariant. One way is to group the binary codes that are rotations of one another (i.e. circular shifts of the binary code). Next, the occurrences of each group are computed and used as feature values. This approach is introduced in [8], and the descriptor is denoted $LBP^{ri}_{N,R}$. For example, when N=8 is used, there are 36 such rotation invariant groups. Since occurrences of rotation groups are considered, the relative distribution within rotation groups is lost. It also means that the $LBP^{ri}_{N,R}$ descriptor achieves rotation invariance by normalizing rotation locally, as described in [23].

Another way of making LBP rotation invariant is introduced in [9, 23]. The occurrences of rotation codes within the rotation groups are not summed, as in $LBP^{ri}_{N,R}$, but Fourier transformed and the resulting power spectrum is used as the feature vector. The descriptor is called LBP histogram Fourier features ($LBP^{HF}_{N,R}$). Since the Fourier features are computed on the global histogram of binary codes in the region/patch investigated, $LBP^{HF}_{N,R}$ achieves rotation invariance globally, and hence, retains the relative distribution within rotation groups [23]. However, the $LBP^{HF}_{N,R}$ descriptor in [9] only considers uniform binary codes (binary codes with at the most two transitions between 0 and 1). It was generalized in [10] to include all binary codes, uniform and non-uniform, and called LBP^{DFT}. Together with the LBP^{DFT} descriptor, the corresponding $ILBP^{DFT}$ descriptor was also introduced in [10].

When having the aforementioned methods at hand, two interesting additional descriptors can be compiled; LTP^{DFT} and $ILTP^{DFT}$. Thus combining the generalized version of the Fourier features from [10], achieving global rotation invariance, with the promising descriptors LTP from [21] and ILTP from [22].

In the tests reported on here, eight descriptors were used, all with N=8 samples on a radius R=1; $LBP_{8,1}$, $LBP^{ri}_{8,1}$, $LBP^{DFT}_{8,1}$, $LTP^{DFT}_{8,1}$, $ILBP_{8,1}$, $ILBP^{ri}_{8,1}$, $ILBP^{DFT}_{8,1}$, $ILTP^{DFT}_{8,1}$. The descriptors are listed in Table 1 together with the length of their respective feature vector. The parameters are selected to describe the textures in their highest scale since the effects of interpolation methods are expected to be most prominent here. $N \neq 9$ is desirable since the dataset has nine equally spaced orientations and they would otherwise line up perfectly. $N \gg 8$ results in feature spaces of very high dimensionality (e.g. $LBP^{DFT}_{16,1} \in \mathbb{R}^{36\,883}$ and $ILTP^{DFT}_{16,1} \in \mathbb{R}^{147\,532}$) and N=8 samples the local neighbourhood at R=1 well.

Table 1: Texture descriptor dimensionality. Length of feature vector using N=8

Descriptor	Dim.	Descriptor	Dim.
LBP $_{8,1}$	256	ILBP $_{8,1}$	511
LBP\(_{8,1}^{\text {ri}}\)	36	ILBP\(_{8,1}^{\text {ri}}\)	71
LBP\(_{8,1}^{\text {DFT}}\)	163	ILBP\(_{8,1}^{\text {DFT}}\)	325
LTP\(_{8,1}^{\text {DFT}}\)	326	ILTP\(_{8,1}^{\text {DFT}}\)	651

CLASSIFICATION PROCEDURE

The interpolation methods and texture descriptors are evaluated by comparing the obtained classification accuracies. A first nearest neighbour (1-NN) classifier with Euclidean metric is used. The 1-NN classifier is used to be able to compare classification results obtained on the same and fair basis. To validate the trained classifier 10-folded cross-validation is performed by randomly assigning each texture sample an index $n \in \{1,2,\ldots,10\}$, creating 10 disjoint subsets with equal number of samples (stratified samples). In the first cross-validation fold, samples with $n \in \{2,3,\ldots,10\}$ will be the training data and samples with n=1 will serve as test data. In the second fold, samples with n=2 will be the test data and the rest is used for training, and so on.

The indices for the cross-validation folds are created once and then kept fixed throughout the experiments. The classification results from the 10 folds are combined into a single confusion matrix estimation and the mean and standard deviation of the classification accuracy is computed.

5 Evaluating interpolation methods and rotation invariance

Before the evaluation a baseline is established for each descriptor by training and testing on the =0° orientation. Table 2 lists the means and standard deviations of the classification accuracies of these baseline tests.

Table 2: Mean classification accuracy at =0°. Standard deviations in brackets

Descriptor	Mean	Std.	Descriptor	Mean	Std.
L B P $_{8,1}$	96.4	(1.2)	*I L B P* $_{8,1}$	95.4	(0.9)
\(\mathrm {LBP^{ri}_{8,1}}\)	97.2	(0.9)	\(\mathrm {ILBP^{ri}_{8,1}}\)	96.9	(0.9)
\(\mathrm {LBP^{DFT}_{8,1}}\)	98.7	(0.7)	\(\mathrm {ILBP^{DFT}_{8,1}}\)	97.5	(0.7)
\(\mathrm {LTP^{DFT}_{8,1}}\)	98.9	(0.5)	\(\mathrm {ILTP^{DFT}_{8,1}}\)	99.6	(0.4)

The descriptors are applied to all the texture samples, rotated by hardware and by each of the five interpolation methods. For the evaluation, the 1-NN classifier is trained on features from the =0° orientation followed by testing on the remaining eight orientations, one by one. The results are shown in Fig. 5. The established baselines are shown as dashed red lines in Fig. 5. One dot in Fig. 5 corresponds to one classification test where the classifier has been trained on =0° orientation and tested on one of the remaining orientations (using 10-folded cross-validation).

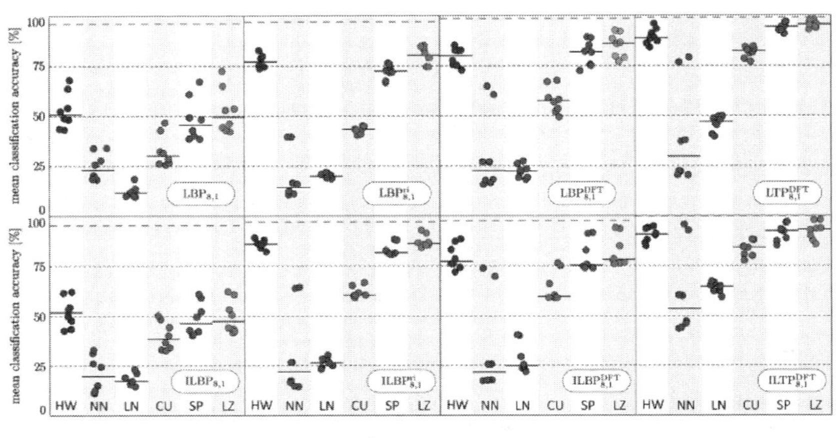

Figure. 5: Rotation invariance test. Plot of mean classification accuracies (in %) when training on the =0° orientation and testing on $\in \{40°, 80°, \ldots 320°\}$ for each descriptor applied on textures rotated by hardware (HW) or nearest neighbour (NN), linear (LN), cubic (CU), B-spline (SP)

and Lanczos 3 (LZ) interpolation. The *horizontal red dashed lines* are the mean accuracies when training and testing on =0°. Each dot corresponds to the mean classification accuracy from one rotation test. The *black horizontal line segments* in each column mark the median in each distribution

Interpolation Method

Figure 5 shows that the obtained classification accuracies differ greatly for the different interpolation methods (Additional file 1). This indicates that the characteristic properties of the textures are retained to a widely varying degree under rotation. Lanczos and B-spline interpolation results in accuracies similar to, or even better than, that of the hardware rotated textures. Nearest neighbour, linear and cubic interpolation all show lower accuracies for all descriptors compared.

For the nearest neighbour rotated textures, some rotation tests (dots in Fig. 5) show relatively high accuracies for several descriptors. To investigate this, Fig. 6 shows the result for the $LBP^{ri}_{8,1}$ descriptor in more detail. Here, each dot is a cross-validation fold in a rotation test. It can be seen that the nearest neighbour has noticeably higher accuracies at =80° and =280°. These two angles correspond to the two dots that often have higher accuracies than the rest for the nearest neighbour interpolation method in Fig. 5. Rotations of a digital image with even multiples of 90° are ideal/lossless. The two orientations of the textures ∈{80°,280° } in the dataset are the two that are closest to even multiples of 90° and, hence, closest to ideal rotations for which nearest neighbour is relatively successful.

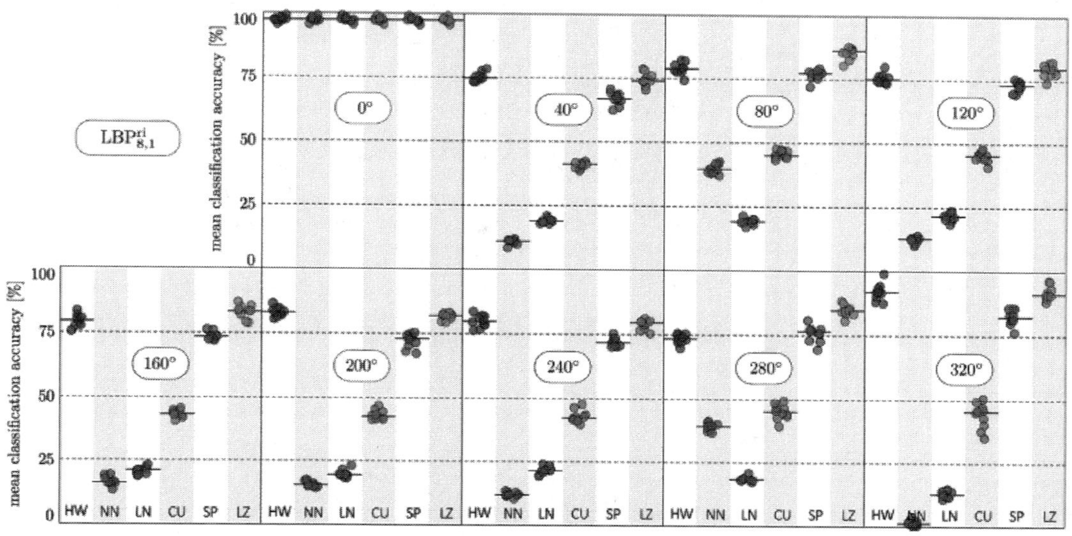

Figure. 6: Rotation invariance test for LBP\(_{8,1}^{\text {ri}}\) in detail. Plot of mean classification accuracies within each cross-validation fold for each orientation for the LBP\(_{8,1}^{\text {ri}}\) descriptor applied on textures rotated by hardware (HW) or nearest neighbour (NN), linear (LN), cubic (CU), B-spline (SP) and Lanczos 3 (LZ) interpolation. The *black horizontal line segments* in each column mark the median in each distribution.

In a few cases, the result achieved using Lanczos and B-spline interpolation exceeds the result obtained using the hardware rotated textures, especially at ∈{80°,280° }, see Fig. 6. This can, to a certain degree, be explained by that, in the case of hardware rotations, the sensor noise is sampled again and again for the different orientations, while the interpolated images all originate from one image with the sensor noise sampled once. The set of images which are rotated by interpolation is hence more homogeneous, and in addition, these two orientations are closest to ideal orientations of the original image, as was discussed above with respect to the higher performance for nearest neighbour interpolation for those two angles. Another explanation can be found by studying the per-class accuracies (data not shown). The classifier runs into problems with class number 12 and 20 for the hardware rotated textures while they are easier to classify in the Lanczos and B-spline interpolated data.

The tests were repeated using N=8 and R=2 for the eight LBP descriptors (data not shown). The interpolation methods appeared in the same order, accuracy-wise, with one exception. The nearest neighbour interpolation was found to perform on a level between linear and cubic interpolation. The descriptors generally achieved higher accuracies in the same order indicating that the selected parameters for the comparison is not optimal if the best overall classification result of the dataset is the objective. However, at R=1, the effects of interpolation methods and difference in the descriptor rotation invariance are more prominent and this is the main focus of this study.

Rotation Invariance

The rotation invariance of the descriptors can be evaluated in Fig. 5. Mean and standard deviation is also reported in Table 3. A truly rotation invariant texture descriptor should only show subtle variations over different orientations of a texture. The classic LBP has obvious problems in the rotation tests and fall far below the baseline. The LBP-versions design to be rotation invariant do achieve higher accuracies than the classic LBP. However, none of the descriptors reach their baseline accuracy.

Table 3: Mean classification accuracy (and standard deviations in brackets) across the eight rotation tests. A rotation test refers to training on =0° and testing on one of the other rotations ≠0°

Descriptor	Mean	Std.	Descriptor	Mean	Std.
$LBP_{8,1}$	52.9	(8.9)	$ILBP_{8,1}$	52.0	(7.3)
$\mathrm{LBP^{ri}_{8,1}}$	77.4	(3.2)	$\mathrm{ILBP^{ri}_{8,1}}$	85.9	(2.2)
$\mathrm{LBP^{DFT}_{8,1}}$	79.3	(4.6)	$\mathrm{ILBP^{DFT}_{8,1}}$	79.4	(6.1)
$\mathrm{LTP^{DFT}_{8,1}}$	89.5	(3.9)	$\mathrm{ILTP^{DFT}_{8,1}}$	90.4	(3.7)

Figure 6 shows that the cross-validation folds give very similar results. This means that the standard deviations over orientations, see Fig. 5, are not due to inadequate validation of the classifier but a genuine variation in descriptor performance.

Reference Methods

To put the rotation invariance of the LBP-based descriptor into perspective, three additional descriptors were evaluated in the same way. The rotation invariant Gabor filter bank achieved a mean accuracy over rotation tests of 90.4 %, Haralick features 75.3 % and MR8 56.6 %. These values should be compared to the mean classification accuracies reported in Table 3 (also showed as black lines in Fig. 5). The Gabor filter bank achieves as good results as the best descriptor tested in the LBP family $\left(\mathrm{ILTP^{DFT}_{8,1}}\right)$ while Haralick achieves just below the local rotation invariant $\mathrm{LBP^{r}_{8,1}i}$ descriptor. Below follows condensed details of how the reference methods were used.

For the Gabor filter bank, definitions and guidelines as described in [7, 24] were used. The frequency ratio was set to $\sqrt{2}$, the number of orientations was set to eight, and the number of frequencies to six. The Gaussian envelope was set to (,)=(3,2), and the kernel size was set to 19×19.

The Haralick features were defined and used as described in [25]. The contrast, correlation, energy and homogeneity measures were computed for four grey-level co-occurrence matrices (four directions) using quantization into 16 grey levels and a distance of two pixels.

MR8 was defined and used as in [26].

COMPUTATIONAL COST

To assess the computational cost of the different interpolation methods and texture descriptors, the relative computation time was measured.

Interpolation Methods

The computational time was measured while rotating the 2500 texture samples to the eight orientations. Figure 7shows the computation time relative to rotating the texture sample using linear interpolation. The two best performing interpolation methods (in terms of retaining texture information) are slightly more computationally expensive; a factor of 1.27 for Lanczos and 1.45 for Spline.

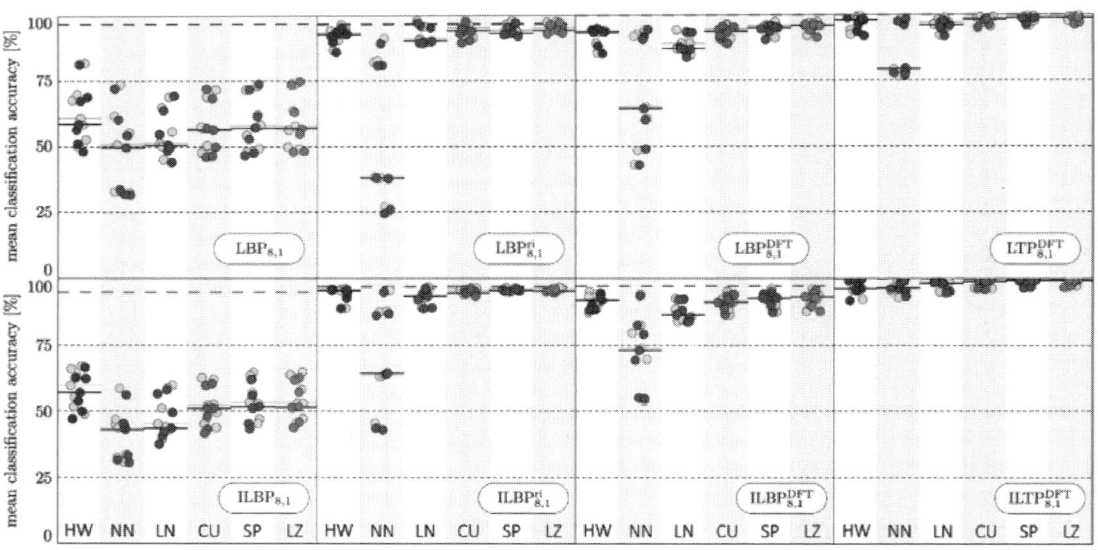

Figure. 7: Computational cost for the different interpolation methods. The costs are given relative to linear interpolation

Texture Descriptors

The computational time for the different descriptors was measured while applying them to the 2500 texture samples of the 0° orientation. Figure 8 shows the computation time relative to the LBP descriptor. The slowest descriptor is the ILTP ^{DFT} being about 2.5 times slower than LBP. However, all the descriptors in the LBP family evaluated in this manuscript are computationally cheap compared to the filter-based texture descriptors used here as reference methods. The Gabor filter bank takes about 37 times as long to compute as the basic LBP. The numbers may vary with implementations and systems but convolving the texture sample with a whole bank of filter kernels is more computationally expensive than computing these LBP-variants.

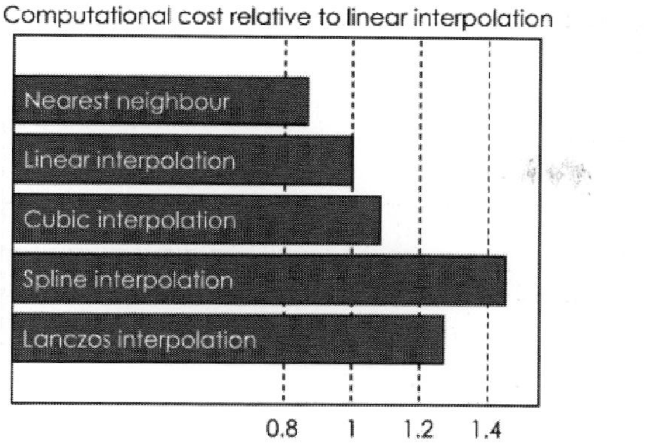

Figure. 8: Computational cost for the texture descriptors. The costs are given relative to the LBP descriptor

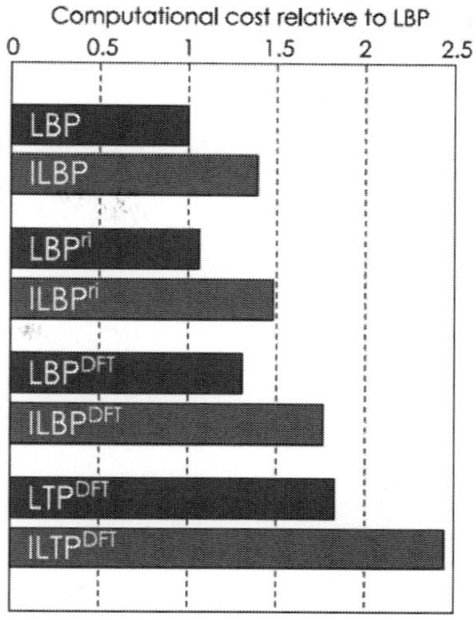

Figure. 9: Rotation invariance when training on two orientations. Plot of mean classification accuracies (in %) when training on the two orientations ∈{0°,40° } and testing on ∈{80°,120°,...320° } for each descriptor applied on textures rotated by hardware (HW) or nearest neighbour (NN), linear (LN), cubic (CU), B-spline (SP) and Lanczos 3 (LZ) interpolation. The horizontal *red dashed lines* are the mean accuracies when training and testing on =0°. The *grey dots* in the background correspond to the mean classification accuracies achieved when the size of the training data is halved. The*horizontal line segments* mark the median in each distribution.

ROTATION REPRESENTATION IN CLASSIFICATION

To investigate the approach of adding virtual training samples by rotating existing training samples, an increasing number of orientations are used in the training data, from one to eight. The classifier is then tested on the remaining orientations. Consequently, the training and test sets are still disjoint, sample and orientation-wise. The results are shown in Fig. 9. Using two orientations in the training data generally results in greatly improved classification accuracies and lower standard deviations. In fact, several of the texture measures approach their baseline. Another striking result is that the choice of interpolation methods is less important. The clear result seen in Fig. 5, that Lanczos 3 and B-spline outperform the other interpolation methods, is drastically reduced when the training data is somewhat more heterogeneous in terms of represented orientations.

In the first test, using training samples ∈{0°,40° }, the training set becomes twice as large. To make sure that the observed improvement originate from better representation of the rotated textures rather than a larger training set, we carried out a complementary test where every other sample was removed from the training set, making it the same size as in the previous tests. These results are shown as grey dots in the background in Fig. 9 and prove that the improvement is not from an increase in the number of training samples.

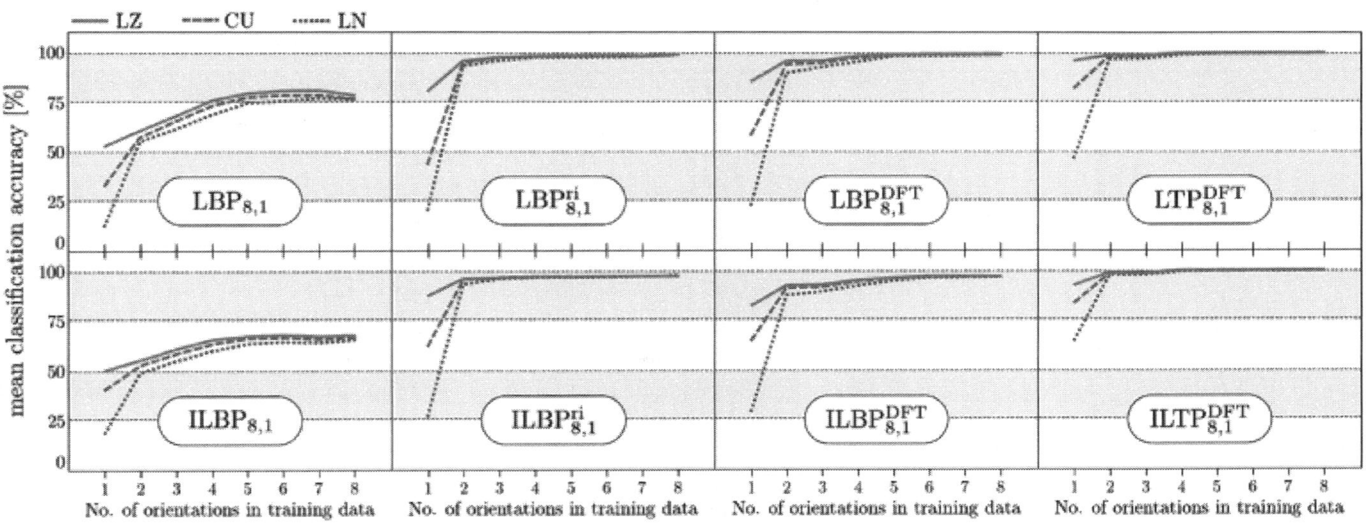

Figure. 10: Increasing the number of rotations in training. Plot of mean classification accuracies with increasing number of Lanczos interpolated orientations in the training data. The additional orientations included in the training data are excluded from the test data

Figure 10 shows how the classification accuracies change when even more orientations are used in the training data, ranging from one (as in the evaluation reported on in Section 4) to eight orientations. The additional orientations are the Lanczos interpolated data, following the scenario of adding "artificial" samples to the training data. It shows that the introduction of a second orientation in the training data is very beneficial in terms of texture recognition. For the less rotation invariant texture descriptors, additional orientations (more than two) further improve the classification accuracies even though the achieved accuracies level out and do not reach the levels of the rotation invariant descriptors. This trend is not as clear for the more rotation invariant descriptors, for which a third and fourth orientation only improves the classification accuracies slightly. Overall, $ILTP_{8,1}^{DFT}$ and $LTP_{8,1}^{DFT}$ are found at the top with mean accuracies between 99 and 100 % when at least four orientations are included in the training data.

CONCLUSIONS

Based on the performed experiments we conclude that

Lanczos 3 interpolation, closely followed by B-splines outperform the other interpolation strategies in all tests. The same levels of texture recognition are achieved using these two interpolation methods as those using hardware-rotated textures.

As expected, the interpolation methods closer to the ideal sinc function retain the texture information best. The commonly used linear and cubic interpolation clearly have shortcomings in preserving texture information in the setting of texture recognition.

The best performing interpolation methods are only slightly more computationally expensive. Lanczos stands out as it is both the best interpolation method in terms of preserving texture information and faster to compute than the second best interpolation method B-spline.

Both the local rotation invariant versions (by counting occurrences of rotation groups), and the global rotation invariant versions (by computing Fourier descriptors of the groups) of the LBP-based texture descriptors are less rotation variant than the classic LBP descriptor but they still suffer from rotation variance.

The LTP^{DFT} and $ILTP^{DFT}$ descriptors perform better than the other tested descriptors. They both show high classification accuracies and low standard deviations over different orientations.

The $ILTP^{DFT}$ achieves the same high classification accuracy as an optimized and rotation invariant Gabor filter bank. The other reference descriptors are inferior to all the tested rotation invariant forms of the LBP descriptors.

Even though the more advanced descriptors in the LBP family are more computationally expensive than the basic LBP, they are all computationally inexpensive compared to common filter bank-based texture descriptors such as Gabor filter banks and MR8.

Including several different orientations of the textures in the training data has great positive impact on the classification accuracies. This also compensates to a large extent the shortcomings of choosing a simple interpolation method. Hence, a simple strategy that should be generally considered!

Linear and cubic interpolation is still commonly used even though they are easily outperformed in terms of retaining texture information. For example, in the image processing software ImageJ, linear interpolation is the default, in GIMP cubic is default (although Lanzos 3 is offered as an option) and MATLAB only supports nearest neighbour, linear and cubic interpolation when rotating images. The software library OpenCV has linear interpolation as default while cubic is optional.

Furthermore, these tests show that the use of rotation invariant texture descriptors may not be enough to achieve rotation invariant texture recognition; representing different orientations of the textures in the training data shows to be very important, even if the extra orientations are artificially generated by rotating already existing samples by means of interpolation.

In the light of these findings, the use of linear interpolation to compute the intensity samples in the local neighbourhoods in all the $LBP_{N,R}$-based descriptors should perhaps be revised.

ACKNOWLEDGEMENTS

The authors would like to thank Cris Luengo for expanding the set of interpolation kernels implemented in DIPimage. This work has been part of the MiniTEM E!6143 project funded by EU and EUREKA through the Eurostars Programme.

REFERENCES

1. X Tan, B Triggs, Enhanced local texture feature sets for face recognition under difficult lighting conditions. IEEE Trans. Image Process. **19**(6), 1635–1650 (2010).
2. K Jafari-Khouzani, H Soltanian-Zadeh, Radon transform orientation estimation for rotation invariant texture analysis. IEEE Trans. Pattern Anal. Mach. Intell. **27**(6), 1004–1008 (2005).
3. B Schölkopf, *Support vector learning* (PhD thesis, Technische Universität Berlin, 1997).
4. J Zhang, T Tan, Brief review of invariant texture analysis methods. Pattern Recogn. **35**(3), 735–747 (2002).
5. Y Xu, H Ji, C Fermüller, Viewpoint invariant texture description using fractal analysis. Int. J. Comput. Vis. **83**(1), 85–100 (2009).
6. M Varma, A Zisserman, A statistical approach to texture classification from single images. Int. J. Comput. Vision. **62**(1-2), 61–81 (2005).
7. F Bianconi, A Fernández, A Mancini, in *Proceedings of the 20th International Congress on Graphical Engineering*. Assessment of rotation-invariant texture classification through gabor filters and discrete Fourier transform (ACMValencia, Spain, 2008).
8. T Ojala, T Mäenpää, Multiresolution Gray-Scale and Rotation Invariant Texture Classification with Local Binary Patterns. IEEE Trans. Pattern Anal. Mach. Intell. **24**(7), 971–987 (2002).
9. T Ahonen, J Matas, C He, M Pietikäinen, in *Proceedings of the 16th Scandinavian Conference on Image Analysis. LNCS*, 5575. Rotation invariant image description with local binary pattern histogram Fourier features (SpringerBerlin Heidelberg, 2009), pp. 61–70.
10. A Fernández, O Ghita, E González, F Bianconi, PF Whelan, Evaluation of robustness against rotation of LBP, CCR and ILBP features in granite texture classification. Mach. Vis. Appl. **22**(6), 913–926 (2011).
11. H Jin, Q Liu, H Lu, X Tong, in *Proceedings of the Third International Conference on Image and Graphics*. Face detection using improved LBP under Bayesian framework, (IEEE Computer Society, WashingtonDC, USA, 2004), pp. 306–309. doi:http://dx.doi.org/10.1109/ICIG.2004.62.

12. A Fernández, M Álvarez, F Bianconi, Texture description through histograms of equivalent patterns. J. Math. Imaging Vis.**45:**, 76–102 (2013).

13. G Kylberg, I-M Sintorn, Kylberg Sintorn Rotation dataset (2015). http://www.cb.uu.se/~gustaf/KylbergSintornRotation/. Accessed 1 Jan 2016.

14. T Ojala, T Mäenpää, M Pietikäinen, J Viertola, J Kyllönen, S Huovinen, in *Proceedings of the 16th International Conference on Pattern Recognition, Quebec, Canada*. Outex - New Framework for Empirical Evaluation of Texture Analysis Algorithms (IEEE Computer SocietyWashington, DC, USA, 2002), pp. 701–706.

15. R Keys, Cubic convolution interpolation for digital image processing. IEEE Trans. Acoust. Speech Signal Process. **29**(6), 1153–1160 (1981).

16. C Lanczos, *Applied Analysis. Prentice-Hall mathematics series* (Prentice-Hall, New Jersey, 1956).

17. CE Duchon, Lanczos Filtering in One and Two Dimensions. J. Appl. Meteor. **18**(8), 1016–1022 (1979).

18. H Hou, H Andrews, Cubic splines for image interpolation and digital filtering. IEEE Trans. Acoust. Speech Signal Process.**26**(6), 508–517 (1978).

19. C Luengo, DIPimage, a Matlab toolbox for scientific image processing and analysis. http://www.diplib.org/. Accessed 1 Jan 2016.

20. T Ojala, M Pietikäinen, D Harwood, A comparative study of texture measures with classification based on featured distributions. Pattern Recogn. **29**(1), 51–59 (1996).

21. X Tan, B Triggs, in *Analysis and Modeling of Faces and Gestures. Lecture Notes in Computer Science*, 4778, ed. by SK Zhou, W Zhao, X Tang, and S Gong. Enhanced local texture feature sets for face recognition under difficult lighting conditions (SpringerSpringer-Verlag Berlin Heidelberg, 2007), pp. 168–182.

22. L Nanni, S Brahnam, A Lumini, A local approach based on a Local Binary Patterns variant texture descriptor for classifying pain states. Expert Syst. Appl. **37**(12), 7888–7894 (2010).

23. G Zhao, T Ahonen, J Matas, M Pietikainen, Rotation-invariant image and video description with local binary pattern features. IEEE Trans. Image Process. **21**(4), 1465–1477 (2012).

24. F Bianconi, A Fernández, Evaluation of the effects of Gabor filter parameters on texture classification. Pattern Recog.**40**(12), 3325–3335 (2007).

25. RM Haralick, K Shanmugam, I Dinstein, Textural Features for Image Classification. IEEE Trans. Syst. Man Cybern. **3**(6), 610–621 (1973).

26. M Varma, A Zisserman, in *Proceedings of the 7th European Conference on Computer Vision, Copenhagen, Denmark. Computer Vision*, 3. Classifying images of materials: Achieving viewpoint and illumination independence (SpringerBerlin, 2002), pp. 255–271.

CHAPTER 3

A SCALE- AND ORIENTATION-ADAPTIVE EXTENSION OF LOCAL BINARY PATTERNS FOR TEXTURE CLASSIFICATION

Sebastian Hegenbart·Andreas Uhl[1]

Department of Computer Sciences, University of Salzburg, Jakob-Haringer Strasse 2, 5020 Salzburg, Austria

ABSTRACT

Local Binary Patterns (LBPs) have been used in a wide range of texture classification scenarios and have proven to provide a highly discriminative feature representation. A major limitation of LBP is its sensitivity to affine transformations. In this work, we present a scale- and rotation-invariant computation of LBP. Rotation-invariance is achieved by explicit alignment of features at the extraction level, using a robust estimate of global orientation. Scale-adapted features are computed in reference to the estimated scale of an image, based on the distribution of scale normalized Laplacian responses in a scale-space representation. Intrinsic-scale-adaption is performed to compute features, independent of the intrinsic texture scale, leading to a significantly increased discriminative power for a large amount of texture classes. In a final step, the rotation- and scale-invariant features are combined in a multi-resolution representation, which improves the classification accuracy in texture classification scenarios with scaling and rotation significantly.

INTRODUCTION

A major challenge in texture classification is dealing with varying camera-scales and orientations. As a result, research focused on scale- and rotation-invariant feature representations has been a hot topic in the last years. Feature extraction methods providing such invariant representations allow to be categorized into four conceptually different categories.

In a theoretically elegant approach, methods of the first category transform the problem of representing features in a scale- and rotation-invariant manner in the image domain, to a possibly easier, but equivalently invariant representation in a suitable transform domain. Pun et al. [1] utilize the Log-Polar transform to convert scaling and rotation into translation, scale- and rotation-invariant features are then computed using the shift invariant Dual-Tree Complex Wavelet Transform (DT-CWT [2]). Jafari-Khouzani et al. [3]propose a rotation-invariant feature descriptor based on the combination of a Radon transform with the Wavelet transform. A general drawback of this class of methods is that scaling can only be compensated at dyadic steps. As an improvement, Lo et al. [4] use a Double-Dyadic DT-CWT combined with a Discrete Fourier Transform (DFT) to construct scale-invariant feature descriptors at sub-dyadic scales. The periodicity of the DFT is also exploited by Riaz et al. [5] to compute scale-

Citation: Sebastian Hegenbart, Andreas Uhl, A scale- and orientation-adaptive extension of Local Binary Patterns for texture classification, Pattern Recognition, Volume 48, Issue 8, August 2015, Pages 2633-2644, ISSN 0031-3203, http://dx.doi.org/10.1016/j.patcog.2015.02.024.

invariant features by compensating the shifts in accumulated Gabor filter responses.

In a more pragmatic approach, methods of the second category achieve scale- and rotation-invariance either explicitly, by a re-arrangement of feature vectors, or implicitly, by selection of suitable transform sub-bands. In general, methods in this class also rely on some sort of image transformation. Lo et al. [6] (using the DT-CWT), Montoya-Zegarra et al. [7] (using the Steerable Pyramid Transform) as well as Han and Ma [8] and Fung and Lam [9] (both relying on Gabor filters responses) are representative approaches of this category. In parallel to the first concept, methods of this class are often limited in the accuracy and amount of compensable scaling and rotation by the nature of the used image transformation.

The obvious, but potentially most devious category, is based on a feature representation with inherent scale- and rotation-invariance. The fractal dimension [10] as measure for the change in texture detail across the scale dimension, is a promising candidate for such a representation. Geometric invariant feature representations based on the temporal series of outputs of pulse coupled neural networks (PCNN) have been used by Ma et al.[11] and Zhan et al. [12]. As a consequence of the inherent scale- and rotation-invariance however, this type of features is likely to have a decreased discriminative power as compared to other feature representations and often requires a generative, model based approach, such as Bag-Of-Words, to be competitive.

The fourth and last category of methods utilizes estimated texture properties to adaptively compute features with the desired invariants. Xu and Chen [13] use geometrical and topological attributes of regions, identified by applying a series of flexible threshold planes. Another large set of methods is based on the response of interest point detectors, such as the Laplacian of Gaussian (LoG [14]), the Harris–Laplace detector[15], Difference of Gaussian (DoG, SIFT [16]), Determinant of Hessian (DoH, SURF [17]) or Wavelet modulus maxima (SIFER [18]) to construct invariant features. Lazebnik et al.[19] apply affine normalization, based on the estimation of local shape and scale at detected interest points, to compute affine invariant features. Hegenbart et al. [20] compute LBP in an affine-adapted neighborhood while Li et al. [21] rely on local responses of the LoG to build a scale-invariant LBP representation. Due to the sparse output of interest point detectors and the stability of selected regions, a feature representation derived from interest points, might not be appropriate for all texture classification scenarios however. Even more, the intrinsic-scale of a large number of textures is inappropriate for a directly adapted computation of discriminative features, due to unsuitably large or small scales. As a consequence, the SIFT, SURF and SIFER feature descriptors are primarily used for tasks in computer vision apart from texture classification. A variation of these methods without scale-selection, based on local descriptors, computed at a dense grid, is generally used for computing features for the classification of textures.

In this work, we present a methodology which combines ideas from the second (alignment of features) and the last category (scale-adaption) to construct a scale- and rotation-invariant LBP feature representation. The method integrates seamlessly into the general computation of LBP, providing a high angular resolution with a fine grained compensation of scaling. Rotation-invariance is achieved by explicit alignment of features at the extraction level, based on a robust global estimate of orientation, using information provided by multi-scale second moment matrices [22]. The distribution of scale normalized Laplacian responses, in a scale-space representation of an image, allows a reliable estimation of the global image scale, which is used for a scale-adaptive feature computation. Based on the estimation of the global scale, intrinsic-scale-adaption is applied to compute features independent of the intrinsic texture scale. This assures the use of suitable LBP-radii, increasing the discriminative power of the feature representation significantly for a large amount of texture classes. In a final step, the rotation- and scale-invariant features are combined in a multi-resolution representation to further improve the discriminative power.

Limitations of LBP with image scaling a nd rotation

The Local Binary Pattern method [23] represents textures as the joint distribution of underlying microstructures, modeled via intensity differences in a pixel neighborhood. Such a neighborhood is defined in relation to a center pixel at position (x,y) as a tuple of nequidistant points on a circle with a fixed radius r. The position of neighbor number k is computed as

$$\eta^{r,n}(k;x,y) = \begin{pmatrix} x + r\cos\left(\frac{2\pi k}{n}\right) \\ y - r\sin\left(\frac{2\pi k}{n}\right) \end{pmatrix}^T. \tag{1}$$

A weighted sum, representing the pixel neighborhood, is computed and interpreted as binary label, based on a sign function sgn(x) mapping to 1 if $x \geq 0$ and 0 else. For a position (x,y) in an image, the standard LBP, based on n neighbors and radius r is computed as

$$\mathbf{LBP}^{r,n}(x,y) = \sum_{k=0}^{n-1} 2^k \operatorname{sgn}\left(I(\eta^{r,n}(k;x,y)) - I(x,y)\right). \tag{2}$$

Finally, the distribution of patterns is represented by a histogram, which is then used, in conjunction with a meaningful distance function, as an LBP feature.

The LBP feature representation has been used in a wide range of texture classification scenarios and has proven to be highly discriminative. A restriction of LBP however, is its sensitivity to affine transformations. As a consequence of the fixed-scale radius and the fixed sampling area dimension of the pixel neighborhood, the locally computed patterns implicitly encode the underlying micro structures of a texture at a scale directly related to the camera-scale of an image. As a result, the LBP feature representation is unable to compensate for different camera-scales. Even more, a rotation of an image is reflected as a circular shift in the individual patterns, which affects the distribution of patterns in a non-linear fashion. As a consequence, the standard LBP feature representation requires either an implicit or explicit alignment of patterns, which is generally done at the encoding level, to compensate for image rotations.

A widely used rotation-invariant encoding of LBP is based on the work of Ojala et al. [24]. The authors construct a rotation-invariant representation at the encoding level by implicit alignment of patterns, representing each individual pattern as the minimal decimal interpretation of all possible bitwise circular shifts of that specific pattern. A major limitation of encoding level based approaches is the highly limited angular resolution. As a consequence, Ojala et al. [24] suggest to combine their rotation-invariant encoding with uniform LBP. This combination however, leads to an even smaller number of individual patterns and a possibly decreased discriminative power of the feature representation. In the same work, the authors propose a multi-resolution representation, which improves the discriminative power of the features, by adding the capability of describing underlying micro structures at multiple scales. The multi-resolution representation however lacks a scale-selection mechanism and is therefore unable to compensate for image scaling.

Li et al. [21] were the first to compute scale-adapted LBP, based on the estimation of local texture scale. The authors use a direct mapping from the estimated local texture scale (in terms of the scale-space) to compute scale-adapted LBP-radii. Rotation-invariance is achieved, based on the methodology proposed by Guo et al. [25], estimating a global orientation on the basis of the computed LBP distribution and using bit alignment on a sub-uniform basis. Unfortunately, using the estimated local image scale as LBP-radius, significantly reduces the reliability of the method. This is a result of computing the features in dependence of the intrinsic texture scale, which is inappropriate for a large number of texture classes (in particular natural textures), due to either very large LBP-radii (low discriminative power) or very tiny LBP-radii (limited possibility of scale-adaption). The proposed scale- and orientation-adaptive (SOA)-LBP, based on prior work[26] and [27], addresses these limitations. The low angular resolution of encoding level based rotation-invariant representations is significantly improved by alignment of patterns at the extraction level, using a robust estimate of global texture orientation. The reliability of the feature representation is greatly enhanced by the means of intrinsic-scale-adaption, allowing the computation of highly discriminative features, independent of a texture's intrinsic-scale.

SCALE-ADAPTIVE LOCAL BINARY PATTERNS

We compute a scale-invariant representation of LBP by appropriate selection of LBP-radii (Section 2.2), based on a global estimate for image scale (Section 2.1). To compensate for the changed spatial extent of image structures due to scaling, we perform Gaussian low-pass filtering in reference to the corresponding scale-adapted LBP-radius, to sample neighbors at the correct scale (Section 2.3).

Estimation of the Global Image Scale

We estimate the global scale of an image utilizing the distribution of scale-normalized Laplacian responses in scale-space. Let $f: \mathbb{R}^2 \mapsto \mathbb{R}$ represent a continuous signal, then the scale-space representation, parametrized in terms of the standard deviation of the Gaussian, $L: \mathbb{R}^2 \times \mathbb{R}_+ \mapsto \mathbb{R}$ is defined by

$$L(\cdot;\sigma) = g(\cdot;\sigma) * f, \qquad (3)$$

with initial condition $L(\cdot;0) = f$. We denote $\sigma \in \mathbb{R}_+$ as the scale parameter (the standard deviation of the Gaussian function g) and represents a convolution operation. The scale-space family L is the solution to the diffusion equation

$$\partial_\sigma L = \sigma\left(\frac{\partial^2 L}{\partial x^2} + \frac{\partial^2 L}{\partial y^2}\right) = \sigma \Delta L. \qquad (4)$$

We construct the scale-space using an exponential spacing of scales $\sigma_i = c\sqrt{2}^{k_i}, k_i \in \{-4, -3.75, \ldots, 7.75, 8\}$ and c=2.1214. The value of c acts as a scaling factor and was initially chosen such that the center scale of the representation corresponds to the LBP-radius 3. We later added a set of larger scales to accommodate for the large intrinsic-scales of natural textures. By using an exponential spacing, we provide a fine grained estimation at small scales and still cover a considerable amount of large scales. Note that as a result of the Gaussian filtering for computing suitable sampling support areas, estimation errors at large scales are not as significant as errors at small scales. The used scale-space parametrization provides a solid foundation for estimating scales in a large number of scenarios. A parametrization specifically optimized for a given problem could potentially improve the accuracy in some cases however.

As a consequence of the sparse output of interest point detectors, scale estimation based on such scale-space extrema has shown to be unreliable for a large number texture classes. Fig. 1 illustrates this by comparing the response distribution of scale-space extrema with the proposed scale estimation function ξ. It can be observed that the sparse nature of interest points significantly limits the reliability of the scale estimation.

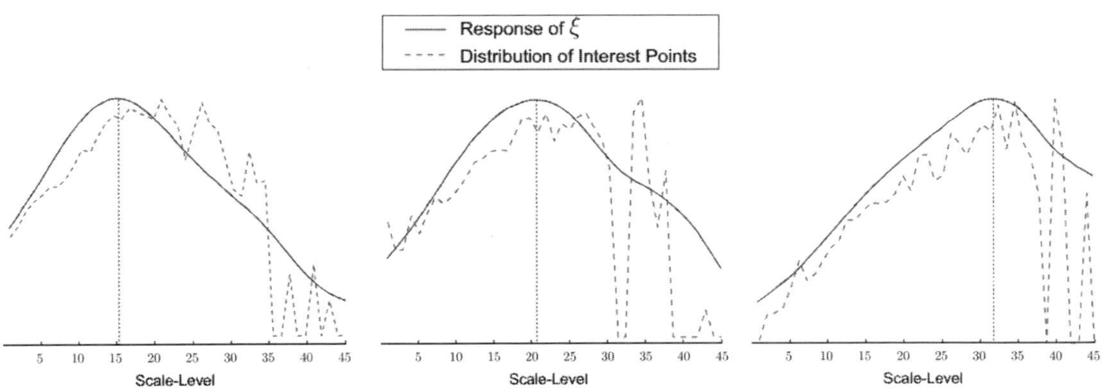

Figure. 1: Normalized response of ξ compared to the normalized response distribution of scale-space extrema.

We therefore use the distribution of the responses of scale-normalized Laplacians in the scale-space representation of an image I , $(\sigma^2|\Delta L(\cdot;\sigma)|$, denoted as $\overline{\Delta I}(\cdot;\sigma)$), computed at all scales in the scale-space, to estimate a global image scale. The scale estimation function ξ is

$$\xi(\sigma_i) = \sum_z \overline{\Delta I}(z;\sigma_i), \tag{5}$$

for $z \in R^2$ corresponding to a Cartesian coordinate on the pixel grid and σi denoting a specific scale-level in the scale-space. To determine the global scale of an image, the first local maximum of ξ is searched, which is then used as seed point for a least-squares Gaussian fit. By using the first local maximum we are capable of consistently estimating the scale of textures exhibiting more than a single dominant global scale. The quality of the estimation is improved by using only data points within a certain offset from the seed point. We use 10 percent of the number of scale-levels in the scale-space as positive and negative offset from the estimated first local maximum to fit the Gaussian function. This value was found during development of the method and has proven to be very stable for various image datasets. The mean value $\tilde{s}$ of the fitted Gaussian function is interpreted as the dominant level in scale-space. The standard deviation u of the fitted Gaussian is used as uncertainty of the estimation. For a given dominant scale-level in scale-space $\tilde{s}_i$, the spatial scale si corresponds to the scale parameter σi in $L(\cdot;\sigma_i)$ (the extent of a spatial structure at scale si is $\sigma_i\sqrt{2}$). Fig. 2illustrates the determination of a global scale by fitting a Gaussian function (dashed red line) to the scale estimation response function ξ (solid blue line).

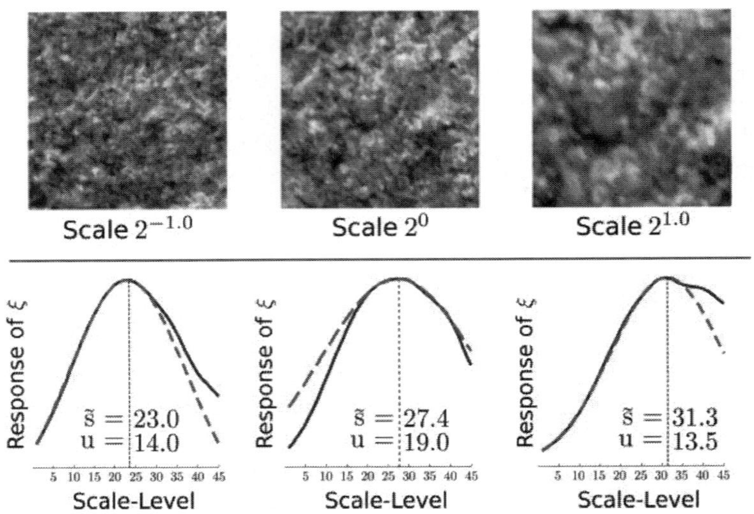

Figure. 2: Estimated scales-levels $\tilde{s}$ with uncertainty u for a texture at three camera-scales. (For interpretation of the references to color in this figure caption, the reader is referred to the web version of this paper.)

The scale estimation method is reliable for the majority of evaluated images but fails completely for a small fraction (approximately 3%). We identify a failed scale estimation by evaluating the uncertainty u. In our implementation, the scale estimation is considered as failed if u, normalized by the number of scale-levels, is greater than a certain thresholdt. In such a case, scale-adapted radii cannot be computed reliably. We therefore fall back to a default, computing the standard LBP with a fixed radius. The value of t was chosen as $20/n$ for n scale-levels (in our case t=0.4082). The specific value for the threshold was found during development and was consequently used across all experiments in this work. Multiple experimental results suggest that this threshold is robust and should generalize well for a large set of different scenarios.

We evaluated the accuracy of the scale estimation for computing scale-adapted LBP-radii, by estimating the global scale of all images in the KTH-TIPS and Kylberg image sets (see Section 5.1) at all 9 scales. Images at the default training scale (2^0) were then used as reference for computing the relative error of scale-adapted LBP-radii compared to the theoretically optimally scale-adapted radius. Fig. 3 presents the relative error (in percent) of scale-adapted LBP-radii, compared to the error of a fixed-scale LBP radius.

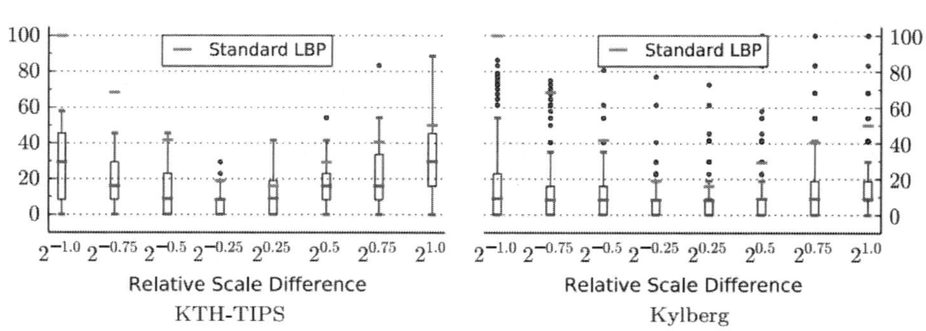

Figure. 3: Relative error (in percent) of scale-adapted LBP-radii.

The results show that the relative errors of scale-adapted LBP-radii are significantly smaller as compared to the fixed-scale LBP-radius. This indicates that the computation of scale-adapted patterns should improve the scale-invariance of the feature representation. Note the general asymmetry of the relative error, which can be observed for the fixed-scale LBP radii.

Intrinsic-Scale-Adaption of the LBP-Radius

The visualized scale of an image in the pixel domain is a function of the camera-scale, which is dependent on intrinsic- and extrinsic-camera parameters such as the focal length, the camera-distance, the image sensor dimensions and resolution, as well as the intrinsic-scale of the texture. The intrinsic-scale of a texture can be interpreted as the spatial extent of its dominant structures. The estimation of the intrinsic-scale is only possible with full knowledge of all camera-parameters. Responses of the scale-normalized LoG attain a maximum if its zeros are aligned with a circular shaped image structure. As a consequence, scales estimated based on the LoG correlate strongly with the visualized scale of the dominant circular shaped structures of a texture. The estimated scale of a texture (using our approach) is therefore highly related to the underlying intrinsic-scale.

Considering that the spatial extent of a circular structure is determined by its diameter, we model the visualized scale of an object using a two dimensional pinhole camera model. For an object with intrinsic-scale ι at distance u to the lens, the scale of the visualized object on the image sensor of a camera with focal length f is given as

$$s = \frac{\iota f}{u} \tag{6}$$

The effective scale of the visualized object in pixels is only dependent on the image sensor format and resolution, which are intrinsic camera-parameters.

An entire category of methods utilizing local texture properties to compute adapted, invariant features (such as Affine Invariant Regions [19] or Li-LBP [21]) are affected negatively by the large variety of intrinsic-scales across texture classes. This is a consequence of using the estimated scale (as combination of the intrinsic- and camera-scale) directly to compute adapted features. Due to unsuitably large or tiny intrinsic-scales for a considerable amount of texture classes, the estimated scales are likely to be inappropriate for computing scale-adapted features. Fig. 4 illustrates how inappropriate intrinsic-scales potentially lead to indiscriminative (too large) LBP-radii after scale-adaption.

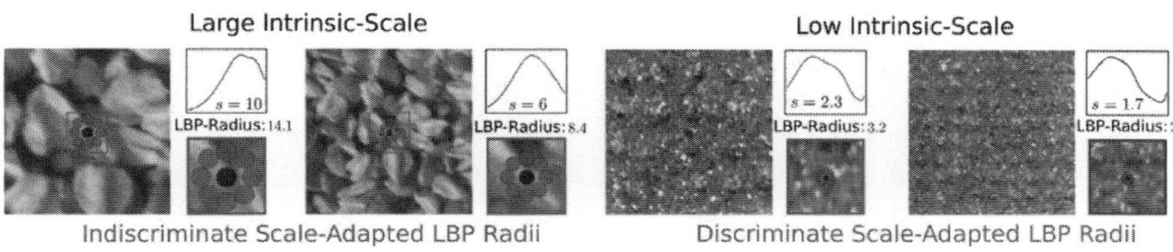

Figure. 4: Impact of intrinsic-scales on scale-adapted LBP-radii.

In this work, we propose a method to compute scale-adaptive LBP at suitable and highly discriminative scales by the means of intrinsic-scale-adaption, which allows scale-adaption based on the camera-scale without actual knowledge of the intrinsic-scale. Considering the quotient of two estimated image scales, either the intrinsic-scales cancel each other out (the images are from the same texture class, hence $\iota_1 \approx \iota_2 = \iota$)

$$\frac{s_1}{s_2} = \frac{\iota f}{u_1}\frac{u_2}{\iota f} \approx \frac{u_2}{u_1} \tag{7}$$

and the quotient is therefore in terms of the camera-scale, or the intrinsic-scales do not match (images are from different texture classes) and the quotient is basically random. By explicit computation of scale-adapted patterns, based on the quotient between the estimated scale of an image and a trained-base-scale, we are able to adapt for unsuitable intrinsic-scales implicitly. Note that this approach assumes that the used cameras have comparable focal lengths, sensor formats and resolutions and the intrinsic-scales within texture classes have moderate variance.

A trained-base-scale, acting as reference for the computation of intrinsic-scale-adapted patterns, is assigned to each texture class in the training data. In particular, we estimate the scales of each image in the training data and use the median of all estimated scales within a texture class as the trained-base-scale of that class. The scale-adapted LBP-radius used for an image with an estimated scale s , in reference to the trained-base-scale $\overline{s_l}$ of texture class l, is then computed as

$$\lambda(s, l, \rho) = \rho \frac{s}{\overline{s_l}}. \tag{8}$$

We define ρ (referred to as base-radius) as the LBP-radius used at the trained-base-scale $\overline{s_l}$. As a trade-off between discriminative power of the representation and the ability of adapting to a large variety of camera-scales, we set ρ=3 as default. This allows for highly discriminative patterns in the case of small relative scale differences and allows to compensate scale differences of up to a factor of 3. Note the linearity of λis a necessary property for scale-invariance. By computing LBP-radii as a function of the quotient of the estimated image scale and a trained-base-scale, the scale-adaptive representation is independent of the intrinsic-scale of the texture. As a consequence, highly discriminative features at suitable LBP-radii can be computed for a much larger set of texture classes.

Our experiments have shown that scale-adapted LBP computed in reference to a wrong trained-base-scale (the wrong texture class), exhibit appropriately the same intra-class variability as compared to the inter-class variability of features computed at matching trained-base-scales (the correct texture class). This is a direct result of the basically random LBP-radii used to compute scale-adapted patterns in such a case. As a consequence, we distinguish between the computation of training features and evaluation features.

The correct class is obviously known for images in the training data as part of the available ground-truth. We therefore compute training features only in relation to the trained-base-scale of the class of each specific image. Concerning images for evaluation, the class labels are unknown. In this case, features are computed in reference to each texture class, with the corresponding trained-base-scale. During classification, only features computed in reference to the same trained-base-scale are compared (see Section 4).

By using this approach we assure that features for training will be computed at suitable discriminative scales, close to the base-radius ρ for a majority of images in the training data. Features for evaluation, computed in reference to the correct trained-base-scale (the same class), benefit from intrinsic-scale-adaption, while evaluation features computed in reference to the trained-base-scale of a different texture class are uninformative due to inappropriate (random) LBP-radii and are insignificant for a later classification.

Adaptive Sampling Support Area Dimension

Scaling of an image changes the spatial extent of textural structures. Therefore the number of pixels covering structural information changes as well. As a consequence, the size of the sampling support area in the LBP neighborhood has to be adapted accordingly. By applying a Gaussian filter, each pixel in the image implicitly encodes information about a circular neighborhood of appropriate spatial scale. The radius of the Gaussian filter for a texture at estimated scale s in relation to a texture class l using base-radius ρ is computed as

$$g_r = \frac{\lambda(s,l,\rho)\pi}{n}, \tag{9}$$

for n defining the number of LBP-neighbors. The Gaussian filter coefficients are then computed such that P percent of the mass of the Gaussian function is covered within the interval [−g_r;g_r] (the kernel is truncated outside the interval limits):

$$\int_{-g_r}^{g_r} e^{-(x^2/2\sigma_g^2)}\,dx = P\int_{-\infty}^{\infty} e^{-(x^2/2\sigma_g^2)}\,dx$$

$$2\int_{0}^{g_r} e^{-(x^2/2\sigma_g^2)}\,dx = P\sigma_g\sqrt{2\pi}$$

$$\sigma_g = \frac{g_r}{\sqrt{2}\,\mathrm{erf}^{-1}(P)}. \tag{10}$$

We chose P to be 0.99 which corresponds to 99% of the mass of the Gaussian function, a value that proved to be robust in a large number of classification scenarios. As the sampling of a Gaussian function with very few sampling points potentially leads to a significant error, we use the error function (erf) using a numerical approach based on Abramowitz and Stegun [28] to improve the stability of sampling the one dimensional Gaussian filters centered at 0

$$G(x;\sigma_g) = \frac{-\mathrm{erf}\left(\frac{x-0.5}{\sigma_g}\right) - \mathrm{erf}\left(\frac{x+0.5}{\sigma_g}\right)}{2}, \tag{11}$$

which are then used in a separable convolution with the analyzed image.

Computation of scale-adapted Local Binary Patterns

The position of LBP-neighbor k, in a scale-adapted computation, in reference to texture class l and an estimated global image scale s, using base-radius ρ with n neighbors is computed as

$$\eta_{l,s}^{\rho,n}(k;x,y) = \begin{pmatrix} x+\lambda(s,l,\rho)\cos\left(\frac{2\pi k}{n}\right) \\ y-\lambda(s,l,\rho)\sin\left(\frac{2\pi k}{n}\right) \end{pmatrix}^T. \tag{12}$$

A Gaussian filter G with the appropriate standard deviation σg (see Eq. (10)) is used to sample neighbors at the correctly adapted spatial scale. Finally, the scale-adapted LBP is computed at position (x,y) with neighborhood $\eta_{l,s}^{\rho,n}$ based on the convolution of image I with G , G, $(I_g = I * G)$, as

$$\mathbf{SA{-}LBP}_{l,s}^{\rho,n}(x,y) = \sum_{k=0}^{n-1} 2^k \,\mathrm{sgn}\left(I_g(\eta_{l,s}^{\rho,n}(k;x,y)) - I_g(x,y)\right). \tag{13}$$

The histogram of patterns computed in reference to the trained-base-scale of texture class l is denoted as Hl and added to the SOA-LBP meta-descriptor of the specific image (see Section 4).

ORIENTATION-ADAPTIVE LBP

To compensate for the non-linear changes of the LBP distribution caused by a rotation of an image, an explicit or implicit alignment of patterns is required. This is generally performed at the encoding level, leading to a low angular resolution. To improve the angular resolution, we perform pattern alignment at the extraction level, which integrates naturally with the scale-adaptive computation of LBP and is based on an estimate of global image orientation.

Estimation of the Global Image Orientation

A main requirement on the orientation estimation in the context of scale-adaptive LBP, is robustness to varying image scales. We therefore utilize multi-scale second-moment-matrices (SMM [22]), computed at the global scale of an image, to estimate a global image orientation. The SMM summarizes the predominant directions of the gradient in a specific area of an image. In contrast to the single-scale SMM, the multi-scale SMM is defined over two scale parameters, the local scale σi as well as the integration scale i . This allows us to estimate the shape of visual structures at appropriate scales, as detected by the scale-estimation algorithm. The integration scale parameter is chosen in relation to the local scale (we use $i = \sqrt{2}\sigma_i$). The local scale parameter is selected as the global scale of the image, using the method described in Section 2.1. The multi-scale SMM of an image at location z∈R^2 is then computed as

$$\mu(z;\sigma_i,i)=\int_{\xi\in\mathbb{R}^2}(\nabla I)(z-\xi;\sigma_i)(\nabla I)^T(z-\xi;\sigma_i)g(\xi;i)\,d\xi. \tag{14}$$

We denote $(\nabla I)(z;\sigma_i)$ as the gradient of the scale-space representation of image I at scale σi and position z. An important property of SMMs in general is positive definiteness. The two (non-negative) eigenvalues of an SMM correspond to the length of the axes of an ellipse (up to some constant factor). The orientation of the eigenvectors correspond to the orientation of the dominant gradient and the orientation perpendicular to the dominant gradient respectively.

To estimate the global orientation of an image I , we compute multi-scale SMMs at a dense grid, corresponding to pixel locations $z\in R^2$. The orientation at a specific location is determined as the angle between the major axis of the ellipse and the vertical axis of the coordinate system (the axes of the image). Due to the ambiguous orientation of the ellipse, we treat all angles modulus π . Hence, the estimated orientation is unambiguous in $[0;\pi]$. We then estimate the global orientation of an image, based on the distribution of local orientations, computed at all coordinates of the sampled grid.

In parallel to the scale estimation method described in Section 2.1, this is done by fitting a Gaussian function to the distribution of local orientations in a least-squares optimization. To improve the accuracy of the estimation, we remove data points with an offset greater than ±15° (a robust, empirically found value that was used successfully on various datasets) from the maximum of the distribution, prior to the fitting process. Finally, the average value of the Gaussian is interpreted as the global orientation, which is used to align the sampling points of the orientation-adaptive LBP.

Fig. 5 illustrates the determination of the global orientation from the local orientation distribution. The dashed red line represents the Gaussian function fitted to the distributions of local orientations (solid blue line) of an image at three different orientations. The numbers centered at each figure present the estimated global orientation of each image.

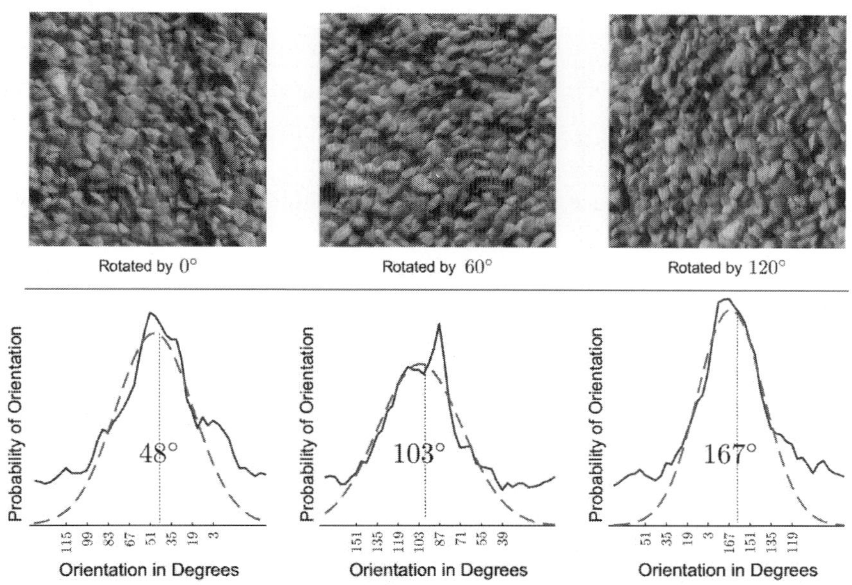

Figure. 5: Estimated orientations for a texture at three orientations. (For interpretation of the references to color in this figure caption, the reader is referred to the web version of this paper.)

To evaluate the accuracy of the orientation estimation method, we computed the absolute error of the estimated orientations (Fig. 6) between a reference image at the default training scale (2^0) and the same image at a different scale and random rotation between 30° and 330° in steps of 30°. The error was evaluated from 891 (81×11) random samples at 8 relative scales using the KTH-TIPS as well as the Kylberg image sets (seeSection 5.1).

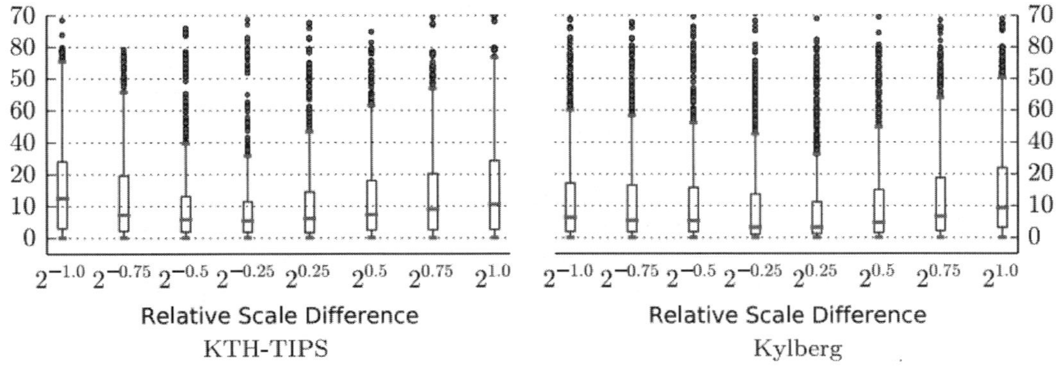

Figure. 6: Absolute errors (in degrees) of the orientation estimation.

The results indicate that the orientation estimation method is robust in respect of image scaling. We see across all scales that the medians of the absolute errors are within a range of 5–10°. Experiments have shown that the standard multi-resolution LBP representation can compensate alignment differences of up to 10°, but fails for orientation differences above. In order to improve the orientation-adaptive representation we apply an error compensation technique based on the accumulation of LBP distributions at multiple orientations.

Orientation Estimation Error Compensation

We found that a distribution of LBP with a small amount of misaligned patterns (a systematic error) will be dominated by the majority of correctly aligned patterns. As a consequence, we accumulate the distribution of LBP based on multiple orientations within an interval of $\pm\Delta o=20°$ of the estimated global orientation o. Experiments show that by using this approach an estimated error of up to 20° can be compensated without a significant loss of discriminative power of the feature representation. Fig. 7 illustrates this error compensation technique.

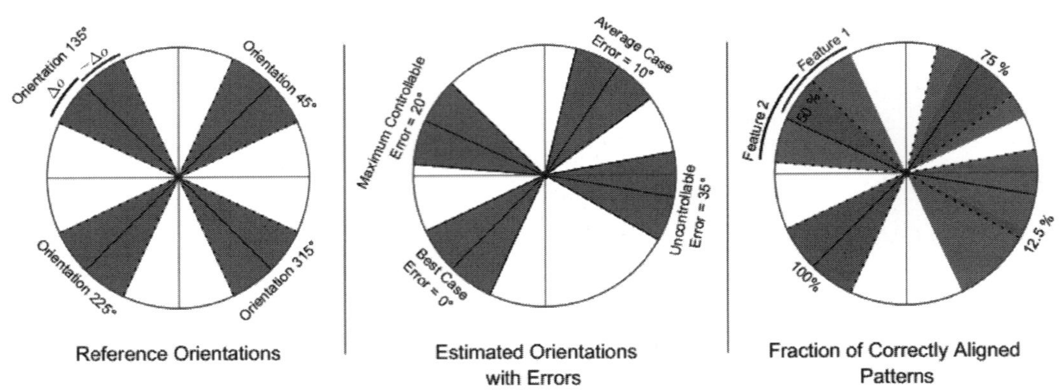

Figure. 7: Orientation estimation error compensation using accumulated pattern distributions.

To improve the reliability of this scheme, we use thresholding to avoid heavy fluctuation of bits due do interpolation artifacts. The modified sign function sgn(x) used in computing the individual patterns therefore requires $x \geq T$ to map to 1. The value of T is selected adaptively based on the Gaussian filtered image Ig, to accommodate for the adapted image properties, as the square root of the standard deviation of all pixel values in Ig.

Computation of Orientation- and Scale-Adaptive LBP (SOA-LBP)

To compute SOA-LBP in reference to a texture class l, estimated global image scale s, global orientation o, base-radius ρ and n neighbors, the position of neighbor k is adapted as

$$\eta_{l,s,o}^{\rho,n}(k;x,y) = \begin{pmatrix} x+\lambda(s,l,\rho)\cos\left(o+\frac{2\pi k}{n}\right) \\ y-\lambda(s,l,\rho)\sin\left(o+\frac{2\pi k}{n}\right) \end{pmatrix}^T . \tag{15}$$

The actual computation of LBP then follows the scheme of the scale-adaptive LBP as depicted in Section 2.4. To accommodate for the ambiguous orientation of multi-scale SMMs, we compute two patterns with initial sample positions at o and $o+\pi$ respectively. Fig. 8 illustrates the computation of scale- and orientation-adaptive LBP schematically. The red sampling points indicate the initial sample positions.

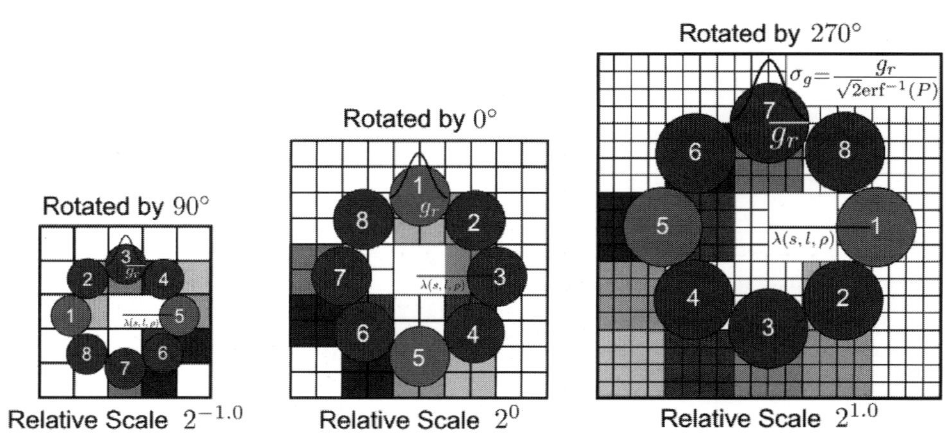

Figure. 8: Schematic computation of scale- and orientation-adaptive LBP. (For interpretation of the references to color in this figure caption, the reader is referred to the web version of this paper.)

SOA-LBP IN A MULTI-RESOLUTION FEATURE REPRESENTATION

The computation of multiple LBP-features (histograms) per image, each in reference to an individual trained-base-scale, requires the construction of a meta-feature-representation for classification. We abstract the set of computed LBP-features per image as a single SOA-LBP meta-descriptor and define a meaningful distance function between a pair of such descriptors. A meaningful distance exists only between LBP-features computed in reference to the same trained-base-scale. As a consequence, we define the distance between LBP-features computed at different trained-base-scales as ∞. Experimentation has shown that LBP-features computed at incorrectly adapted scales generally yield a significantly higher intra-class variability as compared to LBP-features computed at correctly adapted scales. The distance between two meta-descriptors is therefore defined as the minimum distance between all pairs of LBP-features abstracted by the descriptors. For two SOA-LBP meta-descriptors M_1 and M_2, both representing a set of LBP-features, each computed individually in reference to a texture class in the training data $\{H_1, \ldots, H_n\}$, the distance is defined as

$$D(M_1, M_2) = \min\{d(H_l, H_k) \mid H_l \in M_1 \wedge H_k \in M_2\}, \tag{16}$$

with

$$d(H_l, H_k) = \begin{cases} 1 - \sum_{i=1}^{N} \min(H_l(i), H_k(i)) & \text{if } l = k \\ \infty & \text{if } l \neq k. \end{cases} \tag{17}$$

In our implementation the histogram-intersection is used as a measure for similarity. A notable drawback of using the meta-descriptor abstraction is that it does not easily integrate with all classification methodologies. We therefore restrict the experimentation in this work to a classification method that allows for a straight forward integration (a standard k-nearest neighbors classifier).

Algorithm 1.

Selection of valid multi-resolution feature subsets.

Data: Let H^1l and H^2l be the sets of multi-resolution LBP-features
(histograms) computed in reference to texture class l at the base-radii
$\rho = \{\rho_1, \rho_2, \rho_3\}$ for two images with estimated scales s_1 and s_2.
$H_l^1 = \{h^1_{l,\rho_1}, h^1_{l,\rho_2}, h^1_{l,\rho_3}\}$ and $H_l^2 = \{h^2_{l,\rho_1}, h^2_{l,\rho_2}, h^2_{l,\rho_3}\}$
Result: Valid subsets V_1, V_2 of features from H^1l and H^2l.
$V_1 = H_l^1$ and $V_2 = H_l^2$
foreach $\rho_i \in \rho$ **do**
$r_1 = \lambda(s_1, l, \rho_i)$ // intrinsic−scale−adapted LBP−radius of h^1_{l,ρ_i} $r_2 = \lambda(s_2, l, \rho_i)$ // intrinsic−scale−adapted LBP−radius of h^2_{l,ρ_i} **if** $\min(r_1, r_2) < 1$ **or** $\max(r_1, r_2) > 5.44$ **or** $\max(r_1, r_2)/\min(r_1, r_2) > 3$ **then** $V_1 = V_1 \setminus h^1_{l,\rho_i}$ $V_2 = V_2 \setminus h^2_{l,\rho_i}$ **end**
end

Ojala et al. [24] suggest to compute multiple LBP-features, each at separate fixed LBP-radii, to improve the discriminative power of the feature representation. Multi-resolution LBP-features are then created from a set of standard LBP-features by concatenation.

We combine the rotation- and scale-invariant SOA-LBP in a multi-resolution feature representation, to improve the general discriminative power, by reducing the required amount of low-pass filtering for adapting the sampling area and adding the capability of describing underlying microstructures at multiple scales.

Experimental results on various image texture sets suggest that the discriminative power of the multi-scale LBP representation starts to decrease at scales larger than LBP-scale 3 (this corresponds to a radius larger than 5.44 pixels). We therefore consider radii within the interval $[1; 5.44]$ to be the most discriminative. To compute scale-adaptive patterns at multiple resolutions, we use a set of distinct base-radii for intrinsic-scale adaption $\rho = \{\rho_1, \rho_2, \rho_3\} = \{1.5, 3, 4.5\}$, instead of relying on a single base-radius. Hence, a multi-resolution SOA-LBP representation computed in reference to texture class l consists of the set of SOA-LBP-features computed at each of the base-radii and is denoted as $H_l = \{h_{l,\rho 1}, h_{l,\rho 2}, h_{l,\rho 3}\}$.

The specific values for the base radii were chosen to guarantee a high discriminative feature representation for texture image at small scale differences and a minimum amount of required low-pass filtering during scale-adaption for textures at larger scale differences. The values were chosen to augment the default base-radius of 3 in equal steps within the interval of the most discriminative radii.

Considering the small radius $\rho_1=1.5$ as well as the large radius $\rho_3=4.5$ it is likely that either the lower- or the upper-bound on discriminative LBP-radii is violated for a considerable amount of images, which effectively reduces the discriminative power of the multi-resolution representation. We therefore adaptively select the best subset of SOA-LBP-features for constructing the multi-resolution representation during each computation of the distance between two SOA-LBP meta-descriptors (see Algorithm 1).

Once the best subset of SOA-LBP-features is identified for a pair of meta-descriptors, the final multi-resolution representation is constructed by simple concatenation of the normalized histograms. Note that as a consequence of considerably different intrinsic-scales, or a failed scale estimation, the possibility of $V_1=V_2=\emptyset$ exists. In such a case, it is likely that the two SOA-LBP-features represent different texture classes. We consider such a pair of features as incomparable in a scale-adaptive sense and define the distance as ∞.

EXPERIMENTS

We evaluate the proposed SOA-LBP in reference to a set of scale- and orientation-invariant methods, representative for all categories discussed in Section 1. To assess the reliability of the intrinsic-scale-adaption for a large number of textures, we rely on three different images sets for experimentation. We specifically study the scale-invariance properties (Section 5.4) as well as the effects of combined scaling and rotation (Section 5.5). We finally present a runtime performance analysis of the SOA-LBP (Section 5.6) in relation to the compared methods.

IMAGE DATA

We perform the experimentation on three image sets with appropriate characteristics.Table 1 summarizes the most important information about the used data.

Table 1: Information on the image sets used for experimentation

Database	Class-es	Images per scale	Scales	Training scale
CURET	4	184	2	Mixed
KTH-TIPS	9	81	9	2^0
Kylberg	25	500	9	2^0

CURET. The CURET image set contains data with different viewing and illumination conditions. In a four-class classification scenario, textures at two different scales are available as 200×200 pixel images. The scale difference of the textures is reported to be approximately 1.7. As a consequence of the significant amount of signal noise in the CURET data, this image set provides an interesting opportunity to evaluate the effects of noise on the proposed method.

KTH-TIPS. The KTH-TIPS [29] image set consists of images from 10 different materials captured at 9 individual relative scales between $2^{-1.0}$ and $2^{1.0}$ with 9 samples per material. Due to the dimension of the original images of material "cracker" (the texture would only fill half of the images at certain scales), we could not use this class for simulating rotations and consequently removed the class in all experiments, leading to a classification scenario with only 9 classes. Sub-images of size 128×128 pixels were extracted from the center of each image to be consistent with the orientation evaluation experiments.

Kylberg. The Kylberg texture set [30] consists of 28 materials captured at a single camera-scale. The data set contains rotated versions of each image at 30 degree steps within a range of 0–330°. The large image size (576×576 pixels each) allows us to simulate signal scaling without relying on up-sampling, which leads to a reduced amount of unwanted interpolation artifacts. We simulated scaling to match the scales of the KTH-TIPS set such that the scale of the original images is interpreted as the maximum scale$2^{1.0}$ (KTH-TIPS scale 1). Sub-images of size 128×128 pixels were then extracted from the center of the re-scaled images to build the image sets. We created two distinct sets for experimentation, a training set consisting of 20 unique texture patches (types a and b) per material and an evaluation sets comprised of 20 unique texture patches (types c andd) per class. Note that the texture classes rice1 and rice2 as well as stone1, stone2 and stone3, respectively show minimal visual distinction in textural appearance. As a consequence we removed the texture classes rice2, stone2 and stone3 to improve the interpretability of the experiments, leading to a classification scenario with 25 classes.

Compared Feature Extraction Methods

We compare the proposed SOA-LBP to a set of methods, representative for the four categories of scale- and rotation-invariant methods, as discussed in Section 1. We believe that the conceptual properties used by these methods will allow us to establish a comprehensive overview. The used methods are as follows:

Category I. DT-CWT with Log-Polar Transform (Log-Polar[1]).
Category II. Dominant Scale (Dominant Scale[7]).
Category III. Fractal Analysis using Filter Banks (MFS MR8[10]) and Intersecting Cortical Model (ICM[11]).
Category IV. Affine Invariant Regions (Affine Regions[19]) and Fisher vector encoding of dense SIFT descriptors (Dense SIFT[31]). We also compared the method to a standard, multi-resolution LBP with 3 scales (LBP[24]) and the proposed scale-invariant LBP representation of Li et al. (Li-LBP[21]).

Evaluation Protocol and Presentation of Results

We implemented the experiments in a scale-constrained cross-validation scheme to accommodate for the rather small size of the KTH-TIPS image set. The scheme is based on two distinct sets for training and evaluation. Images for training were always selected from a fixed scale (the default training scale, see Table 1), while the scales for evaluation varied according to the specific experiment. This approach allows us to study the characteristics of each method in reference to signal scaling at various scale differences.

Cross-validation was then performed by an iterated random selection (consistent among all methods) of subsets from the training set (75%) and the evaluation set (25%). A standard k-nearest neighbors classifier was used for classification of features extracted from the specific image subsets. The maximum k-value corresponds to the number of images in each class of the training set (at maximum 20). The reported results represent the mean accuracy over all k-values, averaged in a scale-constrained cross validation with 100 iterations.

We report statistical significance on a per-figure basis to improve the readability. Two-tailed Wilcoxon rank-sum tests were performed at a significance level $\alpha = 0.001$ to assess the null-hypothesis that the population median of the cross-validation results obtained with the proposed methodology (SOA-LBP) is equal to the medians of all corresponding methods presented in the specific figure. An arrow pointing upwards (↑) indicates that the null-hypothesis could always be rejected and the SOA-LBP performed significantly better as compared to all corresponding methods in the figure. An arrow pointing to the right (→) indicates that the null-hypothesis could not be rejected at least once but no significant difference could be identified. Finally an arrow pointing downwards (↓) indicates that at least one method performed significantly better as compared to the proposed method. Note that the markers of each plot are slightly displaced on the x-axis to improve the readability of the error-bars, which represent the standard deviations of the individual cross-validation results.

We present the results based on the CURET image set using asymmetric bar charts (Fig. 12). Each side of a bar represents the classification accuracy of a single experiment. The slope of the bar gives an indication of the scale-invariance of each method. The dashed lines represent the average classification accuracies of both experiments. The arrows indicate statistical significance in relation to the SOA-LBP (e.g. an arrow pointing downwards indicates that the specific method performed significantly worse as compared to the proposed methodology).

Studying the Effects of Image Scaling

The first set of experiments is aimed specifically at studying the characteristics of each evaluated method in regard to image scaling. In these experiments, we only use the scale-invariant representation of methods that allow us a selective use of rotation-invariant features. This includes LBP, Li-LBP, SOA-LBP and Dominant Scale. We present the results of the experiments based on the KTH-TIPS, Kylberg and CURET image sets without rotation in Fig. 9, Fig. 10, Fig. 11 and Fig. 12. Images at scale 2^0 were used for training, images at all other available scales were used for evaluation (KTH-TIPS and Kylberg).

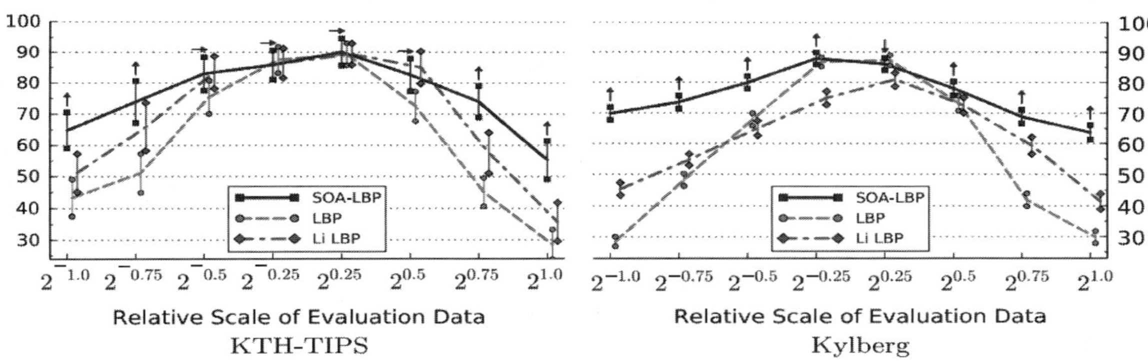

Figure. 9: Classification accuracy (y-axis) for evaluation scales (scaling only).

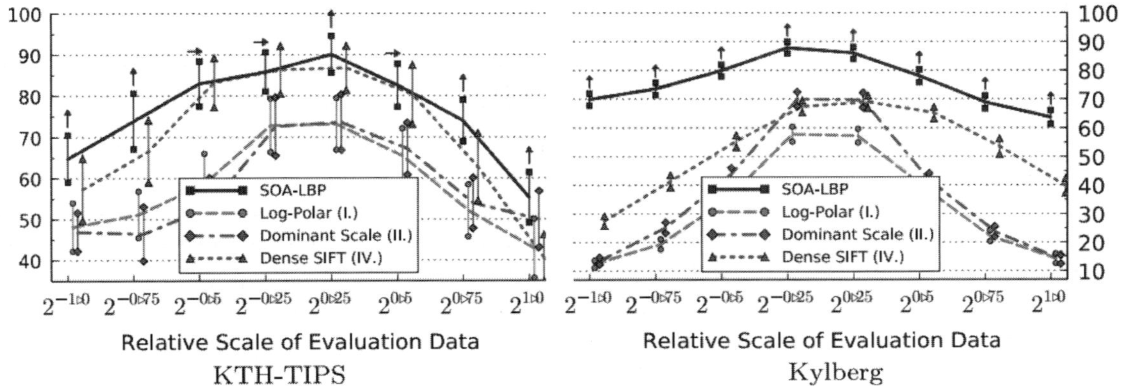

Figure 10: Classification accuracy (y-axis) for evaluation scales (scaling only).

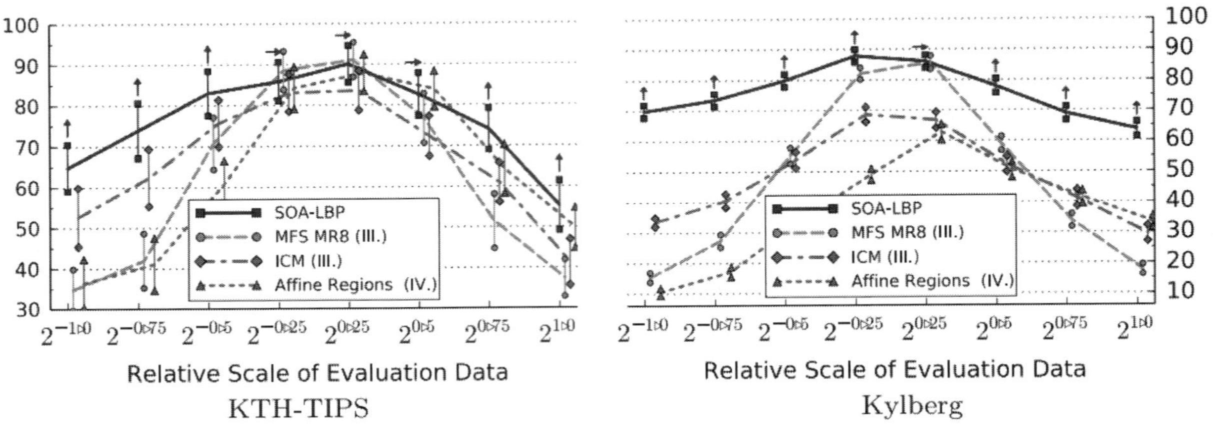

Figure. 11: Classification accuracy (y-axis) for evaluation scales (scaling only).

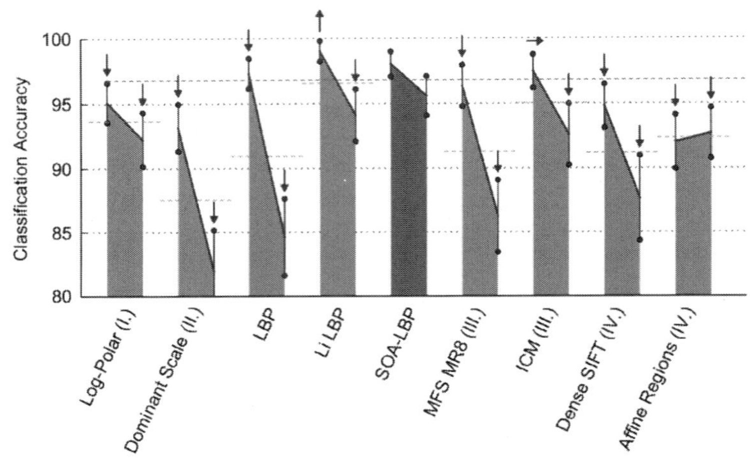

Figure. 12: Classification accuracy of the experiments on the CURET data.

Based on the CURET data, we follow the experimental setup used by Varma and Zisserman [32]. Two separate training sets were constructed. The first training set consists of textures at both scales, while the second training set is based on textures at a single scale. The evaluation set contains textures at both scales. The difference between the two experiments give an indication for the scale-invariance of each method.

Considering the experiments on the KTH-TIPS image set, we observe that the SOA-LBP performs comparably to the majority of evaluated methods, at evaluation scales close to the training scale. No method performed significantly better as compared to the proposed methodology however, which indicates that the multi-resolution SOA-LBP feature representation is competitive in scenarios with minimal to no scaling. In case of large scale differences (starting at $2^{0.75}, 2^{-0.75}$) between the training and evaluation data, the SOA-LBP significantly outperforms all evaluated methods.

In parallel to the experiments on the KTH-TIPS data, the SOA-LBPs performance is significantly better as compared to all evaluated methods at large scale differences considering the Kylberg experiments. In contrary to the previous experiments however, this behavior is already recognized at relative scale differences of $2^{0.5}$ and $2^{-0.5}$. The results indicate that the used multi-resolution representation provides highly discriminative features in the more challenging classification problem provided by the Kylberg set, even at tiny scale differences ($2^{0.25}$, $2^{-0.25}$). The only method that performed significantly better as compared to SOA-LBP was the standard multi-resolution LBP at relative scale $2^{0.25}$, which is caused by a small amount of erroneously estimated image scales of the proposed method. Interestingly, the Li-LBP method performed significantly worse even for small scale differences as compared to the standard LBP as well as the proposed method. We assume this characteristic is caused by the direct mapping from estimated scale to the LBP-radius (the average intrinsic-scale of the Kylberg set is higher as compared to the KTH-TIPS data) in combination with a missing, more powerful, multi-resolution representation.

The experiments on the CURET data indicate a high degree of scale-invariance of the SOA-LBP. Only the Li-LBP method performed significantly better in the experiment without required scale-invariance (mixed training scales). The results on the CURET set show that the SOA-LBP is suited for classification in noisy scenarios, outperforming the majority of evaluated methods.

The experiments indicate that the proposed SOA-LBP provides significantly improved classification accuracies in scenarios with large scale differences. The use of intrinsic-scale-adaption allows the computation of discriminative features for a variety of different textures, while the multi-resolution representation provides highly competitive features even in scenarios with tiny scale differences.

Studying the Effects of Combined Image Rotation and Scaling

The effects of combined rotation and scaling are studied in the second set of experiments. Feature extraction is based on rotated versions of the Kylberg and the KTH-TIPS image sets. Images at scale 2^0 without rotation were used for training, images at all other available scales were used for evaluation. Subsets of the evaluation sets (KTH-TIPS 891 and Kylberg 1250 images), rotated in steps of 30°, in angles between 30° and 330°, were randomly selected (consistently among all methods) for classification. Only methods providing a scale- and orientation-invariant feature representation were evaluated. LBP was used with the rotation-invariant encoding based on uniform patterns[24]. Li-LBP was used with the proposed sub-uniform patterns [21]. The results are presented in Fig. 13 and Fig. 14.

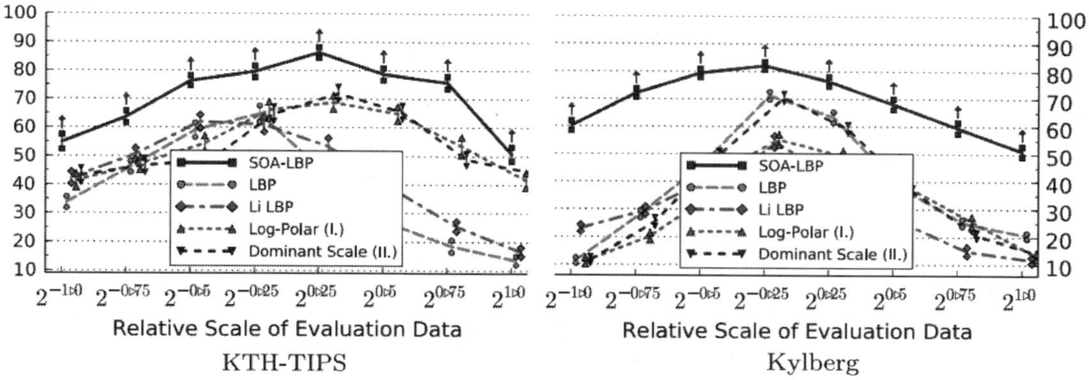

Figure. 13: Classification accuracy for evaluation scales (scaling and rotation).

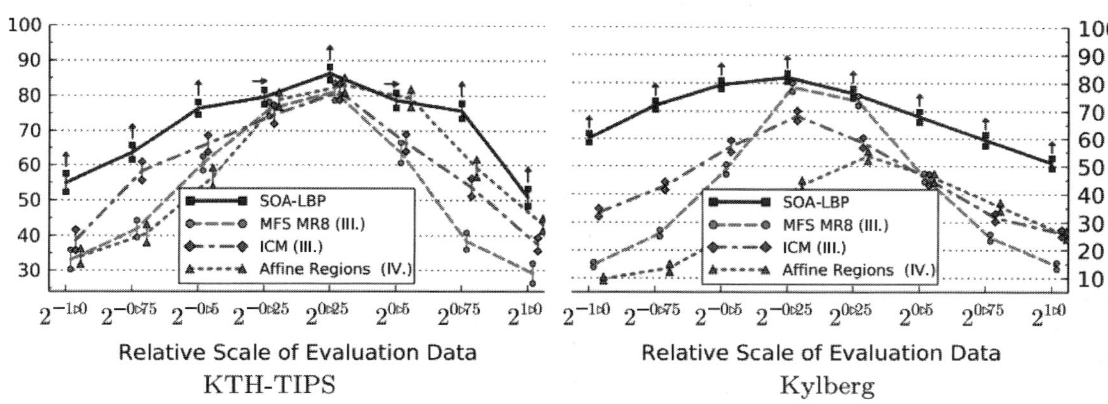

Figure. 14: Classification accuracy for evaluation scales (scaling and rotation).

We observe that the rotation of the images decreased the general accuracy of all methods as compared to the previous experiments. The results show the same trends as recognized in the scaling-only experiments however. Again, the proposed SOA-LBP provides significantly improved classification rates at large scale differences between training and evaluation data and performs highly competitive in scenarios with tiny scale differences. The results indicate that the proposed orientation-adaptive computation is superior as compared to encoding-level based approaches used by LBP and Li-LBP. Interestingly, the Li-LBP method performed worse as compared to the standard LBP method on the Kylberg data even at large scale differences. We assume this is caused by the combination of unsuitable LBP-radii (due to the missing intrinsic-scale-adaption) combined with the less discriminative sub-uniform encoding.

The experiments show that the proposed orientation-adapted computation integrates seamlessly into the scale-adaptive LBP. The results are consistent with the previous experiments (scaling only) and indicate that the extraction-level alignment improves the discriminative power of the features.

Runtime Performance Analysis

To study the computational demand of the proposed method, we analyze the required runtime of all considered methods in a multi-threaded Java implementation (JDK 8), running on an Intel i5-2500k processor at 4.29 GHz (using a higher frequency multiplier than the default). Due to the nature of the Java programming language (JIT-compilation and garbage collection), we report the computational demand per image as an average of the required computation time for 729 images from the KTH-TIPS data set, in a repeated (20 iterations) experiment (Fig. 15). Note that the presented performance should not be considered an exact benchmark, as not all methods have undergone equal amounts of optimization, but is meant to give the reader an idea of the computational complexity of the proposed methodology.

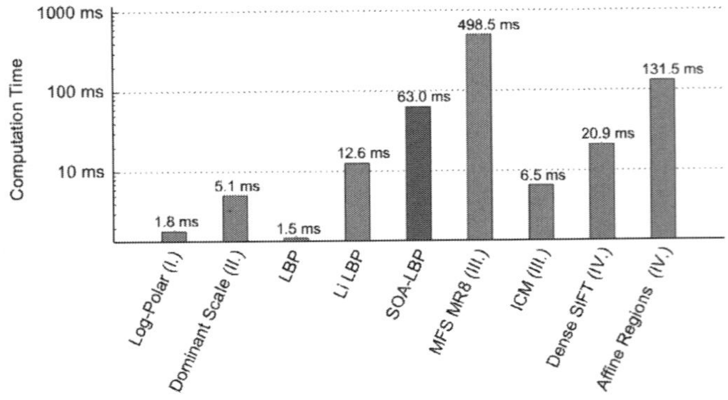

Figure. 15: Average computational time per image (KTH-TIPS).

The results show that the SOA-LBP is considerably slower as compared to the lightweight LBP or the Li-LBP method, which is caused by the increased demand of computing the scales-space, performing scale- and orientation-estimation and the extra amount of feature computation (performing intrinsic-scale-adaption). Considering the improved classification accuracy in environments with varying scales and orientations however, we think that the average computational demand of 63 ms per image is an adequate trade-off. This is even emphasized as the method ranks in the lower middle range among all methods.

CONCLUSION

We presented a generic methodology to compute a scale- and rotation-invariant feature representation based on LBP, by suitable adaption of the LBP neighborhood. The use of intrinsic-scale-adaption allowed the computation of features, independent of the intrinsic-scale of textures, and increased the reliability of the method significantly. This has been shown in experiments based on three different image sets representing a variety of scenarios. The SOA-LBP was significantly superior to all evaluated methods in case of large scale differences. The proposed multi-resolution feature representation was more than competitive in scenarios with tiny scale differences. Experimentation based on the noisy CURET data showed that the proposed methodology provides discriminative and reliable features in difficult scenarios.

Although the computational complexity of the SOA-LBP is significantly higher as compared to the very lightweight LBP, we regard the improved classification accuracies in scenarios with scaling and rotation as an acceptable trade-off for many classification tasks. The proposed methodology is easily applied to a wide variety of LBP based methods [26] and [27], providing a robust scale- and rotation-invariant feature representation.

ACKNOWLEDGMENTS

This work has been supported, in part, by the Austrian Science Fund (Project no. 24366).

REFERENCES

1. C.-M. Pun, M.-C. Lee, Log-polar wavelet energy signatures for rotation and scale invariant texture classification, IEEE Trans. Pattern Anal. Mach. Intell. 25 (5) (2003) 590–603.
2. I. Selesnick, R. Baraniuk, N. Kingsbury, The dual-tree complex wavelet transform, IEEE Signal Process. Mag. 22 (6) (2005) 123–151.
3. K. Jafari-Khouzani, H. Soltanian-Zadeh, Rotation-invariant multiresolution texture analysis using radon and wavelet transforms, IEEE Trans. Image Process. 14 (6) (2005) 783–795.
4. E. Lo, M. Pickering, M. Fratera, J. Arnold, Image segmentation from scale and rotation invariant texture features from the double dyadic dual-tree complex wavelet transform, Image Vis. Comput. 29 (1) (2011) 15–28.
5. F. Riaz, A. Hassan, S. Rehman, U. Qamar, Texture classification using rotationand scale-invariant Gabor texture features, IEEE Signal Process. Lett. 20 (6) (2013) 607–610.
6. E.H.S. Lo, M.R. Pickering, M.R. Frater, J.F. Arnold, Query by example using invariant features from the double dyadic dual-tree complex wavelet transform, in: CIVR, 2009, pp. 1–8.
7. J.A. Montoya-Zegarra, N.J. Leite, R. Torres, Rotation-invariant and scaleinvariant steerable pyramid decomposition for texture image retrieval, in: Proceedings of the XX Brazilian Symposium on Computer Graphics and Image Processing, 2007, pp. 121–128.
8. J. Han, K.-K. Ma, Rotation-invariant and scale-invariant Gabor features for texture image retrieval, Image Vis. Comput. 25 (9) (2007) 1474–1481.
9. K.-K. Fung, K.-M. Lam, Rotation- and scale-invariant texture classification using slide matching of the Gabor feature, in: ISPACS, 2009, pp. 521–524.
10. M. Varma, A. Zissermann, A statistical approach to texture classification from single images, Int. J. Comput. Vis. 62 (1–2) (2005) 61–81.
11. Y. Ma, L. Liu, K. Zhan, Y. Wu, Pulse coupled neural networks and one-class support vector machines for geometry invariant texture retrieval, Image Vis. Comput. 28 (11) (2010) 1524–1529.
12. K. Zhan, H. Zhang, Y. Ma, New spiking cortical model for invariant texture retrieval and image processing, IEEE Trans. Neural Netw. 20 (12) (2009) 1980–1986.
13. Q. Xu, Y.Q. Chen, Multiscale blob features for gray scale, rotation and spatial scale invariant texture classification, in: ICPR, 2006, pp. 29–32.
14. T. Lindeberg, Feature detection with automatic scale selection, Int. J. Comput. Vis. 30 (2) (1998) 79–116.
15. K. Mikolajczyk, C. Schmid, Scale and affine invariant interest point detectors, Int. J. Comput. Vis. 60 (1) (2004) 63–86.
16. D.G. Lowe, Object recognition from local scale-invariant features. in: ICCV, vol. 2, IEEE, Washington, DC, USA, 1999, pp. 1150–1157.
17. H. Bay, A. Ess, T. Tuytelaars, L. Van Gool, Speeded-up robust features (SURF), Comput. Vis. Image Underst. 110 (3) (2008) 346–359.
18. P. Mainali, G. Lafruit, Q. Yang, B. Geelen, L. Gool, R.

Lauwereins, Sifer: scaleinvariant feature detector with error resilience, Int. J. Comput. Vis. 104 (2) (2013) 172–197.

19. S. Lazebnik, C. Schmid, J. Ponce, A sparse texture representation using local affine region, IEEE Trans. Pattern Anal. Mach. Intell. 27 (8) (2005) 1265–1278.
20. S. Hegenbart, A. Uhl, A. Vécsei, G. Wimmer, Scale invariant texture descriptors for classifying celiac disease, Med. Image Anal. 17 (4) (2013) 458–474.
21. Z. Li, G. Liu, Y. Yang, J. You, Scale- and rotation-invariant local binary pattern using scale-adaptive texton and subuniform-based circular shift, IEEE Trans. Image Process. 21 (4) (2012) 2130–2140.
22. T. Lindeberg, Scale-Space Theory in Computer Vision, 1994.
23. T. Ojala, M. Pietikäinen, D. Harwood, A comparative study of texture measures with classification based on feature distributions, Pattern Recognit. 29 (1) (1996) 51–59.
24. T. Ojala, M. Pietikäinen, T. Mäenpää, Multiresolution gray-scale and rotation invariant texture classification with local binary patterns, IEEE Trans. Pattern Anal. Mach. Intell. 24 (7) (2002) 971–987.
25. Z. Guo, L. Zhang, D. Zhang, Rotation invariant texture classification using LBP variance (LBPV) with global matching, Pattern Recognit. 43 (3) (2010) 706–719.
26. S. Hegenbart, A. Uhl, A scale-adaptive extension to methods based on LBP using scale-normalized Laplacian of Gaussian extrema in scale-space, in: ICASSP, 2014, pp. 4352–4356.
27. S. Hegenbart, A. Uhl, An orientation-adaptive extension to scale-adaptive local binary patterns, in: ICPR, 2014, pp. 1–6.
28. M. Abramowitz, I.A. Stegun, Handbook of Mathematical Functions with Formulas, Graphs, and Mathematical Tables, Dover, New York, 1964.
29. E. Hayman, B. Caputo, M. Fritz, J.-O. Eklundh, On the significance of real-world conditions for material classification. in: ECCV, vol. 3024, Springer, Berlin, Heidelberg, 2004, pp. 253–266.
30. G. Kylberg, The Kylberg texture dataset v. 1.0, External Report (Blue Series) 35, Center for Image Analysis, Swedish University of Agricultural Sciences, Uppsala University, Uppsala, Sweden, September 2011.
31. F. Perronnin, C. Dance, Fisher kernels on visual vocabularies for image categorization, in: CVPR, 2007, pp. 1–8.
32. M. Varma, A. Zisserman, A statistical approach to material classification using image patch exemplars, IEEE Trans. Pattern Anal. Mach. Intell. 31 (11) (2009) 2032–2047.

CHAPTER 4

EUCLIDEAN-DISTANCE-BASED CANONICAL FORMS FOR NON-RIGID 3D SHAPE RETRIEVAL

David Pickup·Xianfang Sun[1]·Paul L. Rosin·Ralph R. Martin

School of Computer Science and Informatics, Cardiff University, Queen's Buildings, 5 The Parade, Roath, Cardiff CF24 3AA, United Kingdom

ABSTRACT

Retrieval of 3D shapes is a challenging problem, especially for non-rigid shapes. One approach giving favourable results uses multidimensional scaling (MDS) to compute a canonical form for each mesh, after which rigid shape matching can be applied. However, a drawback of this method is that it requires geodesic distances to be computed between all pairs of mesh vertices. Due to the super-quadratic computational complexity, canonical forms can only be computed for low-resolution meshes. We suggest a linear time complexity method for computing a canonical form, using Euclidean distances between pairs of a small subset of vertices. This approach has comparable retrieval accuracy but lower time complexity than using global geodesic distances, allowing it to be used on higher resolution meshes, or for more meshes to be considered within a time budget.

INTRODUCTION

Content-based 3D shape retrieval is a key research topic, as the large and ever increasing number of available 3D models makes effectively searching for models with a desired shape an increasingly important task. In the case of non-rigid models, where similar objects may appear in many different poses, the task of accurately comparing shapes is an especially challenging problem. One of the most effective approaches to solving this non-rigid retrieval problem is to convert each shape to a *canonical form*, i.e. a standard pose, and perform rigid shape retrieval on those [1] and [2]. Lian et al. [3] gave a method which computes a canonical form (see Fig. 1) for a mesh using the method of Elad and Kimmel [4] to map the geodesic distances between every pair of points on the surface to 3D Euclidean distances using multidimensional scaling (MDS). A view-based method is then used to perform shape retrieval. The drawback of this method is the high,*super-quadratic*, computational cost of geodesic distance computation, which requires the models to be simplified to approximately 2000 vertices to achieve a reasonable run-time.

Citation: David Pickup, Xianfang Sun, Paul L. Rosin, Ralph R. Martin, Euclidean-distance-based canonical forms for non-rigid 3D shape retrieval, Pattern Recognition, Volume 48, Issue 8, August 2015, Pages 2500-2512, ISSN 0031-3203, http://dx.doi.org/10.1016/j.patcog.2015.02.021.

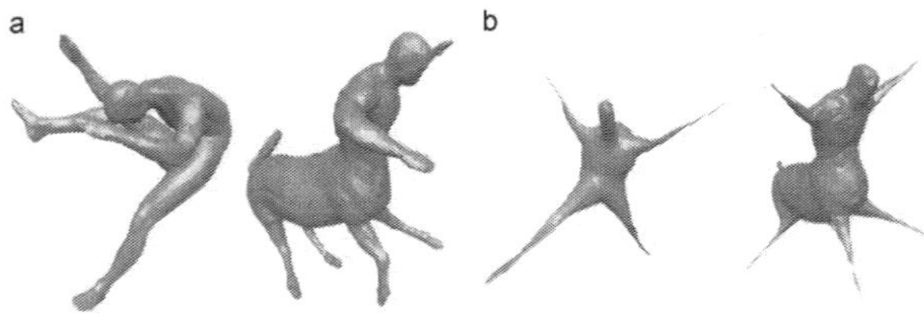

Figure. 1: Canonical forms produced using geodesic distances and multidimensional scaling. The limbs of the models are stretched so that their extremities are distant from one another. (a) Original meshes and (b) Canonical forms.

Our method computes a canonical form without the need for geodesic distances. Instead of mapping geodesic distances to Euclidean distances, we instead maximise the Euclidean distances between a subset of feature points while attempting to preserve the original mesh edge lengths. These feature points are selected based upon the conformal factor of the vertices [5]. This is a pose-invariant measure that represents the amount of local work required to globally transform the mesh into a sphere.

We select $\sqrt{N}$ feature points, where N is the number of mesh vertices, which allows the distances between each pair of feature points to be computed in *linear* time. This reduced computational complexity compared to using geodesic distances means that our algorithm can not only produce canonical forms much faster for simplified meshes, but also allows canonical forms to be computed for higher resolution meshes within a reasonable time. When substituting our canonical forms into Lian et al.'s retrieval method[3], we achieve comparable results to those using Elad and Kimmel's canonical forms [4], but much more quickly. We are also able to produce canonical forms for the dataset used by Lian et al., but without the need to simplify the models first. Another way in which greater speed can be put to use is to allow a larger number of meshes to be compared within a fixed time budget if the search space is a large database.

RELATED WORK

Many works consider rigid shape matching, including ones based on *lightfield* descriptors[6] and *spherical harmonics* [7]. For a detailed review of rigid shape retrieval methods, we refer the reader to two survey papers [8] and [9].

Since our work contributes to non-rigid shape retrieval, we concentrate our review on this topic. Such work is based on shape invariance under isometric transformations. A popular approach compares models based on their local features. Some of these methods directly follow successful methods in computer vision, including *meshSIFT* [10]and *meshHOG* [11]. Sun et al. [12] proposed *heat kernal signatures* (HKS) as a shape descriptor. Inspired by *Video Google* [13], Bronstein et al. [14] used the heat kernel signature and the bags-of-features approach to produce shape signatures for retrieval. Abdelrahman et al. [15] also proposed a similar shape retrieval method based on HKS. Hou et al. [16] used HKS and a novel *bag-of-features graph* method which incorporates spatial relationships of local features. Wang et al. [17] proposed an extension to *spin images* [18] to allow for non-rigid matching, computing intrinsic spin images from a high-dimensional MDS embedding of a 3D model. Both Tabia et al. [19] and Koury et al. [20]describe each detected feature point of a model in terms of a set of local closed curves on its surface.

A second category of techniques uses graphs to represent objects. Hilaga et al. [21] use*multiresolution Reeb graphs* to match the topology of 3D shapes. Sundar et al. [22] use a*thinning* technique to produce an object's *skeleton*, and graph matching to compare skeletons. Sfikas et al. [23] produce a graph representation using geometry-based*discrete conformal factors* [5].

Methods using global information have also been proposed. The *shapeDNA* method of Reuter et al. [24] uses the Laplace–Beltrami operator to provide an isometrically invariant shape descriptor. Smeets et al. [25] and [26] showed that both singular value decomposition and a histogram of the *geodesic distance matrix* can be used as global shape descriptors. They showed that singular value decomposition is especially effective for shape retrieval. Jain and Zhang [27] rely on a *spectral embedding* of a 3D object using the eigenvalues of the geodesic distance matrix. Bronstein et al. [28] directly compute the distance between two non-rigid shapes by formulating the *Gromov–Hausdorff distance*as a multidimensional-scaling-like continuous optimisation problem. Tam and Lau [29]showed that manifold learning can be used to improve the shape retrieval accuracy of existing methods.

Elad and Kimmel [4] proposed computing a canonical form for a mesh using*multidimensional scaling* (MDS) to map the geodesic distances of a mesh into three-dimensional Euclidean distances. Rigid matching can then be performed on the resulting canonical forms. Lian et al. [3] used these canonical forms, along with a *view-based*retrieval method to match non-rigid shapes. It renders depth images of 66 different views of an object, extracts SIFT features from each of these images, and uses the bag of features approach to form a shape descriptor from these features. Recently, Lian et al.[30] created a *feature preserving* canonical form by transforming parts of the original mesh to positions and orientations that correspond to equivalent segments of the mesh's canonical form. Lian et al. [31] have also used a *convexity measurement* of the canonical forms as a global feature for shape retrieval. The drawback of these methods is the*super-quadratic* computational complexity of finding the geodesic distances between all pairs of vertices, requiring the canonical forms to be computed from a simplified version of the original mesh in practice.

Ying et al. [32] gave a parallel geodesic distance algorithm to improve the efficiency of computing these distances. This algorithm still exhibits higher computational complexity than our method described below, but claims an order of magnitude improvement of computation time for meshes containing more than 500,000 faces. However, even if the computation time of the fastest geodesic-distance-based retrieval method we use as a basis for comparison in Section 6 were an order of magnitude faster, our method would*still* be 3.5 times faster than an approach using Ying's method, even for meshes with a resolution of about 10,000 vertices. Unfortunately, as the authors of [32] have not made their code available, we are unable to directly compare computation speeds with a canonical form method using their algorithm. Ultimately, however parallelisation can only lead to a constant factor speed-up, and not a decrease in computational complexity.

A similar approach to ours has been proposed by Wang and Zha [33], who also avoid computing geodesic distances between all pairs of vertices. They also select feature points, but based on local maxima and minima of geodesic distances to the two most*geodesic distant vertices* [34]. They then use the geodesic distances between all pairs of feature points to create target axes used to align local sets of geodesic contours. They compared the result of using their canonical forms for shape retrieval against classical MDS with all-pairs geodesic distances, but not using *least squares MDS*, which has been shown to produce improved results [4] and [3]. Their method is also much more complex than ours, and requires more parameters to be defined by the user. While their method is faster than ours for meshes of approximately 9000 vertices, they did not examine how their method scales to finer meshes. We are unable to determine the exact time complexity of their method, as they omit details of how to minimise certain cost functions. We can deduce however that their geodesic computations have time complexity at least $O(MN \log N)$, where M is the number of feature points, and N is the number of mesh vertices. Our distance calculations have time complexity of $O(N)$, and so will eventually be faster for large enough meshes. Our retrieval results are also more accurate than theirs for each dataset tested later, but these are based on the use of a different retrieval method.

For further insight, we refer the reader to Lian et al. [2] for an in-depth comparison of several non-rigid 3D shape retrieval methods.

PRELIMINARIES

We first give a brief overview of the canonical form computation by Elad and Kimmel [4], which uses least squares multidimensional scaling to map geodesic distances into Euclidean distances. We also give an overview of the shape retrieval method by Lian et al. [3], which successfully uses these canonical forms for non-rigid shape retrieval, as we later substitute our canonical forms into this method in order to evaluate their effectiveness against those of Elad and Kimmel [4].

Multidimensional Scaling for Canonical form Computation.

Multidimensional scaling (MDS) maps a chosen distance measure between pairs of points to Euclidean distances in a low-dimensional space. The use of MDS to compute a pose invariant canonical form of a mesh was first proposed by Elad and Kimmel [4]. They use least squares MDS to map the *geodesic* distances between all points on a 3D mesh to 3D *Euclidean* distances. The geodesic distances are computed using the *fast marching method* of Kimmel and Sethian [35]. They use the *SMACOF* (scaling by maximising a convex function) algorithm [36] to compute the MDS. SMACOF minimises the following functional:

$$S(X) = \sum_{i=1}^{N} \sum_{j=i+1}^{N} w_{i,j}(\delta_{i,j} - d_{i,j}(X))^2, \tag{1}$$

where N is the number of vertices, $w_{i,j}$ are weighting coefficients, $\delta_{i,j}$ is the geodesic distance between vertices i and j of the original mesh, and $d_{i,j}$ is the Euclidean distance between vertices i and j of the resulting canonical mesh X.

This is solved by iteratively computing

$$X_i = V^+ B(X_{i-1}) X_{i-1}, \tag{2}$$

where the elements of matrix $B(X_{i-1})$ are

$$b_{i,j} = \begin{cases} -w_{i,j}\delta_{i,j}/d_{i,j}(X_{i-1}) & i \neq j \text{ and } d_{i,j}(X_{i-1}) \neq 0 \\ 0 & i \neq j \text{ and } d_{i,j}(X_{i-1}) = 0 \end{cases} \tag{3}$$

$$b_{i,i} = \sum_{j=1, j \neq i}^{N} b_{i,j}. \tag{4}$$

The matrix V is given by

$$V = \sum_{i=1}^{N} \sum_{j=i+1}^{N} w_{i,j} A_{i,j}, \tag{5}$$

$$A_{i,j} = (e_i - e_j)(e_i - e_j)^T, \tag{6}$$

where e_i is the i th column of the identity matrix. V^+ is the Moore–Penrose pseudo-inverse of V . The algorithm iterates until $S(X_i) - S(X_{i-1})$ is less than a user specified value ϵ. For more details, see [4].

This canonical form computation is used for 3D shape retrieval by Lian et al. [3]. Due to the super-quadratic computational expense, they first simplify the input meshes to approximately 2000 vertices. In Section 4 we give a method for computing canonical forms which instead of using a *quadratic* number of global *geodesic* distances, only uses the edge lengths of the mesh plus a *linear* number of *Euclidean* distances.

Non-Rigid Shape Retrieval Using Canonical Forms

The state-of-the art method for using canonical forms for non-rigid shape retrieval is due to Lian et al. [3]; their method performed very well in the 2011 Shape Retrieval Competition (SHREC'11) [2] and [37]. They extract a shape descriptor from a model using canonical forms and visual features.

An outline of the steps used is shown in Fig. 2. They first simplify the original model (Fig. 2(a)) to produce a new mesh with approximately 2000 vertices (Fig. 2(b)). They next produce a canonical form of the mesh (Fig. 2(c)), as explained in Section 3.1. It takes too long to process higher resolution meshes without simplification due to the use of all pairs of geodesic distances.

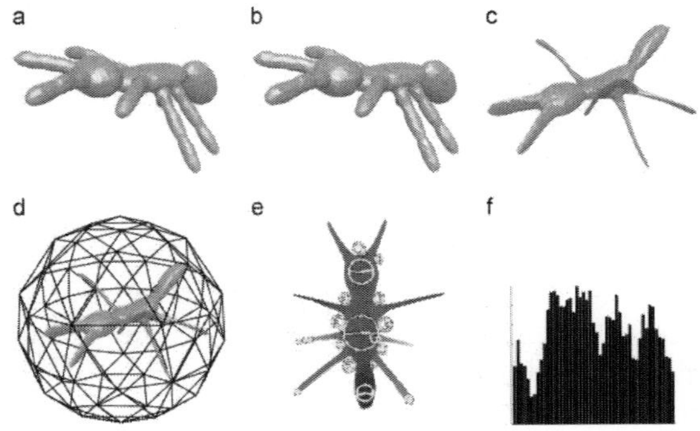

Figure. 2: Computing an object descriptor, according to Lian et al. [3]. (a) Original mesh. (b) Simplify mesh. (c) Compute canonical form. (d) Capture multiple depth images. (e) Extract SIFT features. (f) Construct bag of features.

Once the canonical form has been computed, the position and orientation of the model are normalised using a combination of PCA and rectilinearity [38]. A set of 66 depth images are captured of the model, from viewpoints located at the vertices of a geodesic sphere (Fig. 2(d)). For each of these depth images a set of SIFT features [39] are extracted (Fig. 2(e)). The set of SIFT features is used to generate a histogram (bag-of-features) for each image (Fig. 2(f)).

The distance between two models is computed by comparing the feature histograms for the models using the *maximum dissimilarity histogram intersection distance*.

Our method can be used in this framework to provide alternative canonical forms to those of Elad and Kimmel, significantly speeding up this bottleneck step of the matching algorithm, and eliminating the need to simplify the meshes to such a small number of vertices. We later show that we can compute canonical forms for the models in the SHREC'11 dataset without the need for simplification.

EUCLIDEAN DISTANCE BASED CANONICAL FORM COMPUTATION

Fig. 1 shows two canonical forms produced by Lian et al. [3] from their original meshes. The limbs and similar parts of articulated objects are stretched out, so that their extremities are distant from one another. We achieve this effect more efficiently by maximising the Euclidean distances between certain selected feature points on the mesh, while attempting to preserving the original edge lengths to ensure isometric deformation. The feature points are selected based upon the conformal factors of the vertices [5]. The conformal factor increases along the length of mesh protrusions, which results in high values at the extremities of the mesh.

We first scale the mesh so that the maximum distance of any point on its surface to the centroid of all vertices is one. We then use the method of Ben-Chen and Gotsman [5] to calculate the conformal factor of the mesh as

$$L\Phi = K^T - K^0, \tag{7}$$

where L is the discrete Laplace–Beltrami operator with cotangent weights [40], K^0 is a vector containing the Gaussian curvature for each vertex, and KT is a vector containing a target Gaussian curvature for each vertex.

The Gaussian curvature ^0v at a vertex v is defined as

$$\kappa_v^0 = \begin{cases} 2\pi - \sum_{t \in T_v} \theta_t & v \notin B \\ \pi - \sum_{t \in T_v} \theta_t & v \in B \end{cases}, \tag{8}$$

where $_t$ is the angle subtended at vertex v in triangle t, Tv is the set of triangles connected to v, and B is the set of all vertices on the mesh boundary (if any). The target curvature tv at a vertex v is set to

$$\kappa_v^t = \left(\sum_{i \in V} \kappa_i^0\right) \frac{\sum_{t \in T_v} \mathrm{Area}(t)/3}{\sum_{t \in T} \mathrm{Area}(t)}, \tag{9}$$

where V is the set of all vertices and T is the set of all triangles.

Eq. (7) can be solved for Φ using Cholesky factorisation. An example of the conformal factors of a mesh in various poses is shown in Fig. 3.

Figure. 3: Conformal factors for a human mesh model in varying poses. The conformal factor increases towards the ends of the model's limbs.

To obtain a set of feature points for a mesh with N vertices, we sample the $\sqrt{N}$ vertices which have the largest conformal factors and also satisfy the requirement that they are local maxima. A vertex is defined to be a local maximum if its conformal factor is greater than that of all its neighbours in a 2-ring neighbourhood. Our experiments show that points at the extremities of the mesh are likely to be present within this set of chosen feature points (Fig. 4). Our experiments on a previous dataset [41] show that using 1, 2, or 3-ring neighbourhoods all result in similar final retrieval results; using 2-ring gives marginally better performance. We select $\sqrt{N}$ feature points, as this is the largest number of features we can have while being able to compute the distances between all pairs of feature points in linear time (with respect to the number of mesh vertices). Reducing the number of feature points further would not reduce the overall time complexity of our algorithm, as other aspects of our method also have a linear time complexity. On the other hand, our experiments demonstrate that $\sqrt{N}$ feature points are sufficient to give good shape retrieval in our tests. A further reason for using exactly $\sqrt{N}$ feature points is given later.

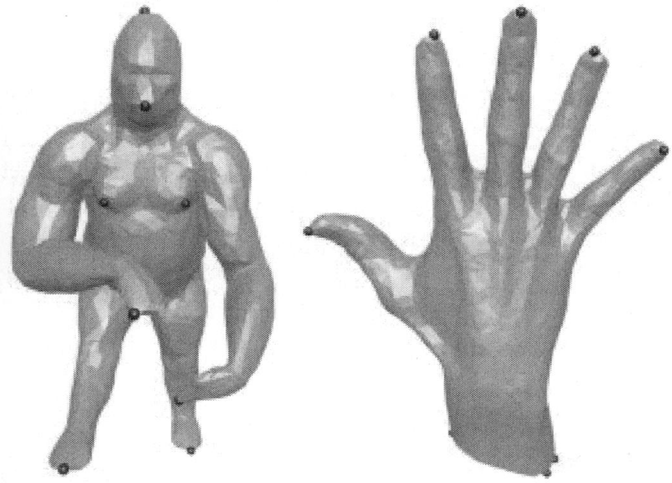

Figure. 4: Feature points selected using conformal factors. The mesh extremities are successfully included in this set.

We compute the canonical form of the mesh by setting the value of $\delta_{i,j}$ in Eq. (1) for all connected vertices i and j equal to the length of the edge connecting them. This aims to preserve the edge lengths of the mesh, to ensure isometric deformation. In order to maximise the distance between feature points, the value of $\delta_{i,j}$ for each pair of the $\sqrt{N}$ sampled vertices is set to a high value . We want this value to be large enough to straighten all the limbs of the model, and our experiments show 10 is large enough. As long as is large enough and the parameter discussed below is optimised accordingly, any value of can be chosen.

If the two vertices i and j are neither a pair of feature points nor connected by an edge, we do not enforce a target distance between them, so $\delta_{i,j}$ and $w_{i,j}$ are both set to zero for such cases. Not having to compute and optimise the distances between these points is crucial in keeping the linear time complexity of our distance calculations. The weights $w_{i,j}$ in Eq. (1) for all i and j that are connected by an edge are set to β/δ_{ij}^2, where is a user defined parameter for preserving edge lengths. (We consider the optimal choice for in the Experiments section later.) We divide by the square of the edge length δ_{ij}^2 so that the distance in Eq. (1) becomes a relative, rather than absolute, difference, making the weighting independent of the length of the edge. The conformal factor is normalised to lie in the interval $[0,1]$, and the entries in the weighting matrix $w_{i,j}$ for each pair of feature points are set to the mean of their conformal factors. This results in vertices which are nearer the ends of long 'limbs' of the object having a higher impact on the resulting canonical form, and avoids stretching out inappropriate parts of the mesh (Fig. 5). The SMACOF algorithm can then be used to minimise Eq. (1) as previously described in Section 3.1.

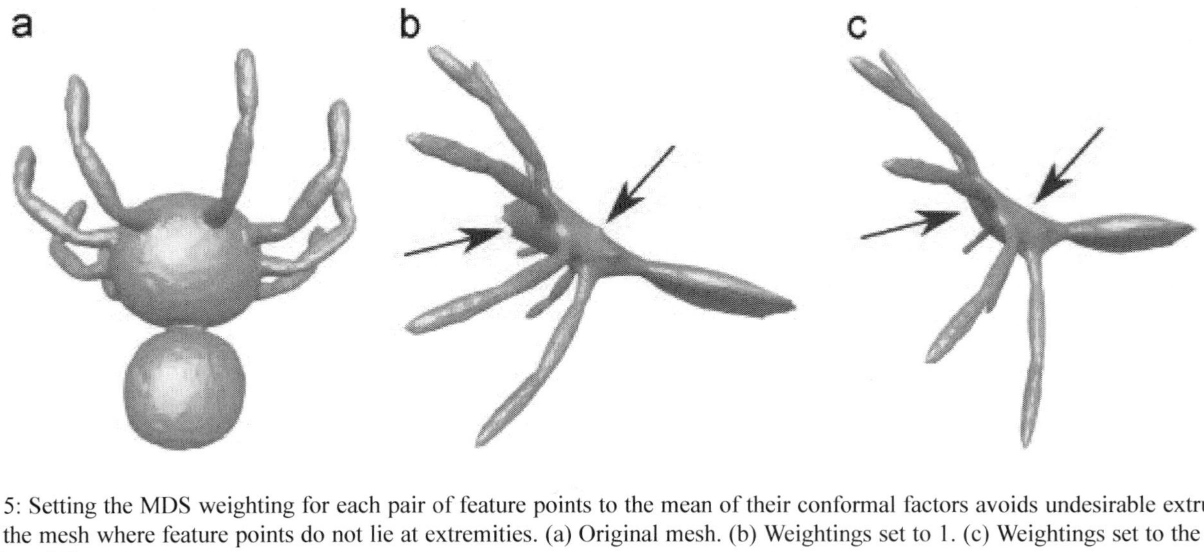

Figure. 5: Setting the MDS weighting for each pair of feature points to the mean of their conformal factors avoids undesirable extrusions at parts of the mesh where feature points do not lie at extremities. (a) Original mesh. (b) Weightings set to 1. (c) Weightings set to the mean of the conformal factors.

In many cases the number of local maxima of conformal factor is less than $\sqrt{N}$. We want the number of feature points to be exactly $\sqrt{N}$ so that the number of edges connecting pairs of feature points grows at the same rate as the number of mesh vertices. This in turn ensures that we can use the same value for the parameter for preserving edge lengths, regardless of mesh resolution. We offer two different solutions to handling this issue. The first is to increase the number of feature points to $\sqrt{N}$ by adding extra randomly selected vertices as feature points.

The second is to separately normalise the weightings $w_{i,j}$ used for pairs of feature points, and for adjacent vertices. We normalise the weights for adjacent vertices by dividing by the total number of edges, and we normalise the feature point pair weights by dividing by the sum of all feature point pair weights. Thus, we may rewrite the final functional to be minimised as

$$S(X) = \sum_{i \in F}\sum_{j \in F, j \neq i} \frac{0.5(\Phi_i + \Phi_j)}{\sum_{i \in F}\sum_{j \in F, j \neq i} 0.5(\Phi_i + \Phi_j)} (\delta_{i,j} - d_{i,j}(X))^2 + \sum_{(i,j) \in E} \frac{\beta}{|E|\,\delta_{i,j}^2} (\delta_{i,j} - d_{i,j}(X))^2, \quad (10)$$

where F is the set of all feature points, E is the set of all edges, and i is the conformal factor of vertex i.

Examples of canonical forms produced by both versions of our method are shown later inFig. 8.

COMPUTATIONAL COMPLEXITY

In this section we analyse the computational complexity of our canonical form computation and compare it to the original method of Elad and Kimmel described in Section 3.1. We first examine the computational complexity of calculating the target distance matrix , used in Eq. (1). Elad and Kimmel used the geodesic distance matrix produced by the fast marching method [35], which has a time complexity of $O(N^2 \log N)$, where N is the number of vertices in the mesh.

Our method instead uses a constant high value for N entries in the distance matrix. These N entries correspond to pairs of feature points. To locate feature points we calculate the conformal factor of each vertex on the mesh, identify at most $\sqrt{N}$ local maxima, and in the first version of our method we add randomly selected vertices if we have fewer than $\sqrt{N}$ local maxima. The conformal factor has several components. Both the Laplace–Beltrami and Gaussian curvature computations are $O(NM)$, where M is the average valence of a vertex on the mesh. Euler's formula tells us that the average valence of a vertex on a triangular mesh is 6 [42]. The computational complexity of locating the feature points is therefore equal to $O(N)$ as M may be considered constant. The computation of the target curvature has a complexity of $O(N+T)$, where T is the number of triangles in the mesh. Again, this is $O(N)$ as the number of triangles is linearly related to the number of vertices. As the matrix L in Eq. (7) has size $N \times N$ but is sparse with $O(N)$ non-zero elements, the linear system can be solved for the conformal factor in$O(N)$ time.

Our method also requires calculation of the edge lengths for all pairs of connected vertices, which has a computational complexity of $O(N)$. Computing the set of feature points, along with all the distance computations is significantly more efficient—linear—than computing the geodesic distances—super-quadratic—required by previous canonical form computations.

Calculating the values of $b_{i,j}$ and $d_{i,j}$ for each iteration of the SMACOF algorithm (Section 3.1) has a computational complexity of $O(N^2)$ when the distance between all pairs of points is used. Our method uses all pairs of $\sqrt{N}$ feature points plus all edge lengths, which lowers the computational complexity to $O(N)$.

EXPERIMENTS

Here we present the results of our experiments. In Section 6.1 we experimentally decide on the best value for the edge preservation weight . Section 6.2 compares the run-time of our method with that of the method by Elad and Kimmel [4] based on MDS and geodesic distances (Section 3.1). Finally we substitute our canonical forms into the shape retrieval method of Lian et al. [3], described in Section 3.2, to analyse their effectiveness for non-rigid shape retrieval.

Parameter Optimization

We used the TOSCA non-rigid models dataset [41] to determine the best value to use for the edge preserving weight . This dataset consists of 148 meshes, divided into 12 classes. Each class consists of the same model in different articulated poses. We performed shape retrieval on this dataset using Lian et al.'s method, but using our canonical forms instead of Elad and Kimmel's. The output of the shape retrieval test is a distance matrix between all pairs of models.

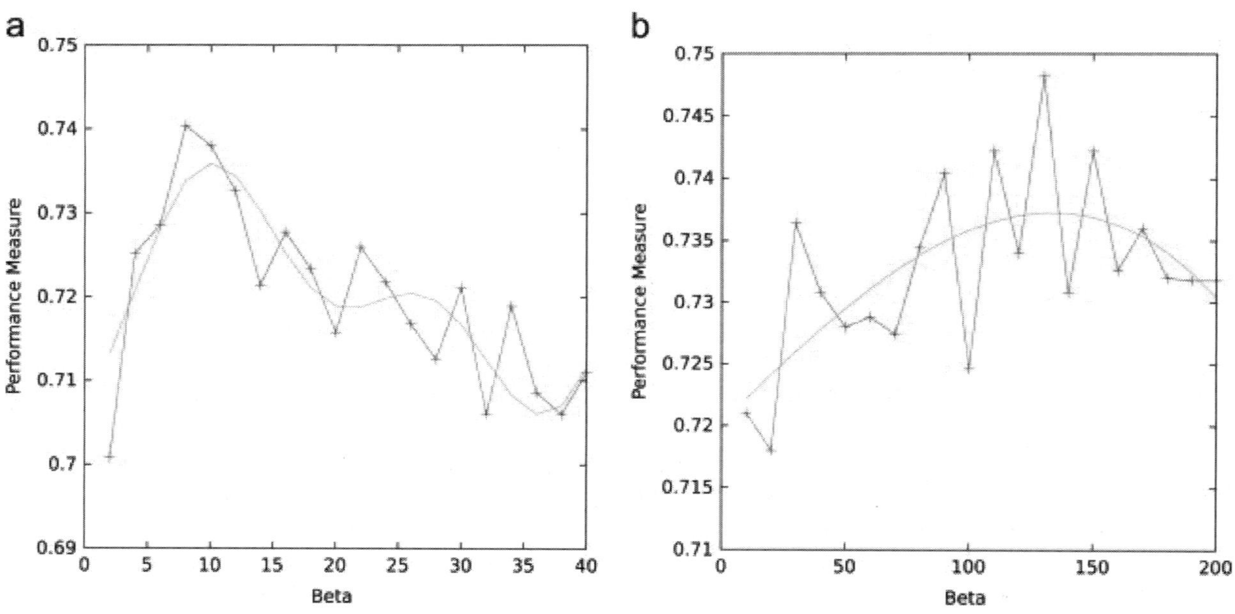

Figure. 6: Shape retrieval performance using our canonical forms for different values for the edge length preserving parameter . (a) Additional random feature points. (b) Normalised weightings.

We performed shape retrieval several times, each time using a different value. We evaluated the distance matrices produced using each of these results using five quantitative measures of how well they perform: nearest neighbour (NN), 1-tier, 2-tier, e-measure, and discounted cumulative gain (DCG); see [2] for a description of these measures and their use in assessing shape retrieval performance. The measures all lie in the range $[0,1]$, and we average over them all and plot the performance against . We do this for both variants of our method which deal differently with meshes of varying resolution (adding random features, or normalising the weights in the MDS formulation) (see Fig. 6). This test shows that =10 is approximately the best choice when adding random features, while performance only drops off by less than 2% if we choose anywhere between 5 and 15. We also come to the same conclusion when studying each statistical measure individually, except for the e-measure which shows high performance for a wider range of parameter values. When using the weight normalisation variant of our method, =130 is the best choice, but again, the method is insensitive to changes in .

Run-Time

The main advantage of our technique over the existing state of the art is that we do not require the expensive computation of geodesic distances to produce canonical forms. In this section, we therefore present a run-time comparison between our method, and a related method using geodesic distances. Our run-time tests were performed on a Linux PC with a 3.2 GHz Intel Core i7-3930k CPU and 32 GB of memory. In Table 1 the timings for computing our canonical forms are compared to the times taken to produce the canonical forms using geodesic distances, for all the models in the SHREC'11 non-rigid dataset [37]. This dataset consists of 600 watertight meshes, with an average of 9300 vertices. The method by Lian et al. [3] simplified these meshes to contain approximately 2000 vertices, before computing their canonical forms using least squares MDS and geodesic distances. The run-time for both our method and the geodesic distance based method is shown for these simplified meshes, but only our method to compute canonical forms could be run on the original meshes in a reasonable enough time for all 600 models. We therefore only tested the run-time when using geodesic distances for one of the original models, which consists of 9300 vertices. The timings for the geodesic distance based method are shown for classical, fast [43] and least squares MDS. We do not include the timings for the simplification process, as they are negligible.

Table 1: Run-time for computing canonical forms for the SHREC'11 dataset our method takes significantly less time, especially for the original meshes. Our method is faster using the original meshes, than the geodesic distance based method using the simplified meshes

Method	All models	Average/model
Geodesic distances (C, simplified meshes)	8.5 h	51 s
Geodesic distances (C, original meshes)	–	21 min[a]
Geodesic distances (F, simplified meshes)	8.5 h	51 s
Geodesic distances (F, original meshes)	–	19 min[a]
Geodesic distances (LS, simplified meshes)	11 h	66 s
Geodesic distances (LS, original meshes)	–	24 min[a]
Our method (simplified meshes)	20 min	2 s
Our method (original meshes)	5.5 h	33 s

[a]Geodesic distances (original meshes) were assessed for a single mesh only due to the length of time needed. Key: C, Classical MDS; F, Fast MDS; LS, Least Squares MDS.

We also compared how the run-time of both our method and using geodesic distances grows as the number of vertices is increased. We selected a single 'typical' model, and created several different versions of it, each simplified to a different number of vertices. A graph showing time taken versus the number of vertices is presented in Fig. 7. When computing canonical forms with geodesic distances, the run-time grows significantly faster than our method as the number of vertices is increased. Our method scales better for higher resolution meshes, as predicted theoretically.

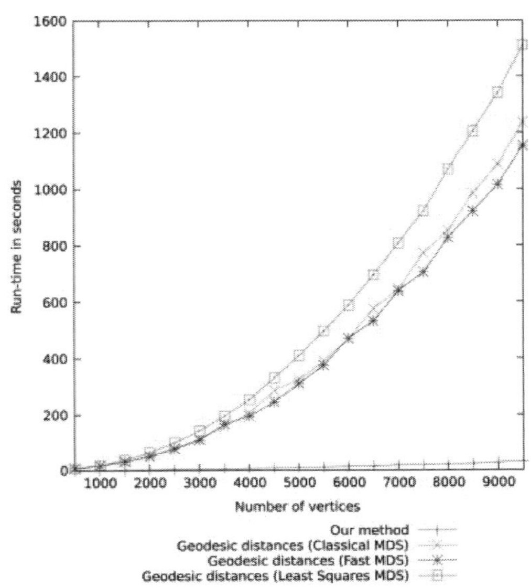

Figure. 7: Run-time for a single model, resampled to different mesh resolutions. Run-times grow significantly faster when using geodesic distances than for our method.

Shape Retrieval

We used the models from the non-rigid track of the 2011 Shape Retrieval Contest (SHREC'11) [37] to evaluate the effectiveness of our method at producing canonical forms suitable for shape retrieval. This dataset consists of 600 watertight meshes, divided into 30 classes. Each class consists of the same shape in 20 different poses. We evaluated the canonical forms using both variants of our method. In experiments using our method with additional random feature points we set =10, and with weightings normalised, we set =130.

Fig. 8 displays a selection of canonical forms produced by our method, for comparison with the canonical forms calculated using least squares MDS and geodesic distances [4]as used by Lian et al. [3] (Section 3.1). The canonical forms using geodesic distances were computed from simplified meshes consisting of approximately 2000 vertices; the original meshes were simplified using MeshLab [44]. For our method, we show both the canonical forms computed using the simplified and original meshes, which on average had approximately 9300 vertices.

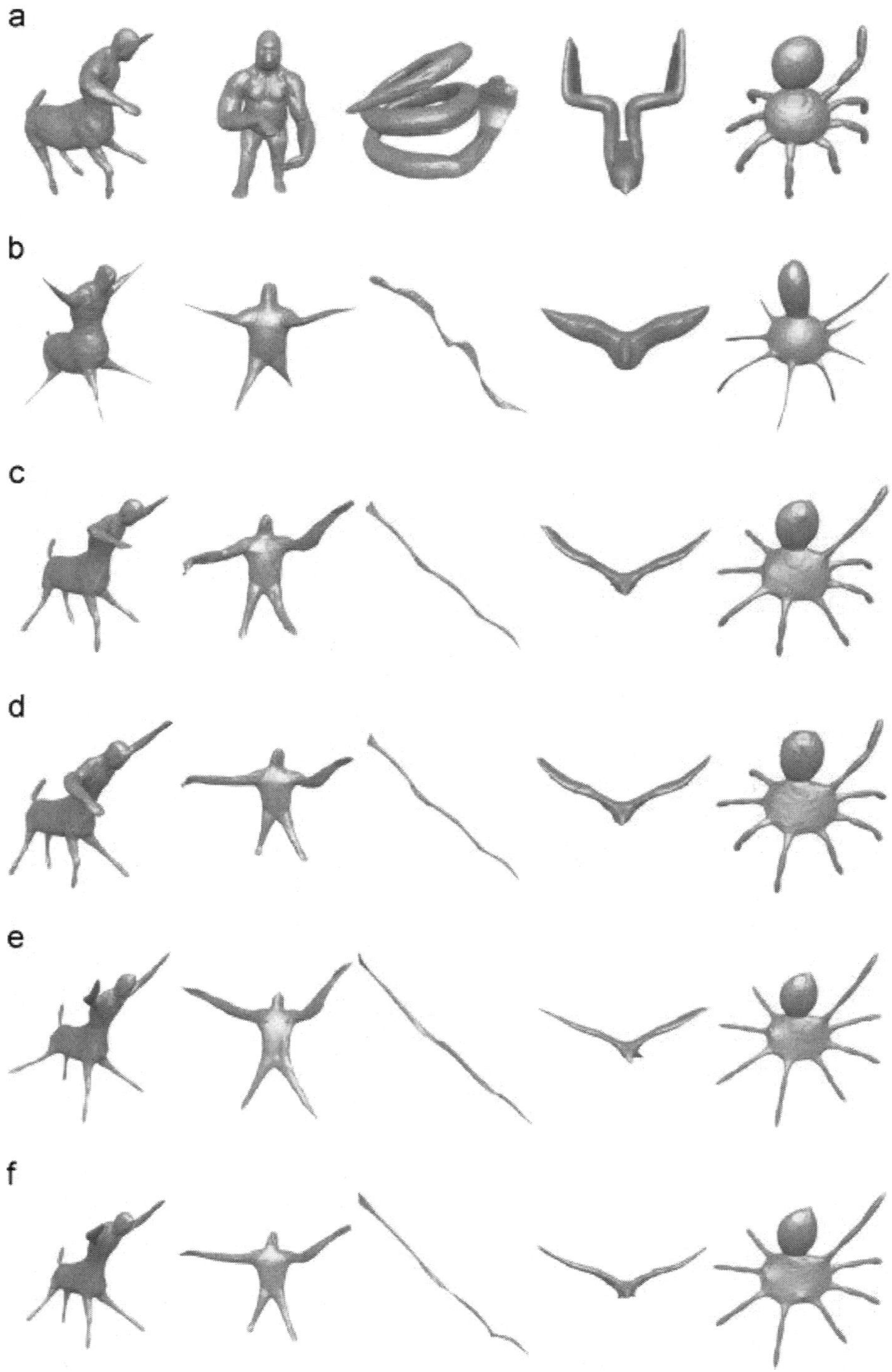

Figure. 8: Comparison of canonical forms for a selection of the SHREC'11 dataset, produced by Elad's method (b), and our methods, both for simplified (c, e) and unsimplified (d, f) meshes. Our method stretches out the limbs of the models, with less distortion along the 'limbs'. Key: S, simplified mesh; O, original mesh; A, additional random features; N, normalised weightings. (a) Original meshes, (b) Canonical forms produced using Least Squares MDS and geodesic distances with simplified meshes, (c) Canonical forms produced using our method (SA), (d) Canonical forms produced using our method (OA), (e) Canonical forms produced using our method (SN) and (f) Canonical forms produced using our method (ON)

Our canonical forms produced using feature points based on conformal factors (Fig. 8(c)) successfully stretch out the 'limbs' of the models, as do the canonical forms produced using geodesic distances (see Fig. 8(b)). Our canonical forms however result in less distortion towards the ends of the extremities. Several canonical forms of the same object, but in different poses, are shown in Fig. 9.

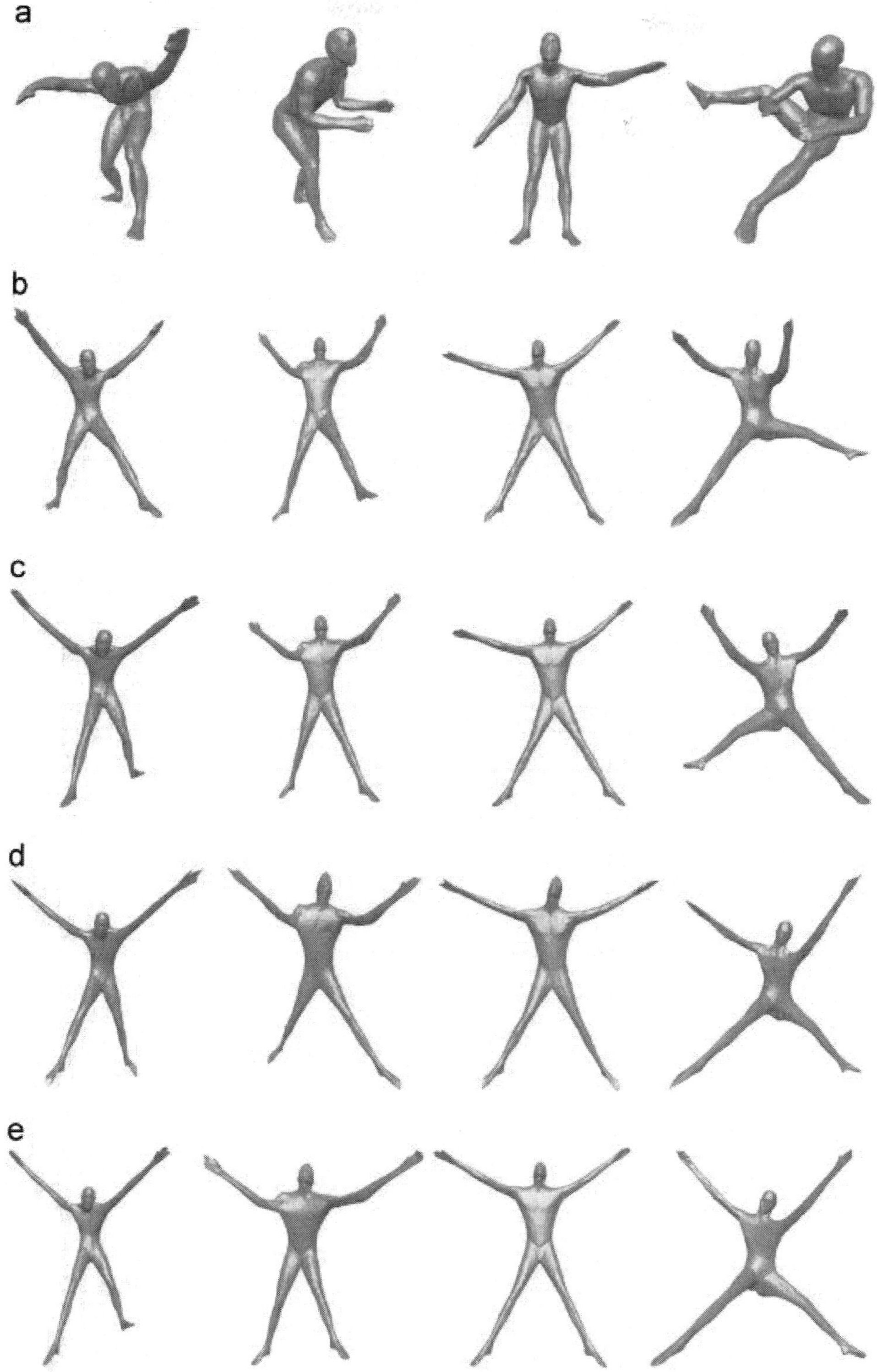

Figure. 9: Canonical forms of the same object initially in different poses, produced by our method, both for simplified (b, d) and unsimplified (c, e) meshes. Key: S, simplified mesh; O, original mesh; A, additional random features; N, normalised weightings. (a) Original meshes, (b) Canonical forms produced using our method (SA), (c) Canonical forms produced using our method (OA), (d) Canonical forms produced using our method (SN) and (e) Canonical forms produced using our method (ON).

We next demonstrate the results of using our method for non-rigid shape retrieval. We computed the distance matrix between all models from the SHREC'11 non-rigid track[37] by substituting our canonical forms into Lian's view-based retrieval method. We compare our results to those from their original method, which uses least squares MDS and geodesic distances to calculate the canonical forms, we also show results for classical and fast MDS. Results based on comparing the original meshes without any canonical form computations are also shown. We evaluate the distance matrices produced using each of these methods using the same five quantitative measures used earlier when tuning the parameter , and analyse them using precision-recall plots. We have also tested our canonical forms using a different view-based method [6], and the results exhibit the same trend. We have therefore not included them in our paper.

Table 2 shows the qualitative measures for each method; higher numbers are better. Using original meshes instead of canonical forms consistently performs worse on all measures, demonstrating the effectiveness of using canonical forms. When using additional random points to allow for varying numbers of feature points, our experiments show that using the original full resolution meshes achieves better results than when the meshes are simplified. The opposite appears true when we instead normalised the weightings of the MDS formulation. The original least squares MDS and geodesic distance method outperforms our method on most of the performance measures, but only by a small amount. Our method outperforms the geodesic based method used with classical or fast MDS.

Table 2: Comparison of retrieval results on the SHREC'11 non-rigid dataset. Our method achieves the same ranking as Lian et al.'s method [3] with geodesic distances. Key: C, Classical MDS; F, Fast MDS; LS, Least Squares MDS; S, simplified mesh; O, original mesh; A, additional random features; N, normalised weightings.

	Method	NN	1-Tier	2-Tier	e-Measure	DCG
Canonical	Original meshes	0.985	0.746	0.863	0.627	0.931
	Geodesic distances (C)	0.987	0.855	0.943	0.691	0.964
	Geodesic distances (F)	0.978	0.795	0.905	0.657	0.945
	Geodesic distances (LS)	0.995	0.913	0.969	0.717	0.982
	Our method (SA)	0.983	0.867	0.966	0.709	0.974
	Our method (OA)	0.988	0.870	0.969	0.710	0.976
	Our method (SN)	0.995	0.875	0.970	0.713	0.977
	Our method (ON)	0.982	0.867	0.964	0.708	0.974
SHREC'11	SD-GDM+meshSIFT	**1.000**	**0.972**	**0.990**	**0.736**	**0.996**
	ShapeDNA	0.992	0.915	0.957	0.705	0.978
	FOG+MRR	0.960	0.881	0.946	0.696	0.959
	BOGH	0.993	0.811	0.884	0.647	0.949
	LSF	0.995	0.799	0.863	0.633	0.943
	BOW-LSD	0.955	0.672	0.803	0.579	0.897
	PatchBOF	0.748	0.642	0.833	0.588	0.837
	HKS	0.837	0.406	0.497	0.353	0.730

Table 2 also shows a comparison with the highest performing submission of each research group to the SHREC'11 non-rigid track. This table shows that our method is ranked in the same position (2nd) as Lian et al.'s method with geodesic distance based canonical forms. Note that the best performing method (SD-GDM+meshSIFT[25] and [10]) also uses all pairs of geodesic distances and therefore suffers from high computational complexity.

Fig. 10 displays precision-recall plots for both the geodesic distance based method, and our method. Again, any canonical form method performs better than using only the original meshes. Our method produced very similar precision-recall plots when using simplified or original meshes. Using least squares MDS with geodesic distances shows a slight advantage over our method.

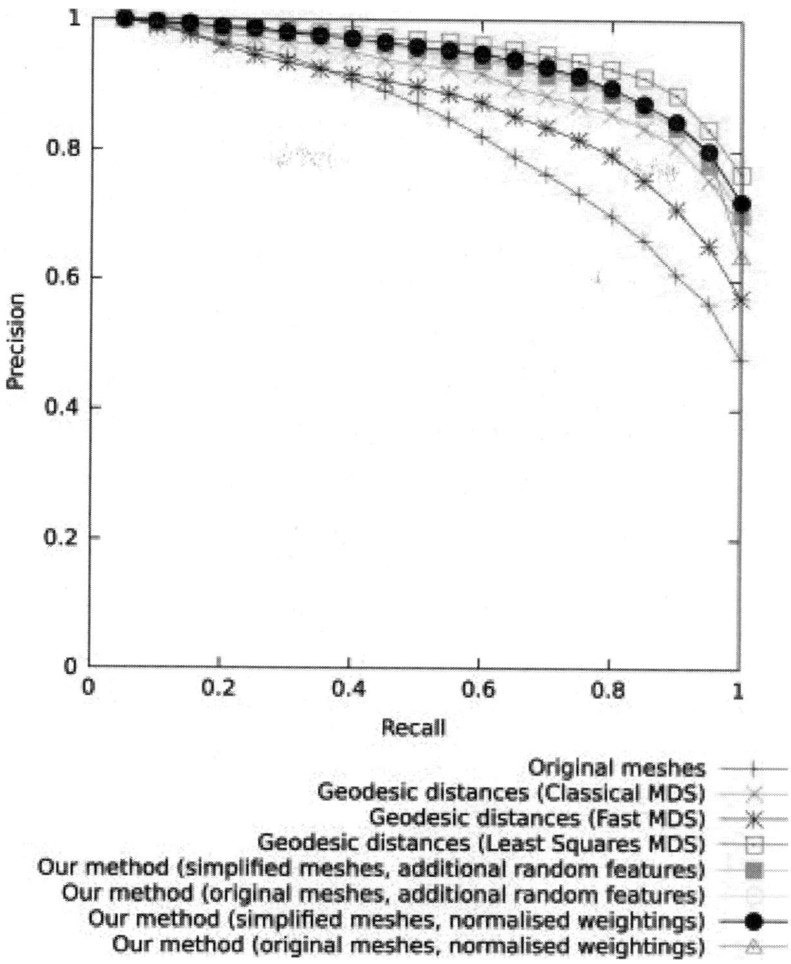

Figure. 10: Precision-recall plots of retrieval results on the SHREC'11 non-rigid dataset.

We have also produced retrieval results for the McGill articulated dataset [45]. This dataset consists of 255 models, made up of 10 different shape classes, with an average mesh resolution of about 14,000 vertices. As with the SHREC'11 dataset, the models were simplified to 2000 vertices to be able to produce canonical forms with the geodesic distance based method. Table 3 shows the retrieval performance of each method, andFig. 11 displays precision-recall plots. On this dataset the use of canonical forms does not show as large an advantage over using the original articulated meshes. This may be because the differences between each class are still mostly visible, even when the meshes are in different poses. Both the classical and fast MDS with geodesic distances perform worse than when using the original articulated meshes. This may be because the large distortions they apply to the mesh cause greater retrieval error than the non-rigid articulations. For this dataset our method with additional random features outperforms using weighting normalisation when using the original full resolution meshes. Table 3 also shows the retrieval results presented by Wang and Zha [33]. Our method produces more accurate results than those which they report, but we use a different retrieval framework to compare canonical forms.

Table 3: Retrieval results for the McGill dataset. Key: C, Classical MDS; F, Fast MDS; LS, Least Squares MDS; S, simplified mesh; O, original mesh; A, additional random features; N, normalised weightings

Method	NN	1-Tier	2-Tier	e-Measure	DCG
Original meshes	0.980	0.744	0.896	0.691	0.940
Geodesic distances (C)	0.961	0.728	0.868	0.678	0.931
Geodesic distances (F)	0.918	0.692	0.860	0.649	0.909
Geodesic distances (LS)	**0.996**	**0.830**	**0.947**	**0.778**	**0.970**
Our method (SA)	0.988	0.771	0.919	0.728	0.951
Our method (OA)	0.973	0.793	0.925	0.737	0.954
Our method (SN)	0.977	0.780	0.909	0.731	0.945
Our method (ON)	0.969	0.761	0.891	0.710	0.940

Method	NN	1-Tier	2-Tier	e-Measure	DCG
Wang and Zha [33]	0.980	0.746	0.895	0.707	0.938

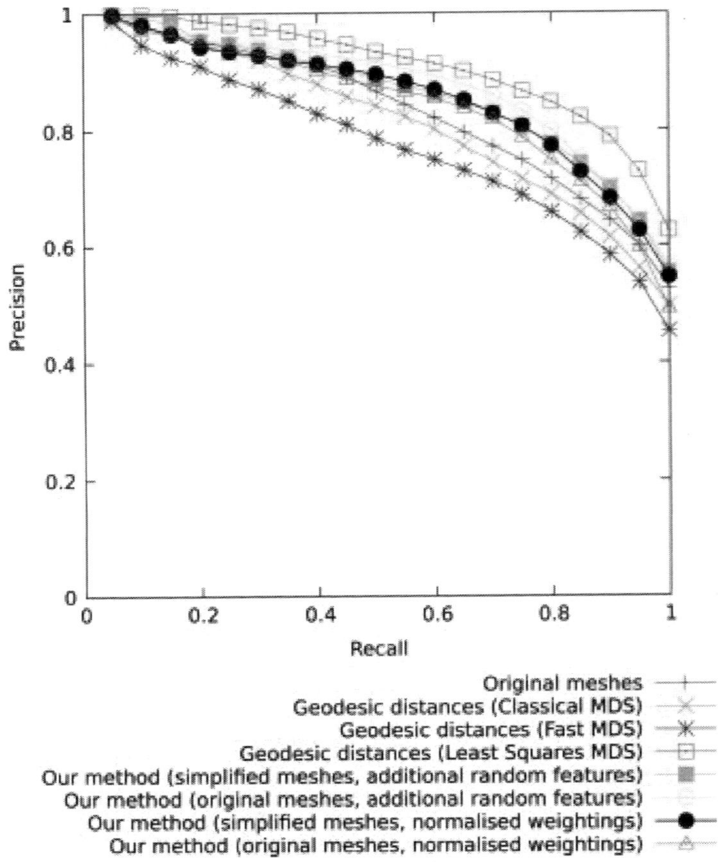

Figure. 11: Precision-recall plots of retrieval results on the McGill articulated dataset.

As our method for computing canonical forms does not distort the local details of the mesh as much as the original geodesic distances method, we would expect our method to perform better when such details are important. We tested this hypothesis on a specialised subset of the SHREC'11 dataset, comprising models in which all share the same basic shape (humanoid), but which vary in local detail. There are 80 models in this subset, split equally into the classes armadillo, gorilla, man, and woman. An example from each of these classes is shown in Fig. 12.

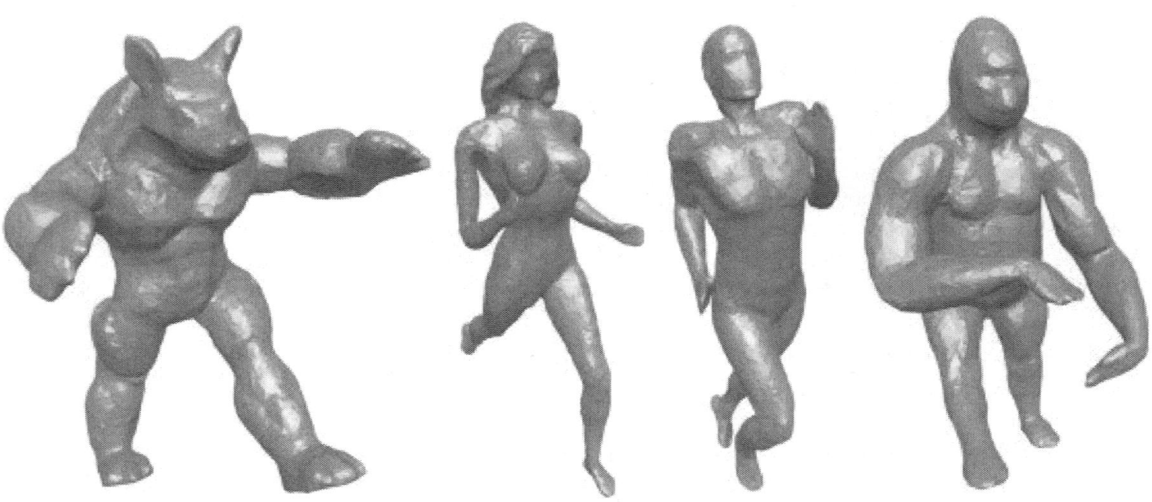

Figure. 12: Examples from the humanoid subset of the SHREC'11 dataset.

Results of running shape retrieval on the humanoid subset are shown in Table 4: our method outperforms the best geodesic distance method on all but one measure, both when using the simplified models with normalised weightings, and when using the full resolution models with additional random features. When using normalised weightings, we achieve a score of 100% for the nearest neighbour performance measure. The least squares MDS geodesic based method performs slightly better overall than the other two runs of our algorithm, but not across all measures. We are able to both outperform best-of-class previous methods on such data, *and* produce results much more quickly.

Table 4: Retrieval results for the SHREC'11 humanoid subset. Variants of our method outperform the original method using MDS and geodesic distances. Key: C, Classical MDS; F, Fast MDS; LS, Least Squares MDS; S, simplified mesh; O, original mesh; A, additional random features; N, normalised weightings

Method	NN	1-Tier	2-Tier	e-Measure	DCG
Original meshes	0.963	0.657	0.870	0.607	0.880
Geodesic distances (C)	0.975	0.691	0.910	0.649	0.919
Geodesic distances (F)	0.925	0.688	0.936	0.658	0.916
Geodesic distances (LS)	0.975	0.761	0.949	0.678	0.928
Our method (SA)	0.913	**0.773**	0.948	0.677	0.938
Our method (OA)	0.963	0.767	0.952	0.679	0.945
Our method (SN)	**1.000**	0.753	**0.959**	**0.685**	**0.948**
Our method (ON)	0.913	0.731	0.938	0.667	0.935

Precision-recall plots for these results are shown in Fig. 13; again results for our method are still very similar for both the original and simplified meshes.

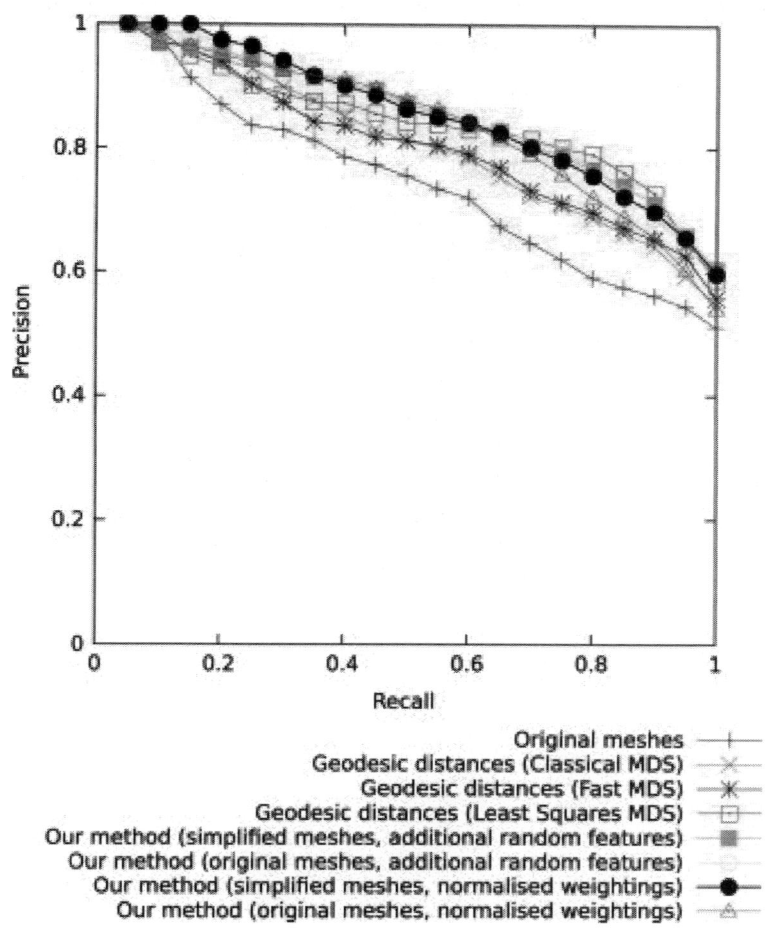

Figure. 13: Precision-recall plots of the retrieval results on the SHREC'11 humanoid subset.

For both datasets, using our method to produce canonical forms using either the original or simplified meshes leads to very similar shape retrieval accuracy. This is likely to be for two reasons. The first is that the feature point locations chosen using conformal factors are stable over a wide range of mesh resolutions, leading to the same standardised pose. The second is that there is a large dissimilarity between the classes in the SHREC'11 non-rigid dataset, and so the different shapes are distinguishable even without the finer details discarded during simplification.

LIMITATIONS

Our method restricts the number of feature points to $\sqrt{N}$, and in the unlikely case that the mesh has more than $\sqrt{N}$ extremities, our method will not be able to detect them all, so sections of the mesh will not be straightened out. All the models in the datasets we have used have fewer than $\sqrt{N}$ extremities, and therefore we have not encountered this problem.

To test how our method deals with noise, we have used the normalised weightings version of our method to produce canonical forms for the SHREC'11 dataset, but with noise added to each of the meshes. The noise was added by randomly displacing each vertex by a small amount using Meshlab [44]. We have also produced canonical forms of these meshes using the geodesic based method by Elad and Kimmel [4]. Fig. 14 shows an example of the effect of noise on the canonical forms for a single mesh. The canonical forms for the meshes with additional noise are still reasonable, but they are not completely consistent with the canonical forms produced from the original meshes. Table 5 shows the retrieval results when using the noisy meshes. The results show a drop in performance for each method, but the results using our method show a larger drop in performance than when using the least squares or classical MDS geodesic distance based methods. It is unclear, however, how much the drop in retrieval performance is caused by deterioration in the canonical forms, or by noise affecting the view-based retrieval method.

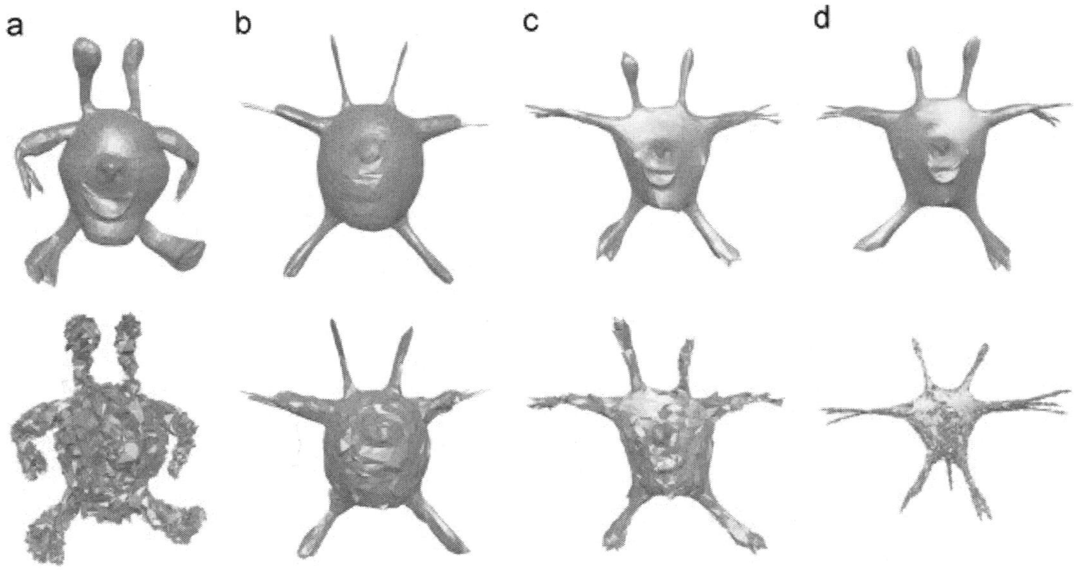

Figure. 14: Canonical forms of the same object with and without additional noise. Key: S, simplified mesh; O, original mesh; N, normalised weightings. Top: No noise, bottom: additional noise. (a) Original meshes, (b) Geodesic distance, (c) Our method (SN) and (d) Our method (ON).

Table 5: Retrieval results for the SHREC'11 dataset, with added noise. The noise has resulted in a decline of retrieval results for all methods. Key: C, Classical MDS; F, Fast MDS; LS, Least Squares MDS; S, simplified mesh; O, original mesh; A, additional random features; N, normalised weightings

Method	NN	1-Tier	2-Tier	e-Measure	DCG
Geodesic distances (C)	0.985	0.802	0.918	0.670	0.948
Geodesic distances (F)	0.920	0.687	0.839	0.599	0.895
Geodesic distances (LS)	**0.992**	**0.863**	**0.960**	**0.701**	**0.970**
Our method (SN)	0.963	0.797	0.909	0.664	0.944
Our method (ON)	0.947	0.736	0.868	0.630	0.918

CONCLUSIONS

We have presented a novel linear-time method for producing canonical forms of meshes for non-rigid shape retrieval. Our method maximises the Euclidean distance between a small number of feature points, while attempting to preserve the original edge lengths. The feature points are chosen based on the conformal factors of the mesh vertices, which concentrates the feature points at the extremities of the model. Our method has lower computational complexity, and is much faster in practice, than methods that require all pairs of geodesic distances to be computed, while resulting in only a small drop in performance for both the SHREC'11 and McGill datasets. We furthermore show that when considering models with a similar basic shape (humanoids), our method can provide the best retrieval performance.

Using the original meshes of both datasets only produces a very small improvement over using simplified versions of the models when using one variant of our method, and produces a small decline in retrieval results for another. This may be because of the large dissimilarity between the classes in these datasets. This means that any finer details of the models, which are lost during simplification, are mostly unnecessary for distinguishing between models of different classes. Our method for finding feature points based on the conformal factors of the mesh vertices is able to find very similar points at different mesh resolutions, therefore the pose of the canonical forms will be similar over a wide range of mesh resolutions. The higher resolution meshes may also exhibit more noise, which could negatively affect the retrieval process.

REFERENCES

1. Z. Lian, A. Godil, T. Fabry, T. Furuya, J. Hermans, R. Ohbuchi, C. Shu, D. Smeets, P. Suetens, D. Vandermeulen, S. Wuhrer, SHREC'10 track: non-rigid 3d shape retrieval, in: Proceedings of the 3rd Eurographics Conference on 3D Object Retrieval, EG 3DOR'10, Eurographics Association, 2010, pp. 101–108.
2. Z. Lian, A. Godil, B. Bustos, M. Daoudi, J. Hermans, S. Kawamura, Y. Kurita, G. Lavoué, H.V. Nguyen, R. Ohbuchi, Y. Ohkita, Y. Ohishi, F. Porikli, M. Reuter, I. Sipiran, D. Smeets, P. Suetens, H. Tabia, D. Vandermeulen, A comparison of methods for non-rigid 3d shape retrieval, Pattern Recognit. 46 (2013) 449–461.
3. Z. Lian, A. Godil, X. Sun, J. Xiao, CM-BOF: visual similarity-based 3d shape retrieval using clock matching and bag-of-features, Mach. Vis. Appl. (2013) 1–20.
4. A. Elad, R. Kimmel, On bending invariant signatures for surfaces, IEEE Trans. Pattern Anal. Mach. Intell. 25 (2003) 1285–1295.
5. M. Ben-Chen, C. Gotsman, Characterizing shape using conformal factors, in: Proceedings of the 1st Eurographics Conference on 3D Object Retrieval, EG 3DOR'08, Eurographics Association, 2008, pp. 1–8.
6. D.-Y. Chen, X.-P. Tian, Y.-T. Shen, M. Ouhyoung, On visual similarity based 3d model retrieval, Comput. Graph. Forum 22 (2003) 223–232.
7. M. Kazhdan, T. Funkhouser, S. Rusinkiewicz, Rotation invariant spherical harmonic representation of 3d shape descriptors, in: Proceedings of the 2003 Eurographics/ ACM SIGGRAPH Symposium on Geometry Processing, SGP '03, Eurographics Association, 2003, pp. 156–164.
8. J. Tangelder, R. Veltkamp, A survey of content based 3d shape retrieval methods, Multim. Tools Appl. 39 (2008) 441–471.
9. P. Shilane, P. Min, M. Kazhdan, T. Funkhouser, The Princeton shape benchmark, in: International Conference on Shape Modeling and Applications, 2004, pp. 167–178.
10. C. Maes, T. Fabry, J. Keustermans, D. Smeets, P. Suetens, D. Vandermeulen, Feature detection on 3d face surfaces for pose normalisation and recognition, in: Fourth IEEE International Conference on Biometrics: Theory Applications and Systems (BTAS), 2010, pp. 1–6.
11. A. Zaharescu, E. Boyer, K. Varanasi, R. Horaud, Surface feature detection and description with applications to mesh matching, in: IEEE Conference on Computer Vision and Pattern Recognition, 2009, pp. 373–380.
12. J. Sun, M. Ovsjanikov, L. Guibas, A concise and provably informative multiscale signature based on heat diffusion, Comput. Graph. Forum 28 (2009) 1383–1392.
13. J. Sivic, A. Zisserman, Video google: a text retrieval approach to object matching in videos, in: Proceedings of Ninth IEEE International Conference on Computer Vision, vol. 2, 2003, pp. 1470–1477.
14. A.M. Bronstein, M.M. Bronstein, L.J. Guibas, M. Ovsjanikov, Shape google: geometric words and expressions for invariant shape retrieval, ACM Trans. Graph. 30 (2011) 1:1–1:20.
15. M. Abdelrahman, M. El-Melegy, A. Farag, Heat kernels for non-rigid shape retrieval: sparse representation and efficient classification, in: Ninth Conference on Computer and Robot Vision (CRV), 2012, pp. 153–160.
16. T. Hou, X. Hou, M. Zhong, H. Qin, Bag-of-feature-graphs: a new paradigm for non-rigid shape retrieval, in: 21st International Conference on Pattern Recognition (ICPR), 2012, pp. 1513–1516.
17. X. Wang, Y. Liu, H. Zha, Intrinsic spin images: a subspace decomposition approach to understanding 3d deformable shapes, in: Proceedings of the International Symposium on 3D Data Processing, Visualization and Transmission, 3DPVT'10, 2010, pp. 17–20.
18. A. Johnson, M. Hebert, Using spin images for efficient object recognition in cluttered 3d scenes, IEEE Trans. Pattern Anal. Mach. Intell. 21 (1999) 433–449.
19. H. Tabia, M. Daoudi, J.-P. Vandeborre, O. Colot, A new 3d-matching method of nonrigid and partially similar models using curve analysis, IEEE Trans. Pattern Anal. Mach. Intell. 33 (2011) 852–858.
20. R. Khoury, J.-P. Vandeborre, M. Daoudi, Indexed heat curves for 3d-model retrieval, in: 21st International Conference on Pattern Recognition (ICPR), 2012, pp. 1964–1967.
21. M. Hilaga, Y. Shinagawa, T. Kohmura, T.L. Kunii, Topology matching for fully automatic similarity estimation of 3d shapes, in: Proceedings of the 28th Annual Conference on Computer Graphics and Interactive Techniques, SIGGRAPH '01, Los Angeles,

ACM, New York, 2001, pp. 203–212.

22. H. Sundar, D. Silver, N. Gagvani, S. Dickinson, Skeleton based shape matching and retrieval, in: Shape Modeling International, 2003, pp. 130–139.

23. K. Sfikas, T. Theoharis, I. Pratikakis, Non-rigid 3d object retrieval using topological information guided by conformal factors, Vis. Comput. 28 (2012) 943–955.

24. M. Reuter, F.-E. Wolter, N. Peinecke, Laplace–Beltrami spectra as shape-DNA of surfaces and solids, Comput.-Aided Des. 38 (2006) 342–366.

25. D. Smeets, T. Fabry, J. Hermans, D. Vandermeulen, P. Suetens, Isometric deformation modelling for object recognition, in: X. Jiang, N. Petkov (Eds.), Computer Analysis of Images and Patterns, Lecture Notes in Computer Science, vol. 5702, Springer, Berlin, Heidelberg, 2009, pp. 757–765.

26. D. Smeets, J. Hermans, D. Vandermeulen, P. Suetens, Isometric deformation invariant 3d shape recognition, Pattern Recognit. 45 (2012) 2817–2831.

27. V. Jain, H. Zhang, A spectral approach to shape-based retrieval of articulated 3d models, Comput.-Aided Des. 39 (2007) 398–407.

28. A. Bronstein, M. Bronstein, R. Kimmel, Efficient computation of isometryinvariant distances between surfaces, SIAM J. Sci. Comput. 28 (2006) 1812–1836.

29. G. Tam, R. Lau, Embedding retrieval of articulated geometry models, IEEE Trans. Pattern Anal. Mach. Intell. 34 (2012) 2134–2146.

30. Z. Lian, A. Godil, J. Xiao, Feature-preserved 3d canonical form, Int. J. Comput. Vis. 102 (2013) 221–238.

31. Z. Lian, A. Godil, P. Rosin, X. Sun, A new convexity measurement for 3d meshes, in: IEEE Conference on Computer Vision and Pattern Recognition (CVPR), 2012, pp. 119–126.

32. X. Ying, S.-Q. Xin, Y. He, Parallel chen-han (pch) algorithm for discrete geodesics, ACM Trans. Graph. 33 (2014) 9:1–9:11.

33. X.-L. Wang, H. Zha, Contour canonical form: an efficient intrinsic embedding approach to matching non-rigid 3d objects, in: Proceedings of the 2nd ACM International Conference on Multimedia Retrieval, ICMR '12, Hong Kong, ACM, New York, 2012, pp. 31:1–31:8.

34. J. Tierny, J.-P. Vandeborre, M. Daoudi, 3D mesh skeleton extraction using topological and geometrical analyses, in: 14th Pacific Conference on Computer Graphics and Applications (Pacific Graphics 2006), 2006, pp. 85–94.

35. R. Kimmel, J.A. Sethian, Computing geodesic paths on manifolds, Proc. Natl. Acad. Sci. 95 (1998) 8431–8435.

36. I. Borg, P. Groenen, Modern Multidimensional Scaling: Theory and Applications, Springer Series in Statistics, Springer, New York, 2005.

37. Z. Lian, A. Godil, B. Bustos, M. Daoudi, J. Hermans, S. Kawamura, Y. Kurita, G. Lavoué, H. V. Nguyen, R. Ohbuchi, Y. Ohkita, Y. Ohishi, F. Porikli, M. Reuter, I. Sipiran, D. Smeets, P. Suetens, H. Tabia, D. Vandermeulen, SHREC'11 track: shape retrieval on non-rigid 3d watertight meshes, in: Proceedings of the 4th Eurographics Conference on 3D Object Retrieval, EG 3DOR'11, Eurographics Association, 2011, pp. 79–88.

38. Z. Lian, P. Rosin, X. Sun, Rectilinearity of 3d meshes, Int. J. Comput. Vis. 89 (2010) 130–151.

39. D. Lowe, Distinctive image features from scale-invariant keypoints, Int. J. Comput. Vis. 60 (2004) 91–110.

40. K. Hildebrandt, K. Polthier, M. Wardetzky, On the convergence of metric and geometric properties of polyhedral surfaces, Geom. Dedic. 123 (2006) 89–112.

41. A.M. Bronstein, M.M. Bronstein, R. Kimmel, Numerical Geometry of Non-rigid Shapes, Monographs in Computer Science, Springer, New York, 2008.

42. X. Sun, E.R. Hancock, Quasi-isometric parameterization for texture mapping, Pattern Recognit. 41 (2008) 1732–1743.

43. C. Faloutsos, K.-I. Lin, Fastmap: a fast algorithm for indexing, data-mining and visualization of traditional and multimedia datasets, in: Proceedings of the 1995 ACM SIGMOD International Conference on Management of Data, SIGMOD '95, ACM, New York, NY, USA, 1995, pp. 163–174. URL: ⟨http://doi. acm.org/10.1145/223784.223812⟩, http://dx.doi.org/10.1145/223784.223812.

44. Meshlab v1.3.0, 2011. URL: ⟨http://meshlab.sourceforge.net⟩.

45. K. Siddiqi, J. Zhang, D. Macrini, A. Shokoufandeh, S. Bouix, S. Dickinson, Retrieving articulated 3-d models using medial surfaces, Mach. Vis. Appl. 19 (2008) 261–275.

CHAPTER 5

APPEARANCE-BASED INDOOR LOCALIZATION: A COMPARISON OF PATCH DESCRIPTOR PERFORMANCE

Jose Rivera-Rubio·Ioannis Alexiou,Anil A. Bharath

Imperial College London, South Kensington Campus, London SW7 2AZ, United Kingdom

ABSTRACT

Vision is one of the most important of the senses, and humans use it extensively during navigation. We evaluated different types of image and video frame descriptors that could be used to determine distinctive visual landmarks for localizing a person based on what is seen by a camera that they carry. To do this, we created a database containing over 3 km of video-sequences with ground-truth in the form of distance travelled along different corridors. Using this database, the accuracy of localization—both in terms of knowing which route a user is on—and in terms of position along a certain route, can be evaluated. For each type of descriptor, we also tested different techniques to encode visual structure and to search between journeys to estimate a user's position. The techniques include single-frame descriptors, those using sequences of frames, and both color and achromatic descriptors. We found that single-frame indexing worked better within this particular dataset. This might be because the motion of the person holding the camera makes the video too dependent on individual steps and motions of one particular journey. Our results suggest that appearance-based information could be an additional source of navigational data indoors, augmenting that provided by, say, radio signal strength indicators (RSSIs). Such visual information could be collected by crowdsourcing low-resolution video feeds, allowing journeys made by different users to be associated with each other, and location to be inferred without requiring explicit mapping. This offers a complementary approach to methods based on simultaneous localization and mapping (SLAM) algorithms.

INTRODUCTION

Satellite-based global positioning systems (GPS) have been available to consumers for many years. When combined with other sensor data, such as terrestrial-based radio signal strength indicators (RSSI), the quality of pedestrian localization within cities, at street level, can be quite reliable. Recently, interest has been gathering for the development of systems for indoor position sensing: we might consider this to be the next challenge in localization systems [15], [17], [28] and [33]. Indoor sensing is likely to require additional infrastructure, such as Bluetooth-based RSSI, or time-of-flight systems. At the time of writing, both of these are reported to be under trial.

Citation: Jose Rivera-Rubio, Ioannis Alexiou, Anil A. Bharath, Appearance-based indoor localization: A comparison of patch descriptor performance, Pattern Recognition Letters, Volume 66, 15 November 2015, Pages 109-117, ISSN 0167-8655, http://dx.doi.org/10.1016/j.patrec.2015.03.003.

Despite this, vision-based navigation systems are under active development. This might be because such systems do not require special markers to be embedded within the environment. However, another reason could be that vision provides an immensely rich source of data, from which estimating position is also possible. For example, in emerging applications [22] such as mobile domestic robots, merely knowing where a robot is not enough: a robot often needs some information about its immediate environment in order to take appropriate decisions. This includes navigating around obstacles and identifying important objects (e.g. pets).

Systems and devices that are designed to help humans to navigate would be improved by incorporating vision as one of the sensing modalities. This is particularly true of systems that are designed to help visually impaired people to navigate (assistive navigation systems). However, for wearable or smartphone-based systems, accuracy and power consumption remain two of the challenges to the reliable and continuous use of computer vision techniques. Visual localization accuracy is affected by several factors, including the techniques used to infer location from visual data. In the case of feature-based SLAM [9], for example, a lack of features, or a highly occluded environment, can reduce accuracy.

Wang et al. [37] have recently suggested an interesting approach to localization based on the principle of identifying landmarks in ambient signals. These ambient signals are acquired from a crowdsourcing-like approach, rather than explicitly mapping out signal strength and WiFi identifiers and appears to offer good performance, with median absolute localization error of less than 1.7 m. Perhaps more importantly, it removes the need to change building infrastructure specifically for localization. One way to strengthen the landmark approach would be to incorporate visual cues, automatically mined from the video data. Theoretically speaking, such an approach might be limited by (i) the quality of the image acquisition, which could be affected by blur, poor focusing, inadequate resolution or poor lighting; (ii) the presence of occlusions to otherwise stable visual landmarks; (iii) visual ambiguity: the presence of visually-similar structures, particularly within man-made environments. We now consider something of an ideal situation in which we can harvest visual signatures from several journeys down the same route; the approach starts with the idea of collecting visual paths, and using the data from these to localize the journeys of users relative to each other, and to start and end points.

VISUAL PATHS

Consider two users, Juan and Mary, navigating at different times along the same notional path. By notional path, we refer to a route that has the same start and end points. An example of indoor notional paths would be the navigation from one office to another, or from a building entrance to a reception point. For many buildings, such notional paths might allow different physical trajectories which could diverge. For example, users might take either stairs or lifts, creating path splits and merges. Such complex routes could be broken down into route (or path) segments, and path segments could contribute to more than one complete notional path.

For any notional path or path segment, both humans and autonomous robots would "experience" a series of cues that are distinctive when navigating along that path. In some instances, however, the cues might be ambiguous, just as they might be for radio signal strength indicators, audio cues and other environmental signals. A vision-based system would need to analyze the visual structure in sequences from hand-held or wearable cameras along path segments in order to answer two questions: which notional path or segment is being navigated, and where along a specific physical path, relative to start and end point, a person might be. We addressed the first of these questions in previous work [30].

Returning to the two-user scenario, let us assume that Juan has been the first to navigate along the path, and has collected a sequence of video frames during his successful navigation. As Mary makes her way along the path, we wish to be able to associate the images taken by Mary with those taken by Juan (see Fig. 1). The ability to do this allows us to locate Mary relative to the journey of Juan from the visual data acquired by both. For only two users, this may seem an uninteresting thing to do. However, imagine that this association is done between not two, but multiple users, and is applied to several physical paths that together form the navigable space of a building. Achieving this association would enable some types of inference to be performed. In particular:

- The visual path data would be a new source of data that could be used for location estimation;
- The association of image locations would allow visual change detection to be performed over many journeys along the same route, made at different times;
- Through non-simultaneous, many-camera acquisition, one could achieve more accurate mapping of a busy space, particularly where moving obstructions might be present;
- Visual object recognition techniques could be applied to recognize the nature of structures encountered along the route, such as exits, doorways and so on.

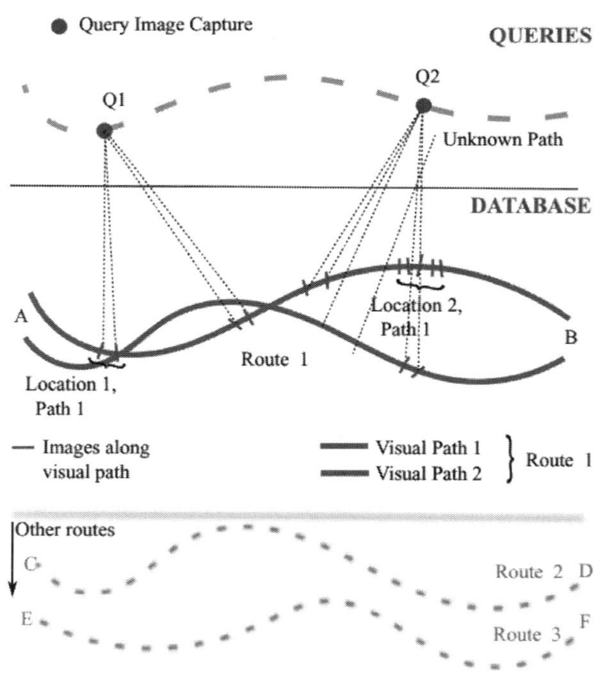

Figure. 1: The idea behind searching across data from navigators of the same physical path: after navigating the space twice, Juan's visual

path data (A, B) is indexed and stored in a database. Mary enters the same space (unknown path), and the images acquired as she moves are compared against the visual path of Juan, providing a journey-centric estimate of location. With many journeys collated, location can be inferred with respect to the pre-collected paths in the database.

Using continuously acquired images provides a new way for humans to interact with each other through establishing associations between the visual experiences that they have shared, independent of any tags that have been applied. The concept is illustrated in Fig. 2(a). In this diagram, four users are within the same region of a building; however, two pairs of users (A,C) and (B,D) are associated with having taken similar trajectories to each other. With a sufficient number of users, one could achieve a crowdsourcing of visual navigation information from the collection of users, notional paths and trajectories.

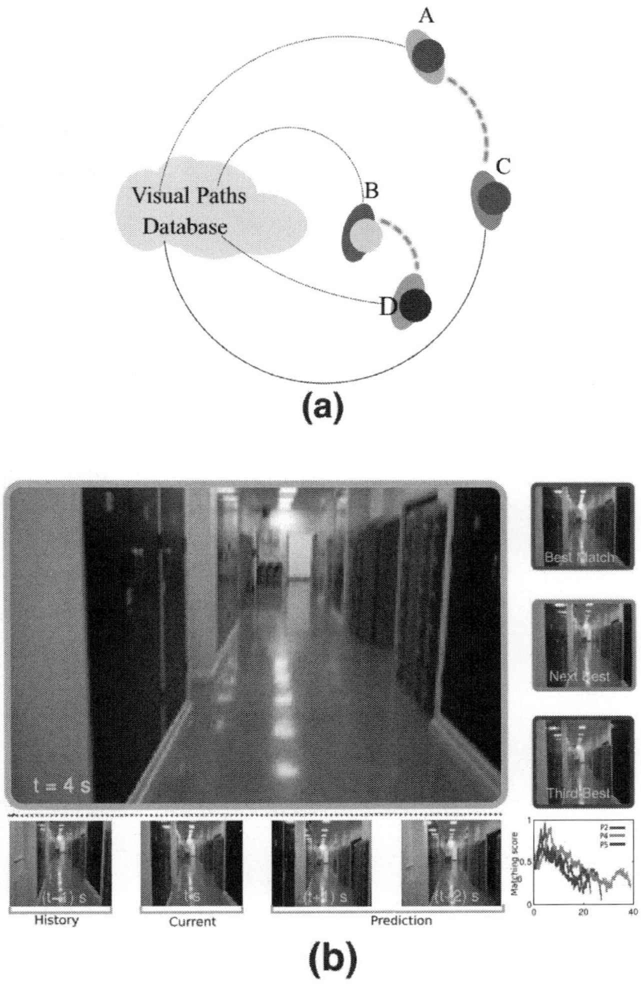

Figure. 2: In (a), we illustrate the concept of using the visual path database to establish rough correspondence in users' locations through their visual experiences. Users (A,C) and (B,D) experience similar visual paths (see the text for details). In (b), the current view captured by a camera and views from the best matchpaths that have been captured through that space, to the immediate right. The first four bottom panels show current, historical, and predicted images, based on the query from the best matching visual path. The right, bottom image shows the similarity scores from other journeys taken along the same notional path. The techniques that enable this type of match to be done are discussed in Section 5.

One intriguing possibility would be to provide information to visually-impaired users. For example, in an assistive system, the visual cues that sighted individuals experience along an indoor journey could be mined, extracting reliable information about position and objects (e.g. exit signs) that are of interest. While other sources of indoor positioning information, such as the locations of radio beacons, can aid indoor navigation, some visual cues are likely to be stable over long periods of time, and do not require extra infrastructure beyond that already commonly installed. Collecting distinctive visual cues over many journeys allows stable cues to be learned. Finally, in contrast to signal-based methods of location landmarks [37], the "debugging" of this type of navigation data—i.e. images, or patches within images—is uniquely human-readable: it can be done simply through human observation of what might have visibly changed along the path. Perhaps most compelling of all, visual path data can be acquired merely by a sighted user sweeping the route with a hand-held or wearable camera.

VISION-BASED APPROACHES TO NAVIGATION

The current state-of-the-art methods for robot navigation make use of simple visual features and realistic robot motion models in order to map, then to navigate. For human navigation, the challenge is slightly greater, due partly to the variability of human motion. Nevertheless, recent progress in simultaneous localization and mapping (SLAM) [24]and parallel tracking and mapping (PTAM) [19] have yielded stunning results in producing geometric models of a physical, recovering geometry and camera pose simultaneously from hand-held devices.

At the same time, being able to recognize certain objects while performing SLAM could improve accuracy, reducing the need for loop closure and allowing better—more reliable—self-calibration [32]. Recognition pipelines in computer vision have recently taken great strides, both in terms of scalability and accuracy. Thus, the idea of collaboratively mapping out a space through wearable or hand-held cameras is very attractive.

Appearance-based navigation, closely related to navigation, has been reported as one of many mechanisms used in biology, and has been explored by various groups in different animals (see, for example, [6], [10] and [13]). Appearance-based approaches can add to the information gained using SLAM-type algorithms. Indeed, in a robust system, we might expect several sources of localization information to be employed. Consider, for example, outdoor navigation in cities: GPS can be combined with WiFi RSSI. Doing so improves overall accuracy, because the errors in these two localization systems are unlikely to be highly correlated over relatively short length scales ($\approx$ 100 m), and would only be trivially correlated (but highly) over longer distances. Localization systems often rely on motion models embedded into tracking algorithms, such as Kalman, extended Kalman [9] filtering, or particle-filtering [27], to infer position. More recently, general purpose graphics processing units (GP-GPUs) have enabled camera position to be quickly and accurately inferred relative to a point cloud by registering whole images with dense textured models [24].

Anecdotal evidence and conversations with UK groups supporting visually-impaired people suggests that no single source of data or single type of algorithm will be sufficient to meet the needs of users who are in an unfamiliar space, or who might suffer from visual impairment. It is likely that a combination of sensors and algorithms is called for.

A Biological Perspective

Research into the mechanisms employed by humans during pedestrian navigation suggests that multisensory integration plays a key role [25]. Indeed, studies into human spatial memory using virtual reality and functional neuroimaging [3] and [4] suggest that the human brain uses a combination of representations to self-localize that might be termed as allocentric and egocentric. The egocentric representation supports identifying a location based on sensory patterns recognized from previous experiences in a given location. Allocentric representations use a reference frame that is independent of one's location. The respective coordinate systems can, of course, be interchanged via simple transformations, but the sensory and cognitive processes underlying navigation in both cases are thought to be different.

The two forms of representation are typified by different types of cells, and, in some cases, different neuronal signal pathways. Within some mammals, such as mice, it appears that a multitude of further sub-divisions of computational mechanisms lie behind location and direction encoding. For example, in the hippocampus, there are at least four classes [14] of encoding associated with position and heading. Hippocampal place cellsdisplay elevated firing when the animal is in a particular location [11]. The environmental cues that affect hippocampal place cells include vision and odour, so the inputs to these cells are not necessarily limited to any one type of sensory input.

Grid cells, on the other hand, show increased firing rates when the animal is present at a number of locations on a spatial grid; this suggests that some form of joint neuronal encoding is at work, and, indeed, there is some evidence that place cell responses arise through a combination of grid cells of different spacing [23]. Boundary cells in the hippocampus appear to encode just that: the distance to the boundaries of a spatial region. This encoding seems to be relative to the direction the animal is facing but independent of the relation between the animal's head and body; they are therefore, examples of an allocentric scheme.

In conclusion, biology seems to employ not only several sensory inputs to enable an organism to locate itself relative to the environment, but also different computational mechanisms. The evidence of these multiple strategies for localization and navigation[14] and [26] motivates the idea for an appearance-based localization algorithm.

The Dataset

A total of 60 videos were acquired from six corridors of the RSM building at Imperial College London. Two different devices were used. One was a LG Google Nexus 4 mobile phone running Android 4.4.2 "KitKat". The video data were acquired at approximately 24–30 fps at two different acquisition resolutions, corresponding to 1280 × 720 and 1920 × 1080 pixels. The other device was a wearable Google Glass (2013 Explorer edition) acquiring data at a resolution of 1280 × 720, and a frame rate of around 30 fps. A surveyor's wheel (Silverline) with a precision of 10 cm and error of ± 5% was used to record distance, but was modified by connecting the encoder to the general purpose input/output (GPIO) pins of a Raspberry Pi running a number of measurement processes. The Pi was synchronized to network time using network time protocol (NTP), enabling synchronization with timestamps in the video sequence. Because of the variable frame rate of acquisition, timestamp data from the video were used to align ground-truth measurements with frames. This data are used to assess the accuracy of estimating positions, and not for any form of training.

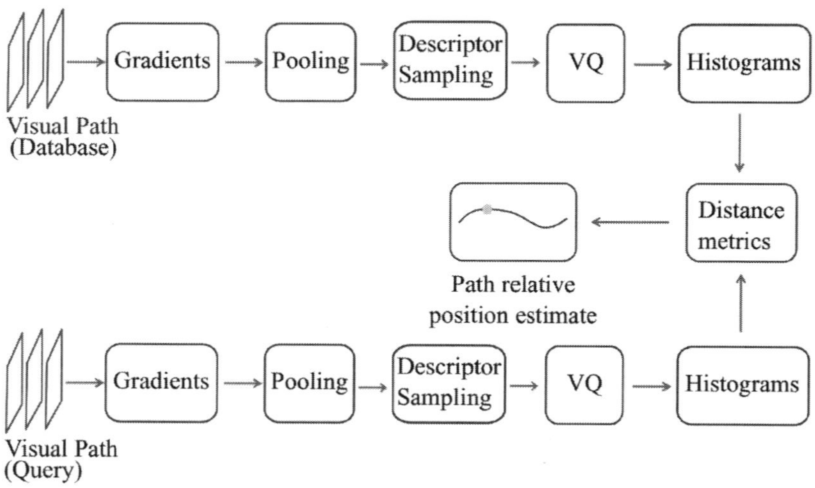

Figure. 3: This diagram illustrates the stages in processing the sequences in the database and for the queries. This diagram does not show the process behind the estimation of ground-truth for the experiments, which is described separately in Section 4. Variants of the gradient operators, pooling operators, quantization and distance metrics are described in Section 5.

In total, 3.05 km of data are contained in this dataset, at a natural indoor walking speed. For each corridor, 10 passes (i.e. 10 separate visual paths) are obtained; five of these are acquired with the hand-held Nexus, and the other five with Glass. Table 1 summarizes the acquisition. As can be seen, the length of the sequences varies within some corridors; this is due to a combination of different walking speeds and/or different frame rates. Lighting also varied, due to a combination of daylight/

nighttime acquisitions, and occasional windows acting as strong lighting sources in certain sections of the building. Changes were also observable in some videos from one pass to another due to the presence of changes (path obstructions being introduced during a cleaning activity) and occasional appearances of people.

Table 1: Detailed summary of the RSM dataset: L length of the corridors and Fr number of video frames. The single frames are representative images from hand-held videos of selected corridors of the RSM building at Imperial College London. The dataset includes both hand-held and wearable camera examples, all containing ground-truth location relative to distance traversed along labelled paths. The grand totals are: L= 3042 km and #Fr = 90,302 frames

	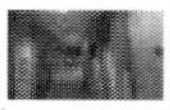				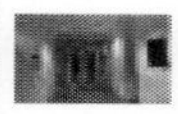	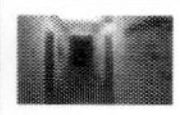
$\bar{L}$(m)	57.9	31.0	52.7	49.3	54.3	55.9
∑L	585.6	312.4	524.2	497.7	562.0	560.4
$\bar{Fr}$	2157	909	1427	1583	1782	1471
∑Fr	19,379	9309	14,638	15,189	16,823	14,964

In total, more than 90,000 frames of video were labelled with positional ground-truth. The dataset is publicly available for download at http://rsm.bicv.org[29].

METHODS: INDEXING

We evaluated the performance of different approaches to query images taken from one visual path against others stored in the database. In order to index and query visual path datasets, we used the steps illustrated in Fig. 3. The details behind each of the steps (e.g. gradient estimation, spatial pooling) are described in the remainder of this section. They include techniques that operate on single frames as well as descriptors that operate onmultiple frames, at the frame level and at the patch level. All the performance evaluation experiments were carried out at low-resolution (208 × 117 pixels) versions of the sequences, keeping bandwidth and processing requirements small.

Frame-Level Descriptor

Based on the use of optical flow in motion estimation [39] and space-time descriptors in action recognition [38] we estimated in-plane motion vectors using a simple approach. We first applied derivative filters along (x, y, t) dimensions, yielding a 2D+t, i.e. spatio-temporal, gradient field. To capture variations in chromatic content from the visual sequence, we computed these spatio-temporal gradients separately for each of the three RGB channels of the pre-processed video sequences. This yielded a 3 × 3 matrix at each point in space. Temporal smoothing was applied along the time dimension, with a support of 11 neighbouring frames. Finally, the components of the matrix were each averaged (pooled) over 16 distinct spatial regions, not very dissimilar to those to be described later in this paper. For each visual path, this yielded 144 signals, of length approximately equal to the video sequences. An illustration of the time series for one visual path is shown in Fig. 4.

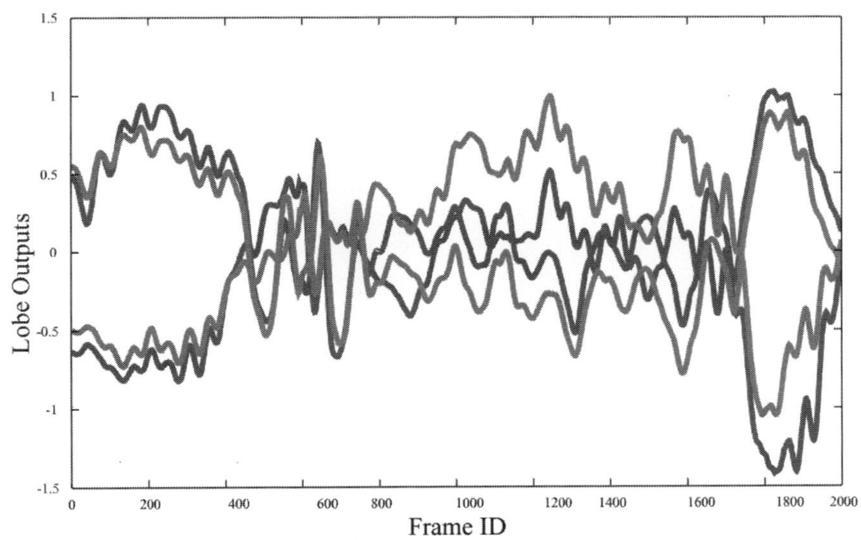

Figure. 4: Four (of 144) representative signals acquired from a visual path; these signals encode changes in red and green channels as a user moves through space. The collection of signal traces at one point in time can be used to build a simple frame-level space-time descriptor: LW_COLOR. The signal amplitudes are spatially pooled temporal and spatial gradient intensities. (For interpretation of the references to color in this figure legend, the reader is referred to the web version of this article.)

At each point in time, the values over the 144 signal channels are also captured into asingle space-time descriptor per frame: LW_COLOR. Our observations from the components of this descriptor are that (a) relative ego-motion is clearly identifiable in the signals; (b) stable patterns of motion may also be identified, though changes in the precise trajectory of a user could also lead to perturbations in these signals, and hence to changes in the descriptor vectors. Minor changes in trajectory might, therefore, reduce one's ability to match descriptors between users. These observations, together with the possibility of partial occlusion, led us to the use of patch based descriptors, so that multiple descriptors would be produced for each frame. These are introduced next.

Patch-Level Descriptors

The patch descriptors can be further divided into two categories: those produced from patches of single frames, and those that are based on patches acquired over multiple frames; the latter are space-time patch descriptors. We explored two distinct single-frame descriptors, and three distinct space-time descriptors. We first describe the single-frame descriptors.

Spatial Patch Descriptors (Single-Frame)

The spatial patch descriptors consist of the Dense-SIFT descriptor [20], [21] and [34] and a tuned, odd-symmetric Gabor-based descriptor. The SIFT descriptor was calculated by dense sampling of the smoothed estimate of $\vec{\nabla} f(x, y; \sigma)$ where f(x, y; σ) represents the scale-space embedding of image f(x, y) within a Gaussian scale-space at scale σ. We used a standard implementation of dense SIFT from VLFEAT [34] with scale parameter,σ ≈ 1, and with a stride length of 3 pixels. This yielded around 2000 descriptors per frame, each describing a patch of roughly 10 × 10 pixels in the frame.

We compared these with another single-frame technique devised in our lab: we used filters that we previously tuned on PASCAL VOC data [12] for image categorization. These consisted of 8-directional, 9 × 9 pixels spatial Gabor filters (k=1,...,8; σ=2). Each filter gives rise to a filtered image plane, denoted **G**k, σ. For each plane, we applied spatial convolution (*) with a series of pooling functions:

$$\mathbf{G}_{k,\sigma} * \Phi_{m,n} \qquad (1)$$

where Φm, n is computed by spatial sampling of the pooling function:

$$\Phi(x, y; m, n) = \exp\left\{-\alpha\left[\log_e\left(\frac{x^2+y^2}{d_n^2}\right)\right]^2 - \beta|\theta-\theta_m|\right\} \qquad (2)$$

with α=4 and β=0.4. The values of m=0,...,7 and n=0,1,2 were taken to construct 8 regions at angles $\theta_m = m\frac{\pi}{4}$ for each of two distances d_1=0.45, d_2=0.6 away from the center of a spatial pooling region in the image plane. For the central region, corresponding to m=0, there was no angular variation, but a log-radial exponential decay. This yielded a total of 17 spatial pooling regions. The resulting 17 × 8 fields were sub-sampled to produce a dense 136-dimensional descriptors, each representing an approximately 10 × 10 pixels image region. This resulted in, again, approximately 2000 descriptors per image frame after the result of Eq. (1) is sub-sampled. This is illustrated in Fig. 5.

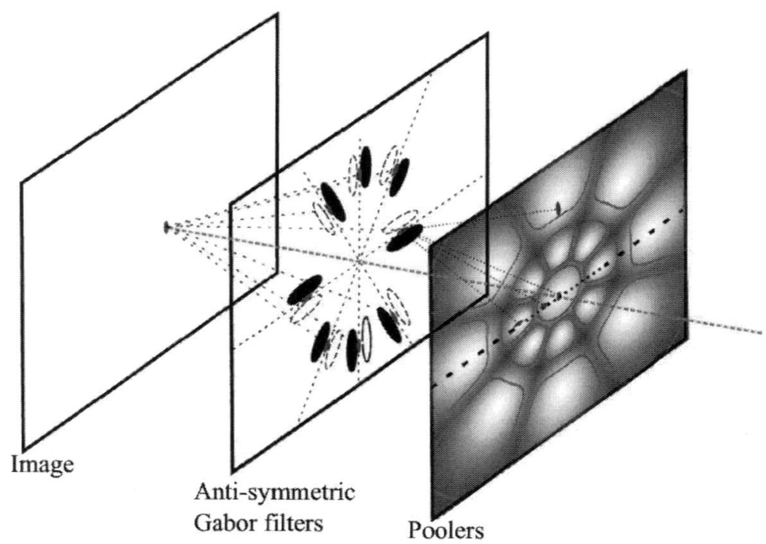

Figure. 5: Orientation selective masks are applied to the video frames by spatial convolution. The outputs of the masks are collected over space using the pooling functions. The outputs of the poolers are sub-sampled over the image plane to produce descriptors for indexing. See the text for further details.

Space-Time Patch Descriptors

Given the potential richness available in the capture of space-time information, we explored three distinct approaches to generate space-time patch descriptors. These approaches all lead to multiple descriptors per frame, and all take into account neighbouring frames in time when generating the descriptor associated with each patch. Additionally, all three densely sample the video sequence. The three methods are (i) HOG 3D, introduced by Kläser et al. [18]; (ii) our space-time, antisymmetric Gabor filtering process (ST_GABOR); and (iii) our spatial derivative, temporal Gaussian (ST_GAUSS) filter.

(i) The HOG 3D descriptor (HOG3D) [18] was introduced to extend the very successful two-dimensional histogram of oriented gradients technique [8] to space-time fields, in the form of video sequences. HOG 3D seeks computational efficiencies by smoothing using box filters, rather than Gaussian spatial or space-time kernels. This allows three-dimensional gradient estimation across multiple scales using theintegral video representation, a direct extension of the integral image idea [36]. The gradients from this operation are usually performed across multiple scales. We used the dense HOG 3D option from the implementation of the authors [18], and the settings yielded approximately 2000 descriptors per frame of video.

(ii) Space-time Gabor (ST_GABOR) functions have been used in activity recognition, structure from motion and other applications [2]. We performed one dimensional convolution between the video sequence and three one-dimensional Gabor functions along either one spatial dimension i.e. x or y, or along t. The one-dimensional convolution is crude, but appropriate if the videos have been downsampled. The spatial extent of the Gabor was set to provide one complete cycle of oscillation over approximately 5 pixels of spatial span, both for the x and y spatial dimensions. The filter for the temporal dimension was set to provide temporal support and one oscillation over approximately 9 frames. We also explored symmetric Gabor functions, but found them less favorable in early performance tests.

After performing three separate filtering operations, each pixel of each frame was assigned a triplet of values corresponding to the result of the each filtering operation. The three values were treated as being components of a 3D vector. Over a spatial extent of around 16 × 16 pixels taken at the central frame of the 9-frame support region, these vectors contribute weighted votes into 13 histogram bins according to their azimuth and elevations, with the weighting being given by the length of the vector. The votes were also partitioned into 17 regions according to the approximate spatial lobe pattern illustrated in Fig. 5, yielding a 221-dimension descriptor.

(iii)A final variant of space-time patch descriptor was designed. This consisted of spatial derivatives in space, combined with smoothing over time (ST_GAUSS). In contrast to the strictly one-dimensional filtering operation used for the space-time Gabor descriptor, we used two 5 × 5 gradient masks for the x and y directions based on derivatives of Gaussian functions, and an 11-point Gaussian smoothing filter in the temporal direction with a standard deviation of 2. Eight-directional quantization was applied to the angles of the gradient field, and a weighted gradient magnitude voting process was used to distribute votes across the 8 bins of a 136-dimensional descriptor. Like the ST_GABOR descriptor, pooling regions were created, similar to those shown in Fig. 5.

Frame-Level Encoding

Our initial conjecture was that whole frames from a sequence could be indexed compactly, using the single-frame descriptor (LW_COLOR). This was found to lead to disappointing performance (see Section 6). For the case of many descriptors-per-frame i.e. descriptors that are patch-based, we have the problem of generating around 2000 descriptors per frame, if dense sampling is used. Thus, we applied vector quantization (VQ) to the descriptors, then used histograms of VQ descriptors, effectively representing each frame as a histogram of words [7]. The dictionary was always built by excluding the entire journey from which queries are to be taken.

Table 2: A summary of the different encoding methods and their relationships to different descriptors the number of elements of each descriptor is also reported (**Dim**)

Method	ST	Dense	Dim	Encoding	Metric
SIFT	No	Yes	128	HA-4000	χ^2
				VLAD-256	Hellinger
SF-GABOR	No	Yes	136	HA-4000	χ^2
				VLAD-256	Hellinger
LW-COLOR	Yes	No	144	N/A	
ST_GABOR	Yes	Yes	221	HA-4000	χ^2
				VLAD-256	Hellinger
ST_GAUSS	Yes	Yes	136	HA-4000	χ^2
				VLAD-256	Hellinger
HOG3D	Yes	Yes	192	HA-4000	χ^2
				VLAD-256	Hellinger

Two different approaches to the VQ of descriptors were taken, one based on standard k-means, using a Euclidean distance measure (hard assignment, "HA"), and one corresponding to the Vector of Locally Aggregated Descriptors (VLAD) [16]. For VLAD, ak-means clustering was first performed. For each descriptor, sums of residual vectors were used to improve the encoding. Further advances to the basic VLAD, which include different normalizations and multiscale approaches, are given by [1]. To compare encodings, either χ^2 or Hellinger distance metrics [35] were used to retrieve results for HA and VLAD encoding approaches respectively. Distance comparisons were performed directly between either hard assigned Bag-of-Words (BoW) or VLAD image encodings arising from collections of descriptors for each frame.

EXPERIMENTS AND RESULTS: PERFORMANCE EVALUATION

The methods for (a) describing spatial or space-time structure, (b) indexing and comparing the data are summarized in Table 2. The choice of parameters was selected to allow (a) as consistent a combination of methods as possible, allowing fair comparisons of the effect of one type of encoding or spatio-temporal operator to be isolated from others (b) to select parameter choices close to other research in the area, e.g. for image categorization, dictionary sizes of $\approx$ 256 and $\approx$ 4000 words are common.

Error Distributions

Error distributions allow us to quantify the accuracy of being able to estimate locationsalong physical paths within the RSM dataset described in Section 4. To generate the error distributions, we did the following:

We started by using the kernels calculated in Section 5.3. One kernel is shown in Fig. 6, where the rows represent each frame from the query pass, and the columns represent each frame from one of the remaining database passes of that corridor. The values of the kernel along a row represent a "score" between a query and different database frames. In this experiment, we associated the position of the best match to the query frame, and calculated the error between this and the ground truth ϵ, in cm. In order to characterize the reliability of such scores, we performed bootstrap estimates of error distributions using 1 million trials. The distribution of the errors gives us a probability density estimate, from which we can get the cumulative distribution function (CDF) $P(x \leq |\epsilon|)$. The outcome is shown in Fig. 8, where only the average across all the randomized samples is shown.

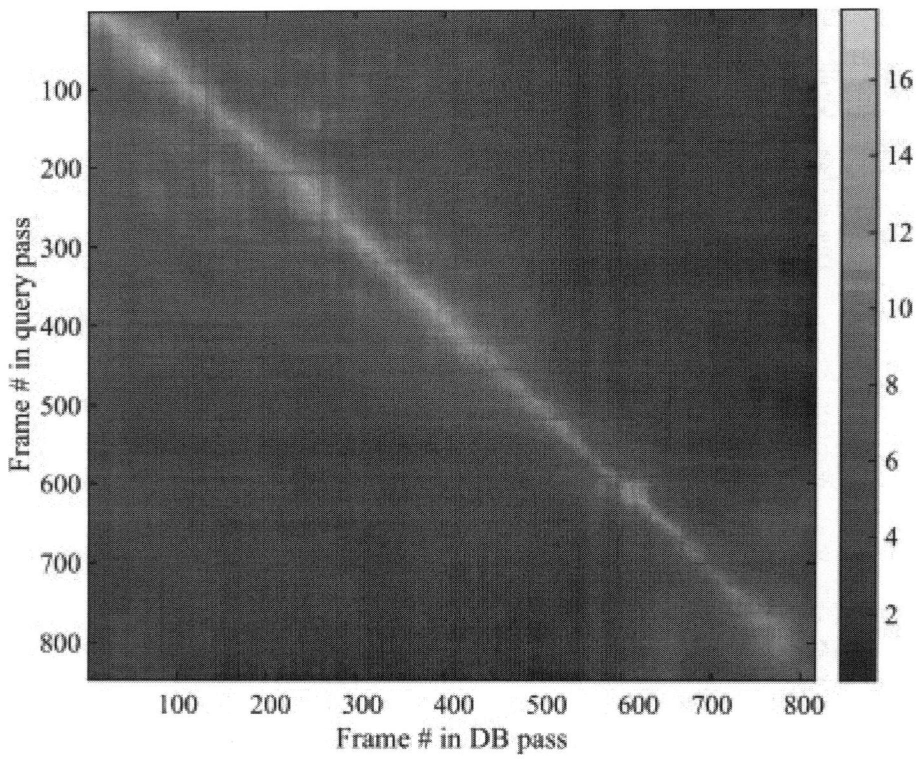

Figure. 6: Example of a χ^2 kernel produced by hard assignment and using the SF_GABOR descriptors when querying with pass 1 of corridor 2 against a database comprised of passes 2–10.

By permuting the paths that are held in the database and randomly selecting queries from the remaining path, we were able to assess the error distributions in localization. Repeated runs with random selections of groups of frames allowed the variability in these estimates to be obtained, including that due to different numbers of paths and passes being within the database.

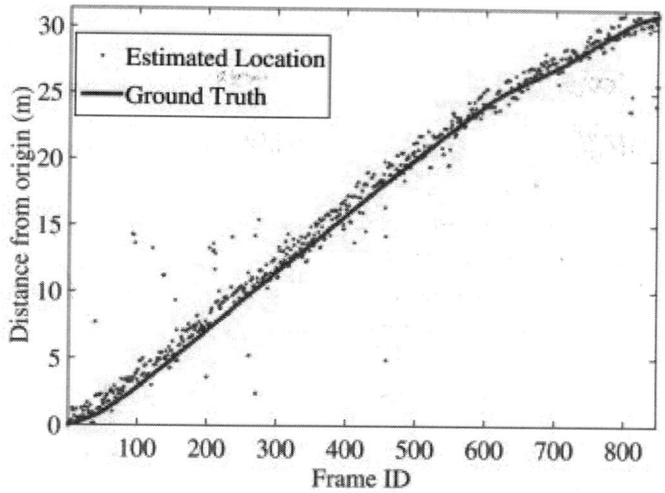

(a) Corridor 3 using Pass 2 acquired with LG Nexus 4. Results for the best spatio-temporal method, ST_GABOR.

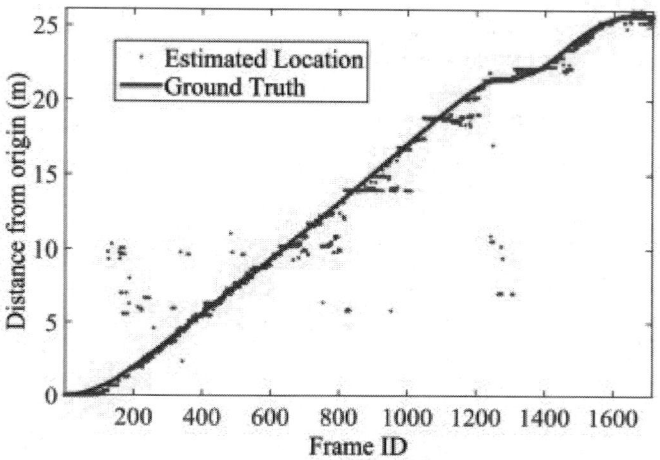

(b) Corridor 3 using Pass 6 acquired with Google Glass. Results for the best single-frame method, SF_GABOR.

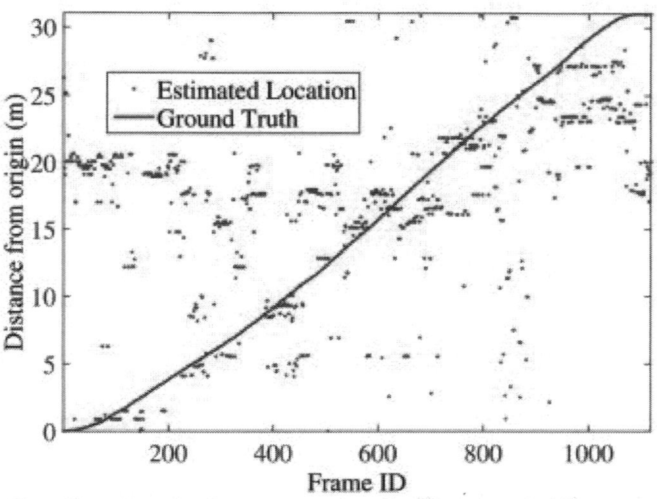

(c) Corridor 2 using Pass 6 acquired with Google Glass. Results for the worst overall method, LW_COLOR.

Figure. 7: Estimated location vs. ground truth. Illustrative examples of good/bad location estimation performance (a) uses the best descriptor and a single-device dataset, (b) uses the best descriptor and a cross-device dataset and (c) uses the worst descriptor, and a multiple-device dataset.

If we consider the idea of crowdsourcing journey information from many pedestrian journeys through the same corridors, this approach to evaluating the error makes sense: all previous journeys could be indexed and held in the database; new journey footage would be submitted as a series of query frames (see Fig. 1).

Localization Error vs Ground-Truth Route Positions

As described in the previous section, by permuting the database paths and selecting, randomly, queries from the remaining path that was left out in the dictionary creation, we can assess the errors in localization along each corridor for each pass, and calculate, also, the average error in localization on a per-corridor basis, or per-path basis. For these, we used the ground-truth information acquired as described in Section 4. Fig. 7provides some examples of the nature of the errors, showing evidence of those locations that are often confused with each other. As can be seen, for the better method (top trace of Fig. 7) while average errors might be small, there are, occasionally, large errors due to poor matching (middle trace). Errors are significantly worse for queries between different devices (see Fig. 7(c)).

Note that we did not use any tracking algorithms, and so there is no motion model or estimate of current location given the previous one. Incorporating a particle filter or Kalman filter should reduce the errors, particularly where there are large jumps within small intervals of time. This deliberate choice allows us to evaluate the performance of different descriptor and metric choices independently.

Performance Summaries

We calculated the average of the absolute positional error (in cm) and the standard deviation of the absolute positional error in a subset of the complete RSM dataset (Table 3). We used a leave-one-journey-out approach (all the frames from an entire journey are excluded from the database). Using bootstrap sampling, we also estimated the cumulative density functions of the error distributions in position, which are plotted in Fig. 8. The variability in these curves is not shown, but is summarized in the last two columns of Table 3 through the area-under-curve (AUC) values. In the best case (SF_GABOR), AUCs of the order of 96% would mean errors generally below 2 m; in the worst (HOG3D), AUCs ≈ 90% would mean errors of around 5 m. These mean absolute error estimates are obtained as we permute the queries, the dictionary and the paths in the database.

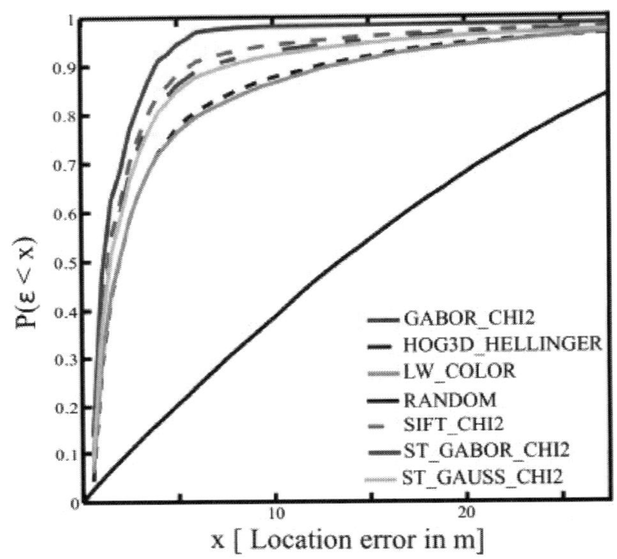

Figure. 8: Comparison of best performing version of all methods. The average CDFs were computed from the distributions generated after the random sampling described in Section 6.1. The results for a random test (RANDOM) were introduced as a "sanity check".

Table 3: Summaries of average absolute positional errors and standard deviation of positional errors for different descriptor types and for different encoding methods (labelled by the corresponding metric used: χ^2 for HA and Hellinger for VLAD). μ_ϵ is the average absolute error, and σ_ϵ is the standard deviation of the error in cm. Single device case and in bold: best and worst AUC

		Error summary (cm)		AUC (%)	
Method	Metric				
		μ_ϵ	σ_ϵ	Min	Max
SF_GABOR	χ^2	130.6	38.8	96.40	**96.75**
SF_GABOR	Hellinger	135.1	46.5	96.29	96.71
ST_GAUSS	χ^2	135.4	44.1	93.61	94.30
ST_GAUSS	Hellinger	144.1	52.4	92.69	93.47
ST_GABOR	χ^2	235.9	86.3	93.97	94.66
ST_GABOR	Hellinger	179.5	62.3	93.98	94.60
SIFT	χ^2	137.5	46.3	94.57	95.14
SIFT	Hellinger	132.7	41.4	94.34	94.95
HOG3D	χ^2	419.6	133.3	**90.89**	91.83
HOG3D	Hellinger	366.5	120.3	91.49	92.37
LW_COLOR	N/A	363.9	113.2	91.42	92.25

Finally, we applied one implementation of the SLAM to this dataset, at the same frame resolution as for the appearance-based localization discussed in this paper. We chose the "EKF Mono SLAM" [5], which uses an extended Kalman filter (EKF) with 1-point RANSAC. We chose this implementation for three reasons: (a) it is a monocular SLAM technique, so comparison with the single-camera approach is fairer; (b) the authors of this package report error estimates—in the form of error distributions; and (c) the errors from video with similar resolutions (240 × 320) to ours were reported as being below 2 m for some sequences [5] in their dataset.

The results of the comparison were surprising, and somewhat unsatisfactory. The challenging ambiguity of the sequences in the RSM dataset, and possibly the low resolution of the queries, might explain the results. The feature detector, a FAST corner detector [31], produced a small number of features in its original configuration. We lowered the feature detection threshold until the system worked on a small number of frames from each sequence. Even with more permissive thresholds, the average number of FAST features averaged only 20 across our experiments. This small number of features led to inaccuracy in the position estimates, causing many of the experimental runs to stop when no features could be matched. The small number of features per frame is also not comparable with the feature density of the methods described in this paper, where an average of 2000 features per frame was obtained for the "dense" approaches. Dense SLAM algorithms might fare better.

DISCUSSION

The performance comparisons shown in the cumulative error distributions of Fig. 8 would seem a fairly natural means of capturing localization performance. Yet, they do not suggest large differences in terms of the AUC metric (Table 3), given the large diversity in the complexity of the indexing methods. However, absolute position estimation errors tell a different story: average absolute errors are as high as 4 m for the worst performing method (HOG3D), and just over 1.3 m for the best performing method (SF_GABOR), if the same camera is used. The best performance compares very favorably with reported errors in positioning from multi-point WiFi signal strength measurements using landmark-based recognition that employs multiple (non-visual) sensing [33]. Indeed, it is very likely that the size of the errors we have observed can be reduced by incorporating simple motion models and a tracker, in the form of a Kalman filter.

A surprising result was that good levels of accuracy were obtained for images as small as 208 × 117 pixels. This suggests that relatively low-resolution cameras can be used to improve the performance of indoor localization systems. Being able to use such low resolutions of image reduces the indexing time, storage, power and bandwidth requirements.

CONCLUSION AND FUTURE WORK

The advent of wearable and hand-held cameras makes appearance-based localization feasible. Interaction between users and their wearable device would allow for new applications such as localization, navigation and semantic descriptions of the environment. Additionally, the ability to crowdsource "visual paths" against which users could match their current views is a realistic scenario given ever improving connectivity.

We evaluated several types of descriptor in this retrieval-based localization scenario, achieving errors as small as 1.30 m over a 50 m distance of travel. This is surprising, given that we used low-resolution versions of our images, and particularly since our RSM dataset also contains very ambiguous indoor scenes.

We are currently working on enlarging the RSM database, by including larger numbers of journeys. A future goal will be to mitigate the effects of partial occlusion between different views of the same physical location. For example, face-detection might be applied to identify when and where people are within the scene acquired along a users' journey; we would avoid generating descriptors that covered these regions of image space. Other movable objects (chairs, trolleys) could also be actively detected and removed from indexing or queries.

The challenges associated with searching across video from multiple devices would still need to be solved. We can see from Section 6 that between-device queries have much higher error than within-device queries. This problem can be solved by either capturing and indexing data from a variety of devices for the same journeys, or by learning a mapping between devices. Another obvious strand of work would be to incorporate information from other sources, such as RSSI indicators, to reduce localization error.

Finally, we are exploring ways to combine the appearance-based technique described in this paper with SLAM and its variants. Doing this would allow geometric models from independent point-cloud sets to be associated with each other, allowing the continuous updating of the models that describe a physical space. Multiple geometric models, acquired from otherwise independent journeys, would support more detailed and reliable descriptions of an indoor, navigable space. It would also allow better interaction between the users of a building with its features, and with each other.

Our long-term goal is to convey the information acquired by sighted users to help people with visual impairment; this would require creating and updating rich descriptions of the visual and geometric structure of a physical space. This could be used in the making of indoor navigational aides, which would be rendered through haptic or audio interfaces, making the planning of journeys easier for the visually impaired.

ACKNOWLEDGEMENTS

Jose Rivera-Rubio's research is supported by an EPSRC DTP scholarship. The dataset hosting is supported by the EPSRC V&L Net Pump-Priming Grant 2013-1. We are also grateful to members of the Imperial College's Biologically Inspired Computer Vision group for their help and collaboration, and to Dr Riccardo Secoli and Dr Luke Dickens for their valuable input and useful discussions.

REFERENCES

1. R. Arandjelovic, A. Zisserman, All about vlad, in: 2013 IEEE Conference on Computer Vision and Pattern Recognition (CVPR), IEEE, 2013, pp. 1578–1585.

2. M. Bregonzio, S. Gong, T. Xiang, Recognising action as clouds of space-time interest points, in: IEEE Conference on Computer Vision and Pattern Recognition, 2009 (CVPR'09), IEEE, 2009, pp. 1948–1955.
3. N. Burgess, Spatial memory: how egocentric and allocentric combine, Trend Cogn. Sci. 10 (2006) 551–557.
4. N. Burgess, E.A. Maguire, J. O'Keefe, The human hippocampus and spatial and episodic memory, Neuron 35 (2002) 625–641.
5. J. Civera, O.G. Grasa, A.J. Davison, J.M.M. Montiel, 1pointt ransac for extended kalman filtering: Application to real-time structure from motion and visual odometry, J. Field Robot. 27 (2010) 609–631.
6. M. Collett, How desert ants use a visual landmark for guidance along a habitual route, Proc. Natl. Acad. Sci. 107 (2010) 11638–11643.
7. G. Csurka, C. Dance, L. Fan, J. Willamowski, C. Bray, Visual categorization with bags of keypoints, in: Workshop on Statistical Learning in Computer Vision, ECCV, Prague, 2004, pp. 1–2.
8. N. Dalal, B. Triggs, Histograms of oriented gradients for human detection, in: IEEE Computer Society Conference on Computer Vision and Pattern Recognition, 2005 (CVPR'05), IEEE, 2005, pp. 886–893.
9. A.J. Davison, Real-time simultaneous localisation and mapping with a single camera, in: IEEE International Conference on Computer Vision (ICCV), IEEE, 2003, pp. 1403–1410.
10. D.A. Dombeck, C.D. Harvey, L. Tian, L.L. Looger, D.W. Tank, Functional imaging of hippocampal place cells at cellular resolution during virtual navigation, Nat. Neurosci. 13 (2010) 1433–1440.
11. G. Dragoi, S. Tonegawa, Selection of preconfigured cell assemblies for representation of novel spatial experiences, Phil. Trans. Roy. Soc. B: Biol. Sci. 369 (2014) 20120522.
12. M. Everingham, L. Gool, C.K.I. Williams, J. Winn, A. Zisserman, The PASCAL Visual Object Classes (VOC) challenge, Int. J. Comput. Vis. 88 (2009) 303–338.
13. S. Fry, R. Wehner, Look and turn: Landmark-based goal navigation in honey bees, J. Exp. Biol. 208 (2005) 3945–3955.
14. T. Hartley, C. Lever, N. Burgess, J. O'Keefe, Space in the brain: How the hippocampal formation supports spatial cognition, Phil. Trans. Roy. Soc. B: Biol.l Sci. 369 (2014) 20120510.
15. R. Huitl, G. Schroth, S. Hilsenbeck, F. Schweiger, E. Steinbach, TUMindoor: An extensive image and point cloud dataset for visual indoor localization and mapping, in: 19th IEEE International Conference on Image Processing (ICIP), IEEE, 2012, pp. 1773–1776.
16. H. Jégou, F. Perronnin, M. Douze, J. Sánchez, P. Pérez, C. Schmid, Aggregating local image descriptors into compact codes, IEEE Trans. Pattern Anal. Mach. Intell. 34 (2012) 1704–1716.
17. W. Kadous, S. Peterson, Indoor maps: The next frontier, in: Google IO, 2013.
18. A. Kläser, M. Marszalek, C. Schmid, A spatio-temporal descriptor based on 3dgradients, in: British Machine Vision Conference (BMVC), British Machine Vision Association (BMVA), 2008, pp. 995–1004.
19. G. Klein, D. Murray, Parallel tracking and mapping on a camera phone, in: 8th IEEE International Symposium on Mixed and Augmented Reality, 2009 (ISMAR'09), IEEE, 2009, pp. 83–86.
20. S. Lazebnik, C. Schmid, J. Ponce, Beyond bags of features: Spatial pyramid matching for recognizing natural scene categories, in: IEEE Computer Society Conference on Computer Vision and Pattern Recognition, 2006, IEEE, 2006, pp. 2169–2178.
21. D.G. Lowe, Object recognition from local scale-invariant features, in: Proceedings of the Seventh IEEE International Conference on Computer Vision, 1999, IEEE, 1999, pp. 1150–1157.
22. A. Möller, M. Kranz, R. Huitl, S. Diewald, L. Roalter, A mobile indoor navigation system interface adapted to vision-based localization, in: Proceedings of the 11th International Conference on Mobile and Ubiquitous Multimedia, ACM, 2012, p. 4.
23. E.I. Moser, E. Kropff, M.B. Moser, Place cells, grid cells, and the brain's spatial representation system, Annu. Rev. Neurosci. 31 (2008) 69–89.
24. R.A. Newcombe, S.J. Lovegrove, A.J. Davison, Dtam: Dense tracking and mapping in real-time, in: 2011 IEEE International Conference on Computer Vision (ICCV), IEEE, 2011, pp. 2320–2327.
25. P. Panagiotaki, S. Lambrey, A. Berthoz, Neural bases and cognitive mechanisms of human spatial memory, Image. Spat. Cogn. 66 (2006) 85–100.
26. B. Poucet, F. Sargolini, E.Y. Song, B. Hangya, S. Fox, R.U. Muller, Independence of landmark and self-motion-guided navigation: a different role for grid cells, Phil. Trans. Roy. Soc. B: Biol. Sci. 369 (2014) 20130370.
27. M.L. Pupilli, a.D. Calway, Real-time camera tracking using a particle filter, in: Proceedings of the British Machine Vision Conference, British Machine Vision Association (BMVA), 2005, pp. 50.1–50.10.
28. M. Quigley, D. Stavens, Sub-meter indoor localization in unmodified environments with inexpensive sensors, in: Intelligent Robots and Systems, IEEE, 2010, pp. 2039–2046.
29. J. Rivera-Rubio, I. Alexiou, A.A. Bharath, RSM dataset, 2014, http://rsm.bicv.org.
30. J. Rivera-Rubio, S. Idrees, I. Alexiou, L. Hadjilucas, A.A. Bharath, Mobile visual assistive apps: Benchmarks of vision algorithm performance, in: A. Petrosino, L. Maddalena, P. Pala (Eds.), Lecture Notes in Computer Science, Springer, Berlin Heidelberg, 2013, pp. 30–40.
31. E. Rosten, T. Drummond, Machine learning for high-speed corner detection, in: European Conference on Computer Vision, Springer, 2006, pp. 430–443.
32. R.F. Salas-Moreno, R.a. Newcombe, H. Strasdat, P.H. Kelly, A.J. Davison, SLAM++: Simultaneous localisation and mapping at the level of objects, 2013

IEEE Conference on Computer Vision and Pattern Recognition (2013) 1352–1359.

33. G. Shen, Z. Chen, P. Zhang, T. Moscibroda, Y. Zhang, Walkie-markie: indoor pathway mapping made easy, in: Proceedings of the 10th USENIX Conference on Networked Systems Design and Implementation, USENIX Association, 2013, pp. 85– 98.

34. A. Vedaldi, B. Fulkerson, VLFeat: An Open and Portable Library of Computer Vision Algorithms, 2008.

35. A. Vedaldi, A. Zisserman, Efficient additive kernels via explicit feature maps, IEEE Trans. Pattern Anal. Mach. Intell. 34 (2012) 480–492.

36. P. Viola, M. Jones, Robust real-time object detection, Int. J. Comput. Vis. 4 (2001) 34–47.

37. H. Wang, S. Sen, A. Elgohary, M. Farid, M. Youssef, R.R. Choudhury, No need to wardrive: unsupervised indoor localization, in: Proceedings of the 10th international conference on Mobile systems, applications, and services, ACM, 2012, pp. 197– 210.

38. H. Wang, M.M. Ullah, A. Klaser, I. Laptev, C. Schmid, Evaluation of local spatiotemporal features for action recognition, in: BMVC 2009-British Machine Vision Conference, BMVA Press, 2009, pp. 124.1–124.11.

39. J. Weickert, A. Bruhn, T. Brox, N. Papenberg, A survey on variational optic flow methods for small displacements, in: Mathematical Models for Registration and Applications to Medical Imaging, Springer, 2006, pp. 103–136.

CHAPTER 6

AN AUTOMATED PATTERN RECOGNITION SYSTEM FOR CLASSIFYING INDIRECT IMMUNOFLUORESCENCE IMAGES OF HEP-2 CELLS AND SPECIMENS

Siyamalan Manivannan,Wenqi Li, Shazia Akbar, Ruixuan Wang, Jianguo Zhang, Stephen J. McKenna

CVIP, School of Computing, University of Dundee, UK

ABSTRACT

Immunofluorescence antinuclear antibody tests are important for diagnosis and management of autoimmune conditions; a key step that would benefit from reliable automation is the recognition of subcellular patterns suggestive of different diseases. We present a system to recognize such patterns, at cellular and specimen levels, in images of HEp-2 cells. Ensembles of SVMs were trained to classify cells into six classes based on sparse encoding of texture features with cell pyramids, capturing spatial, multi-scale structure. A similar approach was used to classify specimens into seven classes. Software implementations were submitted to an international contest hosted by ICPR 2014 (Performance Evaluation of Indirect Immunofluorescence Image Analysis Systems). Mean class accuracies obtained on heldout test data sets were 87.1% and 88.5% for cell and specimen classification respectively. These were the highest achieved in the competition, suggesting that our methods are state-of-the-art. We provide detailed descriptions and extensive experiments with various features and encoding methods.

INTRODUCTION

Antinuclear antibody (ANA) tests are important in the diagnosis and management of autoimmune diseases. These include systemic lupus erythematosus, Sjogren's syndrome, rheumatoid arthritis, polymyositis, scleroderma, Hashimoto's thyroiditis, juvenile diabetes mellitus, Addison disease, vitiligo, pernicious anemia, glomerulonephritis, and pulmonary fibrosis. Immunofluorescene ANA tests have been recommended as the gold standard for ANA testing due to their relatively high sensitivity[1]. Specifically, human epithelial (HEp-2) cell specimens are examined using Indirect ImmunoFluorescence (IIF) imaging [2]. The flow of the IIF procedure includes the following steps: image acquisition, mitosis detection, fluorescence intensity classification and staining pattern recognition. The pattern recognition step is an important one as different patterns are suggestive of different autoimmune diseases. A nucleolar pattern is often associated with scleroderma, for example. Manual analysis of IIF images

Citation: Siyamalan Manivannan, Wenqi Li, Shazia Akbar, Ruixuan Wang, Jianguo Zhang, Stephen J. McKenna, An automated pattern recognition system for classifying indirect immunofluorescence images of HEp-2 cells and specimens, Pattern Recognition, Volume 51, March 2016, Pages 12-26, ISSN 0031-3203, http://dx.doi.org/10.1016/j.patcog.2015.09.015.

is laborious and time-consuming. Furthermore, two or more experts can be required to examine each ANA sample due to inter-observer variability. Aiming to standardize the procedure compared to current manual practice and to reduce workloads, computer aided diagnosis (CAD) systems have been proposed for the analysis IIF images[3] and [4].

This paper describes a system to classify pre-segmented immunofluorescence images of individual HEp-2 cells into six classes (homogeneous, speckled, nucleolar, centromere, golgi, and nuclear membrane) as well as a system to classify HEp-2 specimen images into seven classes (homogeneous, speckled, nucleolar, centromere, golgi, nuclear membrane and mitotic spindle). Instances from these classes are shown inFig. 1 and Fig. 2. These two classification tasks correspond to those used in the contest on *Performance Evaluation of Indirect Immunofluorescence Image Analysis Systems* (I3A)[1] held in conjunction with the 22nd International Conference on Pattern Recognition (ICPR 2014). Our approach to classifying cell and specimen images is based on sets of local features which describe texture and intensity properties of local image regions. Extracted features are encoded via sparse coding and classification is performed using support vector machine (SVM) ensembles.

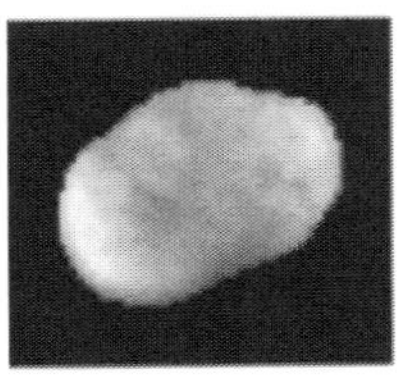

Homoge-
neous

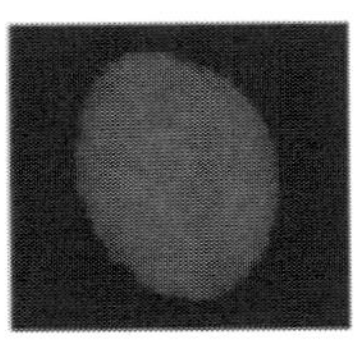

Speckled

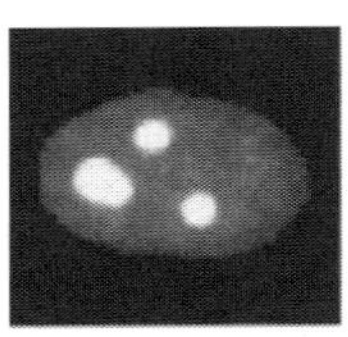

Nucleolar

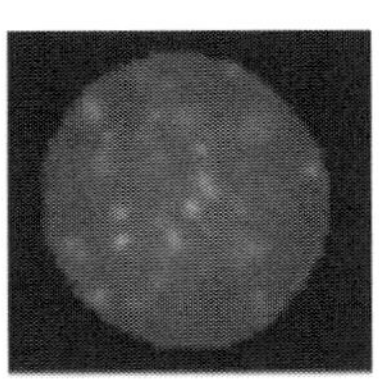

Cen-
tromere

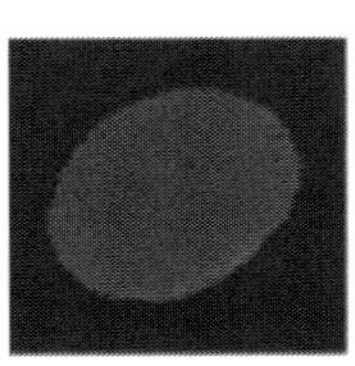

Nuclear
Membrane

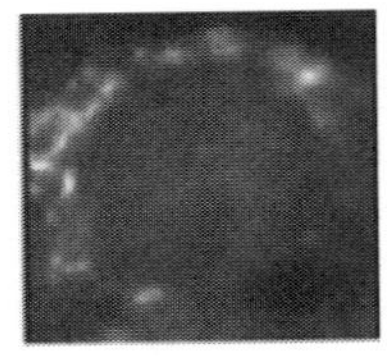

Golgi

Figure. 1: Sample images from the I3A Task 1 dataset (individual cell classification).

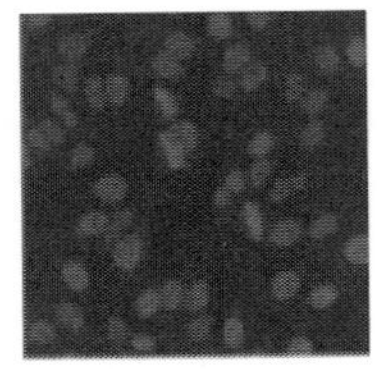

Homoge-
neous

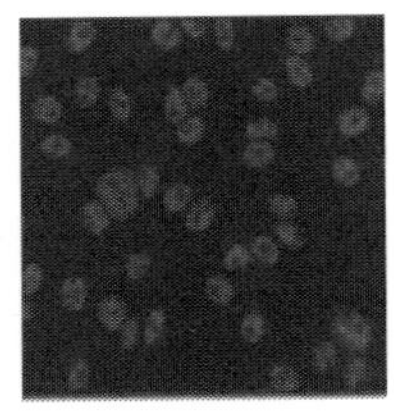

Speckled

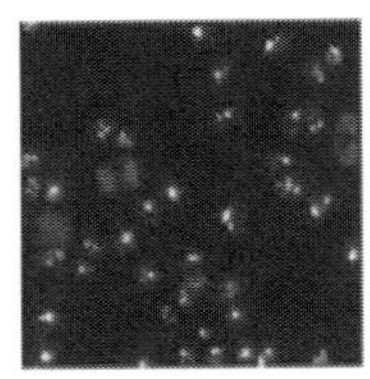

Nucleolar

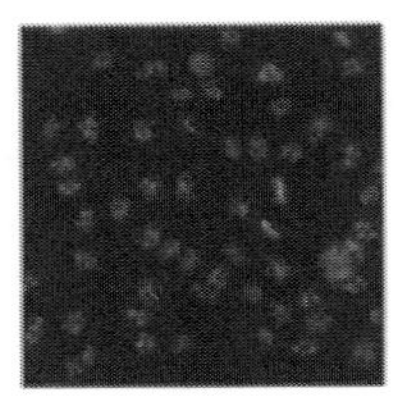

Cen-
tromere

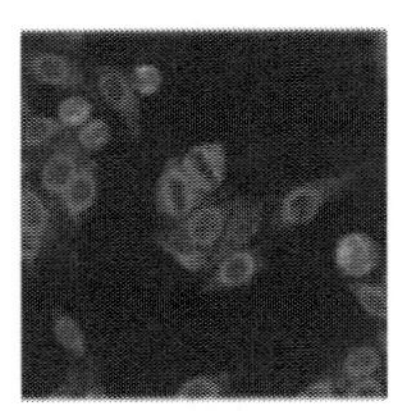

Nuclear
Membrane

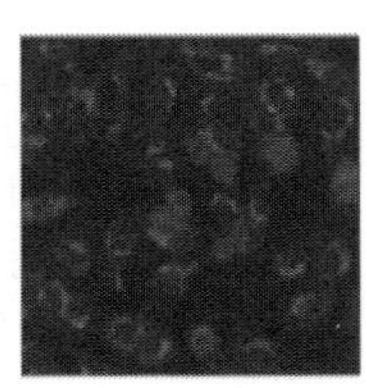

Golgi

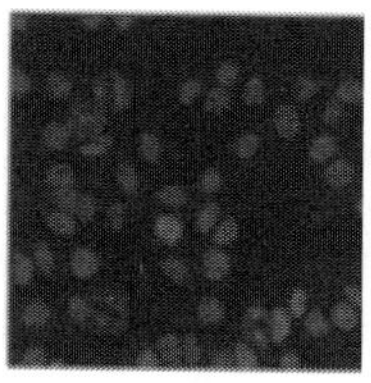

Mitotic Spindle

Figure. 2: Sample images from the I3A Task 2 dataset (specimen classification).

After reviewing competing methods (Section 2) and describing our proposed method in detail (3, 4, 5, 6 and 7), results on the I3A contest datasets are presented (8, 9 and 10). This paper builds on our earlier I3A workshop papers [5] and [6]. It sets the proposed method in the context of related literature, describes it in more detail, presents more extensive experiments to investigate the effect of various components on performance, and summarizes performance in direct comparison with other methods. We present systems tailored to capture cell class-specific properties, leveraging state-of-the-art computer vision techniques. The experiments we report should help guide future developments in this area by providing evidence on the relative contributions of, e.g., feature combinations, data augmentation, cell pyramids, and sparse coding. Systems we proposed and describe in this paper won both tasks at the I3A contest in controlled tests. As such they can serve as benchmarks for researchers developing pattern recognition systems for this important application.

RELATED WORK

This section concisely reviews previous work related to HEp-2 cell image classification in the context of ANA testing. There exists a wider literature on the recognition of fluorescence image patterns characteristic of subcellular structures more generally [7]. However, its review is beyond the scope of this paper.

Perner et al. [8] presented an early attempt at developing an automated HEp-2 cell classification system. Cell regions were represented by a set of basic features extracted from binary images obtained at multiple grey level thresholds. Those features were then classified into six categories by a decision tree algorithm. This approach was further employed and integrated by Sack et al. [9] for identification of positive fluorescence and a set of immunofluorescence patterns. Hsieh et al. [10] performed classification of immunofluorescence patterns using learning vector quantization (LVQ) and various texture features that included grey-level histogram statistics, co-occurrence matrix features, and an estimate of fractal dimension.

The problem of HEp-2 cell classification attracted increased attention among researchers with the benchmarking contests held in conjunction with the International Conference on Pattern Recognition (ICPR) in 2012 [11] (*HEp-2 Cell Classification* [2]) and the International Conference on Image Processing (ICIP) in 2013 [12] (*Competition on Cells Classification by Fluorescent Image Analysis* [3]). A special issue following the ICPR 2012 contest was also organized in the journal Pattern Recognition [4]. Various image descriptors were adopted in those contests, including popular local texture features such as local binary patterns (LBP) and its variants (e.g., multi-resolution LBP), SIFT, summative intensity statistics (e.g., mean and standard deviation) of local or whole image regions, and morphological properties (e.g., eccentricity) [12]. Standard feature encoding methods, in particular bag of words (BoW), were often applied to represent the statistics of local features in the feature space. The majority of the classifiers used were support vector machines although *k*-nearest neighbour classifiers, boosting classifiers, random forest classifiers, and neural networks were also used by some groups[12] and [4]. For more detailed descriptions of the methods submitted to those contests, the reader is referred to the associated reports [11], [12] and [4],11,12].

The I3A contest held in conjunction with ICPR 2014 was the most recent in this series of contests. It received 11 submissions to Task 1 (cell classification) and 7 submissions to Task 2 (specimen classification) [3]. Methods submitted for Task 1 can be classified broadly into two categories: those with feature coding and those without feature coding. The feature coding-based methods basically adopted the popular bag of words framework using either hard coding or sparse coding [13] and [14] and either SVM or boosting classifiers. They differed in their choice of local features. For instance, Ensafi et al. [15] used SIFT and SURF features with sparse coding and max pooling. Theodorakopoulos et al. [16] combined a set of local features, including LBP and rotation-invariant SIFT, with vectors of locally aggregated descriptors (VLAD) [17]. Paisitkriangkrai et al.›s method (as described in [3]) combined features including region covariance and co-occurrence of adjacent LBPs. Gragnaniello et al. [18] used a recently developed local feature called dense scale-invariant descriptors [19] to characterize cell images. The feature coding methods were among the top performing in the contest [3]. The methods without feature coding basically followed a paradigm of global feature representation with a popular machine learning classifier. For example, Roberts› method (as described in [3]) used a set of wavelet-based features with a SVM classifier. Codrescu [20] learned a neural network with finite impulse response filters applied to cell images. Ponomare et al.›s method (as described in [3]) utilized a set of morphological features (such as number of isolated regions in each cell) with a SVM classifier. Taormina et al.›s method (as described in [3]) used an ensemble of nearest neighbour classifiers on 108 features. Those methods generally performed less well in the contest. It is worth noting that Gao et al. [21] used a deep convolutional neural network (CNN) with 8 layers. The last layer was in principle a logistic regression classifier with a soft-max activation function. This method performed reasonably well in the contest. Although deep CNNs have proven very successful on various large-scale image classification benchmarks, they have very large numbers of parameters and it can be difficult to tune their structure to a specific task such as I3A.

Most of the methods used for Task 2 (specimen classification) first performed classification at the individual cell (or sub-region) level using approaches similar to Task 1 and then aggregated the results by majority voting. For instance, Ensafi et al. [15]performed specimen classification by applying voting to the cell classification results for a small number of extracted cells. Similarly, Liu et al. and Paisitkriangkrai et al. (as described in [3]) classified specimen images based on a majority voting over regions where only features within a cell were considered when describing a region. Ponomare et al. [3] applied an unweighted voting scheme for a final classification of the specimen image based on morphological features.

Notable trends when comparing the I3A contest entries to previous work are the use of more advanced hand-crafted local features (e.g. multi-resolution local patterns with cell pyramids as in our entries [5] and [6], and dense scale-invariant descriptors [19]), dataset augmentation, and the deployment of deep learning for automatic feature learning. Our proposed method benefits from a combination of the following factors which we believe contribute to its state-of-the-art performance: complementary multi-resolution features, the use of a specifically designed spatial structure for cell images, sparse coding, and data augmentation.

SYSTEM OVERVIEW

Fig. 3 gives an overview of the system used for generating a feature representation from an image of a cell for input to a classifier. Firstly, each cell image was intensity-normalized. Sets of local features were then extracted and a feature encoding method (e.g. sparse coding) was employed to aggregate the local features into a cell image representation. A two-level cell pyramid was used to capture spatial structure of cell images. Support vector machines were then used to classify images of cells or to classify specimen images containing multiple cells. The following sections describe these system components in detail. Experiments investigating the effect of different system components, feature representations and encodings are then reported.

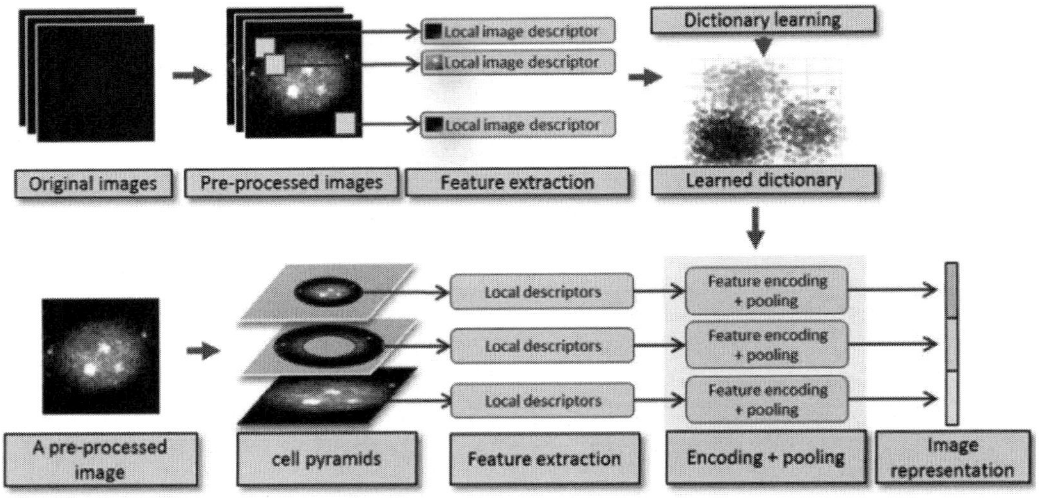

Figure. 3: An overview of the system for generating the image-level feature representation using only one feature type: dictionary learning from training images (first row) and feature encoding to obtain the image-level feature representation (second row). The final image representation is a concatenation of the image level representations obtained by different types of features.

LOCAL FEATURE EXTRACTION

Prior to feature extraction, each cell's image was intensity normalized; specifically, the segmentation mask was dilated (using a 5×5 structuring element) and the intensity values within each cell's dilated mask region were then linearly rescaled so that 2% of pixels in each cell became saturated at low and high intensities (Fig. 4).

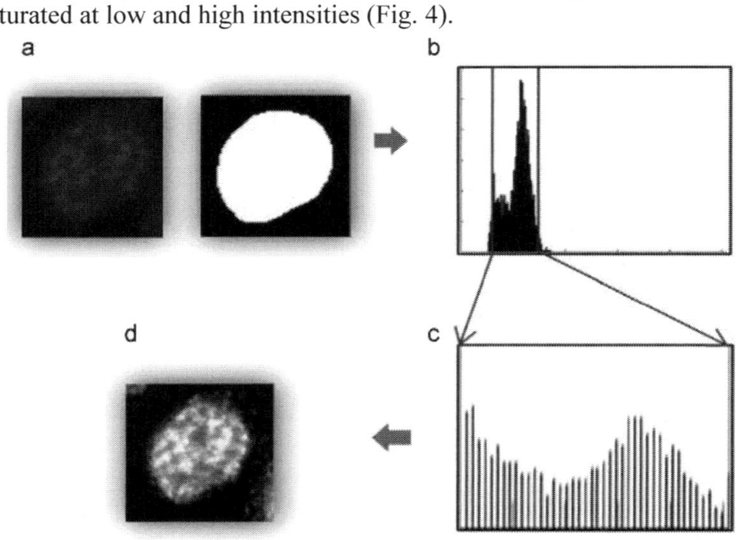

Figure. 4: Image preprocessing: (a) an example cell image and its mask, (b) histogram of intensity values inside cell region in (a), (c) normalized histogram, and (d) preprocessed image.

Local features were extracted densely from each pre-processed cell image. Four types of local feature were considered, namely, Multi-resolution Local Patterns (mLP), Root-SIFT (rSIFT), Random Projections (RP) and Intensity Histograms (IH).

Multi-Resolution Local Patterns

Multi-resolution Local Patterns (mLP) are a multi-resolution adaptation of the local higher-order statistical (LHS) patterns proposed by Sharma et al. [22] for texture classification. LHS is a non-binarized version of the well-known Local Binary Patterns (LBP) descriptor. It operates on a small image neighbourhood of size 3×3. To capture information from a larger neighbourhood and reduce noise effects, we extended LHS from a multi-resolution perspective by employing the sampling patterns described by Mäenpää [23]. This sampling pattern is inspired by the spatial structure of receptive fields in the human retina and has been widely adopted in recently developed visual features in computer vision, e.g., BRISK [24]. Fig. 5 shows an example sampling pattern and the generation of the mLP descriptor. In Fig. 5 the local neighbourhood is quantized radially into three resolutions (radii), and at each resolution a set of (N=8) sampling regions (indicated as circles) are considered. At each sampling point a Gaussian filter is applied, integrating information from the filter's region of support. We call the combination of LHS and these sampling patterns *multi-resolution local patterns*.

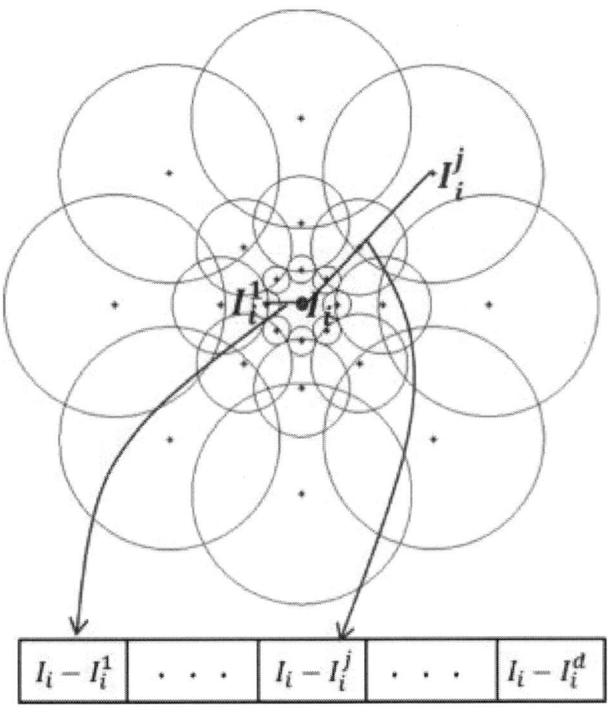

Figure. 5: Generation of multi-resolution local patterns (mLP).

Root-SIFT

Root-SIFT (rSIFT) is a variant of the widely used SIFT descriptor that produces better performance than SIFT on some image matching and retrieval tasks [25]. The standard SIFT descriptor is a histogram representation of local image derivatives and was originally designed to be used with Euclidean distance. Using Euclidean distance to compare histograms often yields inferior performance compared to other measures such as ^{2} or Hellinger for texture classification and image categorization [25]. Therefore, standard SIFT was modified in [25] to create rSIFT such that comparing rSIFT descriptors using Euclidean distance is equivalent to using the Hellinger kernel to compare SIFT vectors.

Random Projection

Random projection (RP) is a simple yet powerful method for dimensionality reduction[26]. It projects patch intensity vectors from the original patch-vector space $R^{D'}$ to a compressed space R^{D} using randomly chosen projection vectors. Such a scheme has been successfully applied to texture image classification [27]. Let $\mathbf{x}$ be a D'-dimensional patch vector and $\hat{\mathbf{x}}$ be its D -dimensional representation in the compressed space. The RP method simply maps these vectors using a $D \times D'$ random projection matrix R, such that

$$\hat{\mathbf{x}} = R\mathbf{x} \qquad (1)$$

Each element in matrix R is sampled from a Gaussian distribution with zero mean and unit variance. The key point of RP is that when projecting the patch-vectors from the original space to the compressed space their relative distances are approximately preserved.

Intensity Histograms

Intensity histograms (IH) were computed from small image patches to represent the local intensity distribution.

FEATURE ENCODING

Feature encoding methods aggregate the local features from an image or image region and play an important role in classification. Four methods for feature encoding were compared, namely bag-of-words (BoW), a sparse coding method (SC), Fisher vectors (FV), and vectors of locally aggregated descriptors (VLAD). A pyramid was used to encode spatial structure. Each of these methods is now described briefly.

Bag-of-Words

Bag-of-Words (BoW) is widely applied as a feature encoding method for medical [2] as well as natural [13] and [28] image classification. In BoW local features sampled from training images are clustered to build a dictionary (codebook). This dictionary represents a set of visual words (or clusters) which are then used to compute a BoW frequency histogram as a feature vector representation of any given image. BoW uses hard quantization where each local image descriptor is assigned to only one visual word.

Sparse Coding

Sparse coding (SC) has shown improved performance over BoW for image classification[28]. In SC each local image descriptor is reconstructed using a sparse weighted combination of visual words. Locality-constrained linear coding (LLC), an efficient variant of sparse coding, utilizes the local linear property of manifolds to project each descriptor into its local coordinate system [14]. Let $X_i \in R^{D \times N_i}$ be a matrix in which each of the $N_i$columns is a D-dimensional local descriptor extracted from an image Ii, i.e. $X_i=[\mathbf{x}_{i1},\mathbf{x}_{i2},\ldots,\mathbf{x}iN_i]$. Given a codebook with M entries, $B=[\mathbf{b}_1,\mathbf{b}_2,\ldots,\mathbf{b}_M]\in R^{D\times M}$, LLC uses the following criterion to compute the codes $C=[\mathbf{c}_{i1},\mathbf{c}_{i2}\cdots \mathbf{c}iN_i]$:

$$\underset{\mathbf{c}}{\operatorname{argmin}} \sum_{j=1}^{N_i} \|\mathbf{x}_{ij} - B\mathbf{c}_{ij}\|^2 + \lambda \|\mathbf{d}_{ij} \odot \mathbf{c}_{ij}\|^2$$
$$\text{s.t.} \quad \mathbf{1}^T\mathbf{c}_{ij} = 1, \quad \forall j \tag{2}$$

where $\odot$ denotes the element-wise multiplication and

$$\mathbf{d}_{ij} = \exp\left(\frac{\text{dist}(\mathbf{x}_{ij}, B)}{\sigma}\right) \tag{3}$$

where $\text{dist}(\mathbf{x}_{ij}, B) = \left[\|\mathbf{x}_{ij}-\mathbf{b}_1\|_2^2, \ldots, \|\mathbf{x}_{ij}-\mathbf{b}_M\|_2^2\right]^T$ and is a decay parameter. A fast approximation to LLC was described in [14] to speed up the encoding process. Specifically, instead of solving (2), the K(<D<M) nearest neighbours of $\mathbf{x}ij$ in B were considered as the local bases $\overline{B}_{ij}$ and a much smaller linear system (Eq. (4)) was solved to get the local linear codes:

$$\underset{\mathbf{c}}{\operatorname{argmin}} \sum_{j=1}^{N_i} \|\mathbf{x}_{ij} - \overline{B}_{ij}\mathbf{c}_{ij}\|^2$$
$$\text{s.t.} \quad \mathbf{1}^T\mathbf{c}_{ij} = 1, \quad \forall j \tag{4}$$

The image representation $\mathbf{z}_i$ of an image Ii is then obtained by pooling the sparse codes associated with the local descriptors. Two kinds of pooling, *max* and *sum*, are applied in the literature for SC. The max-pooling can be defined as $z_i^k = \max \mathbf{c}_{ij}^k,\ j = 1, \ldots, N_i$, and the sum pooling can be defined as $z_i^k = \sum_{j=1}^{N_i} |\mathbf{c}_{ij}^k|$, where, $zik$and $\mathbf{c}_{ij}^k$ are respectively the kth element of $\mathbf{z}_i$ and $\mathbf{c}ij$[29].

Fisher Vectors

Fisher vectors (FV) capture additional information about the distribution of the image descriptors compared to the count (0th-order) statistics in BoW. FV has shown improved performance over BoW and SC for image classification in [30]. In FV, the dictionary is first modelled as a Gaussian mixture model (GMM) $p(\mathbf{x}|\boldsymbol{\Theta})$

$$p(\mathbf{x}|\boldsymbol{\Theta}) = \sum_{m=1}^{M} \pi_m p(\mathbf{x}|\boldsymbol{\mu}_m, \Sigma_m)$$

$$p(\mathbf{x}|\boldsymbol{\mu}_m, \Sigma_m) = \frac{\exp^{-1/2(\mathbf{x}-\boldsymbol{\mu}_m)^T \Sigma_m^{-1}(\mathbf{x}-\boldsymbol{\mu}_m)}}{\sqrt{(2\pi)^d \det(\Sigma_m)}} \tag{5}$$

where $\boldsymbol{\Theta}=(\pi_1,\boldsymbol{\mu}_1,\Sigma_1,\ldots,\pi_M,\boldsymbol{\mu}_M,\Sigma_M)$ are the parameters of the GMM. $\pi_m \in \mathbb{R}^+$ $(\sum_m \pi_m = 1)$, $\boldsymbol{\mu}_m \in \mathbb{R}^D$ and $\Sigma_m \in \mathbb{R}^{D\times D}$ are respectively the weight, the mean and the covariance of the mth Gaussian. GMM uses a soft descriptor-to-cluster assignment:

$$q_m(\mathbf{x}_{ij}) = \frac{\pi_m p(\mathbf{x}_{ij}|\boldsymbol{\mu}_m, \Sigma_m)}{\sum_{l=1}^{M} \pi_l p(\mathbf{x}_{ij}|\boldsymbol{\mu}_l, \Sigma_l)} \tag{6}$$

In FV each cluster is then represented based on the derivative of the GMM with respect to its parameters $\{\boldsymbol{\mu}_m\}$ and $\{\Sigma_m\}$ (1st and 2nd order statistics), i.e.,

$$\mathcal{G}_{\boldsymbol{\mu}_m}^i = \frac{1}{N\sqrt{\pi_m}} \sum_{j=1}^{N_i} q_m(\mathbf{x}_{ij}) \Sigma_m^{-1/2} (\mathbf{x}_{ij} - \boldsymbol{\mu}_m)$$

$$\mathcal{G}_{\Sigma_m}^i = \frac{1}{N\sqrt{2\pi_m}} \sum_{j=1}^{N_i} q_m(\mathbf{x}_{ij}) \left[(\mathbf{x}_{ij} - \boldsymbol{\mu}_m)^T \Sigma_m^{-1} (\mathbf{x}_{ij} - \boldsymbol{\mu}_m) - 1\right] \tag{7}$$

The final image description $\mathbf{z}_i$ is the concatenation of $\mathcal{G}_{\boldsymbol{\mu}_m}^i$ and $\mathcal{G}_{\Sigma_m}^i$ for all $m=1,\ldots,M$, leading to a dimensionality of $2MD$.

Vector of Locally Aggregated Descriptors

Vector of locally aggregated descriptors (VLAD) [17], a simple approximation to FV, uses$k$ -means to learn the dictionary. VLAD uses the 1st-order statistics to represent each cluster; the mth cluster (Qm) representation of an image Ii can be given as

$$\mathbf{v}_m^i = \sum_{\mathbf{x}_{ij} \in Q_m} \mathbf{x}_{ij} - \boldsymbol{\mu}_m \tag{8}$$

The VLAD image representation $\mathbf{z}_i$ is the concatenation of $\mathbf{v}_m^i$ for all $m=1,\ldots,M$ followed by L_2 normalization, leading to a dimensionality of MD.

HEP-2 CELL CLASSIFICATION

From each cell image, each of the four feature types was densely extracted from patches of size 12×12, 16×16, and 20×20 pixels with a step-size of 2 pixels.

SVM Ensemble

Augmenting a classifier's training set with rotated versions of the images may improve classification performance but it also increases memory requirements. Instead we used an ensemble of multi-class one-vs-rest, linear SVMs; the ensemble consisted of four SVMs, one trained on the original training set images, and others trained on images after they were rotated through 90°, 180°, and 270°. The overall system which includes data augmentation as well as the ensemble training is shown in Fig. 6.

At test time, each test image was rotated by 0°, 90°, 180°, and 270°, and each rotated image was then given to the ensemble. This resulted in a set of 16 classification scores for each class (4 rotations×4 SVMs in the ensemble). Scores were treated as probabilities using Platt rescaling [31]. The final classification decision was made by averaging these probabilistic scores and selecting the highest scoring class. Fig. 7illustrates the process of classifying a cell image in detail.

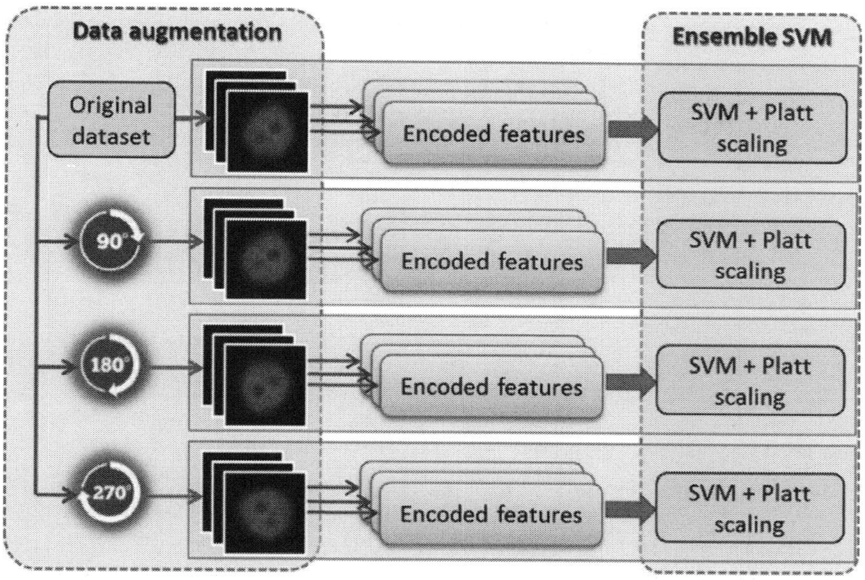

Figure. 6: An overview of the system for data augmentation and SVM ensemble training. Each image can be encoded as shown in Fig. 3.

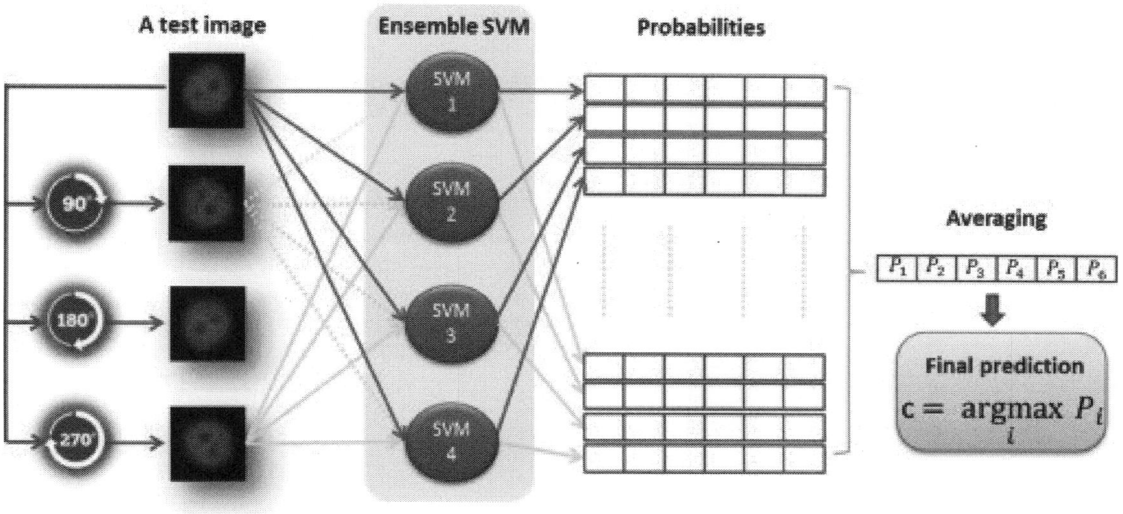

Figure. 7: Testing an image using the SVM ensemble for single cell classification.

Ensemble classification has previously been used for HEp-2 cell classification by Schaefer et al. [32] in the form of a trained linear fusion of classifier outputs. Here the approach we have adopted is based on simply averaging classifier outputs, avoiding the need for training of a fuser. This approach has been used for example to combine the outputs of neural network columns [33] and was shown to perform better than a trained linear combination on handwritten digit recognition [34]. For earlier work on this issue, see e.g., Duin [35].

Cell Pyramids

To capture spatial structure within a cell, a 2-level cell pyramid was used in a similar fashion to the dual-region used by Wiliem et. al. [2] and [36]. Separate dictionaries of size$M$ were learned for each feature type described above. The encoded local features using these dictionaries were pooled to get an image representation. At the first level of the cell pyramid, the encoded features from the whole cell were pooled to get a feature vector of size P (e.g., for BoW, P=M). At the second level, feature vectors were computed from the inner region and from the border region of each cell (see Fig. 3). These three feature vectors were concatenated to give a 3P-dimensional vector. Finally, encoded features from each of the four feature types were concatenated to give a 12P-dimensional vector on which classification was based.

HEP-2 SPECIMEN CLASSIFICATION

Our approach for classifying specimens is similar to the approach we proposed for cell image classification, the main difference being that we extracted features at two sets of scales. Specifically, after preprocessing, two types of local feature, mLP and rSIFT, were extracted. These features were densely extracted from image patches using a patch step-size of 4 pixels. Both *small* patches (12×12 pixels and 16×16 pixels) and *large*patches (48×48 pixels and 64×64 pixels) were used. Intuitively, small patches can capture local properties at cellular level while large patches can capture information about groups of cells. Features from outside the dilated cell masks were discarded.

A separate dictionary of size *M* was learned for each feature type with each group of patches (i.e., one for *small* and one for *large* patches). The image representations using the small-patch dictionary and the large-patch dictionary for each feature type were concatenated. Sparse coding with max-pooling was used.

The dataset provided for Task 2, described in Section 8.2, was augmented by including a 90°-rotated version of each original image. This resulted in a set of 2016 images. Five sub-images were extracted from each image based on the layout shown in Fig. 8(a) with the exception of images in the mitotic spindle class. In mitotic spindle images, metaphase cells in which stained mitotic spindle was apparent were manually identified. Five sub-images were then extracted around those cells, with some random variation, as shown inFig. 8(b). Finally, the 10,080 extracted sub-images were added to the 2016 images, resulting in an augmented dataset of 12,096 images.

a

b

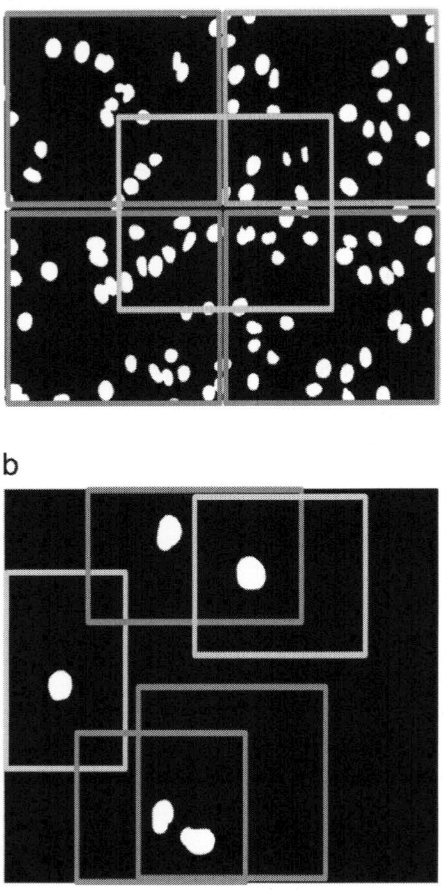

Figure. 8: Sub-images extracted from specimen images for (a) the homogeneous, speckled, nucleolar, centromere, Golgi, and nuclear membrane classes, and (b) the mitotic spindle class. White blobs in the images indicate (a) the segmented regions of individual cells and (b) the manually identified metaphase cells.

A one-vs-rest, multi-class SVM classifier was then trained on the augmented dataset. Since each specimen was imaged at four different locations in the test SVM classifier was applied to each of these four images, resulting in four sets of classification scores per specimen. Scores were treated as probabilities using Platt rescaling [31]. The final classification decision was made by averaging these probabilistic scores and selecting the class with the highest average score.

EXPERIMENT SETUP

Implementation Details

The following parameter settings were used for different feature types:

- mLP: a 3-resolution version with 8 sampling points at each resolution was used as shown in Fig. 5. The parameters of the Gaussian filters at each sampling point were selected as in [23].
- RP: The dimension D' of each linearized patch was reduced to D =300 whenever $D'>300$.
- IH: Local intensity histograms of 256 bins were used.

The public library, vlfeat[37], was used for dictionary learning and feature encoding. For SC, we used the implementation of LLC from [14] with 10 nearest neighbours (K=10). K-means with 300,000 randomly selected instances of each type of local feature was used to build the dictionaries for BOW, SC and VLAD methods.

In all the reported experiments we used the L_2 and power normalizations [30] to normalize the final image representation $\mathbf{z}_i$ of an image *Ii*, given by

$$\mathbf{z}_i \leftarrow \frac{\operatorname{sign}(\mathbf{z}_i)|\mathbf{z}_i|^{1/2}}{\|\mathbf{z}_i\|_2} \tag{9}$$

where $|\mathbf{z}_i|^{1/2}$ applies the square root to each component of $\mathbf{z}_i$.

We used the LIBLINEAR [38] implementation for the SVM classifiers. The code[4] from the authors of [39] was used for Platt scaling. We sampled equal number of positive and negative instances from the training set when learning the Platt calibration [31].

Mean Class Accuracy (MCA) was used as one of the evaluation metrics, as required metric by the I3A contest. It is defined as

$$\text{MCA} = \frac{1}{K}\sum_{k=1}^{K} \text{CCR}_k \tag{10}$$

where CCRk is the correct classification rate for class k and K is the number of classes.

Datasets

In reported experiments, the Task 1 and Task 2 datasets from the I3A contest were used. These were collected between 2011 and 2013 at the Sullivan Nicolaides pathology laboratory, Australia. For each task, a set of training images was provided to the contest participants. Submitted systems were then evaluated on a separate hidden test set which was privately maintained by the contest organizers and not released to the participants. The results obtained on the contest's hidden test set by our entries are reported in9.9 and 10.4. This paper also reports cross-validation results on the contest training sets.

The Task 1 dataset consists of 68,429 images of individual cells extracted from 419 patient positive sera (approximately 100–200 cell images per patient serum) along with their binary segmentation masks. 13,596 images were available during training. The remaining 54,833 images were used for the hidden test set to evaluate performance of systems submitted to the contest. The specimens were automatically photographed using a monochrome high dynamic range cooled microscopy camera. Cell images are approximately 70×70 pixels in size. The dataset has six pattern classes: homogeneous, speckled, nucleolar, centromere, nuclear membrane, and golgi. An example image from each of the six classes is given in Fig. 1.

The Task 2 dataset consists of uncompressed, monochromatic images of 1001 patient sera with positive ANA test. Each specimen was photographed at four different locations (four images per specimen). A total of 1008 images from 252 specimens were made available (approximately 25% of the data) while the remaining images were retained by the organizers for testing. Each image was 1388×1040 pixels and cell masks were obtained based on an automatic segmentation for each image. The dataset has seven pattern classes: homogeneous, speckled, nucleolar, centromere, nuclear membrane, Golgi and mitotic spindle. An example image from each of the seven classes is given inFig. 2.

The distribution of classes for both tasks is shown in Table 1. The homogeneous, speckled, nucleolar and centromere classes represent common ANA patterns whilst the other three classes are less common.

Table 1: Distribution of classes in Task 1 and Task 2 training datasets

Class	Task 1	Task 2
Homogeneous	2494	212
Speckled	2831	208
Nucleolar	2598	200
Centromere	2741	204
Nuclear membrane	2208	84
Golgi	724	40
Mitotic Spindle	–	60

Experimental results are presented in the following two sections. For cell classification (Section 9), we first used the I3A contest training set to explore the effect that different aspects of the system had on performance. These aspects included feature types, encoding methods, cell pyramids, and data augmentation. Subsequently, we used a leave-one-specimen-out setting to test the ability to classify cells from previously unseen specimens. Finally, we include the results reported on the held-out contest test data for the proposed system as well as for the other contest entries (Section 9.9). For specimen classification (Section 10), we report results comparing performance using different features. We then report results on cross-specimen and cross-image generalization. Finally, we include the results reported on the held-out contest test data (Section 10.4).

CELL CLASSIFICATION RESULTS (TASK 1)

Experiment 1: Comparison of Different Features and Encoding Methods

We compared the performance of different features and encoding methods on the Task 1 training dataset. Two-fold cross-validation experiments were carried out, and each was repeated 10 times. Fig. 9 reports the MCAs for different dictionary sizes. rSIFT gave a slightly better performance than the other features. IH gave the worst results. For all encoding methods, larger dictionaries gave higher MCA. SC with sum pooling always gave better performance than other encoding and pooling methods. For all features except IH, FV performed better than VLAD indicating that the additional (2nd order) information it captured was useful. When the dictionary size was 64, FV obtained similar MCA to SC with sum pooling with a dictionary size of 4000, but with an increased feature dimensionality. For example, using rSIFT the dimensionality of an FV image representation was 16,384 compared to 4000 using SC with sum pooling.

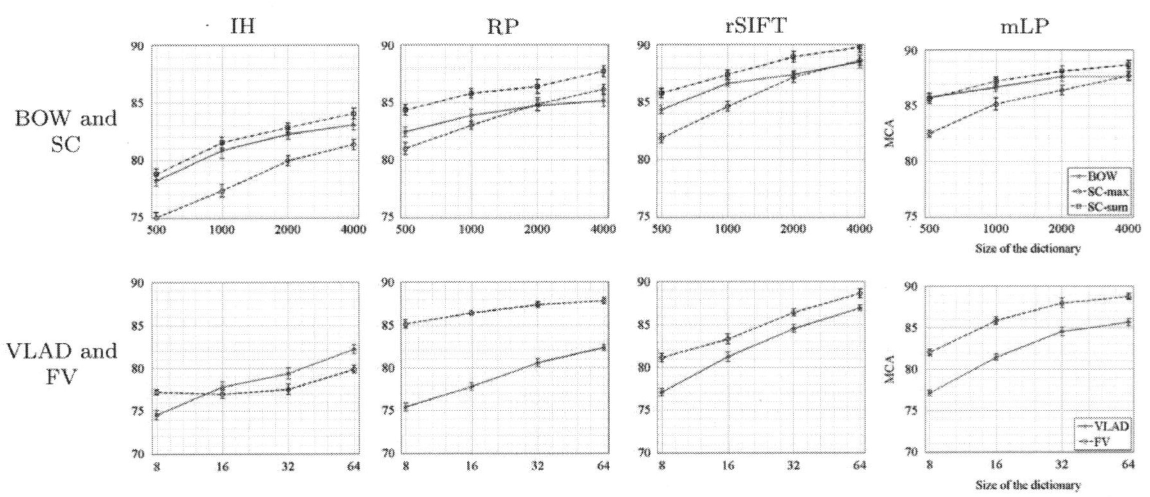

Figure. 9: Performance of various features with different encodings (size of the dictionary vs MCA).

Experiment 2: Combinations of Features

We investigated the performance of combinations of different features. We used BoW and SC encodings for this purpose as they gave better performance than VLAD and FV in Experiment 1. The dictionary size was fixed to 1500. Table 2 reports the results (see columns 2–4). Similar performance was observed using BoW and SC when combining all four types of feature. An improvement of more than 3% was obtained when combining other features with rSIFT (Fig. 9 and Table 2 columns 2–4). Table 5 reports the confusion matrix obtained when combining all the features and encoding with SC and max-pooling. The Golgi class was the least accurately classified; about 8% of Golgi images were misclassified as nucleolar.

Table 2: Two-fold cross-validation results (MCA±std) for different feature combinations with and without CPM and data augmentation (dictionary size of 1500)

Features	Original dataset without CPM			Original dataset with CPM			Augmented dataset with CPM		
	BOW	SC-sum	SC-max	BOW	SC-sum	SC-max	BOW	SC-sum	SC-max
rSIFT	90.4±0.4	91.0±0.4	90.2±0.4	91.1±0.4	92.0±0.5	91.9±0.4	93.6±0.4	94.1±0.3	94.0±0.5
+ mLP rSIFT	89.6±0.3	90.6±0.3	89.7±0.4	90.6±0.4	91.9±0.4	91.6±0.3	93.1±0.3	93.9±0.3	93.7±0.3
+ RP rSIFT	91.0±0.4	91.2±0.3	89.9±0.4	92.6±0.4	93.2±0.3	92.7±0.4	94.2±0.3	94.3±0.3	94.1±0.3
+ IH all	92.6±0.3	93.1±0.5	92.6±0.4	93.6±0.4	94.1±0.4	94.1±0.4	95.2±0.3	95.2±0.2	95.2±0.2

Table 3: Different performance measures for classification based on 2-fold cross-validation (all features, SC, max pooling, dictionary size of 1500)

Method	MCA	Accuracy	Precision	Recall	Fscore
Original dataset without CPM	92.6±0.4	97.7±0.1	0.933±0.003	0.925±0.004	0.929±0.003
Original dataset with CPM	94.1±0.4	98.1±0.1	0.941±0.003	0.944±0.003	0.942±0.002
Augmented dataset with CPM	**95.2±0.2**	**98.4±0.1**	**0.949±0.003**	**0.952±0.003**	**0.950±0.003**

Experiment 3: Effect of Cell Pyramids

To improve classification accuracy, particularly of the Golgi class, we incorporated spatial structure into the feature encoding process via cell pyramids (CPM). Table 2reports the performance of different feature combinations *with* and *without* CPM using BoW and SC approaches (see columns 5–7). When combining all the features and using CPM, the overall MCA was improved by about 1%. In particular, CPM improves the classification accuracy of the Golgi images by about 3% (see Table 5 and Table 6).

Table 4: Computational time (in sec. averaged over 500 cell images) required for different descriptors for feature extraction and encoding (SC, max pooling, dictionary size of 1500)

Features	Original dataset without CPM			Original dataset with CPM			Augmented dataset with CPM		
	Feature extract	Feature encode	Total	Feature extract	Feature encode	Total	Feature extract	Feature encode	Total
mLP	0.02	0.47	0.49	0.02	0.88	0.91	0.10	3.54	3.64
IH	0.79	0.54	1.33	0.79	1.06	1.85	3.16	4.15	7.31
rSIFT	0.06	0.55	0.61	0.06	1.02	1.08	0.23	4.19	4.42
RP	0.55	0.49	1.04	0.54	0.93	1.46	2.16	3.72	5.88

Table 5: Confusion matrix obtained using all features combined, SC with max pooling, and dictionary size of 1500. (Neither CPM nor data augmentation were used here.)

	Homo.	Spec.	Nucl.	Cent.	NuMe.	Golgi
Homo.	93.5	05.1	00.3	00.1	00.8	00.1
Spec.	04.6	90.6	01.7	02.2	00.6	00.3
Nucl.	01.0	02.1	94.8	01.0	00.6	00.5
Cent.	00.2	03.4	01.6	94.5	00.1	00.1
NuMe.	02.2	01.1	00.8	00.1	95.1	00.6
Golgi.	01.3	01.3	07.7	01.3	01.5	87.0

Table 6: Confusion matrix obtained using all features combined, SC with max pooling, dictionary size of 1500, and CPM. (No data augmentation was used here.)

Homo.	Spec.	Nucl.	Cent.	NuMe.	Golgi
94.7	04.2	00.3	00.1	00.6	00.1
04.0	91.9	01.4	01.7	00.8	00.2
00.9	01.7	95.5	00.9	00.6	00.5
00.1	02.8	01.5	95.5	00.0	00.1
01.9	00.8	00.6	00.1	96.0	00.6
00.9	00.8	05.2	00.6	01.7	90.9

Experiment 4: Effect of Data Augmentation

We investigated the effect of augmenting the training set by including rotated images as explained in Section 6. An ensemble SVM was used for classification. Augmenting the dataset improved the classification accuracy (see Table 2 columns 8–10 vs. columns 5–7, and Table 7 vs. Table 6). When combining all the features, the overall MCA was further improved by about 1%. BoW and SC gave similar performance; the MCA saturated at about 95%.

Table 7: Confusion matrix obtained using all features combined, SC with max pooling, dictionary size of 1500, CPM, and data augmentation

Homo.	Spec.	Nucl.	Cent.	NuMe.	Golgi
95.5	03.3	00.3	00.1	00.5	00.3
04.1	92.1	01.3	01.4	00.8	00.4
00.8	01.2	96.2	00.5	00.5	00.8
00.1	02.3	01.5	96.0	00.0	00.1
01.7	00.4	00.6	00.1	96.5	00.8
00.4	00.3	03.1	00.1	01.0	95.0

Different Measures for Classification

In experimental results reported above, MCA was used as the performance measure. This measure was specified for use in the I3A contest. Here we report results using various alternative measures, specifically accuracy, precision, recall, and F-score. Different measures summarize performance in different ways; e.g., accuracy ignores class imbalance thus is biased towards classes that have more examples in the dataset. On the other hand, MCA gives equal importance to all the classes. We direct the interested reader to [40] for detailed explanation and definitions of these measures for multi-class classification. Table 3 reports performance using these measures. These results suggest consistent conclusions regardless of the measure used. Adding CPM improved performance. CPM with data augmentation gave the best performance by all measures.

Computational Time for Feature Extraction and Encoding

Table 4 reports comparisons of the computational time required for feature extraction and encoding in order to compute the cell-level representations. This was by far the most time consuming part of the proposed system. These timings were obtained using Matlab 2014b and Windows 7 running on a machine with a Core i7 processor and 8 GB RAM. IH took more time than other features while resulting in lower MCA (see Fig. 9). On the other hand, mLP took the least time and resulted in competitive MCA. When all feature types were used along with data augmentation and CPM, the system took approximately 21s to compute the cell-level representation for one image.

Experiment 5: Leave-one-Specimen-Out

The above experiments disregarded the identities of the specimens from which cells had been extracted. To test the generalization performance of our system across different specimens, we conducted an experiment in a *leave-one-specimen-out* setting. Specifically, we used the specimen IDs to split the data into training and validation sets. Since 83 different specimens were available, we used images from 82 specimens for training in each split, and the images from the remaining specimen for testing. In this experiment we used the combination of all feature types, the augmented dataset, CPM, SC, max-pooling, and dictionary size of 1500. Table 8 reports the confusion matrix. An MCA of 81.1% was obtained. The Golgi class had poor results (66.7%). This class exhibits high intra-class variability and was poorly represented in the available data set; only 4 Golgi specimens were in the training set.

Table 8: Confusion matrix for leave-one-specimen-out experiment. (All features, CPM, data augmentation, SC, max pooling, dictionary size of 1500.)

Homo.	Spec.	Nucl.	Cent.	NuMe.	Golgi
81.8	14.8	00.8	00.2	02.0	00.4
09.0	75.5	03.7	10.6	00.8	00.4
01.1	03.4	89.4	02.5	01.3	02.3
00.3	10.7	03.4	85.4	00.0	00.2
05.8	01.9	01.5	00.0	87.9	02.8
04.8	02.1	17.4	01.5	07.5	66.7

Experiment 6: Performance on Images Extracted from Task 2 Dataset

We also made use of cell images segmented from the Task 2 dataset. (We did not use the *mitotic spindle* images in the experiment reported in this section.) An automatic procedure was used to select cells from the Task 2 dataset given the segmentation masks provided with that dataset. Firstly, all disjoint regions were identified in the segmentation mask images using connected component analysis. Secondly, eccentricity values were calculated for each connected component. Finally, low-eccentricity components that could be bounded by an 80×80 square with which no other component overlapped were selected. Approximately 5000 isolated cells were selected in this way. This is illustrated in Fig. 10 where red bounding boxes denote cell images that were extracted.

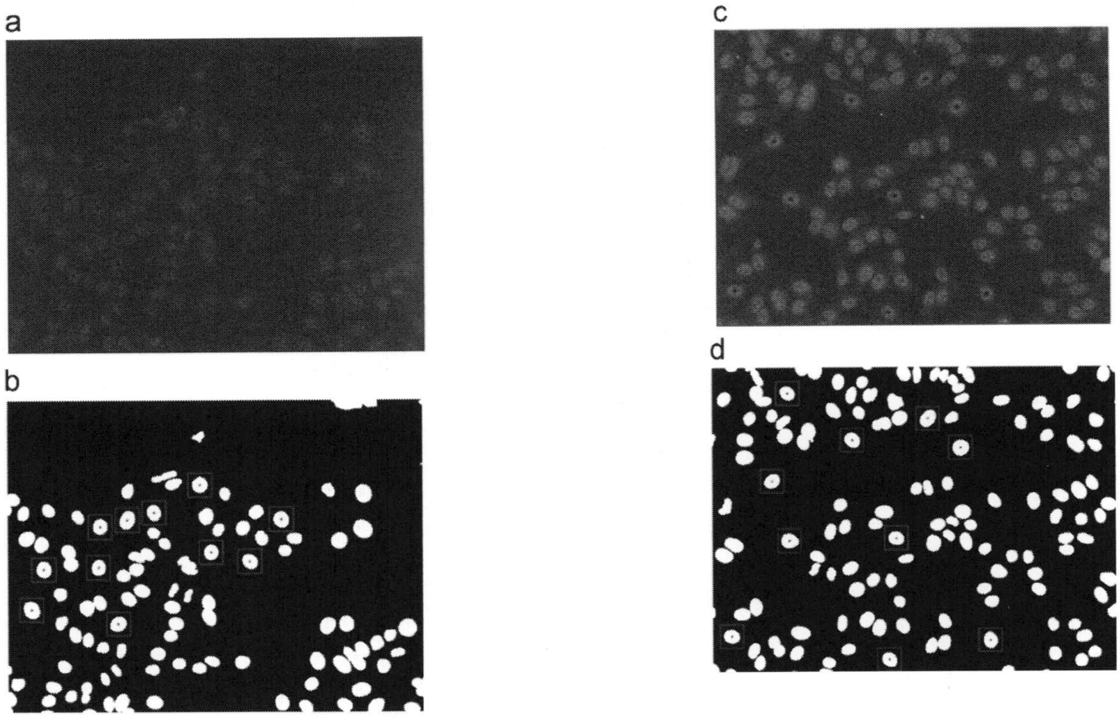

Figure. 10: Sample specimen images from I3A Task-2 dataset. The bounding boxes indicate the cell images which are automatically extracted from these specimen images. (a) A centromere specimen. (b) The segmentation mask for (a). (c) A speckled specimen. (d) The segmentation mask for (c).

We trained an ensemble classifier using all the images from the Task 1 training dataset and then tested it on the cell images extracted from the Task 2 dataset. We used the combination of all feature types with the augmented dataset, CPM, SC and max-pooling (dictionary size of 1500). The results are reported in Table 9; an MCA of 86% was obtained.

Table 9: Confusion matrix of the system trained on Task 1 images and tested on the cell images extracted from Task 2 (SC with max pooling, dictionary size of 1500)

	Homo.	Spec.	Nucl.	Cent.	NuMe.	Golgi
Homo.	65.4	28.5	01.5	00.0	03.8	00.7
Spec.	04.5	90.8	00.6	01.7	02.2	00.2
Nucl.	01.2	01.9	95.7	00.0	00.3	00.9
Cent.	00.1	11.1	06.5	82.0	00.2	00.1
NuMe.	03.7	01.9	00.3	00.0	92.0	02.2
Golgi.	00.0	01.4	03.1	00.2	05.1	90.2

Performance on the Test Dataset

We submitted two systems to the I3A contest for Task 1; the first system used only data made available in the Task 1 training set; the second system trained on a data set consisting of the Task 1 training set together with the additional 5000 cell images extracted from the Task 2 training set (see Section 9.8). Both systems used all the features together with SC (max-pooling, dictionary size 1500), the rotated versions of the images, and CPM.

Fig. 11 reports the MCAs obtained by all of the methods submitted to the contest on the Task 1 test set. Our first submission which made use of only the Task 1 training data obtained an MCA of 84.2%, higher than all the other teams' entries. Our second submission which used additional data (cells extracted from the Task 2 dataset) achieved an MCA of 87.1%. The next best entry, that of Gragnaniello et al. [18], obtained an MCA of 83.6%. Table 10 reports confusion matrices from our method and the method of Gragnaniello et al. The reader is referred to the I3A report [3] for detailed results of other entries.

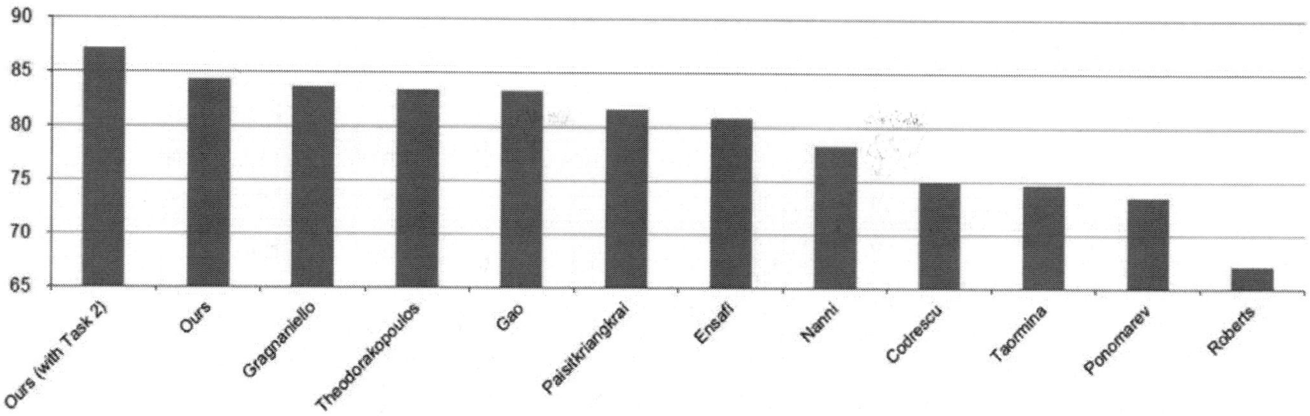

Figure. 11: The MCA at cell level attained by each method on the test set of Task 1.

Table 10: Confusion matrices for the proposed method and that of Gragnaniello et al. [18] on the Task 1 test set

(a) Proposed method (trained with cell images from Task 1 and Task 2 training sets)

	Cent.	Golgi	Homo.	Nucl.	NuMe.	Spec.
Cent.	97.5	00.1	00.5	00.8	00.2	00.9
Golgi.	00.1	82.0	05.4	03.6	08.2	00.7
Homo.	00.2	00.8	82.6	05.7	04.5	06.1
Nucl.	00.8	00.5	01.4	94.8	01.3	01.3
NuMe.	00.1	00.6	04.9	00.7	92.2	01.4
Spec.	10.3	00.5	12.0	02.0	01.5	73.6

(b) Gragnaniello et al. [18]

	Cent.	Golgi	Homo.	Nucl.	NuMe.	Spec.
Cent.	95.5	00.4	00.2	01.2	00.1	02.7
Golgi.	00.0	71.8	04.7	07.3	14.6	01.6
Homo.	00.1	00.8	78.6	04.9	08.1	07.6
Nucl.	00.8	01.6	02.0	92.5	01.7	01.5
NuMe.	00.1	00.8	03.1	00.9	93.3	01.8
Spec.	13.4	00.7	11.1	02.7	02.2	70.0

Examples of Correctly and Incorrectly Classified Cells

Fig. 12, Fig. 13, Fig. 14, Fig. 15, Fig. 16 and Fig. 17 show examples of cells from each class that were correctly and incorrectly classified. Some of the misclassified images are particularly noisy, e.g., the misclassified Golgi images in Fig. 17. In most other cases of misclassification shown, visual inspection reveals qualitative similarity to the class whose label was assigned to it.

a

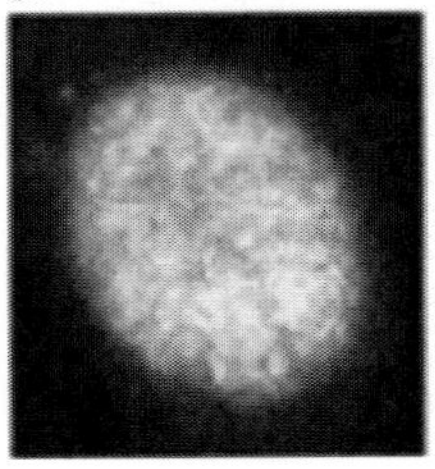

b

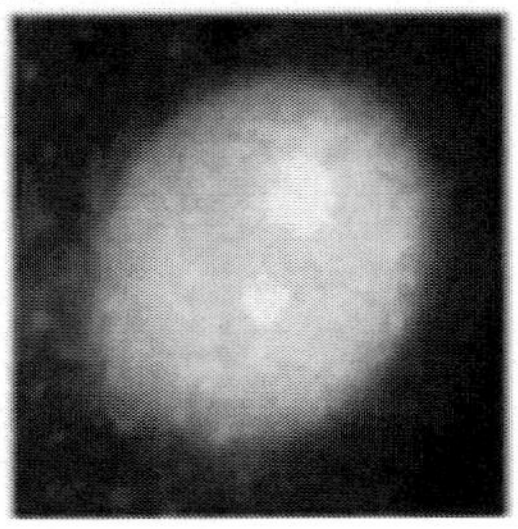

c

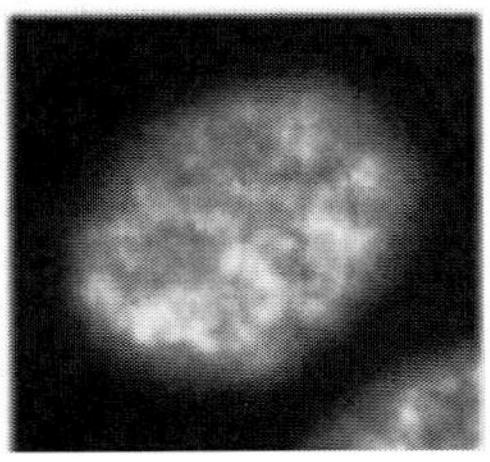

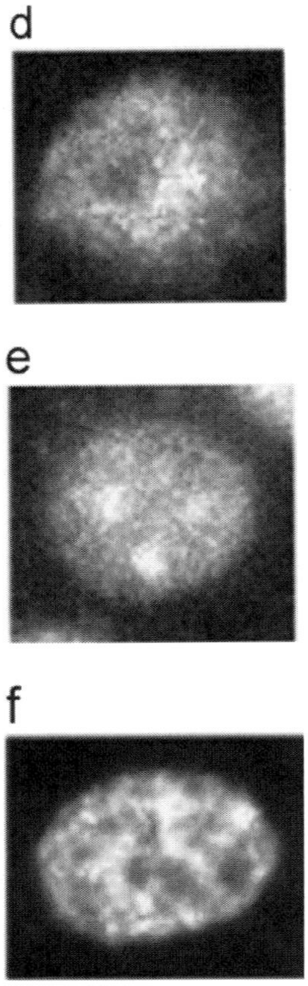

Figure. 12: Examples of *Homogeneous* cells that were (a)–(c) correctly classified, (d) classified as Spec., (e) classified as Nucl., and (f) classified as NuMe.

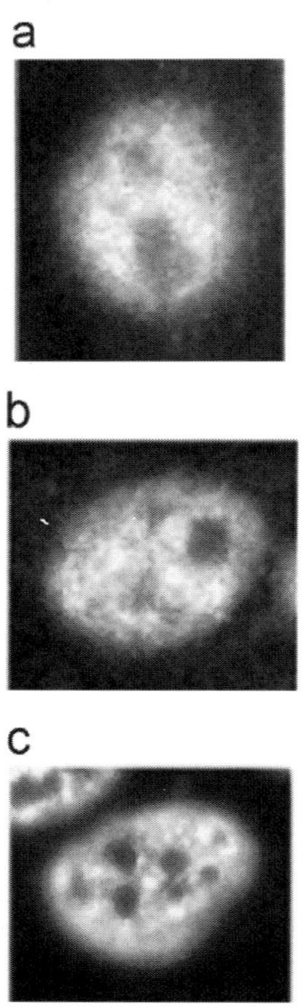

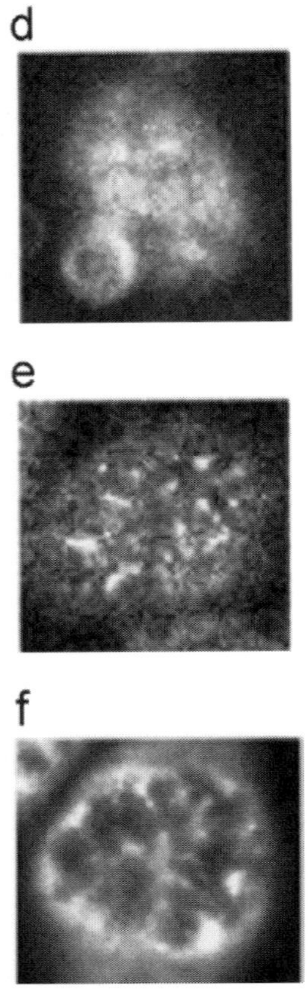

Figure. 13: Examples of *Speckled* cells that were (a)–(c) correctly classified, (d) classified as Homo., (e) classified as Cent., and (f) classified as Golg.

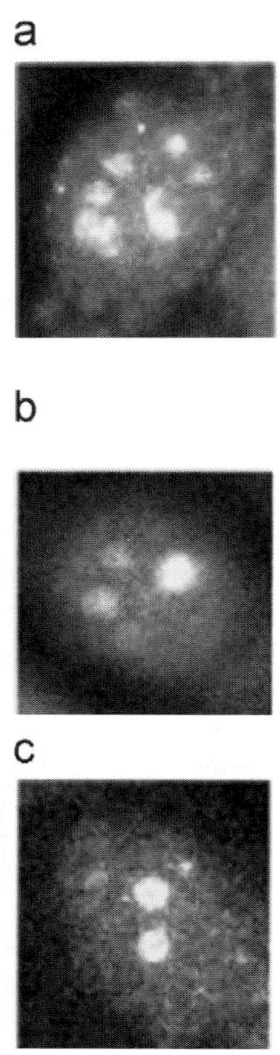

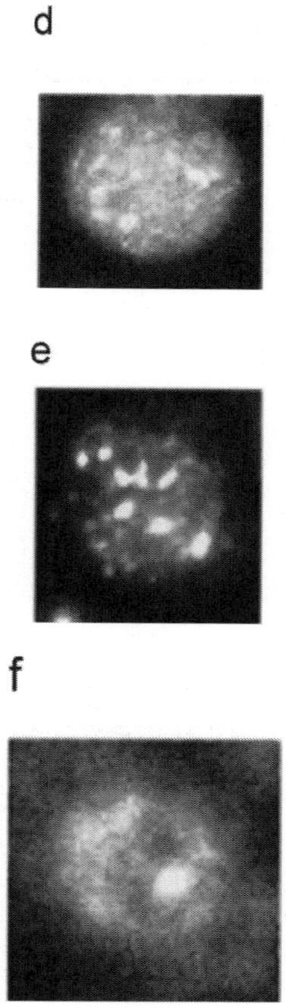

Figure. 14: Examples of *Nucleolar* cells that were (a)–(c) correctly classified, (d) classified as Homo., (e) classified as Cent., and (f) classified as Spec.

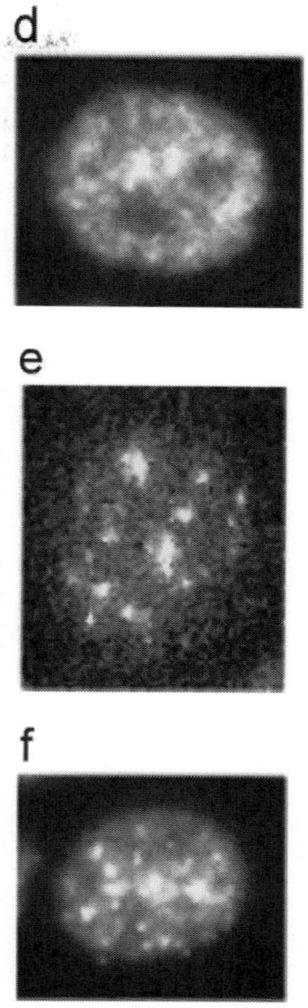

Figure. 15: Examples of *Centromere* cells that were (a)–(c) correctly classified, (d) classified as Homo., (e) classified as Nucl., and (f) classified as Spec.

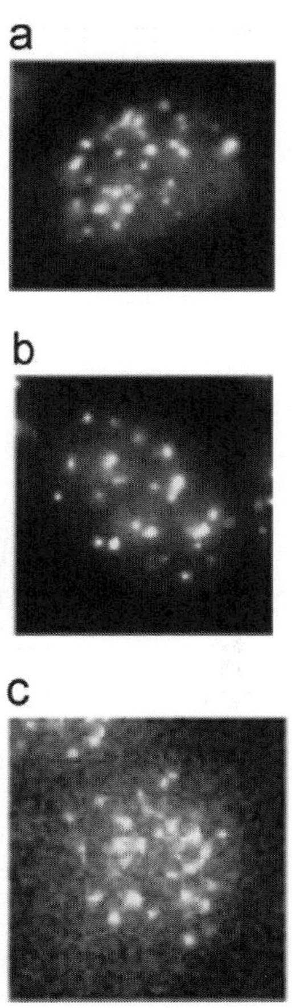

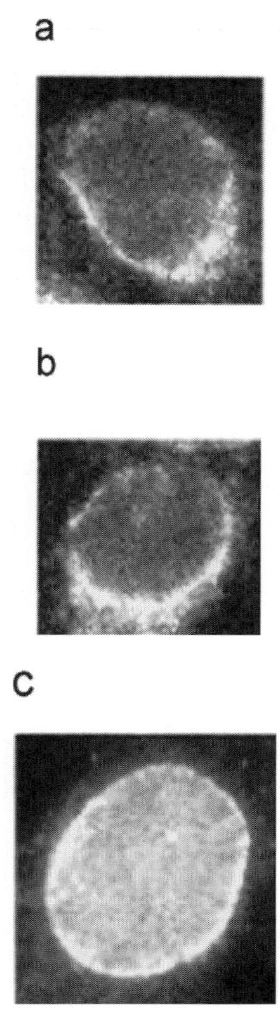

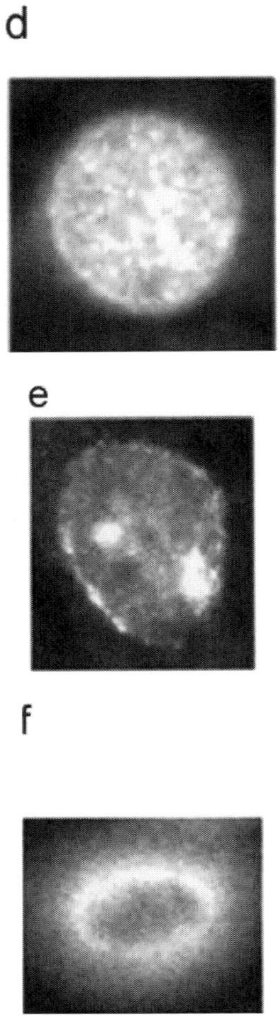

Figure. 16: Examples of *Nuclear Membrane* cells that were (a)–(c) correctly classified, (d) classified as Homo., (e) classified as Nucl., and (f) classified as Spec.

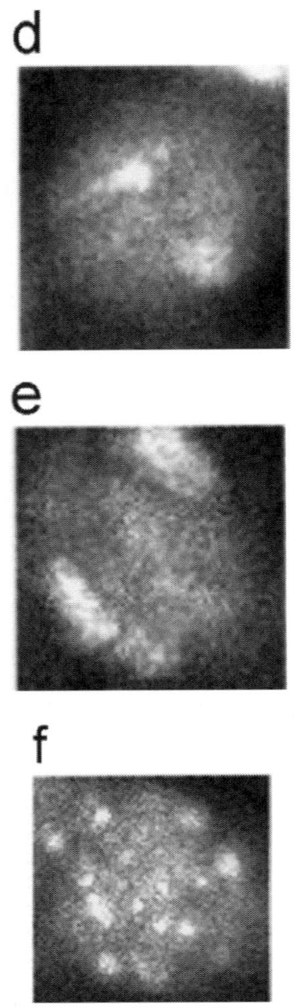

Figure. 17: Examples of *Golgi* cells that were (a)–(c) correctly classified, (d) classified as Nucl., (e) classified as NuMe., and (f) classified as Spec.

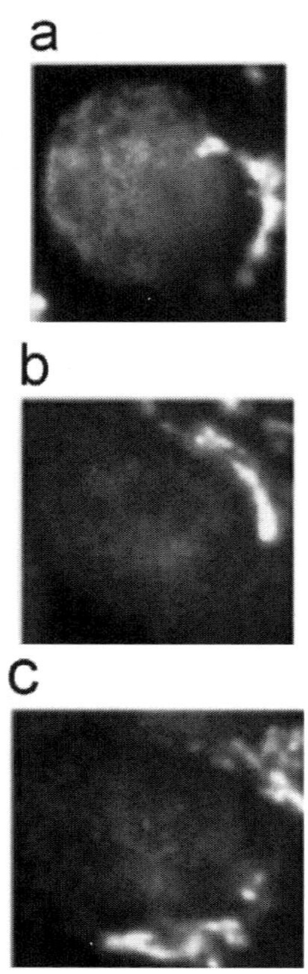

SPECIMEN CLASSIFICATION RESULTS (TASK 2)

Experiment 1: Comparison of Different Features

We performed five-fold cross-validation experiments to compare the performance obtained when using different features for Task 2. The dictionary size was fixed to 5000. Table 11 reports the MCA for each feature type as well as their combination. mLP with larger patch sizes outperformed other features. rSIFT with larger patch sizes gave the worst result. Combining features together resulted in an improved MCA of 89.9%. Table 12 shows the confusion matrix.

Table 11: Classification performance using different features based on five-fold cross-validation. (All=histogram of rSIFT plus histogram of mLP, with *smaller* and *larger* patch sizes.)

Feature type	MCA(%)
mLP $(12\times12, 16\times16)$	85.63
rSIFT $(12\times12, 16\times16)$	85.62
mLP $(48\times48, 64\times64)$	87.34
rSIFT $(48\times48, 64\times64)$	82.20
All	89.93

Table 12: Confusion matrix based on 5 fold-cross-validation for Task 2 specimen image classification

	Homo.	Spec.	Nucl.	Cent.	Golg.	Nume.	Mits.
Homo.	86.8	09.4	01.9	00.0	00.0	01.9	00.0
Spec.	01.9	96.1	00.0	00.0	00.0	01.9	00.0
Nucl.	00.0	00.0	98.0	02.0	00.0	00.0	00.0
Cent.	00.0	00.0	00.0	100.0	00.0	00.0	00.0
Golg.	00.0	00.0	00.0	00.0	100.0	00.0	00.0
Nume.	00.0	00.0	00.0	00.0	00.0	95.2	04.8
Mits.	26.7	06.7	00.0	00.0	00.0	13.3	53.3

Experiment 2: Leave-one-Specimen-out Experiments

A leave-one-specimen-out experiment was carried out using the specimen IDs provided to split the data into training and validation sets. Since 252 different specimens were available, we used images and their rotated versions from 251 specimens for training in each split. Table 13 reports the confusion matrix. Accuracy of 100% was obtained for the centromere and Golgi classes. The mitotic spindle class had the lowest accuracy being confused mostly with the homogeneous class. An overall MCA of 89.9% was obtained.

Table 13: Confusion matrix for leave-one-specimen-out experiment

	Homo.	Spec.	Nucl.	Cent.	Golg.	Nume.	Mits.
Homo.	88.7	09.4	00.0	00.0	00.0	01.9	00.0
Spec.	03.8	94.2	00.0	00.0	00.0	01.9	00.0
Nucl.	00.0	00.0	98.00	02.00	00.0	00.0	00.0
Cent.	00.0	00.0	00.0	100.00	00.0	00.0	00.0
Golg.	00.0	00.0	00.0	00.0	100.0	00.0	00.0
Nume.	00.0	00.0	00.0	00.0	00.0	95.2	04.7
Mits.	26.7	06.7	00.0	00.0	00.0	13.3	53.3

Experiment 3: Classification of Individual Images taken at Different Locations

In the above two experiments (10.1 and 10.2), the classification decision for each specimen was made based on averaging the classification scores (probabilities) of its four images taken from different locations as explained in Section 7. In this experiment we consider each of the four images separately and classify them individually. A leave-one-specimen-out experiment was performed, where at each iteration a classifier was trained on the images (and their rotated versions) obtained from 251 specimens, and tested on each of the four images of the test specimen. An MCA of 87.9% was obtained, 2% lower than the best accuracy (89.9%) obtained in the experiment reported in Section 10.2.

Performance on the Test Dataset

Fig. 18 reports the MCAs obtained by each of the submitted methods on the Task 2 test set. Our method achieved an MCA of 88.5%, outperforming all the other methods submitted. The second placed method, that of Liu et al. (described in [3]), achieved an MCA of 86.1%. Table 14 reports the confusion matrices of our method and that of Liu et al.

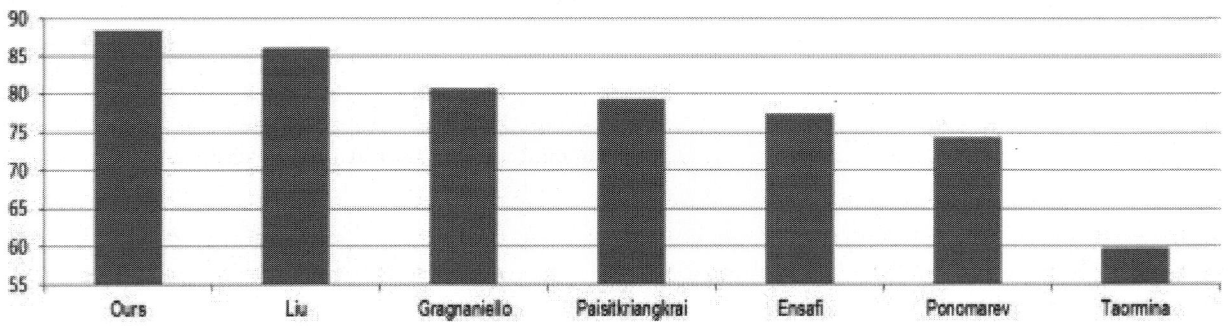

Figure. 18: The MCA at specimen level obtained by each method on the test set of Task 2.

Table 14: Task 2 confusion matrices for (a) our method, and (b) the methods of Liu et. al. (a) Our method (b) Liu et. al

(a)

	Cent.	Golgi	Homo.	Nucl.	NuMe.	Spec.	MitSp
Cent.	98.7	00.0	00.0	00.7	00.0	00.0	00.7
Golgi.	00.0	80.8	00.0	03.9	03.9	11.5	00.0
Homo.	00.0	00.0	93.0	00.0	00.0	01.9	05.1
Spec.	00.0	00.0	18.2	61.4	02.3	11.4	06.8
Nucl.	00.0	01.3	00.7	00.7	96.0	00.7	00.7
NuMe.	00.0	00.0	01.6	00.0	00.0	98.4	00.0
MitSp	00.0	00.0	07.0	00.6	00.0	01.3	91.1

(b)

	Cent.	Golgi	Homo.	Nucl.	NuMe.	Spec.	MitSp
Cent.	98.7	00.7	00.0	00.0	00.0	00.0	00.7
Golgi.	00.0	80.8	07.7	03.9	03.9	03.9	00.0
Homo.	00.0	00.0	93.0	00.0	00.0	01.3	05.7
Spec.	00.0	02.3	34.1	52.3	02.3	06.8	02.3
Nucl.	00.0	00.0	00.7	00.7	98.0	00.7	00.0
NuMe.	00.0	00.0	06.5	00.0	00.0	91.9	01.6
MitSp	00.0	00.0	11.4	00.6	00.0	00.0	88.0

CONCLUSION AND RECOMMENDATIONS

In this paper we explained in detail our winning entries for both Task 1 (cell image classification) and Task 2 (specimen image classification) of the I3A contest. To more fully understand the contributions of the components of our classification systems, we empirically studied the local feature extraction and encoding methods, as well as data augmentations applied in training the systems. We found that (1) the combination of different features outperforms individual feature types; (2) for cell classification SC performs better than BOW. FV and VLAD could achieve similar accuracies to BOW and SC but only with feature representations of much higher dimensionality; (3) adding spatial information from the cell images via the use of cell pyramids can improve classification performance for cell images (an improvement of ~3% was observed for Golgi images); and (4) augmenting the training set by the use of rotated training images further improves the classification performance. When combined, these aspects differentiate our entries from other entries in the competition. They are important factors to consider in building state-of-the-art cell and specimen image classification systems.

System design choices were guided by using mean class accuracy as the measure of classification performance. (This was the measure specified for use in the I3A contest.) However, experiments with alternative measures suggested similar conclusions. Future contests could be usefully enhanced by investigating the statistical and clinical significance of differences in performance between competing methods. One way to go about assessing statistical significance is bootstrap sampling, as recently incorporated in the PASCAL Visual Object Classes Challenge [41]. It would also be informative to quantify the reliability of the *ground truth* labels, especially given the visual similarity of many of the misclassified examples to their assigned classes (see Fig. 12, Fig. 13,Fig. 14, Fig. 15, Fig. 16 and Fig. 17). The features we used were pre-defined (hand-crafted). We believe that future work could further explore learning adaptive feature extractors for cell and specimen classification.

ACKNOWLEDGEMENTS

We thank the ICPR-I3A contest organizers, Brian C. Lovell and Arnold Wiliem (University of Queensland, Australia), Gennaro Percannella and Mario Vento (University of Salerno, Italy), for making the data and test results available for this research. S. Manivannan and R. Wang were funded by 2011–2016 EU FP7 ERC project "CODIR: colonic disease investigation by robotic hydrocolonoscopy", collaborative between the Universities of Dundee (PI Prof. Sir. A. Cuschieri) and Leeds (PI Prof. A. Neville). S. Akbar was funded by the UK Engineering and Physical Sciences Research Council (Grant EP/J500392/1) and W. Li was funded by the overseas student scholarship of the University of Dundee.

REFERENCES

1. P.L. Meroni, P.H. Schur, ANA screening: an old test with new recommendations, Ann. Rheum. Dis. 69 (8) (2010) 1402.
2. A. Wiliem, Y. Wong, C. Sanderson, P. Hobson, S. Chen, B. C. Lovell, Classification of human epithelial type 2 cell indirect immunofluoresence images via codebook based descriptors, in: IEEE Workshop on Applications of Computer Vision (WACV), IEEE, Piscataway, 2013, pp. 95–102.

3. B.C. Lovell, G. Percannella, M. Vento, A. Wiliem, Performance Evaluation of Indirect Immunofluorescence Image Analysis Systems, Technical Report, ICPR Workshop, 2014.
4. P. Foggia, G. Percannella, A. Saggese, M. Vento, Pattern recognition in stained HEp-2 cells: where are we now?, Pattern Recognit. 27 (2014) 2305–2314.
5. S. Manivannan, W. Li, S. Akbar, R. Wang, J. Zhang, S. J. McKenna, HEp-2 cell classification using multi-resolution local patterns and ensemble SVMs, in: 1st Workshop on Pattern Recognition Techniques for Indirect Immuno- fluorescence Images (I3A), IEEE, Piscataway, NJ, 2014, pp. 37–40, http://dx.doi. org/10.1109/I3A.2014.1.
6. S. Manivannan, W. Li, S. Akbar, R. Wang, J. Zhang, S. J. McKenna, HEp-2 specimen classification using multi-resolution local patterns and svm, in: 2014 1st Workshop on Pattern Recognition Techniques for Indirect Immuno-fluorescence Images (I3A), IEEE, Piscataway, NJ, 2014, pp. 41–44, http://dx.doi. org/10.1109/I3A.2014.20.
7. X. Chen, M. Velliste, R.F. Murphy, Automated interpretation of subcellular patterns in fluorescence microscope images for location proteomics, Cytom. Part A 69A (2006) 631–640.
8. P. Perner, H. Perner, B. Muller, Mining knowledge for HEp-2 cell image classification, Artif. Intell. Med. 26 (1–2) (2002) 161–173.
9. U. Sack, S. Knoechner, H. Warschkau, U. Pigla, F. Emmrich, M. Kamprad, Computer-assisted classification of HEp-2 immunofluorescence patterns in autoimmune diagnostics, Autoimmun. Rev. 2 (5) (2003) 298–304.
10. T.-Y. Hsieh, Y.-C. Huang, C.-W. Chung, Y.-L. Huang, HEp-2 cell classification in indirect immunofluorescence images, in: 7th International Conference on Information, Communications and Signal Processing, Piscataway, NJ, 2009, pp. 1–4.
11. P. Foggia, G. Percannella, P. Soda, M. Vento, Benchmarking HEp-2 cells classi- fication methods, IEEE Trans. Med. Imaging 32 (10) (2013) 1878–1889.
12. P. Hobson, G. Percannella, M. Vento, A. Wiliem, International Competition on Cells Classification by Fluorescent Image Analysis, Technical Report, ICIP, 2013. ⟨http://nerone.diiie.unisa.it/contest-icip-2013/ICIP2013_report.pdf⟩.
13. J. Zhang, M. Marszałek, S. Lazebnik, C. Schmid, Local features and kernels for classification of texture and object categories: a comprehensive study, Int. J. Comput. Vis. 73 (2) (2007) 213–238.
14. J. Wang, J. Yang, K. Yu, F. Lv, T. Huang, Y. Gong, Locality-constrained linear coding for image classification, in: IEEE Conference on Computer Vision and Pattern Recognition (CVPR), IEEE, Piscataway, NJ, 2010, pp. 3360–3367.
15. S. Ensafi, S. Lu, A. Kassim, C. L. Tan, A bag of words based approach for classification of HEp-2 cell images, in: 1st Workshop on Pattern Recognition Techniques for Indirect Immunofluorescence Images (I3A), 2014, pp. 29–32.
16. I. Theodorakopoulos, D. Kastaniotis, G. Economou, S. Fotopoulos, HEp-2 cells classification using morphological features and a bundle of local gradient descriptors, in: 1st Workshop on Pattern Recognition Techniques for Indirect Immunofluorescence Images (I3A), 2014, pp. 33–36.
17. H. Jégou, M. Douze, C. Schmid, P. Pérez, Aggregating local descriptors into a compact image representation, in: 2010 IEEE Conference on Computer Vision and Pattern Recognition (CVPR), IEEE, 2010, pp. 3304–3311.
18. D. Gragnaniello, C. Sansone, L. Verdoliva, Biologically-inspired dense local descriptor for indirect immunofluorescence image classification, in: 1st Workshop on Pattern Recognition Techniques for Indirect Immuno- fluorescence Images (I3A), 2014, pp. 1–5.
19. I. Kokkinos, M. Bronstein, A. Yuille, Dense Scale Invariant Descriptors for Images and Surface, Technical Report 7914, INRIA, 2012.
20. C. Codrescu, Quadratic recurrent finite impulse response MLP for indirect immunofluorescence image recognition, in: 1st Workshop on Pattern Recognition Techniques for Indirect Immunofluorescence Images (I3A), 2014, pp. 49–52.
21. Z. Gao, J. Zhang, L. Zhou, L. Wang, HEp-2 cell image classification with convolutional neural networks, in: 1st Workshop on Pattern Recognition Techniques for Indirect Immunofluorescence Images (I3A), 2014, pp. 24–28.
22. G. Sharma, S. ul Hussain, F. Jurie, Local higher-order statistics (LHS) for texture categorization and facial analysis, in: European Conference on Computer Vision (ECCV), 2012, pp. 1–12.
23. T. Mäenpää, The local binary pattern approach to texture analysis: extensions and applications (Ph.D. thesis), University of Oulu, 2003.
24. S. Leutenegger, M. Chli, R. Y. Siegwart, BRISK: Binary robust invariant scalable keypoints, in: IEEE International Conference on Computer Vision (ICCV), IEEE, Piscataway, NJ, 2011, pp. 2548–2555.
25. R. Arandjelovic, A. Zisserman, Three things everyone should know to improve object retrieval, in: IEEE Conference on Computer Vision and Pattern Recognition (CVPR), 2012, pp. 2911–2918.
26. E. Bingham, H. Mannila, Random projection in dimensionality reduction: applications to image and text data, in: ACM SIGKDD International Conference on Knowledge Discovery and Data Mining, ACM, New York, NY, USA, 2001, pp. 245–250.
27. L. Liu, P. Fieguth, Texture classification from random features, IEEE Trans. Pattern Anal. Mach. Intell. 34 (3) (2012) 574–586.
28. J. Yang, K. Yu, Y. Gong, T. Huang, Linear spatial pyramid matching using sparse coding for image classification, in: IEEE Conference on Computer Vision and Pattern Recognition (CVPR), 2009, pp. 1794–1801.
29. Y.-L. Boureau, N. Le Roux, F. Bach, J. Ponce, Y. LeCun, Ask the locals: multi-way local pooling for image

recognition, in: IEEE International Conference on Computer Vision (ICCV), 2011, pp. 2651–2658.

30. F. Perronnin, J. Sánchez, T. Mensink, Improving the fisher kernel for large-scale image classification, Computer Vision—ECCV LNCS, 6316, Special Publication, Crete, Greece, 2010.

31. J.C. Platt, Probabilistic outputs for support vector machines and comparisons to regularized likelihood methods, Advances in large margin classifiers, MIT Press, Cambridge, MA, USA (2000), p. 61–74.

32. G. Schaefer, N.P. Doshi, B. Krawczyk, HEp-2 cell classification using multidimensional local binary patterns and ensemble classification, in: Second IAPR Asian Conference on Pattern Recognition, 2013, pp. 951–955.

33. D.C. Ciresan, U. Meier, J. Schmidhuber, Multi-column deep neural networks for image classification, in: CVPR, 2012.

34. U. Meier, D.C. Ciresan, L.M. Gambardella, J. Schmidhuber, Better digit recognition with a committee of simple neural nets, in: International Conference on Document Analysis and Recognition, 2011, pp. 1135–1139.

35. R.P.W. Duin, The combining classifier: to train or not to train?, in: International Conference on Pattern Recognition, 2002, pp. 765–770.

36. A. Wiliem, C. Sanderson, Y. Wong, P. Hobson, R.F. Minchin, B.C. Lovell, Automatic classification of human epithelial type 2 cell indirect immuno- fluorescence images using cell pyramid matching, Pattern Recognit. 47 (7) (2014) 2315–2324.

37. A. Vedaldi, B. Fulkerson, VLFeat: an open and portable library of computer vision algorithms, in: Proceedings of the International Conference on Multimedia (MM), ACM, New York, NY, USA, 2010, pp. 1469–1472.

38. R.-E. Fan, K.-W. Chang, C.-J. Hsieh, X.-R. Wang, C.-J. Lin, LIBLINEAR: a library for large linear classification, J. Mach. Learn. Res. 9 (2008) 1871–1874.

39. H.-T. Lin, C.-J. Lin, R.C. Weng, A note on Platt's probabilistic outputs for support vector machines, Mach. Learn. 68 (3) (2007) 267–276.

40. M. Sokolova, G. Lapalme, A systematic analysis of performance measures for classification tasks, Inf. Process. Manag. 45 (4) (2009) 427–437.

41. M. Everingham, S.M. Ali Eslami, L. van Gool, C.K.I. Williams, J. Winn, A. Zisserman, The PASCAL visual object classes challenge: a retrospective, Int. J. Comput. Vis. 111 (1) (2015) 98–136.

CHAPTER 7

CONTEXTUAL ANOMALY DETECTION IN CROWDED SURVEILLANCE SCENES

Michael J.V. Leach[a], Ed.P. Sparks[b], Neil M. Robertson[a]

[a] Heriot-Watt University, Edinburgh Campus, Edinburgh, Scotland, United Kingdom

[b] Roke Manor Research, Romsey, Hampshire, United Kingdom

ABSTRACT

This work addresses the problem of detecting human behavioural anomalies in crowded surveillance environments. We focus in particular on the problem of detecting subtle anomalies in a behaviourally heterogeneous surveillance scene. To reach this goal we implement a novel unsupervised context-aware process. We propose and evaluate a method of utilising social context and scene context to improve behaviour analysis. We find that in a crowded scene the application of Mutual Information based social context permits the ability to prevent self-justifying groups and propagate anomalies in a social network, granting a greater anomaly detection capability. Scene context uniformly improves the detection of anomalies in both datasets. The strength of our contextual features is demonstrated by the detection of subtly abnormal behaviours, which otherwise remain indistinguishable from normal behaviour.

INTRODUCTION

As a society we have the need to monitor public and private space in order to prevent criminal behaviour and identify security threats. The scale at which surveillance is undertaken and the density of information in video results in a huge amount of data – the analysis of which using human resources is often prohibitively expensive. The solution is to automate human surveillance (Makris and Ellis, 2005). Due to advances in pedestrian detection and robust tracking long term human centred tracks are becoming more prevalent (Kalal and Matas, 2009 and Felzenszwalb et al., 2010). It is becoming plausible to autonomously profile the behaviour of a single, or multiple, humans over time. An abnormal event in automated surveillance is one which has a low statistical representation in the training data (Loy, 2010). Our approach is motivated by this definition with an emphasis upon contextual information as a method of creating separation between otherwise only subtly distinct behaviours. A good behaviour representation should encode the dataset in such a way that homogeneous clusters of behaviour can be segmented from the heterogeneous mass of data. Equally a poor behaviour representation is incapable of measuring the distinction between desired subgroups of data. Subtle behaviours provide a greater challenge because the information required to segment them from the greater set

Citation: Michael J.V. Leach, Ed.P. Sparks, Neil M. Robertson, Contextual anomaly detection in crowded surveillance scenes, Pattern Recognition Letters, Volume 44, 15 July 2014, Pages 71-79, ISSN 0167-8655, http://dx.doi.org/10.1016/j.patrec.2013.11.018.

is not directly measurable. Subtle behaviours can be handled in the following two ways; firstly by measuring more relevant information which better segments the data into homogeneous subsets, or secondly by implementing a better suited model which is capable of fitting the nuances of the data domain. In this research we tackle the former point; inspired by work in Scene Modelling (Makris and Ellis, 2005) and Social Signal Processing (Cristani and Raghavendra, 2012) we demonstrate the extraction and use of high level surveillance information which provides a *contextual* basis to identify subtly abnormal behaviour. Simple surveillance scenes may not contain much contextual information, in fact at its simplest a surveillance scene can be said to have only one contextual state. In such cases a simple trajectory matching algorithm may be appropriate to detect outlier behaviour. However, a dynamic or crowded surveillance scene may be heterogeneous, and thus behaviour in one context may not be representative of behaviour in a different context. In any non-trivial surveillance scene contextual information such as scene region, social context, periodic events, and entry or exit points impact the dynamics of behaviour (Lan and Wang, 2010). We can use this contextual information to provide further means of segmenting abnormal behaviours from the mass of data, and perhaps provide the means to segment subtle behaviours from the mass of data. For a more general discussion on contextual anomaly detection see (Chandola, 2009 and Song et al., 2007).

With this work we demonstrate the significance of inferring social links between people in a surveillance application. We provide further validation of the growing trend in automatic scene understanding, additionally providing a novel approach. Furthermore we demonstrate a novel social context based anomaly detection procedure. We evaluate our systems capability to detect *subtle* behavioural anomalies within a complex and crowded human surveillance scene. Our main contributions are a novel method of acquiring scene structure information in surveillance, the development of a novel mutual information social group metric, and the demonstration that social and scene contextual information is effective in combination at anomaly detection.

Related Work

We focus upon social and scene region contextual knowledge as a means of improving the detection of subtle behavioural anomalies. The scene regions provides an understanding of portions of the scene in which we would expect normal behaviours to be different from other areas (Makris and Ellis, 2005). Previous approaches such as Li et al. develop a scene segmentation method which divides the scene into regions based upon behavioural dissimilarity (Li and Xiang, 2008). Similarly, Chen Loy segments a scene into spatial regions of similar behaviour by virtue of behaviour correlation (Loy, 2010). This work introduces a second line of contextual scene knowledge: temporal state. This contextual information is particularly apt for the traffic junction, in which behaviour is clearly temporally segmented in short time intervals. However, it is far less applicable to many human surveillance environments where the periodicity of behaviour is far less structured, if at all. Wang et al. uses a Dual Hierarchical Dirichlet Process to cluster behaviours spatially, learning both observation and trajectory clusters simultaneously (Wang and Teck Ma, 2008). The second source of contextual information we use is social context. Social Context grants the ability to learn the distinction between normal behaviour for groups and individuals independently. The social model provides an additional benefit; it ensures that the behaviour of each individual is analysed in reference to people external to the same social group. Thus a homogeneous group of individuals all acting abnormally cannot be self-justifying. Furthermore social information enables us to create likelihood dependencies between individuals in a social group. Thus if one individual in a group is behaving abnormally the expectation of other group members behaving abnormally goes up. To estimate social groupings Ge et al. uses a proximity and velocity metric to associate individuals into pairs, iteratively adding additional individuals to groups using the Hausdorff distance as a measure of closeness (Ge and Ruback, 2009). Yu et al. implements a graph cuts based system which uses the feature of proximity alone (Yu et al., 2009). However modelling social groups by positional information alone is perilously primitive and prone to finding false social connections when individuals are within close proximity due to external influences such as queuing. Oliver et al. uses a Coupled HMM to construct a priori models of group events such as Follow-reach-walk together, or Approach-meet-go separately (Oliver and Pentland, 1998). Certain actions are declared group activities and thus groups can be constructed from individuals via mutual engagement in a grouping action. Robertson and Reid utilise gaze direction in order to determine whether individuals are within each other's field of view (Robertson, 2011). Gaze direction is significant as it departs from the use of motion features alone by taking into account visual interest (Farenzena and Tavano, 2011). For a comprehensive and complete review of the emerging field of social signal processing see the work of Cristani (Cristani and Raghavendra, 2012).

METHOD

The extraction of pedestrian trajectories from surveillance video is non-trivial, particularly when there is occlusion and crowding. It is not our goal to develop a novel low level feature extractor and for that reason we rely upon the large amount of research in computer vision already devoted to producing tracking solutions. Extracting pedestrian trajectories requires two main stages: detection of pedestrians, and tracking of targeted pedestrians. Detection is achieved using the Felzenszwalb part based detector (Felzenszwalb et al., 2010). Tracking of human targets in the image plane is achieved with the use of the Predator TLD tracker (Kalal and Matas, 2009). We track the heads of pedestrians in the crowded PETS-2007 scene, see Fig. 1(a). for the second dataset, the Oxford data, we use the published tracking results provided by Benfold (Benfold and Video, 2011). We select the TLD tracker due to high performance amongst state of the art trackers (Kalal and Matas, 2010) and utilise its capability to learn a target model and discriminate between potential targets in a crowded surveillance scene. The pedestrian tracking performance of the TLD tracker is extensively tested against alternative recent tracking procedures in the author's paper (Kalal and Matas, 2010).

(a)

(b)

Figure. 1: We illustrate here the tracked dataset PETS-2007 (a), and tracked Oxford data (b). The PETS-2007 data presents a challenging crowded environment and contains far less structure in the apparent motion of individuals in the scene. In contrast the Oxford data contains very structured trajectory information, and is sparsely populated. Our social context extraction is geared towards crowded scenes such as the PETS-2007 data, however this presents a harder surveillance challenge.

Scene context: Building upon the work of Makris and Ellis (2005) our scene model consists of four potential regions: Traffic lanes, idle areas, convergence/divergence regions, and general area. Convergence and divergence is synonymous as there is no temporal direction. Each region is defined to isolate a different dynamic of a scene, and is captured as a relation between the direction, speed, persistence (the number of frames a trajectory last for), and energy and entropies of trajectories through the scene. For each of the four potential regions a heat map is constructed on the ground plane and a threshold segments positive regions from negative. Scene regions are mutually exclusive of each other. We define each of the four scene context regions as follows:

Traffic lanes: A traffic lane represents an area of the scene which contains a high number of trajectories in a structured motion. The traffic region is defined as:

$$T_{xy} = \frac{N_{xy}}{\bar{N}} \frac{1}{-\sum P(\theta_{xy})log(P(\theta_{xy}) + \frac{1}{\pi}\sum\sqrt{(\theta_{xy}-\bar{\theta}_{xy})^2}} \quad (1)$$

where θ is a histogram of directions populated by all target trajectories to go through region x, y in the scene. The numerator Nxy gives the number of trajectories through the location x, y, and $\bar{N}$ gives the mean number of trajectories for any given location. High scoring traffic locations coincide with regions displaying a high number of trajectories, low directional entropy and low trajectory energy.

Idle regions: The idle region captures the area of the scene which hold enough evidence of near stationary trajectories that the region is considered a legitimate place to remain idle.

$$I_{xy} = \frac{T_{xy}}{\bar{T}} \frac{v_{xy}}{\sum\sqrt{(v_{xy}-\bar{v}_{xy})^2} + \sum v_{xy}} \quad (2)$$

The mean temporal persistence Txy provides the mean numbers of frames that trajectories persist for in the region x, y, this coefficient is balanced by the denominator $\bar{T}$ the mean number of frames for all regions. The speeds of trajectories observed in location x, y is denoted by histogram *v*. We define likely idle regions as those with a high mean temporal persistence, low speed and low speed energy.

Convergence divergence areas: These areas of the scene are responsible for imposing a force which brings trajectories together or releases them allowing them to diverge. Typically such regions are appended to the ends of a traffic lane.

$$C_{xy} = \frac{\frac{1}{\pi}\sum\sqrt{(\theta_{xy}-\bar{\theta}_{xy})^2}}{-\sum P(\theta_{xy})log(P(\theta_{xy})} \quad (3)$$

where θ is the histogram of direction observed at x, y. We define the convergence region by a high directional energy low directional entropy region. Thus a structured splitting of trajectories over a region would be considered a likely candidate for a convergence or divergence region.

General Area: having scored the scene with the above region definitions we normalise the region intensity maps between [0,1], and apply a threshold to segment active regions. The remaining area of the scene not classified as any of the above regions is considered the general area. The interpretation of the general area is as the region which does not impose any influence on the motion vector of tracked pedestrians.

Social context. The basis of our social model is the premise that a high degree of shared trajectory information implies a social dependence between two individuals. Our social model is geared towards effective detection of social groups in a moving crowd. Crowded surveillance provides an environment in which socially connected individuals are more likely to move together, and thus display more similar trajectory information. The more entropic the underlying motion of the crowd is the more salient similar trajectories will be. For an illustration of typical social pairs see Fig. 2(b).

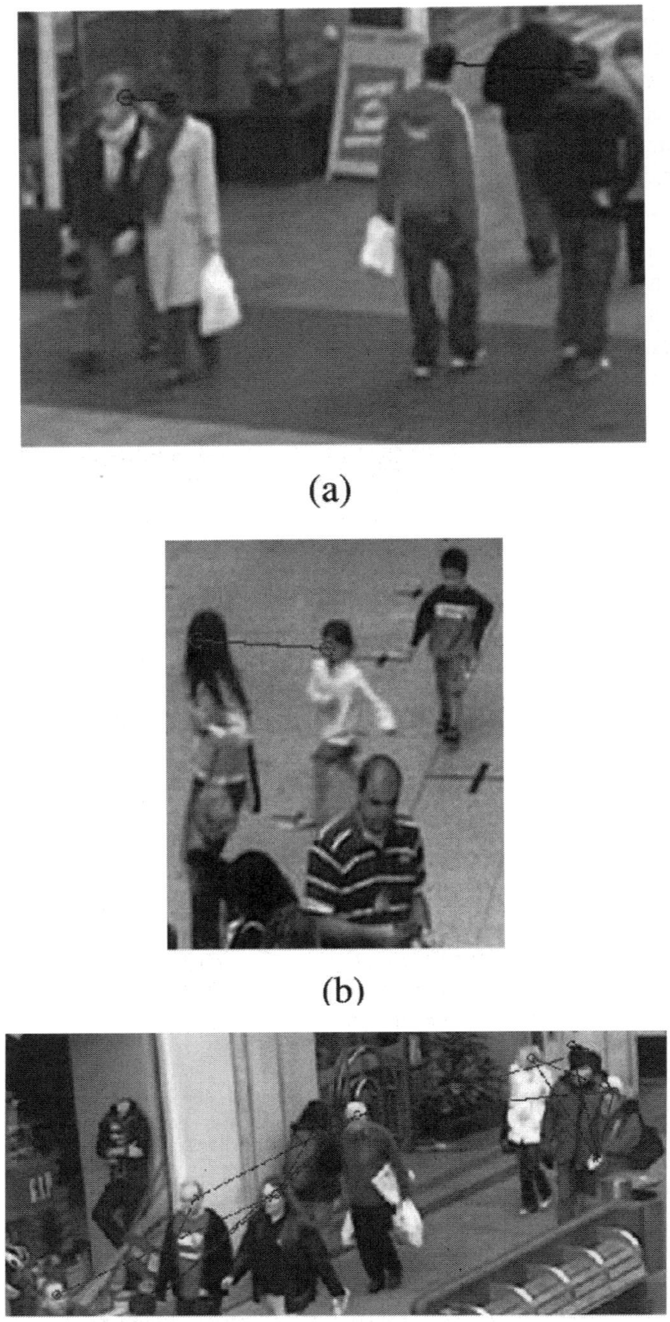

(a)

(b)

(c)

Figure. 2: An example of social grouping from the Oxford data (a) and the PETS-2007 data Scene 04 (b) derived using our social connection strength metric. Both (a) and (b) show a true positive result. (c) demonstrates a failure mode.

We use a novel metric to identify the strength of pair-wise social connections consisting of the weighted product of multiple features. We identified 4 features as effective at detecting pair connections between two individuals: the mutual information of direction ($I\Theta ijt$), the mutual information of speed ($IVijt$), the proximity between two individuals ($\Delta Pijt$) and the temporal overlap ratio between two individuals (τijt). We train a set of weighting variables $\alpha_{\Delta P}$, α_{IV}, $\alpha_{I\Theta}$, α_{τ} which weight each feature in the social metric based upon the classification score of each feature independently on the ground truth training data. The feature weights are distributed proportional to each features classification score. The features which compose the pairing metric are defined as:

$$\Delta P_{ijt} = \alpha_{\Delta P} e^{-\frac{\frac{1}{N}\sum_n |S_{it}-S_{nt}| + \frac{1}{N}\sum_n |S_{jt}-S_{nt}|}{2|S_{it}-S_{jt}|}} . \quad (4)$$

For 2 tracked individuals i and j at frame t where Sij is the distance between trajectory $i$and j at time t . The proximity between any two individuals ΔP is scaled by the distance between i and j to the set of all other individuals N in the scene. Thus we incorporate a measure of scene density which places a bias upon pairs being closer together in denser areas, and allows pairs to drift apart in sparse areas.

$$\Delta \tau_{ijt} = \alpha_T e^{-\frac{|T_i - T_j|}{2T_{ij}}} \quad (5)$$

where τijt is the temporal overlap ratio between i and j up to the current frame t, which is to say the ratio of time both individuals have existed contemporaneously to total time of existence, thus rewarding individuals who enter and exit the scene at similar times. T_i, and T_j is the frame length of trajectory i and j respectively, and Tij is the number of frames in which both i and j have coexisted.

Whilst $\Delta P ijt$ and $\Delta\tau ijt$ are direct measures of trajectory statistics it is important to note that both $IVijt, I\Theta ijt$ are more complex in nature. We use mutual information (MI) instead of the Euclidean distance as it handles non-linear and non-Gaussian random variables effectively and provides a principled method of comparing orthogonal feature dimensions. We define the Gaussian distributions of speed $P(v)$ and direction $P(\theta)$ as the Maximum Likelihood Estimation (MLE) derived from the most recent 1 s of trajectory data. The joint probability is calculated as the MLE Gaussian for the combined data of both person i and j over the last second. The mutual information between individual i and $j$is calculated for a number of temporal offsets thus permitting an individual reaction time to the trajectory it has dependence upon. Thus we calculate the mutual information between each individual with set time offsets of 10 frames consecutively forwards and backwards, and take the maximal mutual information for all time offsets.

$$IV_{ijt} = -\alpha_{IV}\sum_{b} P(v^i(b))\log_2(P(v^i(b))) - \alpha_{IV}\sum_{b} P(v^j(b))\log_2(P(v^j(b))) + \alpha_{IV}\sum_{b} P(v^{ij}(b))\log_2(P(v^{ij}(b))). \quad (6)$$

Where v^i is the MLE distribution over speed for person i over the most recent time window. The mutual information calculation for direction $I\Theta ijt$ is structured identically to the above, replacing the MLE speed distribution v^i with the MLE direction distribution θ^i.

Each feature is used independently to classify pair connections between tracked individuals and scored with against the ground truth classification. We observed that the features of proximity between two individuals (ΔP) and the temporal overlap ratio between two individuals ($Tijt$) present a significant ability to classify pairs in the test data. The overall performance is improved with the inclusion of the mutual information measures for direction and speed, see Fig. 3. Whilst the individual features of mutual information speed and direction provide better classification we find there is a lack of correlation with the true positives exemplified by the Euclidean features of proximity and temporal overlap in this dataset. In this dataset the impact is a slightly reduced true positive rate. However we select the mutual information metric over Euclidean distance as it is a more principled method and scores better than the Euclidean features.

To measure the overall social connection strength between two individuals we utilise the pairwise strength in the previous step in the following way. A trajectory of length T frames consists of T tuples (S,v,θ) for 2D ground plane position vector S, speed scalar v and direction of trajectory in radians θ. We can calculate the pair strength at frame T between any two individuals i and j, for $i,j \in N$ where N is the set of all individuals in the scene for all frames. The social connection strength κ between two individuals i and j at time T is:

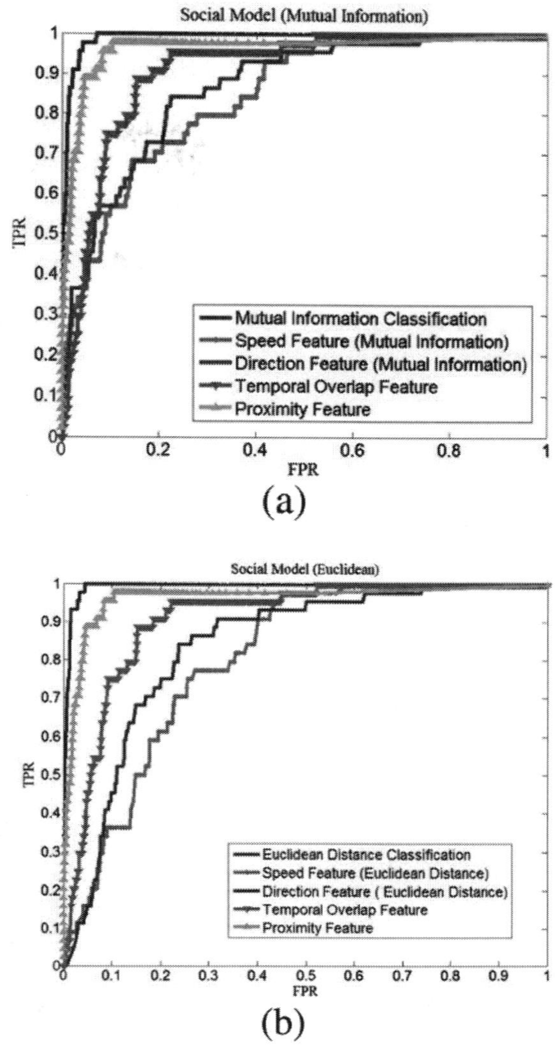

Figure. 3: A comparison of the features which comprise the Mutual information social model (a) and for comparison the Euclidean distance equivalent (b) both trained upon the PETS 2006 dataset and tested upon the PETS 2007 data set. The proximity and temporal overlap in both metrics are identical. The critical difference is in the speed and direction information. We observe that the mutual information speed and direction metrics outperform the Euclidean distance feature metrics in overall true positive classification.

$$\kappa_{ijt} = \frac{1}{T}\sum_{t}^{T} IV_{ijt} I\Theta_{ijt} \Delta P_{ijt} \tau_{ijt}, \quad (7)$$

τ_{ijt}, IV_{ijt}, $I\Theta_{ijt}$, ΔP_{ijt} are the temporal overlap, mutual information for speed, mutual information for direction and proximity difference between person i and j, as detailed in the feature Eqs. (4), (5) and (6). We classify the social state S, for $S=\{0,1\}$, by applying social strength threshold λ which is set empirically from the training data. Connections between individuals which score higher than λ are considered socially connected, providing the binary social context state used in the anomaly detection stage.

Anomaly detection. Anomaly detection splits into three distinct segments: the *behaviour ontology*, the method for *calculating normality* of observations, and the algorithm for*detecting anomalies*.

Behaviour Ontology: Our behaviour ontology is represented by a four part feature vector$x = R^4$, consisting of a bivariate motion component [speed, persistence], and the two contextual states [social state, scene region]. Speed is measured in meters per second on the ground plane, and social state is a binary state describing whether the individual is part of a social group or not. The persistence of an individual is a measure in frames of how long an individual has remained in the scene for. Lastly, the scene region identifies the scene context region in which the individual resides, denoted by a numerical identifier. For an individual with trajectory length T frames we have T feature vector observations. The observations are accumulated to a discrete 4 dimensional feature space representing a 4D histogram, termed the behaviour profile X_i, for individual i . Defined in this way X_i consists of a feature distribution from a large number of observations. The advantage to this is that it hides short-term measurement noise resulting in a behaviour ontology which is more robust. Furthermore, as measurement noise is often correlated rather than Gaussian white noise, the order independent nature of the behaviour profile X_i overcomes the appearance of anomalies that arise from structured noise. Our behaviour profile provides flexible temporal scaling of behaviours; something DBNs struggle with, however it results in the loss of time series information which may reduce the descriptive capacity of the ontology.

Normality of behaviour observations: As our approach is unsupervised anomalies are discovered due to their contrasting nature to previously observed behaviour. Much work to date has focused upon a frequency based analysis to determine the normality of behaviour observations. However, frequency-based anomaly detection suffers under the following assumption: that the normality of any observed behaviour is proportional to the relative frequency of observations of the behaviour. Whilst we can expect abnormal events to be rare, it is not the case that normal events are all frequent, and proportionally represented. We wish to distinguish here between the *normality* of a behaviour and the*expectation* of a behaviour. The expectation of a behaviour is how likely it is to occur next, whereas the normality of a behaviour is how permitted the behaviour is in the scene; how legitimate it is. A frequency based analysis reveals expectation of each behaviour to occur next, not the intrinsic normality of the behaviour itself, thus missing the mark. We instead implement a Nearest Neighbour method to search for supporting evidence for an observation from others within the data. The normality of any behaviour is based upon its distance to the nearest K instances of supporting evidence *not* the frequency of observation for that behaviour.

Whilst a nearest neighbour approach could be expected to segment out anomalies with strong contrary motions, a subtle anomaly may not be distant from the set of normal behaviour with regard to the majority of features. A subtle anomaly may be abnormal for only a subset of features, and furthermore only when seen in the context of another feature. For example the speed is abnormal only when seen in the context of a specific scene region, rather than the speed and scene region both being independently abnormal. As such we need to assign a normality score to each feature in context of each other feature, independently of every other feature, a step critical to detecting subtle differences between behaviours. This step enables us to see context dependent distinctions between behaviours which when viewed in the full feature space are too subtle to impact a distance calculation. To represent each feature in the context of another we reduce our 4D histogram feature space to a set of 1D feature distributions Y_n^{f1f2} detailing the distribution of feature f1 given the currently observed value for feature f2 for person n at frame t . For a feature vector x_i with dimensionality D there are $D^2 - D$ feature context pairs covering each $\{f_1, f_2\}$ feature pairing, when $f_1 \neq f_2$. In our 4D feature space 12 individual feature pairs are assessed at each frame for each individual, each representing a different observation given context pairing. To reduce the dimensionality of X_i to 1 for a particular feature context pair we sum the distribution X_i for all dimensions f in the set of dimensions F where $f_1 \neq f_2$ resulting in a 2D joint distribution Y_n of observation feature f_1 and context feature f_2. We then take a further step reducing the 2D distribution to the target 1D distribution by taking the distribution through the current context feature value $f_2(i)$ only. Thus our resulting distribution Y_n^{f1f2} details the distribution of observed feature values for observation feature dimension $f_1$given the context feature state $f_2(i)$. An example of which would be the distribution of the speed feature given the scene feature of idle region.

We apply the Nearest Neighbour (NN) function to distribution Y_n^{f1f2} and the set of all distributions Y to determine the nearest neighbour Y_m^{f1f2} to Y_n^{f1f2} for each possible feature context pairing $\{f_1, f_2\} \in F$. The Nearest Neighbour distance metric specified is the Bhattacharyya coefficient. The nearest neighbour distance metric for feature context pair $\{f_1, f_2\}$ is thus defined as:

$$B(Y_n, Y_m) = \sum_h \sqrt{Y(h)_n^{f1f2} Y(h)_m^{f1f2}} \tag{8}$$

where we sum over all histogram bins h for feature dimension f_1. Thus given a feature vector for individual $n \in N$ at frame $t \in T$ we find the nearest neighbour m where $\{m \in N : n \neq m\}$.

$$NN(Y_n) = \{Y_m \in Y | \forall Y_p \in Y : B(Y_n, Y_m) \geqslant B(Y_n, Y_p)\} \tag{9}$$

The nearest neighbour equation specifies m the index of the least distant behaviour profile of n for feature context pair $\{f_1, f_2\}$ and B the resultant Bhattacharya coefficient. As the Bhattacharyya coefficient is a measure of similarity, scoring more similar distributions higher, the NN finds the greatest Bhattacharyya coefficient to distribution Y_n from the set of all distributions Y given the feature context pair $\{f_1, f_2\}$, we then recombine the independent feature context pairs to generate a single value for the abnormality coefficient $A(n,t)$ for person n, at frame t. The abnormality coefficient of behaviour at frame t for person n is the least supported feature pairing; the lowest similarity to the nearest neighbour:

$$A(n,t) = argmin_{f1f2} B(Y_n^{f1f2}, Y_m^{f1f2}) \tag{10}$$

A consequence of segmenting subgroups is that an observation may be the only member of a context defined sub group. Ideally in operation an active learning methodology would be implemented to determine the normality of an observation in a new area of the behaviour space. However, in our application we chose to suspend judgment of new instances of behaviour, specifying that no evidence of an alarm is not an alarm. It would be equally valid to select the opposite, the effect of which would be to place a bias upon highlighting rare behaviour.

Anomaly detection: Threshold μ upon $A(n,t)$ separates anomalies from normal observations and in effect represents the sensitivity of the system. If we seek to detect only anomalies then μ represents the expectation of abnormal behaviour in the sequence. For the end user μ represents a constant surveillance workload for the operator. Variable μ can be either set by the operator or defined empirically in an additional training phase. Anomalies $A(n,t)$ at frame t for person n are classified by:

$$A(n,t) = \delta(A(n,t)) = \begin{cases} 1, & A(n,t) < \mu \\ 0, & A(n,t) \geqslant \mu \end{cases} \tag{11}$$

Based upon the assumption that there is dependence between the behaviour of individuals within the same social group we utilise the social contextual information in an additional two ways. Firstly we ensure that the behaviour of each individual is only analysed in reference to people external to their social group. Thus a behaviourally homogeneous group of individuals all acting abnormally cannot be self-justifying. We enforce this by removing the indexes of individuals from the same social group from the nearest neighbour calculation for individuals in that group. Secondly, social information enables us to propagate the expectation of an anomaly through the entire social group. In this way each member of a social group at any given frame has the highest anomaly score for all individuals in that group. Thus if one individual in a group is behaving abnormally all group members are equally as abnormal. We do not implement any post process alarm filtering. We justify the exclusion of this process as it may obscure the change in accuracy resulting from the inclusion and exclusion of contextual information.

EXPERIMENT

We wish to evaluate whether social and scene region contextual knowledge improves the detection of behavioural anomalies and permits the detection of subtle behavioural anomalies. We now detail the results of an anomaly detection experiment on the PETS 2007 dataset with the inclusion and exclusion of contextual information. Furthermore we test against a state of the art behaviour anomaly detection system which is itself designed to detect subtle anomalies.

The publicly available PETS 2007 dataset (PETS2007 and accessed, 2012) offers a source of multi camera real world surveillance footage. The datasets consists of 8 sequences each captured from 4 different viewpoints. We consider the PETS 2007 data to be a crowded scene. The data contains a total of 573 individuals over 11902 frames, averaging 24 people in the scene at any given frame in a space measuring 16.2 meters by 7.2 meters. Behavioural anomalies in this dataset are characterised by strong motion abnormality such as a group running across part of the scene, or subtle anomalies such as a single individual standing still in a busy area, or a group loitering amongst a crowd. We specifically chose this data due to its behavioural complexity for anomaly detection. The second dataset selected is the Oxford dataset. The Oxford data contains 430 tracked pedestrians over 4500 frames. There are an average of 15 individuals in any given frame, with a minimum of 5 and a maximum of 29. We consider this data as sparsely populated. The trajectory motion in the Oxford data is far more structured; the vast majority of individuals travel at walking pace in one of two directions. We select the second dataset, the Oxford data, to test our social context approach for failure modes. In the Oxford data the trajectories of socially unconnected pedestrians are often very similar, and often close in proximity - giving the appearance of social connectivity. We expect this will produce false positive social context information. We evaluate upon 3 non-sequential videos from the PETS 2007 selected due to the ground truth behaviour abnormalities present. PETS Scene 02 consists of 4500 images, Scene 04 is 3500 images long, and Scene 07 is 3000 images in length. All three are imaged at 25 fps. The single scene from the Oxford dataset is captured at 25 fps and 4500 frames in length. each sequence is treated individually. We apply the tracking procedure outlined earlier upon the jpeg the format images with no other pre-processing.

Scene segmentation. We found well defined regions for the idle, divergence and traffic region in the PETS data which fit with the intuitive interpretation of the scene. For clarity we illustrate the scene segmentation, see Fig. 4. The Oxford data held well defined areas for the traffic region and the divergence region. However the idle region hardly featured. This finding fits with the highly structured nature of the Oxford data in which there are very few stationary tracks. As our approach is data driven, scene regions are defined by virtue of being a tool for segmenting the behaviour space rather than fitting an intuitive interpretation of scene regions.

(a)

(b)

(c)

Figure. 4: (a), (b) and (c) illustrate the automatic scene segmentation we arrived at using the all trajectories from the PETS-2007 datasets. Each unique scene context is designated by a colour; Idle region – Red, Traffic region – Blue, and Divergence region – Green. Areas of the scene not included in either scene region class do not have sufficient supporting evidence to be classified and as such remain blank. (For interpretation of the references to color in this figure legend, the reader is referred to the web version of this article.)

Social context. We test the social context classification against an independently constructed ground truth for social connections. The training data (PETS 2006) consisted of 28 people with 14 true positive unique social connections between them of varying strength. The test data (PETS 2007) contains 152 tracked individuals, 44 social connections. Classifying social connections in the PETS 2007 data using parameters trained in the PETS 2006 data achieved a true positive detection rate (TPR) of 0.92 and a false positive rate (FPR) of 0.092, see Fig. 3(a). There are a greater number of false positive social connections in the Oxford data. The optimal result found 0.412 TPR and 0.0149 FPR. However beyond this true positive rate the false positives escalated greatly.

Anomaly detection. To demonstrate the impact context information has upon anomaly detection we determine the accuracy in four states: no contextual information, only scene context, only social context and with both types of contextual information. A comparison is made of the TPR and FPR, for detection of groundtruth anomalies. See Table 1 for a full list of anomalies. For examples of subtle anomaly detection see Fig. 5. The anomaly ground truth reveals 12 behavioural anomalies in the PETS 2007, and 3 anomalies over 4500 frames in the Oxford data. In both the PETS and Oxford data we vary the μ threshold from 0 to 1 in small increments to adjusts the systems sensitivity to unlikely observation. Fig. 6(a)–(c) demonstrates the anomaly detection success in the PETS 2007 dataset. Fig. 7 illustrates the results on the Oxford data.

Table 1: The behavioural anomalies in PETS 2007 (3 sequences) and Oxford Data. (1), (2) and (3) occur due to a group standing on the left of the scene looking around and suddenly dispersing in different directions. Anomalies (4) and (5) occur due to two individuals entering the scene, turning a corner and then suddenly turning around and leaving in the same place they entered. (6) is a known ground truth behavioural anomaly. One of the participants in the PETS 2007 experiment purposefully loiters in a busy scene. (6), (7) and (8) are all members of a small group of 3 running through the scene, from the top to the bottom of the scene. (9), (10), (11), and (12) are four more instances of known ground truth anomalies. Two individuals purposefully loiter in the scene whilst another two suspiciously switch baggage. In the Oxford data, anomaly (13) is due to the unique behaviour of the individual interacting with a bin in the scene. Anomaly (14) captures an individual entering the scene at the bottom and loitering in the middle. Anomaly (15) captures a women meandering slowly through the scene

PETS 2007 (Scene s00)	Id	Start	End
Unusual group behaviour	1	1	2656
Unusual group behaviour	2	1	2419
Unusual group behaviour	3	1	2714
Abrupt you turn in busy area	4	2627	2928
Abrupt you turn in busy area	5	2604	2928
PETS 2007 (Scene s02)			
ground-truth loitering	6	160	4497
PETS 2007 (Scene s04)			
Running through scene	6	109	275
Running through scene	7	130	290
Running through scene	8	148	322
Bag swap, unusual motion	9	1	3496
Bag swap, unusual motion	10	1	3496
ground-truth loitering	11	1	2596
ground-truth loitering	12	497	1726
Oxford Data			
Motion + interaction with scene	13	3554	4349

PETS 2007 (Scene s00)	Id	Start	End
Loitering	14	3867	4500
Abnormally slow movement	15	2382	3454

(a)

(b)

(c)

Figure. 5: Illustrated here is three examples of anomalies detected by our system in the PETS 2007 data set. (a) shows two true positives with a false positive in the bottom left corner. The anomalies in (a) refer to anomaly Id: 6 and 7 in Table 1. In (b) two examples of loitering are detected, anomaly Id: 11 and 12. In (c) loitering is detected, Anomaly Id: 9, and 10.

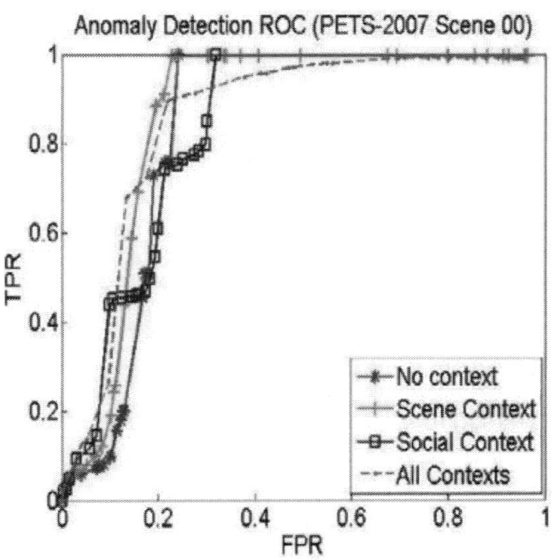

(a)

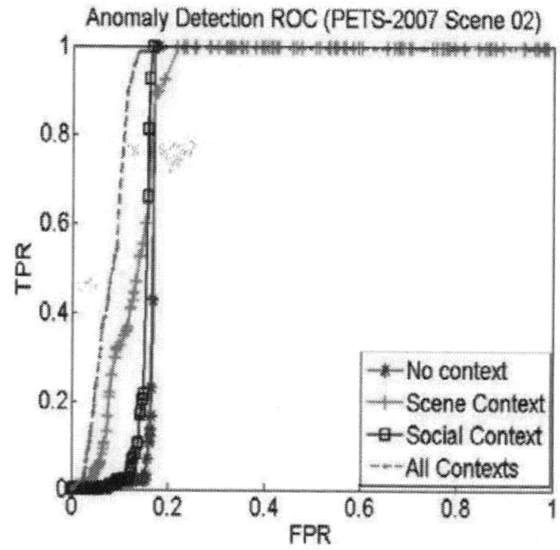

(b)

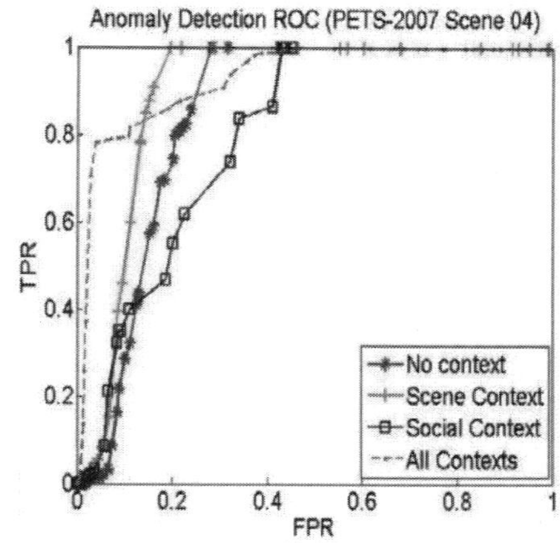

(c)

Figure. 6: ROC charts for Anomaly Detection classification, with a comparison of different contextual setups. (a) shows the results from PETS-2007 Scene 00, (b) from PETS-2007 Scene 02, and (c) from PETS-2007 Scene 04.

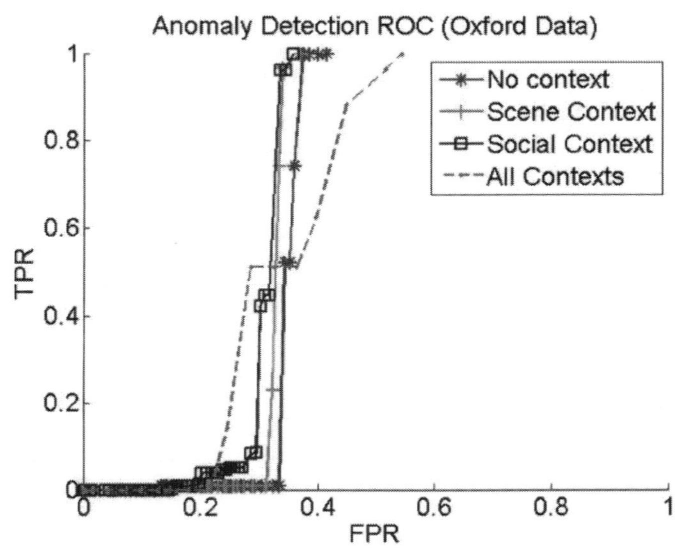

Figure. 7: The anomaly detection results on the Oxford Dataset. we test upon the Oxford data to test for a failure mode in the social model.

EVALUATION

The final TPR and FPR classification results with the inclusion of both types of context are affected by three factors above the no-context baseline. Firstly, the inclusion of scene context, the inclusion of social context, and impact of propagating anomalies through a social group and denying self-justifying social groups. In the three PETS-2007 datasets we observe that the addition of scene context improves the TPR over FPR detection of anomalies over all datasets in comparison to the no-context baseline. This is most significantly observed in Scene 04, Fig. 6(c). The inclusion of social context alone into the PETS-2007 data demonstrates a reduction in anomaly detection capacity in Scene 02,Fig. 6(c). PETS-2007 Scene 02 shows only a minor improvement. The significant result is that with the inclusion of both social context and scene context the TPR is improved above the TPR of scene context inclusion alone. This is due to the inclusion of the capability introduced by the social context to deny self-justifying groups and propagate anomalies within social groups. Particularly in PETS Scene 04, we observe that by propagating low likelihood scores throughout the group the bulk of true positive anomalies are discovered earlier, reducing the FPR from 0.2 to 0.03, see Fig. 6(c). The overall classification score with both social and scene context for all PETS-2007 data is shown in Fig. 8. We recorded a drop in the false positive rate of 0.13 for the optimal classification rate of 0.78 when applying the social and scene context.

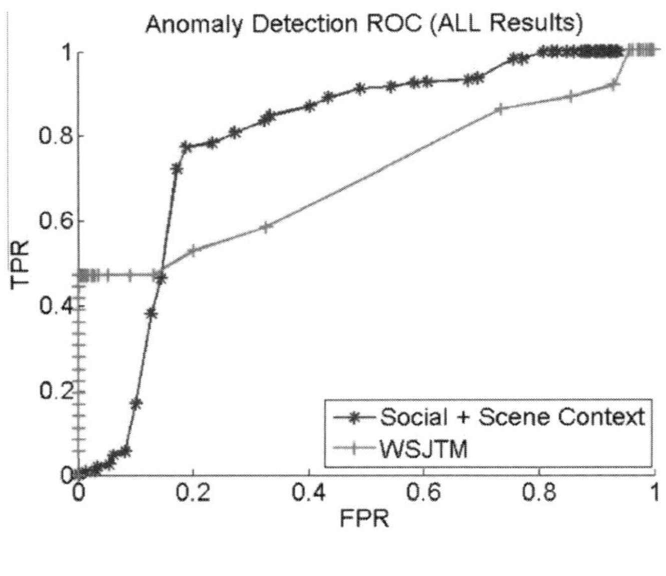

Figure. 8: A comparison between the Weakly Supervised Joint Topic model and our context aware method on the challenging PETS-2007 dataset. We trained and tested against all PETS-2007 data for both datasets.

In the Oxford data set the use of context information does not appear to raise the ability to detect anomalies significantly. We believe this to be due to the highly structured simple nature of the Oxford data. There is in effect very little contextual information to leverage our method upon. The false positive social connections in the Oxford data has not adversely affected use of social context, however, the inclusion of denying self-justifying groups, and propagating anomalies through social groups has a notable negative impact. The impact of denying self-justifying groups in the presence of false positive social groups is to remove potential training data, thus increasing the probability of false positive anomaly alarms. We observe this failure mode in the Oxford data, see Fig. 7which reflects our original prediction that our social model, geared towards crowds, would present a failure mode in the highly structured motion of Oxford data. To further test our approach we applied our context aware algorithm to maritime AIS shipping data in Southampton Harbour. The social context depicted mutual dependencies such as tugs pulling ships and convoy behaviour. Scene context was directly comparable. We achieved a true positive anomaly detection rate of 0.98 with a false positive rate of 0.17 over 66 h of data. However as the focus of our approach is computer vision we do not discuss the results further in this work.

In the PETS-2007 data anomalies such as loitering are subtle behavioural anomalies as the trajectories of these behaviours are very similar to a large number of legitimate behaviours in the scene, particular in the queuing areas. Because motion alone is not sufficient to define the behaviour as an anomaly we require extra contextual information to segment these subtle behaviours from the main body of data, particularly the scene context. The output of our system is displayed in Fig. 5. Images (a) through (c) show correct identification of anomalies. Image (a) shows an example of a context independent anomaly: running through the scene. Image (b) shows two examples of context dependent anomalies. The motion features pertaining to the anomaly are common within the entire scene, requiring scene context for them to be detected as anomalies.

To see our anomaly detection system in reference to the state of the art we include an implementation of the Weakly Supervised Joint Topic Model (WSJTM) proposed and developed by T. Hospedales, Jian Li, Shaogang Gong and Tao Xiang. We select the WSJTM as it is designed specifically to detect *subtle* abnormal behaviour similar in style to our own work. Furthermore, it is based upon a different behaviour representation whilst its use of positional information makes it comparable to our scene contextual information. For a detailed account of this work see (Hospedales and Xiang, 2011). We use the code provided by the author to make the comparison. The results from our own and the WSJTM procedure can be seen in Fig. 8. We find that the WSJTM outperforms our method at low TPR and FPR rates. However the results sharply fall off as it is incapable of segmenting a range of anomalies from the challenging PETS-2007 data. The WSJTM is capable of finding gross motion anomalies better than our method however it fails to detect subtle anomalies such as loitering. We observe that our method achieves a better overall TPR over FPR.

CONCLUSION

We successfully demonstrated the capability to detect anomalies based upon contextual information and trajectories in two scenes, presenting distinctly different behavioural environments. The application of social context provides a improvement in anomaly detection in the crowded PETS-2007 data. However, failure of the social model can result in a negative impact upon anomaly detection, as witnessed in the Oxford dataset. We found that our context aware method performs significantly better than the equivalent method without contextual information; reducing the false positive rate from 0.2 to 0.03. We show an overall true positive classification rate of 0.78 over 0.19 false positives on the PETS-2007 data, a reduction in the false positive rate of 0.13 due to the inclusion of contextual information. We conclude that in a crowded scene the application of social context to prevent self-justifying groups and propagate anomalies is highly relevant. Scene context uniformly improved the detection of anomalies in both datasets, and provided the ability to detect subtle context dependent anomalies. The metric for comparing behaviours in this work can be interchanged with other state of the art methods; the implication being that contextual information, particularly scene regions could be complimentary used with other anomaly detection systems revealing subtle anomalies that otherwise may be missed.

ACKNOWLEDGEMENTS

This work was supported by funding from the EPSRC Industrial Doctorate Centre in Optics and Photonics Technologies (EP/G037523/1), and the EPSRC/MOD University Defence Research Collaboration in Signal Processing (EP/J015180/1).

REFERENCES

1. Benfold, I.R.B., 2011. Stable multi-target tracking in real-time surveillance video.Chandola, V., 2009. Anomaly detection: a survey, ACM Comput. Surv.
2. Cristani, A.D.B.V.M.M., Raghavendra, R., 2012. Human behavior analysis in video surveillance: a social signal processing perspective, Neurocomputing.
3. Farenzena, L.B.D.T.M., Tavano, A., 2011. Social Interactions by visual focus of attention in a three-dimensional environment, Expert Syst.
4. Felzenszwalb, D.M.P., Girshick, R., Ramanan, D., 2010. Object detection with discriminatively trained part based models, IEEE Trans. Pattern Anal. Mach. Intell. 32.
5. Ge, R.T.C.W., Ruback, B., 2009. Automatically detecting the small group structure of a crowd. In: IEEE Workshop on Applications of Computer Vision.
6. Hospedales, S.G.T., Xiang, T., 2011. Identifying rare and subtle behaviours: a weakly supervised joint topic model, IEEE Trans. Pattern Anal. Mach. Intell.
7. Kalal, K.M.Z., Matas, J., 2009. Online learning of robust object detectors during unstable tracking. In: Third Online Learning for Computer Vision Workshop, Kyoto, Japan, IEEE CS.
8. Kalal, K.M.Z., Matas, J., 2010. P-N learning: bootstrapping binary classifiers by structural constraints. In: 23rd IEEE Conference on Computer Vision and Pattern Recognition.
9. Lan, W.Y.G.M.T., Wang, Y., 2010. Beyond actions: discriminative models for contextual group activities, Adv. Neural Inf. Process. Syst.
10. Li, J.L.J., Xiang, T., 2008. Scene segmentation for behaviour correlation. In: European Conference on Computer Vision.
11. Loy, C.C., 2010. Activity understanding and unusual event detection in surveillance videos (Ph.D. thesis). Queen Mary University of London.
12. Makris, D., Ellis, T., 2005. Learning semantic scene models from observing activity in visual surveillance, IEEE Trans. Syst. Man Cybern.
13. Oliver, B.R.N., Pentland, A., 1998. Statistical Modelling of human interactions. In: CVPR Workshop on Interpretation of Visual Motion.
14. PETS2007, 2012. <http://www.cvg.rdg.ac.uk/PETS2007/data.html>, accessed 24/09/ 2012.
15. Robertson, I.D.R.N.M., 2011. Automatic reasoning about causal events in surveillance video, EURASIP J. Image and Video Process.
16. Song, C.J.X., Wu, M., Ranka, S., 2007. Conditional anomaly detection, IEEE Trans. Knowl. Data Eng. 19.
17. Wang, G.N.E.G.X., Teck Ma, K., 2008. Trajectory analysis and semantic region modeling using a nonparametric Bayesian model, Comput. Sci. Artif. Intell. Lab.Technical Report.
18. Yu, T., Lim, S., Patwardhan, K., Krahnstoever, N., 2009. Monitoring, recognising and discovering social networks, IEEE Comput. Vision and Pattern Recognit.

CHAPTER 8

IDENTIFYING DEFICITS OF VISUAL SECURITY METRICS FOR IMAGES

Heinz Hofbauer·Andreas Uhl

University of Salzburg, Department of Computer Sciences, Austria

ABSTRACT

Visual security metrics are deterministic measures with the (claimed) ability to assess whether an encryption method for visual data does achieve its defined goal. These metrics are usually developed together with a particular encryption method in order to provide an evaluation of said method based on its visual output. However, visual security metrics themselves are rarely evaluated and the claim to perform as a visual security metric is not tied to the specific encryption method for which they were developed. In this paper, we introduce a methodology for assessing the performance of security metrics based on common media encryption scenarios. We systematically evaluate visual security metrics proposed in the literature, along with conventional image metrics which are frequently used for the same task. We show that they are generally not suitable to perform their claimed task.

INTRODUCTION

The claim of visual security metrics (security metrics for brevity) is usually the ability to assess the functionality of an encryption method based on the output of the encryption of visual data. In particular, the evaluation of an encryption method is only based on the visual output (i.e., the ciphertext), which is either an image or video. While such metrics are often created in conjunction with a specific encryption method and tested, if at all, only for this encryption method, the claim to perform as a security metric is usually universal. Furthermore, regular image quality metrics, such as the frequently used PSNR and SSIM, are also utilized in the literature to evaluate encryption methods [1], [2],[3] and [4].

The problem with the evaluation of security metrics is the fact that there is no established testing methodology. Thus, even if security metrics are tested, the test is usually based on the evaluation procedures for regular image metrics, which are not sufficient to establish whether a method is applicable in the context of encryption.

Regarding cryptographic security, Shannon›s work [5] shows that the highest level of security is reached when applying a secure cipher to a redundancy free plain text. Current image/video codecs exploit redundancy for compression and thus we can consider a bit stream to be an almost redundancy free

Citation: Heinz Hofbauer, Andreas Uhl, Identifying deficits of visual security metrics for images, Signal Processing: Image Communication, Volume 46, August 2016, Pages 60-75, ISSN 0923-5965, http://dx.doi.org/10.1016/j.image.2016.05.001.

plain text in the sense of Shannon. Consequently, for maximal security, the encryption of the entire bit stream with a state-of-the-art cipher, such as AES, would suffice ("conventional encryption"). Lookabaugh and Sicker [6] showed that selective encryption is sound and demonstrated its relation to Shannon's work. However, [7] showed that side information can compromise security.

However, there are application scenarios which make it necessary to move away from full encryption. Methods which do not utilize full encryption of the underlying data are called Lightweight/Soft/Partial/Selective Encryption. Specifically, selective encryption is the application of an assumed secure encryption method to a *selected* part of the plain text. In selective encryption the *encryption* part is assumed to be secure, e.g., by using AES. The final security of the selective encryption comes then from the *selection* part. What is evaluated in order to gauge the final security of the selective encryption is, to what extent the information left in plain text can be used to reconstruct an image or video.

Furthermore, an attack on the selective encryption method does not come from attacking the *encryption*, but from attacking the *selection*. This is usually done by using knowledge about the original format of the image/video. An attack is usually based on removing the negative impact on quality by the, essentially random, signal introduced by the encryption. This is typically done by replacing the random signal by a signal which introduces the least amount of error into the final decoding. In order to do so, very specific knowledge about the containing format has to be exploited, and there is usually only a single method to go about this, i.e., *the attack*.

Another, (implicit) assumption about the selective encryption method under test is the*format compliance*. Format compliance requires an encrypted bitstream to be decodable by a standards compliant decoder. In other words, format errors may not be introduced by the encryption. Thus, in the following, when we refer to an encrypted image/video we mean the image or video that results from decoding an unattacked encrypted bitstream.

The notion of security in selective encryption is different from the traditional notion of security: First, we knowingly leave information in plain text to retain format compliance; second, the focus is on content security not information security, i.e., the content should be secure (to some defined extent), while information about the content might be allowed to leak. In order to be able to discuss the exact notion of security in such non-conventional encryption schemes, we need to distinguish distinct application scenarios of encryption schemes for visual data:

Confidentiality encryption:

Means MP security (message privacy). The formal notion is that if a system is MP-secure an attacker cannot efficiently compute any property of the plain text from the cipher text [8]. This can only be achieved by the conventional encryption approach.

Content confidentiality:

This is a relaxation of confidential encryption. Side channel information may be reconstructed or left in plaintext, e.g., header information, packet length, but the visual content must be secure in the sense that it must not be intelligible/discernible [9].

Sufficient encryption:

Means we do not require full security, just enough security to prevent abuse of the data. The content must not be consumable due to high distortion (e.g., for DRM systems) by destroying visual quality to a degree which prevents a pleasant viewing experience or destroys the commercial value. This implicitly refers to message quality security (MQ), which requires that an adversary cannot reconstruct a higher quality version of the encrypted material than specified for the application scenario [10].

Perceptual/transparent encryption:

Means we want consumers to be able to view a preview version of the video but in a lower quality while preventing them from seeing a full version. As an example: this can be used in a pay per view scheme where a lower quality preview version is available from the outset to attract the viewers interest, q.v., [11]. The difference between sufficient and transparent is the fact that there is no minimum quality requirement for sufficient encryption. Encryption schemes which can do sufficient encryption cannot necessarily ensure a certain quality and are thus unable to provide transparent encryption.

Given these different application scenarios it is clear that depending on the goal, a security metric has to fulfill different roles. For example, under the assumption of sufficient encryption, a given security metric would have to evaluate which quality is low enough to prevent a pleasant viewing experience. In contrast, for the transparent encryption case, a metric not only has to assess whether the quality of an image or video is low enough, but also whether the quality is high enough to be useful to attract interest. When it comes to content confidentiality the question of quality is no longer applicable. Content confidentiality requires that image content must not be identified by human or automated recognition. This requirement also has to be maintained for any part of the image. Image metrics, in general, do not deal with such questions but rate the overall image quality, the question of intelligibility is usually not covered at all. A drastic example would be an image where only a small part of the image is partly visible. Classical metrics would judge the whole image and consequently would attribute a high security, even though a part of the image is still recognizable which contradicts content confidentiality. Still, it has to be pointed out that content confidentiality can have different forms. To prevent a face recognition scheme from working properly it is sufficient to protect any facial information in a surveillance video, while humans could still be identified in such a video by using gait recognition. Furthermore, if the appearance of a person has to be concealed entirely, a much stronger extent of protection (i.e., higher security) is required. Finally, confidential encryption cannot be solely assessed with security metrics since the scope goes beyond assessing security based on the visual appearance only. Furthermore, we should note that the application of security metrics on video is performed at a frame by frame basis in the literature. We will adopt this model but should note that for the discussion of confidential encryption motion data is of importance, e.g., in [12] it was shown that a replacement attack combined with motion information can reveal the content of a scene even though the visual content of every frame is encrypted.

Consequently, depending on a given application scenario different properties are required from a security metric and different approaches to construct such a metric might perform better or worse for some applications scenarios. This dependence on the evaluation goal of a security metric is hardly ever discussed in the papers introducing a metric. Sufficient and transparent encryption scenarios have a clear and distinct link to the traditional notion of (low) visual quality, while it is highly questionable or at least doubtful if content confidentiality can be assessed by the classical quality notion. While the lack of relation to spatial areas of most security metrics could be compensated in the design to provide locally varying results, the lack of relation to intelligibility in general can probably not be easily resolved.

For both, security metrics and regular image metrics, in the literature we do not find any evaluation whether a given metric can perform the claimed function or how such an evaluation correlates to actual security. However, for regular image metrics it is well known that the correlation with human observations over the full range of possible quality (from high to low quality) does not imply a good performance on a given subset. More specifically, it was pointed out recently that most image metrics perform very poorly for the low quality range ([13]–using the low quality end of the LIVE database). For security metrics, not even this question has been covered so far.

In this paper, we will try to remedy this situation by giving an overview of requirements regarding security goals and formulating these requirements into a testing methodology. Based on this methodology we will evaluate the various security metrics in the literature as well as applicable conventional image metrics. However, we will not deal with every application scenario equally explicitly. We will only make a first step to cover the content confidentiality scenario. The main reason for this is a lack of ground truth. It is not obvious how to generate ground truth for this scenario since there is a disparity between how an image metric works and what is necessary to evaluate content confidentiality. Image metrics, and as an extension security metrics, measure the quality of an image respective to human judgement. This works well for high quality images but suffers for low quality images where human observers can have difficulties differentiating between the severity of an impairment. Thus the methodology to systematically generate ground truth based on human observation needs to be changed for content confidentiality which is not in the scope of this paper. On the other hand, for the image quality-related scenarios (sufficient and transparent encryption), ground truth data is available, in the form of image impairment databases with mean opinion scores (MOS) based on a number of human observations.

In the following we will motivate and introduce a methodology which allows to measure desirable traits of a security metric. This methodology can, and should, be used to asses newly developed metrics.

To motivate this, we will give an overview of security metrics used in the literature inSection 2. We will not analyse why metrics fail and the presented counterexamples should be seen as a proof by contradiction rather than a exhaustive list of failings. Note that this is a question of generality, we do not dispute that certain metrics are a good fit for specific encryption types.

In Section 3 we will describe what is expected of security metrics and the methodology how security metrics can be evaluated. This is done by providing certain desirable traits of a security metric, a motivation for the traits and a formal description of the traits to allow measuring them.

In Section 4 we will present the evaluation of state-of-the-art metrics and measure their ability to perform as security metrics. This is done based on the traits introduced inSection 3 and shows the usefulness of these traits to find shortcomings of current metrics and to prevent the same shortcomings in future metrics.

Section 5 will conclude the paper and give a list of state-of-the-art reference metrics and their usability as security metrics.

OVERVIEW OF SECURITY METRICS

In this section a brief overview of security metrics will be given. The metrics discussed are taken from the recent literature and are specifically designed to ascertain whether the image quality after encryption is sufficiently reduced. The metrics given in this section are discussed as general security metrics, i.e., not limited to the specific method with which they were designed together. References to the original work will be given for each security metric as well as some examples where the metric fails to assess given example images as would be expected from a general security metric. The SSIM and PSNR are also included in this overview. Even though they were not designed to be security metrics, they are frequently used as such, e.g., [3], [14], [15] and [16] (SSIM), [4], [17],[18] and [19] (PSNR) and [1] and [2] (SSIM and PSNR).

Peak Signal to Noise Ratio (*PSNR*): The peak signal-to-noise ratio (PSNR) is still widely used because it is unrivaled in speed and ease of use. The PSNR is a quality metric, meaning a high metric score reflects a high quality, which gives a score in the range$[0,\infty]$. However, it is also well known that the correlation to human judgement is somewhat lacking even for high and medium quality [20]. Fig. 1 illustrates the performance of the PSNR metric on samples from the IVC-SelectEncrypt [21] database (see Section 3).

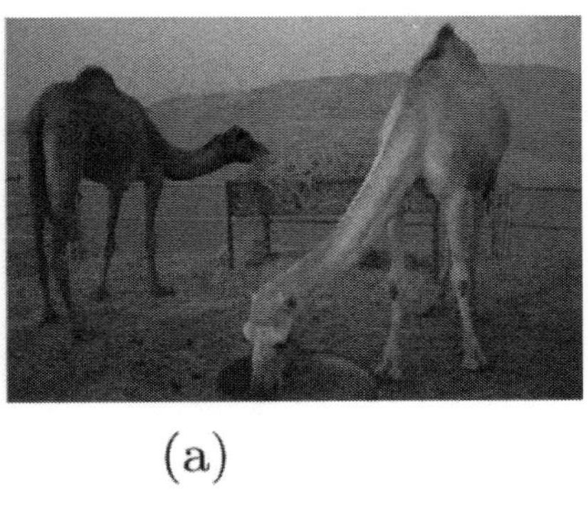

(a)

(b)

(c)

Figure. 1: PSNR metric scores for images from the IVC-SelectEncrypt database. According to PSNR images (a) and (b) are of the same quality and (c) is of much higher quality, i.e., less secure than (a) and (b).

Structural Similarity Index Measure (*SSIM*): The structural similarity index measure (SSIM) [22] extracts three separate scores from the image and combines them into the final score. First the visual influence is calculated locally then luminance, contrast and structural scores are calculated globally. These separate scores are then combined with equal weight to form the SSIM score. The SSIM is a quality metric, a high metric score reflects a high quality, which gives a score in the range $[0,1]$. Fig. 2 illustrates the performance of the SSIM metric on samples from the IVC-SelectEncrypt database.

(a)

(b)

(c)

Figure. 2: SSIM metric scores for images from the IVC-SelectEncrypt database. According to SSIM images (a) and (b) are of the same quality and (c) is of much lower quality, i.e., more secure than (a) and (b).

Edge Similarity Score (*ESS*): The edge similarity score (ESS) was introduced in [23] and uses localized edge direction information to compare two images. The ESS is a quality metric, a high metric score reflects a high quality, which gives a score in the range $[0:1]$.Fig. 3 illustrates the performance of the ESS metric on the foreman sequence when encryption according to [24] is applied in comparison to white noise.

(a)

(b)

(c)

(d)

Figure. 3: ESS comparison for frame 80 of the foreman sequence (a). ESS judges the white noise (c) to be of higher quality than the residual information from the encrypted frame (d). In order to show the amount of information actually retained in the encrypted frame a post processed version is also shown (b).

Luminance Similarity Score (*LSS*): The luminance similarity score (LSS) was introduced in [23] and uses localized luminosity information to compare two images. The LSS is a quality metric, a high metric score reflects a high quality, which gives a score in the range$[-8.5:1]$. Fig. 4 illustrates the performance of the LSS metric on the foreman sequence when encryption according to [25] is applied in comparison to noise.

(a)

(b)

(c)

Figure. 4: LSS comparison for a frame of the foreman sequence (c). LSS judges the noise (a) to be of higher quality than the residual information from the encrypted frame (b).

Neighborhood Similarity Degree (*NSD*): The neighborhood similarity degree metric (NSD), introduced in [26], uses local pixel similarity correlation between original and impaired image. The NSD depends on two parameters, one to define the region for pixel similarity correlation (d) and one to define the similarity threshold (m). The parameters $m$and d were set to the same values as in the experiments in [26], i.e., m=5, d =3, border extension is done by repeating the last border pixel. The NSD is an impairment metric, a high metric score reflects a low quality, which gives a score in the range $[0:1]$. Fig. 5illustrates the performance of the NSD metric on samples from the IVC-SelectEncrypt database.

(a)

(b)

(c)

Figure. 5: NSD metric scores for images from the IVC-SelectEncrypt database. According to NSD images (a) and (b) are of the same quality and (c) is of much lower quality, i.e., more secure than (a) and (b).

Local Entropy (*LE*): The local entropy metric was introduced in [27] (LE), it is a no reference metric operating only on an impaired image. The LE metric uses the average of normalized localized entropy scores, on 8×8 blocks, as image quality predictor. The LE is an impairment metric, a high metric score reflects a low quality, which gives a score in the range $[0:1]$. Fig. 6 illustrates the performance of the LE metric on samples from the IVC-SelectEncrypt database.

(a)

(b)

(c)

Figure. 6: LE metric scores for images from the IVC-SelectEncrypt database. According to LE images (a) and (b) are of the same quality and (c) is of much lower quality, i.e., more secure than (a) and (b).

Local Feature Based Visual Security (*LFBVS*): The local feature based visual security metric (LFBVS) was introduced in [28] and utilizes localized edge and luminance features which are combined and weighted according to error magnitude, i.e., error pooling. The LFBVS is an impairment metric, a high metric score reflects a low quality, which gives a score in the range $[0:1]$. Fig. 7 illustrates the performance of the LFBVS metric on the silent sequence when encryption according to [24] is applied in comparison to white noise.

(a)

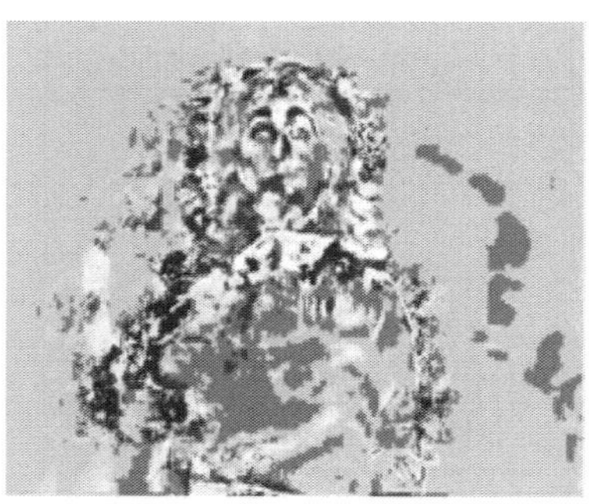

(b)

(c)

(d)

Figure. 7: LFBVS comparison for frame 80 of the silent sequence (a). LFBVS judges the residual information from the encrypted frame (d) to be about the same as the information contained in white noise (c). In order to show the amount of information actually retained in the encrypted frame a post processed version is also shown (b).

All the given image metrics, with the exception of the local entropy metric (LE), are full reference metrics, meaning they utilize information from the original and comparison (encrypted) image to calculate an assessment of the visual similarity. The local entropy metric by Sun et al. is a no reference metric, i.e., it utilizes only the impaired image to judge the resulting quality. By measuring entropy, LE can also be interpreted to assess the encrypted image compared to random noise (which exhibits maximal LE). Since all of the given security metrics are proposed to be general we will not differentiate between full- and non-reference metrics in the following but compare them solely on the task they are supposed to solve.

From the description of the various security metrics it can be seen that a wide range of approaches exist, from metrics targeting signal properties, e.g., PSNR which targets noise, to LE which targets local entropy, metrics which use higher level information, e.g., NSD or ESS which use a form of object detection (mostly based on edges), to metrics which use information about the HVS to improve their performance, e.g., SSIM or LFBS which use simulation of the fovea centralis and error pooling respectively. However, for every security metric the accompanying figure demonstrates an obvious fault in the performance of the given metric. Such examples can be found for every metric of course, the question we will try to answer in the following sections is whether the demonstrated fault is singular or systemic.

In order to gauge the effectiveness of a dedicated security metric it is useful to compare them to regular metrics. To facilitate a fair comparison, three recent metrics, in addition to PSNR and SSIM, are chosen and included in the evaluation. The local edge gradient image metric (LEG) [29] shows a good correlation with human judgement and is reasonably fast to compute such that it can be used even under time constraints.

The visual information fidelity (VIF) [30] and CPA1 [31] outperform the SSIM and LEG in regard to correlation with human judgement but are a lot slower to compute [29]. A second reason to include those metrics is to gauge whether they can be used as security metrics. The LEG and VIF are quality metrics with a score in the range $[0,1]$, and the CPA1 is an impairment metric with a score in $[0,\infty]$.

EVALUATION METHODOLOGY

In this section we outline (1) our evaluation methodology, (2) the reason to use this methodology, and (3) the application scenario(s) which can be assessed by employing a certain methodology. A discussion of the desired outcome from these tests for security metrics is also provided. This section is the guideline of how security metrics and image metrics are evaluated for the use as security metrics in Section 4.

Application Domain

Selective encryption methods, in a majority of cases, aim at format-compliant encryption, i.e., encryption in such a way that the media can be decoded by a standards compliant decoder without error. Frequently, security metrics are applied on a direct decoding of the encrypted bitstream. This can have adverse effects since encryption introduces noise which can hide plain text data and consequently a security metric might judge that an encryption method is more secure then it actually is. There are a number of options how a security metric can be applied, as illustrated in Fig. 8. All security metrics under evaluation are applied on a direct decoding of the encrypted bitstream (which we will denote as the *encrypted domain*) by the authors in the corresponding original papers. Other options would be to attack the encryption method in a way which does not break it but reduces the obfuscation of the plain text data in the encrypted domain (denoted*extracted* in the figure). Such attacks usually utilize knowledge about the bitstream rather than the encryption method (other than location). Typical attacks would be error concealment and replacement attacks against selective encryption schemes [32]. In general, such attacks replace the encrypted content with bitstream elements that are statistically known to introduce less noise than the random elements produced by encryption. Another possibility would be to utilize post processing to further help the metric detect residual information (denoted *processed* in the figure).

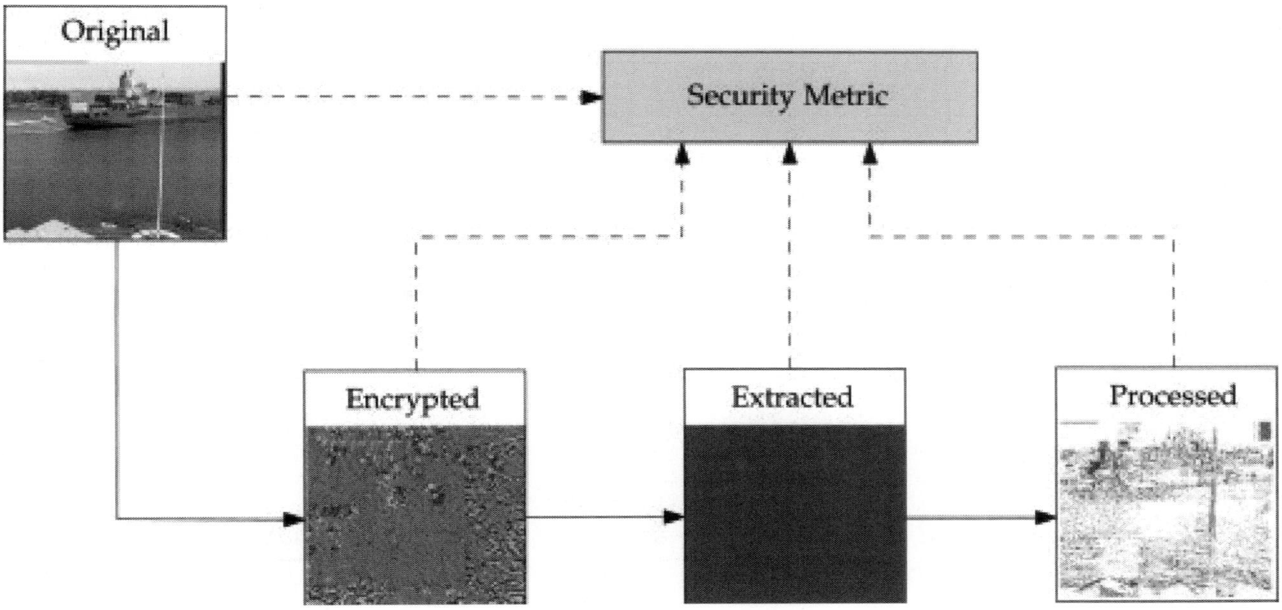

Figure. 8: The possible domains which can be used by a metric to compare to an original image. Either the direct output of an encrypted bit stream for format compliant encryption or an extraction of the plain text data which minimizes the disruptive effect of the encrypted data on the resulting bit stream. Another possibility would be post processing to further accent the residual plain text data contained in an image.

In the evaluation we only handle the difference between security metrics applied directly in the encrypted domain versus application in the extracted domain. The post processing step is only provided to better highlight the remaining information in an image. It cannot be directly used as an application domain since the post processing step is either specific to the encryption, in which case it should be included in the attack, or specific to the metric, in which case it should be included in the security metric. Post processing, in general, can influence different metrics in various ways, an example of this is given in Fig. 9 where post processing increases ESS and, at the same time, decreases the SSIM. This means that it is never actually possible to determine an optimal postprocessing method which would be required in a sensible assessment.

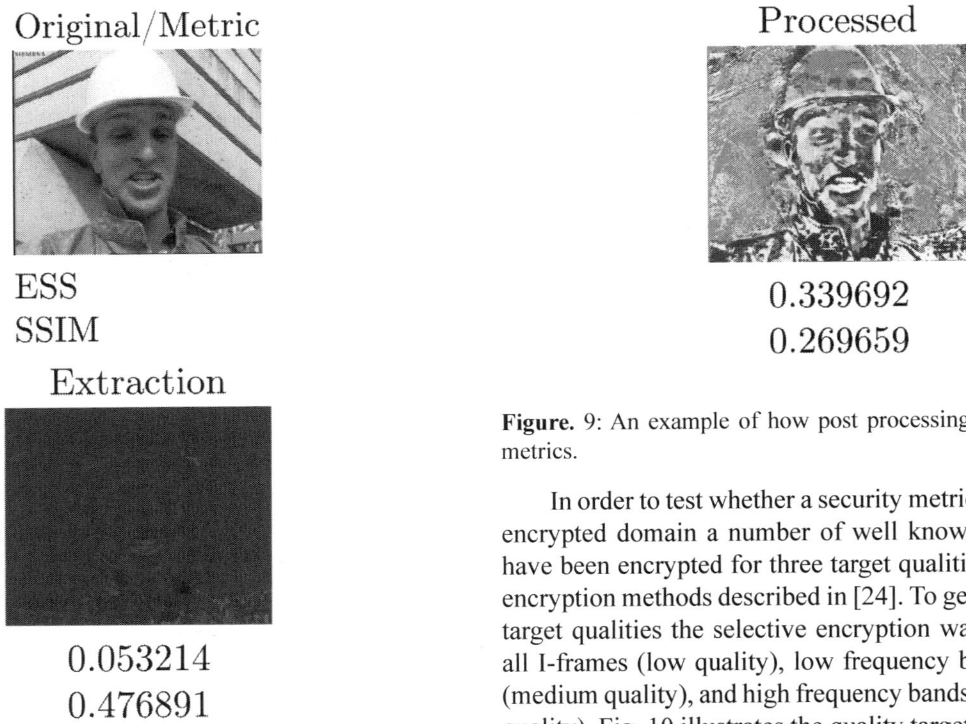

Figure. 9: An example of how post processing influences different metrics.

In order to test whether a security metric can operate in the encrypted domain a number of well known video sequences have been encrypted for three target qualities, utilizing EZBC encryption methods described in [24]. To generate the different target qualities the selective encryption was applied to either all I-frames (low quality), low frequency bands of all frames (medium quality), and high frequency bands of all frames (high quality). Fig. 10 illustrates the quality targets of the encryption process. The targets where chosen to contain high, medium and low residual information. Under the assumption that a security metric can operate in the encrypted domain it should be able to reliably order the encrypted frames of each sequence for every sequence from highest to lowest quality.

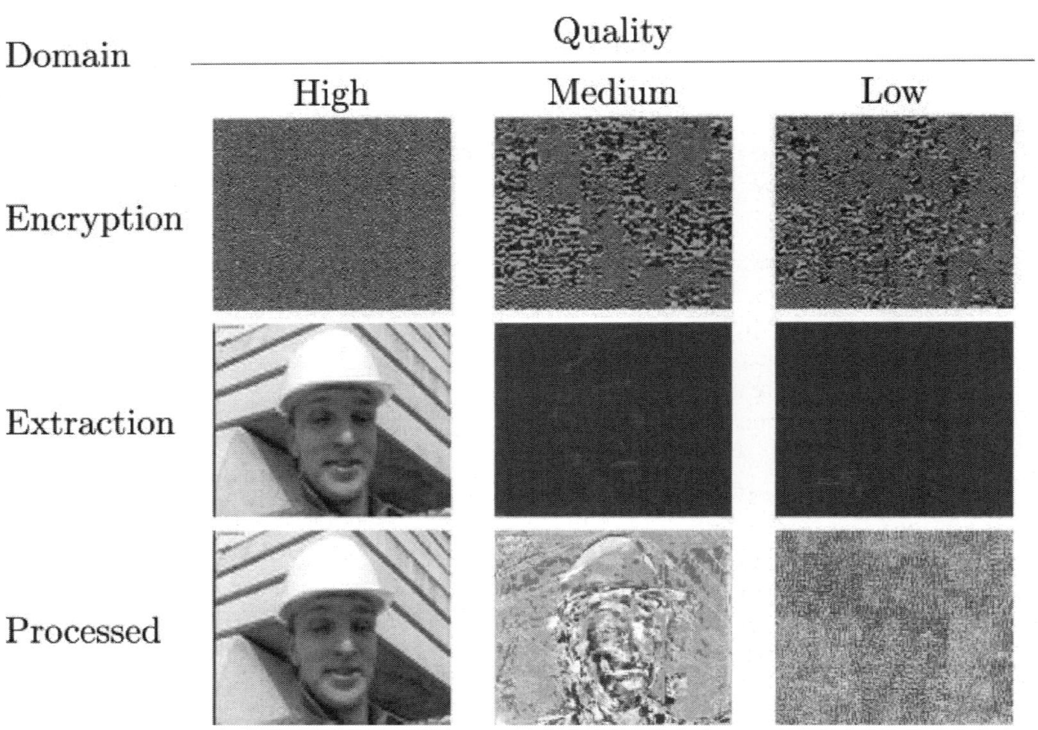

Figure. 10: A sample of the quality ordering test set based on the foreman sequence. Samples from the high, medium and low quality sequences are shown in the encryption, extraction and processed domain.

We generate two sets of ground truth, one for the high-medium and one for the medium-low quality comparison, containing ordered (based on quality) tuples for each frame. For a sequence S containing frames $s^1, \ldots, s^{N_S}$ and high (SH), medium (SM) and low (SL) quality versions we generate two sets of ordered (by quality) tuples $D_{HM}(S) = \{(s_H^1, s_M^1), \ldots, (s_H^{N_S}, s_M^{N_S})\}$, and likewise for *DML*.

Then a tested metric should also generate two sets with ordered tuples based on the estimated quality provided by the metric $M(o, c) \mapsto \mathbb{R}: Q_{HM}(S, M) = \{q(M, s^1, s_H^1, s_M^1), \ldots, q(M, s^{N_S}, s_H^{N_S}, s_M^{N_S})\}$,

, where

$$q(M, o, c_1, c_2) = \begin{cases} (c_1, c_2) & \text{if } M(o, c_1) \geq M(o, c_2), \\ (c_2, c_1) & \text{otherwise,} \end{cases}$$

and likewise for $QML(S, M)$.

Based on these four sets the ability of a metric to order in a given domain can be estimated by

$$O_M(S) = \frac{1}{2}\frac{|Q_{HM}(S, M) \cap D_{HM}(S)|}{|D_{HM}(S)|} + \frac{1}{2}\frac{|Q_{ML}(S, M) \cap D_{ML}(S)|}{|D_{ML}(S)|}. \quad (1)$$

values of 1 (100% correct ordering) or 0 (0% correct orderings, i.e., 100% correct reverse ordering) show a high ability of the metric to order correctly in the given domain. The difference between 1 and 0 is that image metrics can either measure similarity of images, i.e., quality metrics, resulting in a normal ordering or the difference between images, i.e., impairment metrics, resulting in a reverse ordering. Results around 0.5 are akin to random decisions and reflect the inability of the tested metric to perform in this domain.

Furthermore, since the low quality range chosen is in (or at least close to) the domain of content confidentiality, i.e., the rightmost column in Fig. 10 does not exhibit intelligible visual content, this setting also serves as an indication whether an image metric might be useful for content confidentiality. While a good performance on this evaluation does not necessarily mean a image metric is qualified for content confidentiality, a low performance is a strong indicator that the metric is unfit for this task. Based on the information which parts of the data have been encrypted and the entirely evident differences in visual appearance, ground truth is out of question here.

Note also that this is a proof by contradiction. That is, if a metric claims to be able to generally work in the encrypted domain and we can demonstrate that there are cases where it does not, then this claim has been falsified.

Correspondence to HVS: Confidence

Besides the encryption application scenarios where a certain quality is required (sufficient and transparent encryption), further examples for the importance of the quality notion are watermarking where the resulting quality should not be below a certain threshold, and of course, lossy compression. However, the notion of quality in this cases is not as straightforward as it seems. On the one hand we use the term quality in the context of the human visual system (HVS), i.e., how a person consuming the content would judge the quality. On the other hand, the term quality can refer to the score returned by a (security) metric which is tied to the quality in the HVS sense. This relation is not exact and it is not inherently clear how to choose a metric which correlates to the HVS quality which is targeted, although in practice Algorithm 1 is usually applied.

Algorithm 1.

Method for finding a target metric score based on a target HVS quality.

1: Chose a source image.
2: Alter the image until it fits the perceived target quality.
3: Apply the security metric on the altered and original image, the resulting metric score is the target quality.

While this results in a target quality which can be used, we know nothing about how well this score actually reflects the human judgement, since it is well known that the correlation between human judgement and image metrics is not perfect. In other words, how confident can we be in the choice of image metric score in relation to the perceived quality?

In order to evaluate this, well known databases which contain impaired and encrypted images and the perceived quality, in the form of mean opinion scores (MOS), will be used. The databases contain a set of points p representing impaired images with associated values pv for metric value and pd for MOS value, ordered from lowest to highest quality. Based on a target MOS quality score D two values can be calculated, Fig. 11 illustrates this.

Zero false negative: $Vmin(D)$ refers to the metric value for which the following holds: $p_d > D \Rightarrow p_v > Vmin(D)$. That is if the metric score is below $Vmin(D)$ we are sure that the perceived quality is below the MOS quality score (D).

Zero false positives: $Vmax(D)$ refers to the metric value for which the following holds: $p_v > Vmax(D) \Rightarrow p_d > D$. That is if the metric score is above $Vmax(D)$ we are sure that the perceived quality is above the MOS quality score (D).

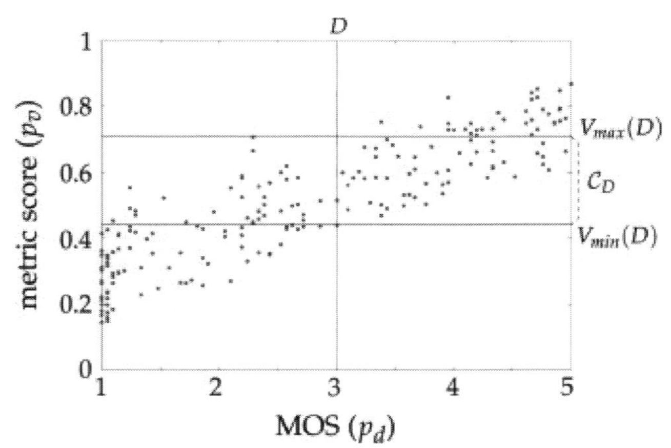

Figure. 11: Illustrations of zero false negative $Vmin(D)$, zero false positives $Vmax(D)$ and confidence C_D for a MOS value of D=3 is shown based on the IVC-SelectEncrypt database and the LEG metric.

This also means that if a target metric quality score pvt is obtained as given by Algorithm 1we are assured that $V_{min}(D) \geq p_v^t \geq V_{max}(D)$.

Thus we can define the confidence C_D for a metric score based on a given perceived quality D as $C_D := |Vmax(D) - Vmin(D)|$. A confidence score over the full perceived quality range can be given as

$$C = \frac{1}{\#S} \sum_{D \in S} C_D,$$

where *S* is the set of distinct MOS samples from the database.

Also note that we can interpret C as the average over C_D, $\mu_{D \in S}(C_D)$, and consequently we can also calculate the standard deviation, $\sigma_{D \in S}(C_D)$. The reason for calculating is to estimate how stable the confidence range is over the whole range of visual quality. This has to be taken into account since it is well known that image metrics exhibit different correlation to human judgement depending on the quality range, e.g.,[13].

For security metrics, and image metrics in general, the lower $\mu(C_D)$ and $\sigma(C_D)$ the better Algorithm 1 can be used to estimate a target image quality metric score.

Furthermore, since the signal is reduced to statistical components it is also of interest which shape the signal takes in conjunction with $\mu(C_D)$ and $\sigma(C_D)$. The shape, together with the monotonicity (see Section 3.3), can be used to indicate a possible application scenario for a security metric, essentially whether the security metric can be used for all quality ranges or only on high/low quality applications.

By shape of the signal we mean the distribution of outliers, where we define outlier based on the *z* score [1] of a data point D as

$$z_D = \frac{C_D - \mu(C_D)}{\sigma(C_D)}.$$

we will define *high outliers* as outliers with $z_D < -1$ and likewise *low outliers* as outliers with $z_D > 1$, indicating a higher and lower confidence respectively. Based on the distribution of high and low outliers we can specify the shape of the signal as follows.

- A signal is *stable* if there are no outliers. That is, if $-1 \geq z_D \geq 1$ holds for all $D \in S$.
- A signal is *biased* if it consist of two clearly separable parts, where one exhibits a good confidence score (C_D small) and the other a bad confidence score (C_D large). We say the signal is biased towards the quality where the good confidence score is located, i.e., where the metric is performing well.

That is, A signal is biased if there exists a *Dt* such that $z_D < -1 \Rightarrow D < D_t$ and $z_D > 1 \Rightarrow D > D_t$ or $z_D < -1 \Rightarrow D > D_t$ and $z_D > 1 \Rightarrow D < D_t$. If a low *D* indicates a high quality we specify the shape to be *biased towards high quality* if $z_D < -1 \Rightarrow D < D_t$ or *biased towards low quality* if $z_D > 1 \Rightarrow D < D_t$. If a low *D* indicates low quality the definition is switched accordingly.

- A signal which is neither stable nor biased is considered *unstable*.

Correspondence to HVS: Monotonicity

What is required from image metrics in general is monotonicity with regard to human observations. That is, if an image metric decides that image A is of better quality than image B a human observer should also prefer image A over image B. This is akin to correlation but since the human visual system is not a linear system regular linear correlation is meaningless. Thus in order to ascertain the correlation of an image metric and human observations the notion of monotonicity is utilized. Rank order correlation, which essentially judges the monotonicity of the signals, is most often used, usually in the form of Spearman's rank order coefficient (SROC) [33] or Kendall Tau ( ) [34].

Hofbauer and Uhl [13] pointed out that the correlation of an image metric over the full quality range does not imply that a high correlation is achieved for the low quality range. This is especially important for security metrics since certain application scenarios specifically target the low quality range of images, e.g., sufficient encryption. We cannot confine the evaluation to the low quality range since there are also applications for higher quality, i.e., transparent encryption. Also note that this is a dual property to the confidence in the sense that for the confidence we evaluate the relation of choosing a MOS value and evaluating the range of metric scores which can potentially fall onto this MOS value. Monotonicity is evaluated on specific sets of impairment and looks at how well an increase in metrics score reflects an increase in the MOS.

To properly evaluate security metrics for all encryption scenarios we will evaluate them using a high quality, a low quality and a full quality range dataset. The reason to also include the high quality range is to be inclusive in terms of possible application scenarios. For example the upgrade to high definition quality from PAL/NTSC quality is just as valid in terms of application scenarios as from hand held quality to PAL/NTSC quality. As basis for the evaluation we will use well known databases which contain mean opinion score of human judgement over different impairments. The SROC will be used for evaluation purpose, as is current best practice for metric evaluation.

From security metrics we would expect a high correlation with human judgement for the low quality range. While the low quality range is often the target of encryption some transparent encryption schemes could target a higher quality, consequently, a good correlation with human judgement on the high quality range is also desirable.

EVALUATION

In this section we present the results of the evaluation process as detailed in Section 3. Each evaluation contains a short description of the test data, the actual data from the evaluation and a discussion. We remark that the discussion is focussed on the evaluation methodology rather than the details of why and how a metric fails a test. Discussing the latter would require an in-depth description of how the metrics work, together with the explicit details about the test set which is out of scope of this paper.

With respect to implementations, we used our own code[2] for LSS, ESS, LE, NSD, LFBVS, SSIM and PSNR. For VIF, we used the implementation from the "MeTriX MuX Visual Quality Assessment Package",[3] version 1.1. For CPA1, we used the MATLAB implementation provided by Florent Autrusseau.[4]

Evaluation of the Application Domain

In order to evaluate the extraction versus encrypted domain applicability of metrics, we used a number of standard sequences: akiyo, bus, coastguard, container, flower, foreman, mobile, news, silent, tempete and waterfall.[5] The ordering, as discussed inSection 3.1 above is performed on a frame by frame basis and averaged over all the frames in a given sequence. Additionally we provide the average over all sequences in order to simplify the comparison. Table 1 shows the results for the encryption and extraction domain. The optimal result would be for a metric to perform equally well independent of the application domain.

Table 1: Results for the ordering given by Eq. (1) (Section 3.1) of high versus medium and medium versus low quality sequences in the encryption and extraction domain the averages over all the frames in a sequence as well as the average over all sequences is shown per image metric.

	LEG	SSIM	LSS	ESS	PSNR	LFBVS	LE	NSD	VIF	CPA1
Results for the encryption domain (*QM*)										
akiyo	0.168	0.500	0.496	0.441	0.527	0.500	0.609	0.481	0.449	0.496
bus	0.398	0.383	0.402	0.332	0.555	0.516	0.895	0.426	0.383	0.250
coastguard	0.500	0.391	0.340	0.434	0.582	0.590	0.754	0.422	0.508	0.254
container	0.246	0.484	0.207	0.391	0.348	0.500	0.777	0.563	0.586	0.035
flower	0.191	0.430	0.527	0.410	0.445	0.488	0.996	0.398	0.488	0.395
foreman	0.289	0.481	0.492	0.410	0.508	0.504	0.547	0.453	0.500	0.481
mobile	0.340	0.481	0.500	0.305	0.457	0.512	0.688	0.492	0.453	0.473
news	0.430	0.500	0.395	0.418	0.742	0.500	0.500	0.457	0.363	0.496
silent	0.270	0.332	0.242	0.102	0.250	0.602	0.992	0.375	0.320	0.148
tempete	0.414	0.481	0.500	0.395	0.500	0.500	0.742	0.481	0.449	0.465
waterfall	0.422	0.492	0.488	0.461	0.715	0.508	0.539	0.512	0.492	0.496
average	0.334	0.450	0.417	0.373	0.512	0.520	0.731	0.460	0.454	0.363
Results for the extraction domain (*QM*)										
akiyo	1.000	1.000	0.516	0.941	1.000	0.500	0.500	0.500	0.945	0.094
bus	0.996	1.000	0.512	0.969	0.992	0.500	0.500	0.461	0.981	0.234
coastguard	1.000	1.000	0.508	0.992	0.996	0.457	0.500	0.246	1.000	0.008
container	1.000	1.000	1.000	0.941	0.000	0.500	0.500	0.500	1.000	0.004
flower	1.000	1.000	0.953	0.926	0.063	0.418	0.500	0.473	0.984	0.094
foreman	1.000	1.000	0.531	0.996	0.500	0.500	0.500	0.500	0.996	0.000
mobile	0.984	1.000	0.922	0.977	0.434	0.500	0.500	0.410	0.984	0.020
news	1.000	1.000	0.500	0.996	1.000	0.500	0.500	0.500	0.996	0.000
silent	1.000	1.000	0.902	0.981	0.578	0.500	0.500	0.500	1.000	0.000
tempete	1.000	1.000	0.570	0.992	1.000	0.500	0.500	0.481	1.000	0.000
waterfall	0.957	0.992	0.695	0.828	0.930	0.500	0.500	0.246	0.981	0.008
average	0.994	0.999	0.692	0.958	0.681	0.489	0.500	0.438	0.988	0.042

From the overall averages it can be clearly seen that the performance in the encrypted domain is worse than in the extraction domain with the exception of LE. While LE performs better in the encrypted domain than in the extraction domain, the performance is still very low. In the extraction domain, most metrics still perform poorly; only LEG, SSIM, VIF, CPA1 and, to a lesser degree, ESS exhibit good performance. Aside from the latter, all other security metrics exhibit extremely poor performance.

This allows to conclude that *application in the encrypted domain should not be performed(although, it is routinely done in all corresponding papers)*.

When looking closer at the detail information from the extraction domain, some interesting effects can be observed. While the performance of LEG, SSIM and VIF are consistently good, we notice that there are cases where even a metric that shows good performance can have problems. In the actual case the performance of the ESS is significantly reduced for the waterfall sequence and the CPA1 shows poor performance on the bus sequence. It stands to reason that there are other, untested, sequences which would lead to similar reduced performance for the LEG, SSIM and VIF.

Evaluation of Confidence

In order to evaluate the confidence, databases with either pertinent content, i.e., encrypted images, or a large dataset with distortions are optimal. The IVC-SelectEncrypt database [21] contains various instances of JPEG 2000 transparent encryption, using different encryption techniques, and is a useful tool for evaluating confidence with respect to encrypted images. It is the only database available containing encrypted visual data and corresponding MOS. The test sets contained in the IVC-SelectEncrypt database (and their abbreviation) are traditional encryption (trad), truncation of the code stream (trunc), window encryption without error concealment (iwind_nec), window encryption with error concealment (iwind_ec), and wavelet packet encryption (res), for detailed information see [35]. However, the IVC-SelectEncrypt database has a rather small set of impairments, i.e., five per test set, and is focused on JPEG 2000 only.

In order to get a more diverse view on the confidence of metrics we additionally utilize the LIVE database [36] to supplement the IVC-SelectEncrypt database of encrypted images. While the LIVE database does not contain encrypted images, the quality range of the images reaches from high to low quality; this makes it at least relevant for transparent encryption where a certain target quality is required. Furthermore, images in the low quality range of the LIVE database exhibit strong distortions which can be equated to encrypted images in the sense that strong distortions mask a lot of the visual information. Consequently, the distorted images can be used to assess how well a metric can identify information contained in a distorted/encrypted image; this is exactly the property that we want from security metrics. An example of these strong distortions is shown in Fig. 12which contains an encrypted image from the IVC-SelectEncrypt database as well as heavily distorted versions of images from the LIVE database. These examples illustrate that the LIVE database contains not only images which are similar to the IVC-SelectEncrypt database in terms of content masking, but also images which are clearly in the quality realm of sufficient encryption. The test sets contained in the LIVE database (and their abbreviation in plots and figures) are JPEG 2000 compression (jp2k), JPEG compression (jpeg), white noise (wn), Gaussian blur (gblur), and bit errors in JPEG2000 bit stream transmission over a simulated fast fading Rayleigh Channel (fastfading), for detailed information see [37].

Figure. 12: The lowest quality images of each test set in the LIVE database as well as one of the lowest quality images from the IVC-SelectEncrypt database.

What we tried to do is to break it down the complex relation of the confidentiality as far as possible, to facilitate comparison, without losing too much information. The result is that the measures μ and σ combined with the signal shape actually contain a lot of information. The purpose of the detailed discussion in this section is to relate to the reader the implications carried by the 'scores' μ, σ and signal shape.

Fig. 13 shows the detailed evaluation of confidence on the LIVE database. For each metric, the figure shows a scatter plot of MOS and metric values, the bounding curves$V_{min}(D)$ and $V_{max}(D)$ as well as a plot of the local confidence value C_D. Table 2 lists the confidence scores $\mu(C_D)$ and $\sigma(C_D)$ for each metric on the LIVE database. In order to make the confidence scores comparable a pseudo normalization was used. Bounded metrics are normalized and unbounded metrics, e.g., PSNR, are normalized by mapping the range of occurring metric scores into $[0,1]$. For the calculation of the signal shape the leading and trailing 10% of the MOS range where not taken into account because, at the high and low ends of the MOS, the difference in metric scores is limited due the boundary of the metric range, see Fig. 13. This would result in false outliers. The shape was calculated, similar to $\mu(C_D)$ and $\sigma(C_D)$, on pseudo normalized local confidence value C_D. In Fig. 13, outliers are indicated by down and up arrows for high and low quality outliers, respectively.

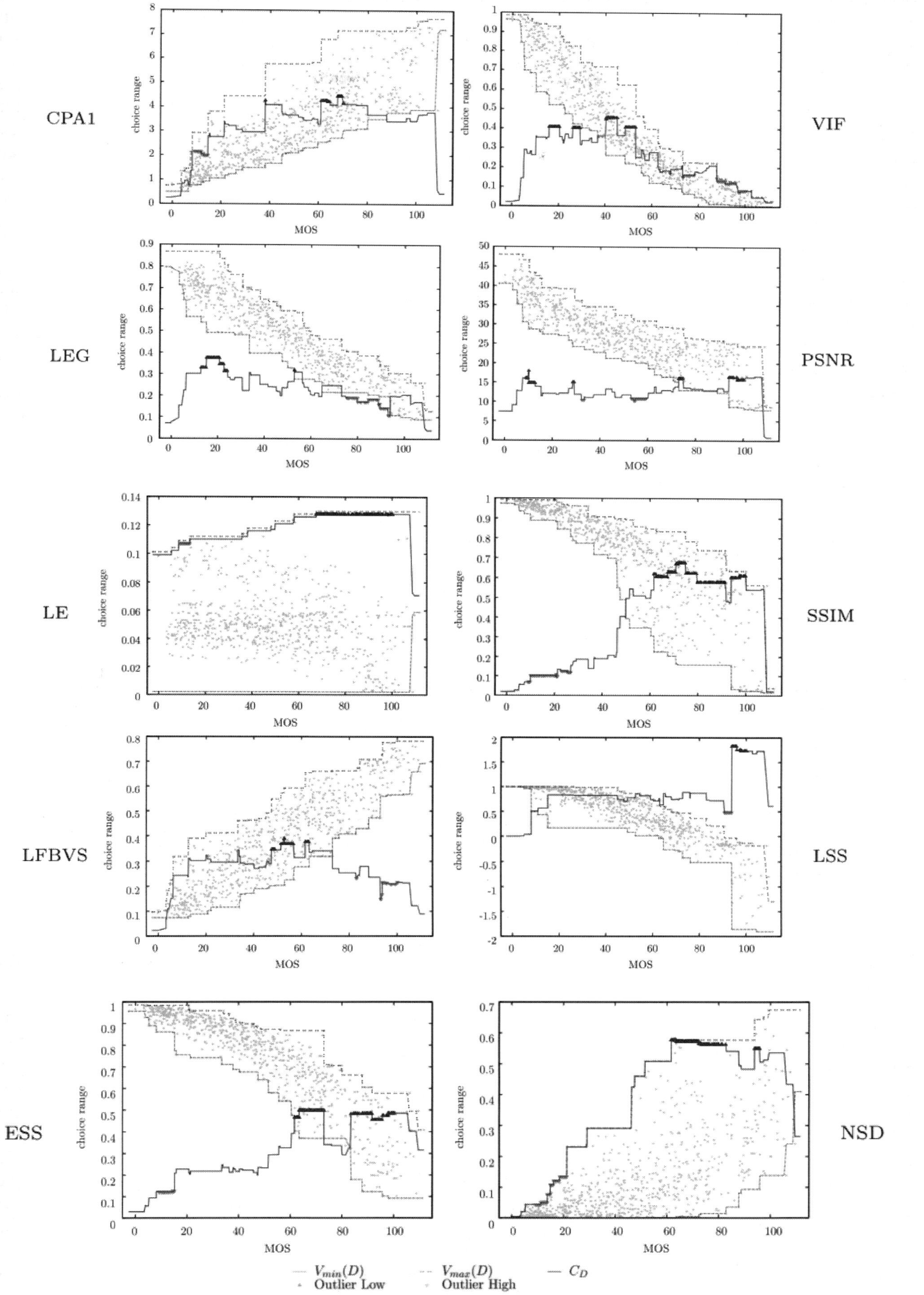

Figure. 13: Confidence plots for the LIVE database for different image metrics. The plot shows the scatter plot for the MOS and metric score pairs, the plot of *Vmin(D)*, *Vmax(D)* and C_D.

Table 2: Average and standard deviation of normalized confidence and signal shape on the LIVE database

	SSIM	LEG	VIF	CPA1	LSS	ESS	LFBVS	LE	NSD	PSNR
$\mu(C_D)$	0.357	0.291	0.285	0.431	0.415	0.300	0.370	0.906	0.537	0.265
$\sigma(C_D)$	0.225	0.070	0.110	0.109	0.159	0.133	0.071	0.069	0.277	0.038
Signal shape	Bias high	Bias low	Bias low	Bias high	Bias high	Bias high	Bias low	Bias high	Bias high	Unstable

For the IVC-SelectEncrypt database the accordant plot and confidence scores are given in Fig. 14 and Table 3, respectively. Notice that while a MOS score of 0 is high quality on the LIVE database, a MOS score of 0 represents low quality on the IVC-SelectEncrypt database.

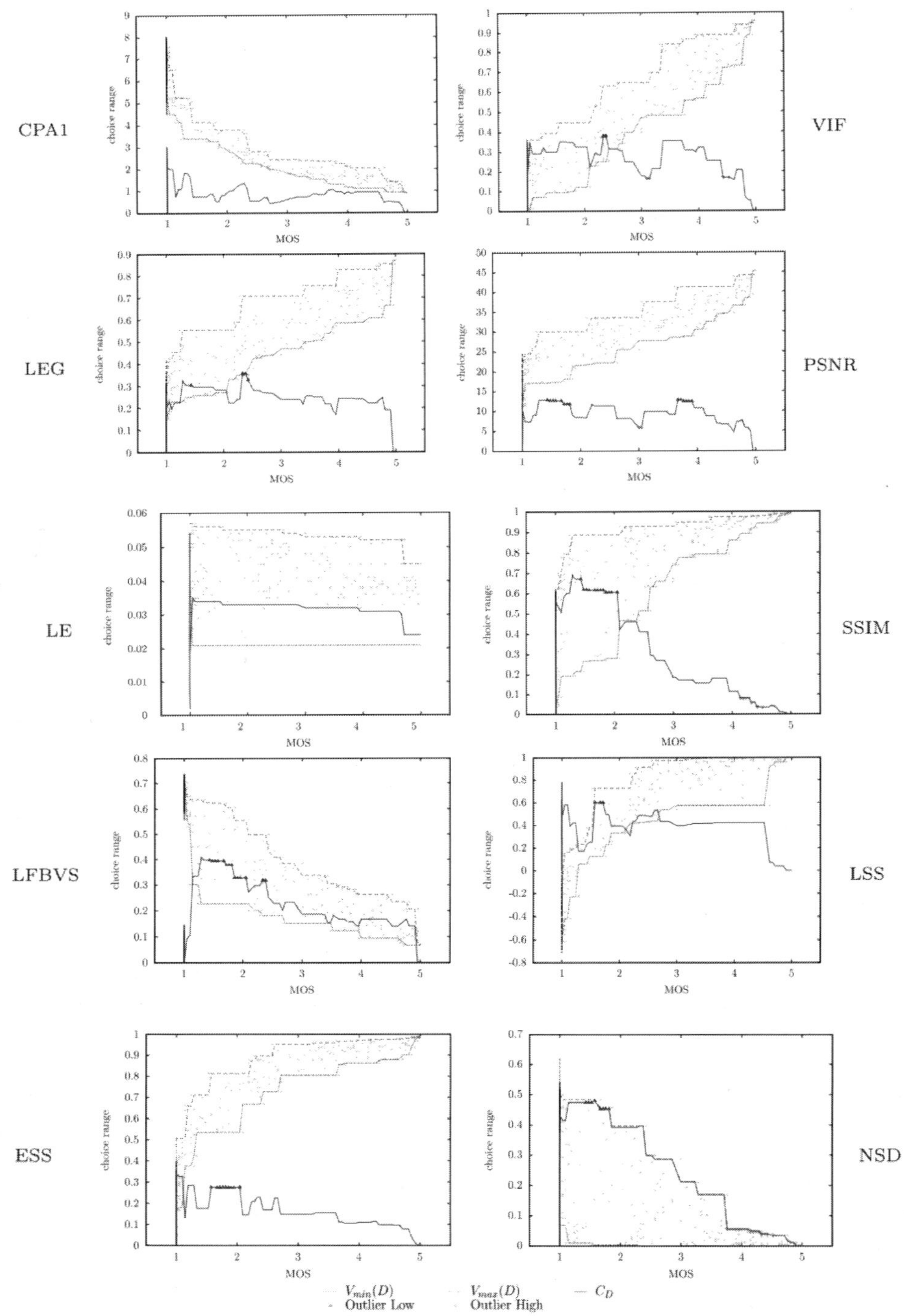

Figure. 14: Confidence plots for the IVC-SelectEncrypt database for different image metrics. The plot shows the scatter plot for the MOS and metric score pairs, the plot of *Vmin(D)*, *Vmax(D*) and C_D.

Table 3: Average and standard deviation of normalized confidence and signal shape on the IVC-SelectEncrypt database

	SSIM	LEG	VIF	CPA1	LSS	ESS	LFBVS	LE	NSD	PSNR
$\mu(C_D)$	0.319	0.268	0.277	0.119	0.374	0.168	0.273	0.540	0.394	0.196
$\sigma(C_D)$	0.226	0.077	0.098	0.066	0.173	0.090	0.139	0.107	0.274	0.063
Signal shape	Bias high	Bias high	Bias high	Stable	Bias high	Bias high	Bias high	Stable	Bias high	Unstable

When it comes to confidence we can safely state that none of the metrics show good performance overall, i.e., none of the metrics are stable with a low $\mu(C_D)$ and $\sigma(C_D)$. However, for certain test sets there are metrics with good performance, e.g., CPA1 on the IVC-SelectEncrypt test set shows exceptionally high confidence and is stable. The CPA1 metric on the LIVE database, however, shows extremely poor performance.

Let us consider the relation between shape, $\mu(C_D)$ and $\sigma(C_D)$. The notional average performance is given by $\mu(C_D)$. The amount of deviation from the notional average is given by $\sigma(C_D)$. The way the deviation is distributed is given by the signal shape. Contrariwise, $\sigma(C_D)$ indicates the magnitude of the shape, i.e., assuming constant$\mu(C_D)$ a biased signal will be biased to a higher degree if $\sigma(C_D)$ is higher.

Compare LEG and SSIM as an example: While LEG shows better overall $\mu(C_D)$ and$\sigma(C_D)$, SSIM is much more biased than LEG, i.e., SSIM outperforms LEG where its bias is. Contrary to that, LEG outperforms SSIM outside of the bias. This behavior is only identifiable when $\sigma(C_D)$ is considered in conjunction with shape. In particular, even though the LEG is biased, due to the small $\sigma(C_D)$ we can deduce that the bias is far smaller than the bias of the SSIM (which exhibits a high $\sigma(C_D)$).

A similar behavior can be observed for unstable shapes. A metric with an unstable shape and a high $\sigma(C_D)$ will have much more severe outliers than one with a low $\sigma(C_D)$. This is nicely reflected in the PSNR: Although unstable, the magnitude of the outliers should be relatively small since the PSNR shows a low $\sigma(C_D)$. This is exactly what we see in the plots of Fig. 13 and Fig. 14.

For a stable shape, $\mu(C_D)$ becomes more important since it shows where the stable part of the confidence lies. The LE metric on the IVC-SelectEncrypt database is a prime example of this: While it is stable, the actual confidence score shows that it is stable in the sense that it exhibits poor performance over the whole quality range.

Regarding confidence values and shape, it can be seen from Table 2 and Table 3 that can exhibit non-uniform behavior over different test sets. If this is the case, when calculating the overall performance scores, the worst value should be taken into account.

What is also noticeable from the two tables is the fact that the evaluated image and security metrics are more often biased towards the high quality range. Indeed, on the IVC-SelectEncrypt database which is the actual encryption database, not a single metric is biased towards the low quality range. Furthermore, the metrics biased towards the low quality range on the LIVE database, i.e., LEG, VIF and LFBVS, are all biased towards high quality on the IVC-SelectEncrypt database, and should thus be considered unstable overall.

To sum up the findings regarding the confidence of the metrics we can state the following: *First*, LE, LSS and CPA1 show extremely poor performance overall. While CPA1 performs exceptionally well on the IVC-SelectEncrypt database, it performs poorly on the LIVE database; consequently, the overall performance of CPA1 is not good.*Second* , LEG, VIF, ESS, LFBVS and PSNR – while not stable – exhibit a low $\sigma(C_D)$and are thus closest to being considered good metrics over the full quality range. However, in each case the $\mu(C_D)$ is relatively high overall (at least from a security standpoint) and thus could use some improvement, or replacement. *Third*, SSIM and NSD show a strong bias towards the high quality range. While the confidence for these metrics is not good overall, the confidence over the high quality range is actually quite good. Consequently, when the application scenario is known to target the higher quality range, e.g., in transparent encryption, these metrics should be considered.

Monotonicity for Low Quality Images

The monotonicity of an image metric over the MOS for the full quality range is what defines the quality of an image metric. However, for transparent/sufficient encryption it is more important that a metric has good monotonicity properties on the lower quality range than on the full quality range. It is well known, c.f. [13], that this can be a problem so it is necessary to study this in more detail.

The absolute SROC values for the test sets of the LIVE database are given in Table 4(a) for the full quality range, the low quality range and the high quality range. In the table,*high* means SROC >0.9 (marked bold); a SROC score of <0.5 means *unsatisfactory*(underlined). For the IVC-SelectEncrypt database the same information is listed in Table 4(b).

Table 4: Spearman's rank order correlation (SROC) for full, high and low quality range

(a) SROC for the LIVE database

	fast-fading	gblur	jp2k	jpeg	wn
Full quality range					
SSIM	0.942	0.903	0.936	0.946	0.962
LEG	0.971	0.966	0.945	0.960	0.960
VIF	0.965	0.972	0.968	0.984	0.985
CPA1	0.881	0.927	0.958	0.962	0.984
LSS	0.843	0.916	0.953	0.970	0.965
ESS	0.933	0.888	0.929	0.949	0.886
PSNR	0.891	0.782	0.895	0.881	0.985
LFBVS	0.932	0.937	0.853	0.920	0.891

(a) SROC for the LIVE database

	fast-fading	gblur	jp2k	jpeg	wn
LE	0.241	0.236	0.076	0.766	0.228
NSD	0.815	0.803	0.741	0.806	0.657
High quality range					
SSIM	0.517	0.815	0.710	0.879	0.875
LEG	0.652	0.792	0.638	0.761	0.734
VIF	0.632	0.782	0.797	0.920	0.891
CPA1	0.681	0.689	0.761	0.889	0.840
LSS	0.504	0.762	0.754	0.825	0.700
ESS	0.473	0.618	0.694	0.744	0.827
PSNR	0.517	0.662	0.676	0.801	0.869
LFBVS	0.289	0.707	0.393	0.651	0.572
LE	0.295	0.127	0.228	0.136	0.143
NSD	0.028	0.760	0.275	0.370	0.674
Low quality range					
SSIM	0.662	0.487	0.635	0.339	0.802
LEG	0.893	0.872	0.617	0.699	0.804
VIF	0.937	0.920	0.646	0.829	0.911
CPA1	0.614	0.669	0.657	0.402	0.897
LSS	0.788	0.732	0.647	0.595	0.817
ESS	0.877	0.847	0.397	0.735	0.570
PSNR	0.611	0.408	0.510	0.046	0.897
LFBVS	0.863	0.849	0.391	0.702	0.659
LE	0.268	0.338	0.070	0.720	0.290
NSD	0.731	0.582	0.427	0.668	0.231

(b) SROC for the IVC-SelectEncryptdatabase

	iwindec	iwindnec	resolution	tradition	truncation
Full quality range					
SSIM	0.925	0.954	0.887	0.968	0.879
LEG	0.869	0.956	0.876	0.966	0.863
VIF	0.937	0.969	0.767	0.982	0.954
CPA1	0.925	0.953	0.906	0.967	0.959
LSS	0.888	0.907	0.955	0.948	0.948
ESS	0.868	0.894	0.933	0.906	0.860
PSNR	0.909	0.910	0.916	0.965	0.889
LFBVS	0.756	0.930	0.916	0.943	0.878
LE	0.136	0.579	0.562	0.550	0.201
NSD	0.698	0.922	0.790	0.757	0.523
High quality range					
SSIM	0.652	0.863	0.714	0.975	0.835
LEG	0.713	0.852	0.643	0.920	0.610
VIF	0.729	0.907	0.143	0.885	0.662
CPA1	0.683	0.890	0.000	0.868	0.934
LSS	0.705	0.945	0.393	0.865	0.975
ESS	0.809	0.846	0.571	0.679	0.794

(b) SROC for the IVC-SelectEncryptdatabase

	iwindec	iwindnec	resolution	tradition	truncation
PSNR	0.636	0.857	0.500	0.879	0.830
LFBVS	0.521	0.863	0.250	0.698	0.723
LE	0.095	0.489	0.464	0.431	0.019
NSD	0.251	0.736	0.179	0.236	0.602
Low quality range					
SSIM	0.191	0.694	0.644	0.680	0.695
LEG	0.141	0.823	0.490	0.652	0.181
VIF	0.518	0.732	0.823	0.913	0.832
CPA1	0.400	0.628	0.897	0.592	0.706
LSS	0.523	0.240	0.875	0.386	0.679
ESS	0.422	0.411	0.551	0.609	0.549
PSNR	0.251	0.474	0.688	0.663	0.541
LFBVS	0.188	0.562	0.507	0.416	0.022
LE	0.386	0.004	0.152	0.166	0.272
NSD	0.100	0.655	0.290	0.589	0.291

In order to better compare the difference in high, low and overall, rank order correlation for a given test set and database is illustrated in Table 5 for the LIVE and IVC-SelectEncrypt database. For each combination of test set and metric, the graphical entry displays the range of possible SROC scores as background, with SROC=0 at the bottom and SROC=1 at the top. The light gray background bar shows the SROC value for the full quality range while the smaller bars indicate the SROC score for the high quality (left) and low quality range (right). What can be directly seen is that the overall performance of a metric does not imply good performance for either the high or low quality range, although most metrics tend to perform better on the high quality range. In some cases, a metric can even perform better on a limited quality range than over the whole range of quality. For example, the VIF performs better for the low quality end of resolution encryption on the IVC-SelectEncrypt database than for the whole range of resolution. In other cases, the performance over the full quality range is drastically reduced over the high and low parts of the quality range. For example, the VIF performs poorly on the high quality range for resolution encryption on the IVC-SelectEncrypt database. For other test sets, the impact of either low or high quality is minor and most likely due to the reduced number of samples from the database, e.g., the VIF performs well for white noise distortion (from the LIVE database) irrespective of the quality range. This in essence shows that the actual performance of an image metric is dependent on the distortion type as well as the quality range. For most test sets and image metric combinations, the SROC over the full quality range can at most be used as an upper bound for the limited quality cases. However, there are exceptions, e.g., VIF for the low quality range of the resolution test set performs better than on the overall quality range for the same test set.

Table 5: Visual representation of Spearman's rank order correlation ($|SROC| \in [0,1]$) for the LIVE Image Quality Assessment and IVC-SelectEncrypt databases for full quality range (light gray), low quality range (orange bar on the right) and high quality range (green bar on the left side)

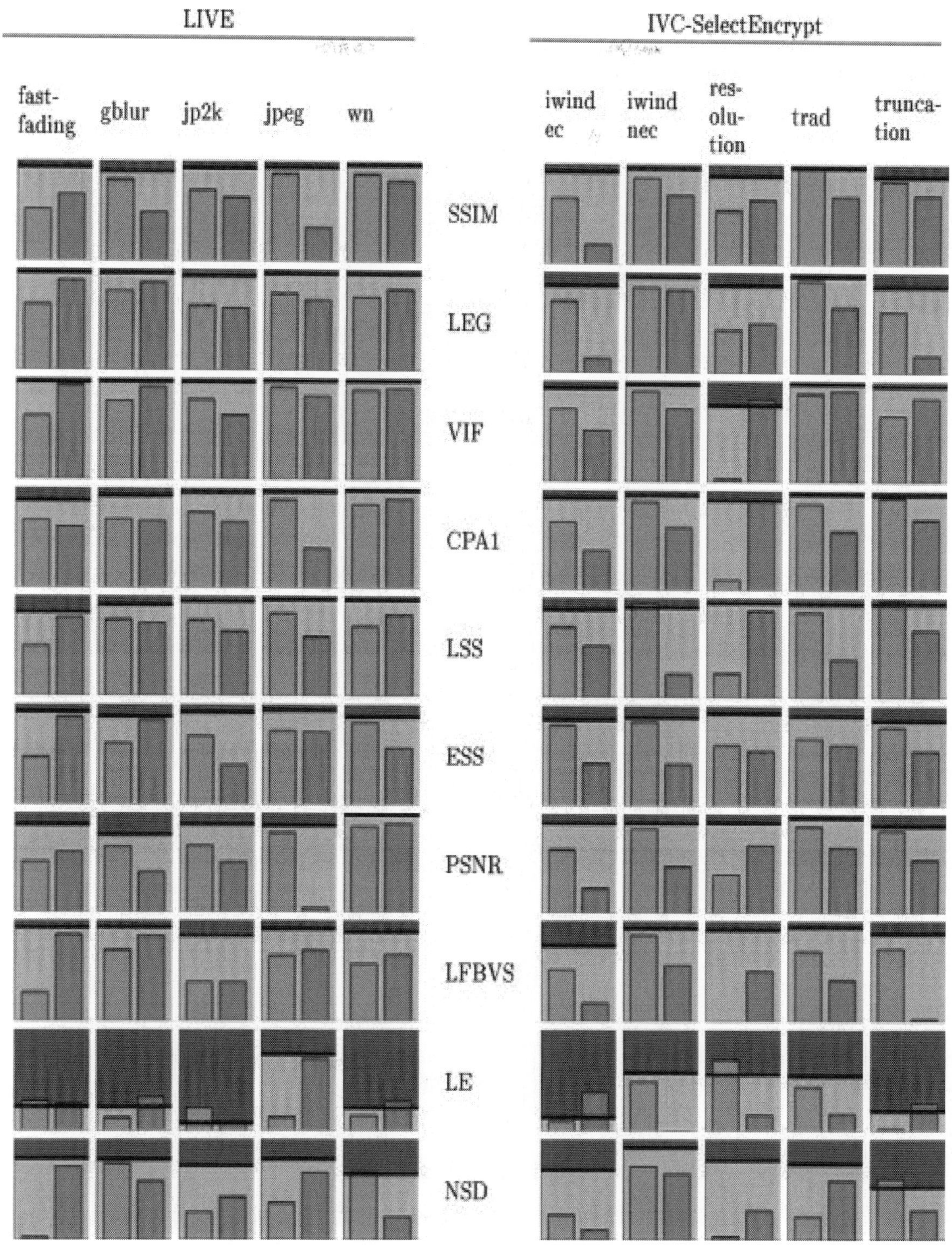

With regard to security metrics and their performance on the low quality range, there is no noticeable difference in behavior to regular image metrics. Interestingly, though, for each test set the best performance over the low quality range can be found among the traditional image metrics.

Comparing the high and low quality ranges, we can see that the low quality range has a far higher number of unsatisfactory SROC scores. This shows that image metrics tend to perform better at differentiating the different strength in distortion for high quality images. A high SROC score over the whole quality range indicates that the metric can differentiate between high and low quality images. In essence, image metrics which perform well on the overall quality range can still be utilized to identify sufficient encryption, even though a metric which exhibits good performance in the range of the quality threshold should be preferred. For transparent encryption, where the goal is to find the best image below a certain threshold, monotonicity in the chosen quality range becomes more important. In particular, the target quality is a lot closer to the threshold, thus a high monotonicity (expressed by a high SROC) is required.

Another interesting aspect of the high versus low quality test is the fact that none of the tested metrics exhibits a better performance on the high or low quality range for all test sets. Consequently, the metrics cannot be reduced to a single SROC score and a bias towards either high or low. There are cases where the performance over both high and low quality is far worse than for the whole quality range, e.g., LFBVS on the jp2k test set. Thus, in order to evaluate whether an image metric can be used as a security metric, tests regarding low and high quality performance have to be conducted.

Summing up the monotonicity tests, we can state the following: *First*, VIF, CPA1, SSIM, LSS and LEG perform best over the whole quality range. LFBVS, PSNR and ESS also show a good behavior, while NSD and especially LE perform poorly.

Second, for the high quality range, most quality metrics still show a decent performance. However, only SSIM, LEG, and PSNR exhibit no unsatisfactory performance in a single test set.

Third, for the low quality range, only VIF exhibits good performance over all test sets. All other metrics have at least two test sets where their performance is unsatisfactory.

CONCLUSION AND FUTURE WORK

We have outlined an evaluation method for security metrics based on practical application scenarios and considerations. These methods were used to evaluate (1) state-of-the-art security metrics, (2) image metrics which are used as security metrics as well as (3) state-of-the-art image metrics. A summary of the evaluation and a basic evaluation score is given in Table 6. To simplify and give a clear comparison between metrics, we assign a single score to each metric based on its performance in the various evaluation steps. In particular, for each step, we assign a score of 1 for desired behavior and a penalty of −1 for a clear failure. The final score is the sum of the individual test scores. In Table 6, desired behavior is indicated in bold and failure in italics.

Table 6: Summary of evaluation

	SSIM	LEG	VIF	CPA1	LSS	ESS	PSNR	LFBVS	LE	NSD
Application domain										
Encryption	*0.450*	0.334	*0.454*	0.363	*0.417*	0.373	*0.512*	*0.520*	0.731	*0.460*
Extraction	**0.999**	**0.994**	**0.988**	**0.042**	0.692	**0.958**	0.681	*0.489*	*0.500*	*0.438*
Confidence on the LIVE database										
$\mu(C_D)$	0.357	**0.291**	**0.285**	0.431	0.415	0.300	**0.265**	0.370	*0.906*	*0.537*
$\sigma(C_D)$	0.225	**0.070**	0.110	0.109	0.159	0.133	**0.038**	**0.071**	**0.069**	*0.277*
Signal shape	Bias high	Bias low	Bias low	Bias high	Bias high	Bias high	Bias low	Bias high	Bias high	*Unstable*
Confidence on the IVC-SelectEncrypt database										
$\mu(C_D)$	0.319	**0.268**	**0.277**	**0.119**	0.374	**0.168**	**0.196**	**0.273**	*0.540*	0.394
$\sigma(C_D)$	0.226	**0.077**	**0.098**	**0.066**	0.173	**0.090**	**0.063**	0.139	0.107	*0.274*
Signal shape	Bias high	Bias high	Bias high	Stable	Bias high	Bias high	Bias high	Stable	Bias high	*Unstable*
Low quality SROC on the LIVE database										
fastfading	0.662	0.893	**0.937**	0.614	0.788	0.877	0.611	0.863	*0.268*	0.731
gblur	*0.487*	0.872	**0.920**	0.669	0.732	0.847	*0.408*	0.849	*0.338*	0.582
jp2k	0.635	0.617	0.646	0.657	0.647	*0.397*	0.510	*0.391*	*0.070*	*0.427*
jpeg	*0.339*	0.699	0.829	*0.402*	0.595	0.735	*0.046*	0.702	0.720	0.668
wn	0.802	0.804	**0.911**	0.897	0.817	0.570	0.897	0.659	*0.290*	*0.231*
Low quality SROC on the IVC-SelectEncrypt database										
iwind ec	*0.191*	*0.141*	0.518	*0.400*	0.523	*0.422*	*0.251*	*0.188*	0.386	*0.100*

	SSIM	LEG	VIF	CPA1	LSS	ESS	PSNR	LFBVS	LE	NSD
iwind nec	0.694	0.823	0.732	0.628	*0.240*	*0.411*	*0.474*	0.562	*0.004*	0.655
resolution	0.644	*0.490*	0.823	0.897	0.875	0.551	0.688	0.507	*0.152*	*0.290*
trad	0.680	0.652	**0.913**	0.592	*0.386*	0.609	0.663	*0.416*	*0.166*	0.589
truncation	0.695	*0.181*	0.832	0.706	0.679	0.549	0.541	*0.022*	*0.272*	*0.291*
Comparison score, −1 or +1 for insufficient or good performance, −1 for conflict in signal shape										
Score	*−3*	1	**6**	0	*−3*	0	*−2*	*−5*	*−11*	*−12*

For the *application domain* a sorting error of less than 10% is considered good, while a performance around 50% (±10%) is considered a failure. For the *confidence*, a score for $\mu\pm$ of 0.3±0.1 is considered good, while a result of 0.5±0.25 is considered a failure. Regarding signal shape, we desire stable signals. The signal shape will also be penalized if it does not agree on the two test sets. For the *quality and SROC* , a value ≤ 0.5 is a failure while an SROC greater than 0.9 is desirable. From the final scores inTable 6, we conclude that none of the security metrics are fit to perform as general purpose security metrics.

Regarding transparent and sufficient encryption, the LE and NSD metrics especially show that metrics engineered to fit a certain application scenario cannot claim generality. Furthermore, SSIM and PSNR, which are frequently used as security metrics, also perform poorly. Most state-of-the-art image metrics hardly perform the security metric task adequately. Only VIF, apart from a borderline confidence score and stability, demonstrates good performance.

Regarding content confidentiality we cannot make a strong statement due to lack of ground truth for recognizability tests. However, the performance in the encrypted domain, during the evaluation of the application domain, gives a strong indication that none of the tested image metrics can perform the task of evaluating the content confidentiality.

The presented methodology for the evaluation of security metrics should be used to asses newly developed metrics to prevent similar shortcomings as those shown by the investigated metrics. Table 6 gives a concise overview of the state-of-the-art metrics, and can be used as a basis for comparison for new image and security metrics.

Future Work

The inability to properly evaluate image metrics with regard to content confidentiality naturally leads to the conclusion that more data is required. This also holds true for the lower quality ranges with respect to regular metrics, which would undoubtedly benefit from a dataset that is specifically designed for high impairment cases. In future work, we will gather ground truth data for content confidentiality (and will also design protocols how to properly capture human assessment for these data sets) and extend the work presented in this paper to properly encompass content confidentiality. Furthermore, we intend to gather more data on the low quality range as well, in order to better evaluate security metrics for transparent and sufficient encryption.

ACKNOWLEDGMENTS

This work was partially supported by the Austrian Science Fund, project no. P27776.

REFERENCES

1. S.-K. Au Yeung, S. Zhu, B. Zeng, Quality assessment for a perceptual video encryption system, in: 2010 IEEE International Conference on Wireless Communications, Networking and Information Security (WCNIS), 2010, pp. 102– 106, http://dx.doi.org/10.1109/WCINS.2010.5541898.
2. M. Khan, V. Jeoti, A. Malik, Perceptual encryption of JPEG compressed images using DCT coefficients and splitting of DC coefficients into bitplanes, in: 2010 International Conference on Intelligent and Advanced Systems (ICIAS), 2010, pp. 1–6.
3. E.J. Wharton, K. Panetta, S. Agaian, Simultaneous encryption/compression of images using alpha rooting, in: 2008 Data Compression Conference (DCC 2008), 25–27 March 2008, Snowbird, UT, USA, IEEE Computer Society, 2008, p. 551ff, https://www.computer.org/csdl/proceedings/dcc/2008/3121/00/ 3121a551-abs.html.
4. S. Lian, Efficient image or video encryption based on spatiotemporal chaos system, Chaos Solitons Fractals 40 (5) (2009) 2509–2519, http://dx.doi.org/ 10.1016/j.chaos.2007.10.054, URL ⟨http://www.sciencedirect.com/science/arti cle/pii/S0960077907009277⟩.
5. C.E. Shannon, Communication theory of secrecy systems, Bell Syst. Tech. J. 28 (1949) 656–715.
6. T.D. Lookabaugh, D.C. Sicker, Selective encryption for consumer applications, IEEE Commun. Mag. 42 (5) (2004) 124–129.
7. A. Said, Measuring the strength of partial encryption schemes, in: Proceedings of the IEEE International Conference on Image Processing (ICIP'05), vol. 2, 2005, pp. 1126–1129, http://dx.doi.org/10.1109/ICIP.2005.1530258.
8. M. Bellare, T. Ristenpart, P. Rogaway, T. Stegers, Format-preserving encryption, in: Proceedings of Selected Areas in Cryptography, SAC '09, vol. 5867, SpringerVerlag, Calgary, Canada, 2009, pp. 295–312.
9. T. Stütz, A. Uhl, A survey of H.264 AVC/SVC encryption, IEEE Trans. Circuits Syst. Video Technol. 22 (3) (2012) 325–339.

10. T. Stütz, A. Uhl, Efficient format-compliant encryption of regular languages: block-based cycle-walking, in: B.D. Decker, I. Schaumüller-Bichl (Eds.), Proceedings of the 11th Joint IFIP TC6 and TC11 Conference on Communications and Multimedia Security, CMS '10, IFIP Advances in Information and Communication Technology, vol. 6109, Springer, Linz, Austria, 2010, pp. 81–92.

11. Q. Li, I.J. Cox, Using perceptual models to improve fidelity and provide resistance to valumetric scaling for quantization index modulation watermarking, IEEE Trans. Inf. Forensics Secur. 2 (2) (2007) 127–139.

12. H. Hofbauer, A. Uhl, Selective encryption of the MC-EZBC bitstream and residual information, in: 18th European Signal Processing Conference, 2010 (EUSIPCO-2010), Aalborg, Denmark, 2010, pp. 2101–2105.

13. H. Hofbauer, A. Uhl, Visual quality indices and low quality images, in: IEEE 2nd European Workshop on Visual Information Processing, Paris, France, 2010, pp. 171–176.

14. Y. Zhou, K. Panetta, S. Aagaian, Partial multimedia encryption with different security levels, in: 2008 IEEE Conference on Technologies for Homeland Security, 2008, http://dx.doi.org/10.1.1.152.725, URL ⟨http://biron.usc.edu/ sagetong/resume.html⟩.

15. B. SaiChandana, S. Anuradha, A new visual cryptography scheme for color images, Int. J. Eng. Sci. Technol. 2 (6) (2010) 1998–2000.

16. L. Dubois, W. Puech, J. Blanc-Talon, Confidentiality metrics and smart selective encryption for HD H.264/AVC videos, in: 2013 Proceedings of the 21st European Signal Processing Conference (EUSIPCO), 2013, pp. 1–5.

17. L. Dubois, W. Puech, J. Blanc-Talon, Smart selective encryption of cavlc for h.264/avc video, in: 2011 IEEE International Workshop on Information Forensics and Security (WIFS), 2011, pp. 1–6, http://dx.doi.org/10.1109/WIFS.2011. 6123130.

18. W. Wang, D. Peng, H. Wang, H. Sharif, An adaptive approach for image encryption and secure transmission over multirate wireless sensor networks, Wirel. Commun. Mob. Comput. 9 (3) (2009) 383–393, http://dx.doi.org/10.1002/wcm.550.

19. Z. Shahid, W. Puech, Visual protection of HEVC video by selective encryption of CABAC binstrings, IEEE Trans. Multimed. 16 (1) (2014) 24–36, http://dx.doi.org/10.1109/TMM.2013.2281029.

20. Q. Huynh-Thu, M. Ghanbari, Scope of validity of PSNR in image/video quality assessment, Electron. Lett. 44 (13) (2008) 800–801.

21. F. Autrusseau, T. Stuetz, V. Pankajakshan, Subjective quality assessment of selective encryption techniques, ⟨http://www.irccyn.ec-nantes.fr/autrusse/ Databases/⟩, 2010.

22. Z. Wang, A. Bovik, H. Sheikh, E. Simoncelli, Image quality assessment: from error visibility to structural similarity, IEEE Trans. Image Process. 13 (4) (2004) 600–612.

23. Y. Mao, M. Wu, Security evaluation for communication-friendly encryption of multimedia, in: Proceedings of the IEEE International Conference on Image Processing (ICIP'04), IEEE Signal Processing Society, Singapore, 2004.

24. H. Hofbauer, A. Uhl, Selective encryption of the MC EZBC bitstream for DRM scenarios, in: Proceedings of the 11th ACM Workshop on Multimedia and Security, ACM, Princeton, New Jersey, USA, 2009, pp. 161–170.

25. H. Hofbauer, T. Stütz, A. Uhl, Selective encryption for hierarchical MPEG, in: H. Leitold, E. Markatos (Eds.), Communications and Multimedia Security, Proceedings of the 10th IFIP International CMS 2006 Conference, Lecture Notes on Computer Science, vol. 4237, Springer Verlag, Heraklion, Crete, 2006, pp. 151– 160.

26. Y. Yao, Z. Xu, J. Sun, Visual security assessment for cipher-images based on neighborhood similarity, Informatica 33 (2009) 69–76.

27. J. Sun, Z. Xu, J. Liu, Y. Yao, An objective visual security assessment for cipherimages based on local entropy, Multimed. Tools Appl., 53 (1) (2011) 75–95, http://dx.doi.org/10.1007/s11042-010-0491-5.

28. L. Tong, F. Dai, Y. Zhang, J. Li, Visual security evaluation for video encryption, in: Proceedings of the International Conference on Multimedia, MM '10, ACM, New York, NY, USA, 2010, pp. 835–838, http://dx.doi.org/10.1145/1873951. 1874091.

29. H. Hofbauer, A. Uhl, An effective and efficient visual quality index based on local edge gradients, in: IEEE 3rd European Workshop on Visual Information Processing, Paris, France, 2011, p. 6pp.

30. H.R. Sheikh, A.C. Bovik, Image information and visual quality, IEEE Trans. Image Process. 15 (2) (2006) 430–444.

31. M. Carosi, V. Pankajakshan, F. Autrusseau, Towards a simplified perceptual quality metric for watermarking applications, in: Proceedings of SPIE, Multimedia on Mobile Devices, vol. 7542, SPIE, San Jose, CA, USA, 2010.

32. T. Stütz, A. Uhl, On JPEG2000 error concealment attacks, in: Advantages in Image and Video Technology: Proceedings of the 3rd Pacific-Rim Symposium on Image and Video Technology, PSIVT '09, Lecture Notes in Computer Science, Springer, Tokyo, Japan, 2009, pp. 851–861.

33. C. Spearman, The proof and measurement of association between two things, Am. J. Psychol. 100 (3/4) (1904) 441–471.

34. M.G. Kendall, A new measure of rank correlation, Biometrika 30 (1/2) (1938) 81–93.

35. T. Stütz, V. Pankajakshan, F. Autrusseau, A. Uhl, H. Hofbauer, Subjective and objective quality assessment of transparently encrypted JPEG2000 images, in: Proceedings of the ACM Multimedia and Security Workshop (MMSEC '10), ACM, Rome, Italy, 2010, pp. 247–252.
36. H.R. Sheikh, Z. Wang, L. Cormack, A.C. Bovik, LIVE image quality assessment database release 2, 2006 ⟨http://live.ece.utexas.edu/research/quality⟩.
37. H.R. Sheikh, M.F. Sabir, A.C. Bovik, A statistical evaluation of recent full reference image quality assessment algorithms, IEEE Trans. Image Process. 15 (11) (2006) 3440–3451.

CHAPTER 9

NEW INEXACT EXPLICIT THRESHOLDING/SHRINKAGE FORMULAS FOR INVERSE PROBLEMS WITH OVERLAPPING GROUP SPARSITY

Gang Liu, Ting-Zhu[1,2] Huang[2], Xiao-Guang[3] Lv and Jun Liu[2]

[1] College of Applied Mathematics, Chengdu University of Information Technology

[2] Research Center for Image and Vision Computing/School of Mathematical Sciences, University of Electronic Science and Technology of China

[3] School of Science, Huaihai Institute of Technology

ABSTRACT

The least-square regression problems or inverse problems have been widely studied in many fields such as compressive sensing, signal processing, and image processing. To solve this kind of ill-posed problems, a regularization term (i.e., regularizer) should be introduced, under the assumption that the solutions have some specific properties, such as sparsity and group sparsity. Widely used regularizers include the ℓ_1 norm, total variation (TV) semi-norm, and so on. Recently, a new regularization term with overlapping group sparsity has been considered. Majorization minimization iteration method or variable duplication methods are often applied to solve them. However, there have been no direct methods for solving the relevant problems due to the difficulty of overlapping. In this paper, we proposed new inexact explicit shrinkage formulas for one class of these relevant problems, whose regularization terms have translation invariant overlapping groups. Moreover, we apply our results to TV deblurring and denoising problems with overlapping group sparsity. We use alternating direction method of multipliers to iteratively solve them. Numerical results verify the validity and effectiveness of our new inexact explicit shrinkage formulas.

INTRODUCTION

The least-square regression problems or inverse problems have been widely studied in many fields such as compressive sensing, signal processing, image processing, statistics, and machine learning. Regularization terms with sparse representations (for instance, the ℓ_1 norm regularizer) have been developed into an important tool in these applications, and a list of methods have been proposed [1, 2, 3, 4]. These methods are based on the assumption that signals or images have a sparse representation, that is, only containing a few nonzero entries. The corresponding task is to solve the following problem

$$\min_{\mathbf{z}} \quad \|\mathbf{z}\|_1 + \frac{\beta}{2}\|\mathbf{z}-\mathbf{x}\|_2^2, \tag{1}$$

where $\mathbf{x} \in \mathbb{R}^n$ is a given vector, $\left(\sum_{i=1}^{n} |z_i|^p\right)^{\frac{1}{p}}$ with $(p = 1, 2)$ represents the ℓ_p norm of vector z. The first term of (1) is called the regularization term, the second term is called the fidelity term, and $\beta>0$ is the regularization parameter.

To further improve the solutions, recent studies suggested to go beyond sparsity and took into account additional

Citation: Gang Liu, Ting-Zhu Huang, Xiao-Guang Lv and Jun Liu, New inexact explicit thresholding/ shrinkage formulas for inverse problems with overlapping group sparsity, DOI 10.1186/s13640-016-0118-5.

information about the underlying structure of the solutions [1, 2, 5]. In particular, a wide class of solutions which have specific "group sparsity" structure are considered. In this case, a group sparse vector can be divided into groups of components satisfying (a) only a few of groups contain nonzero values and (b) these groups are not needed to be sparse. If we insert such group vectors into a matrix as row vectors of the matrix, this matrix will only have few nonzero rows and these rows may be not sparse. This property is called "group sparsity" or "joint sparsity," and many literature have considered these new sparse problems [1, 2, 3, 4, 5, 6]. These problems can be formulated as the following expression.

$$\min_{\mathbf{z}} \quad \sum_{i=1}^{r} \|\mathbf{z}[i]\|_2 + \frac{\beta}{2}\|\mathbf{z}-\mathbf{x}\|_2^2, \qquad (2)$$

where $|z_i|$ is the absolute value of z_i, and $\mathbf{z}[i]$ is the ith group of z with $\mathbf{z}[i]\cap\mathbf{z}[j]=\emptyset$ and $\bigcup_{i=1}^{r}\mathbf{z}[i]=\mathbf{z}$. "Group sparsity" solutions from (2) have better representation and have been widely studied both for convex and nonconvex cases [1, 3, 6, 7, 8, 9].

More recently, overlapping group sparsity (OGS) have been considered [1, 3, 10, 11, 12, 13, 14, 15, 16, 17, 18, 19]. These methods are based on the assumption that signals or images have a special sparse representation with OGS. The task is to solve the following problem

$$\min_{\mathbf{z}} \quad \|\mathbf{z}\|_{2,1} + \frac{\beta}{2}\|\mathbf{z}-\mathbf{x}\|_2^2, \qquad (3)$$

where $\|\mathbf{z}\|_{2,1}=\sum_{i=1}^{n}\|(z_i)_g\|_2$ is the generalized $\ell_{2,1}$ norm. Here, each $(z_i)_g$ is a group vector containing s(called group size) elements that surrounding the ith entry of $\mathbf{z}$. For example, $(z_i)_g=(z_{i-1},z_i,z_{i+1})$ with s=3. In this case, $(z_i)_g$, $(z_{i+1})_g$, and $(z_{i+2})_g$ contain the (i+1)th entry of $\mathbf{z}$, z_{i+1}, which means "overlapping" different from the form of group sparsity in (2). In particular, if s=1, the generalized $\ell_{2,1}$ norm degenerates into the original ℓ_1 norm, and the relevant regularization problem (3) degenerates to (1).

To be more general, researchers considered the weighted generalized $\ell_{2,1}$ norm $\|z\|_{w,2,1}=\sum_{i=1}^{n}\|w_g\circ(z_i)_g\|_2$ (we only consider that each group has the same weight, which means translation invariant) instead of the former generalized $\ell_{2,1}$ norm. The task can be extended to

$$\min_{\mathbf{z}} \quad \|\mathbf{z}\|_{w,2,1} + \frac{\beta}{2}\|\mathbf{z}-\mathbf{x}\|_2^2, \qquad (4)$$

where w_g is a nonnegative real vector with the same size as $(z_i)_g$ and "$\circ$" is the point-wise product or Hadamard product. For instance, $w_g\circ(z_i)_g=((w_g)_1 z_i,(w_g)_2 z_{i+1},(w_g)_3 z_{i+2})$ with s=3 as the former example. In particular, the weighted generalized $\ell_{2,1}$ norm degenerates into the generalized $\ell_{2,1}$ norm if each entry of w_g equals to 1.

The problems (3) and (4) have been considered in [1, 14, 15, 16, 17, 18, 19]. They solve the relevant problems by using variable duplication methods (variable splitting, latent/auxilliary variables, etc.). For example, Deng et al. in [1] introduced a diagonal matrix G in their method. However, matrix G is not easy to find and would break the structure of the coefficient matrix, which makes the difficulty of solving solutions under high-dimensional vector cases. Moreover, it is difficult to extend this method to the matrix case.

Considering the matrix case of the problem (4), we can get

$$\min_{A} \quad \|A\|_{W,2,1} + \frac{\beta}{2}\|A-X\|_F^2, \qquad (5)$$

where $X,A\in\mathbb{R}^{m\times n}$, $\|A\|_{W,2,1}=\sum_{i=1}^{m}\sum_{j=1}^{n}\|W_g\circ(A_{i,j})_g\|_F$. Here, each $(A_{i,j})_g$ is a group matrix containing $K_1\times K_2$ (called group size) elements that surround the (i,j)th entry of A. It is defined by

$$W_g\circ(A_{i,j})_g = \begin{bmatrix} (W_g)_{1,1}A_{i-l_1,j-l_2} & (W_g)_{1,2}A_{i-l_1,j-l_2+1} & \cdots & (W_g)_{1,K_2}A_{i-l_1,j+r_2} \\ (W_g)_{2,1}A_{i-l_1+1,j-l_2} & (W_g)_{2,2}A_{i-l_1+1,j-l_2+1} & \cdots & (W_g)_{2,K_2}A_{i-l_1+1,j+r_2} \\ \vdots & \vdots & \ddots & \vdots \\ (W_g)_{K_1,1}A_{i+r_1,j-l_2} & (W_g)_{K_1,2}A_{i+r_1,j-l_1+1} & \cdots & (W_g)_{K_1,K_2}A_{i+r_1,j+r_2} \end{bmatrix} \in\mathbb{R}^{K_1\times K_2},$$

where $l_1=\lfloor\frac{K_1-1}{2}\rfloor$, $l_2=\lfloor\frac{K_2-1}{2}\rfloor$, $r_1=\lfloor\frac{K_1}{2}\rfloor$, $r_2=\lfloor\frac{K_2}{2}\rfloor$ (with $l_1+r_1+1=K_1$ and $l_2+r_2+1=K_2$) and $\lfloor x\rfloor$ denotes the largest integer less than or equal to x. In particular, if $K_2=1$, this problem degenerates to the former vector case (4). If $K_1=K_2=1$, this problem degenerates to the original ℓ_1 regularization problem (1) for the matrix case.

Much more recently, Chen et al. [10] considered the problem (5) when $(W_g)_{ij}\equiv 1$ for $i=1,\cdots,K_1$, $j=1,\cdots,K_2$. They used an iterative algorithm to solve this problem based on the principle of majorization minimization (MM). Their experiments showed that their method on solving the problem (3) is more efficient than other variable duplication methods (variable splitting, latent/auxilliary variables, etc.) [1, 14, 15, 16, 17, 18, 19]. Although their method is efficient for computing the solution, it may cost more time since the solution is obtained after many iterations rather than a direct step computation.

To our knowledge, there have been no direct methods for solving the relevant problems (3), (4), and (5). In this paper, we propose new inexact explicit shrinkage formulas for solving them directly, which are direct methods, and faster or more convenient than other methods, for instance, the most recently MM iteration method [10]. This new direct method can save more time than the MM method in [10] while getting very similar solutions. Numerical results are given to show the effectiveness of our new explicit shrinkage formulas. Moreover, the new method can be used in application for solving the subproblem in many other OGS problems, such as the wavelet regularizer with OGS in compressive sensing and the total variation (TV) regularizer with OGS in image restoration [11, 12]. For example, we will apply our results to image restoration using TV with OGS in this work similarly as [11, 12]. Moveover, we expend the works, only considering ATV in [11, 12] to both ATV and ITV (more detail can be referred to Section 4). Numerical results will show that our method can save more time with getting very similar results.

The outline of the paper is as follows. In Section 2, we deduce our explicit shrinkage formulas in detail for the OGS problems (3), (4), and (5). In Section 3, we propose some extensions for these shrinkage formulas. In Section 4, we apply our results to image deblurring and denoising problems with OGS TV. Numerical results are given in Section5. Finally, we conclude this paper in Section 6.

OGS SHRINKAGE

Classic Shrinkage

In this subsection, we will review the original shrinkage formulas and their principles, since our new explicit OGS shrinkage formulas are based on them. Firstly, we give the following definition.

Definition 1.

Define shrinkage mappings Sh_1 and Sh_2 from RN×R+RN×R+ to RNRN by

$$Sh_1(\mathbf{x},\beta)_i = \operatorname{sgn}(x_i)\max\left\{|x_i| - \frac{1}{\beta}, 0\right\}, \tag{6}$$

$$Sh_2(\mathbf{x},\beta) = \frac{\mathbf{x}}{\|\mathbf{x}\|_2}\max\left\{\|\mathbf{x}\|_2 - \frac{1}{\beta}, 0\right\}, \tag{7}$$

where the expression is taken to be zero when the second factor is zero, and "sgn" represents the signum function indicating the sign of a number. More precisely, sgn(x) =0 if x=0, sgn(x) =−1 if x<0, and sgn(x) =1 if x>0.

The shrinkage (6) is known as soft thresholding and occurs in many algorithms related to sparsity since it is the proximal mapping for the ℓ_1 norm. The shrinkage (7) is known as high-dimensional shrinkage formula.

Now, we consider solving the following problems

$$\min_{\mathbf{z}} \quad \|\mathbf{z}\|_p + \frac{\beta}{2}\|\mathbf{z}-\mathbf{x}\|_2^2, \ p = 1, 2. \tag{8}$$

The minimizer of (8) with p=1 is the following equation.

$$\arg\min_{\mathbf{z}} \quad \|\mathbf{z}\|_1 + \frac{\beta}{2}\|\mathbf{z}-\mathbf{x}\|_2^2 = Sh_1(\mathbf{x},\beta). \tag{9}$$

Due to the additivity and separability of both the ℓ_1 norm and the square of the ℓ_2 norm, Eq. (9) can be deduced easily by the following formula:

$$\min_{\mathbf{z}} \quad \|\mathbf{z}\|_1 + \frac{\beta}{2}\|\mathbf{z}-\mathbf{x}\|_2^2 = \sum_{i=1}^{n}\min_{z_i}|z_i| + \frac{\beta}{2}|z_i - x_i|^2. \tag{10}$$

The minimizer of (8) with p=2 is the following equation.

$$\arg\min_{\mathbf{z}} \quad \|\mathbf{z}\|_2 + \frac{\beta}{2}\|\mathbf{z}-\mathbf{x}\|_2^2 = Sh_2(\mathbf{x},\beta). \tag{11}$$

This formula is deduced by the Euler equation of (8) with p=2. In particular, without considering **x**=**0** or **z**=**0**, we obtain the following Euler equation.

$$\beta(\mathbf{z}-\mathbf{x}) + \frac{\mathbf{z}}{\|\mathbf{z}\|_2} \ni \mathbf{0}, \tag{12}$$

$$\left(1 + \frac{1}{\beta}\frac{1}{\|\mathbf{z}\|_2}\right)\mathbf{z} - \mathbf{x} \ni \mathbf{0}. \tag{13}$$

We can easily get that the necessary condition is that the vector **z** is parallel to the vector **x**. That is, $\frac{\mathbf{z}}{\|\mathbf{z}\|_2} = \frac{\mathbf{x}}{\|\mathbf{x}\|_2}$. Substituting into (13) and considering **x**=**0** or **z**=**0**, we obtain the formula (7). More details can be referred to [20, 21].

Our new explicit OGS shrinkage formulas are based on these observations, especially the properties of additivity, separability, and parallelity.

Remarks 1.

The problem (2) is easy to be solved by a simple shrinkage formula, which is not used in this work. More details can be referred to [1, 22, 23].

The OGS Shrinkage Formulas

Now, we focus on the problem (3) firstly. The difficulty of this problem is "overlapping." Therefore, we must take some special techniques to avoid "overlapping." That is the point of our new explicit OGS shrinkage formulas.

It is obvious that the first term of the problem (3) is additive and separable. So if we find some relative rules such that the second term of the problem (3) has the same properties with the same variable as the first term, one approximate solution of (3) can be easily found similar as classic shrinkage.

Assuming periodic boundary condition (PBC) (boundary condition (BC) is necessary in OGS functionals) is used here, we observe that each entry z_i of the vector **z** would appear exactly s times in the first term. Therefore, to hold on the uniformity of vectors **z** and **x**, we need to multiply the second term by s. To maintain the invariability of the problem (3), after some manipulations, we have

$$\begin{aligned} f_m(\mathbf{z}) &= \min_{\mathbf{z}}\|\mathbf{z}\|_{2,1} + \frac{\beta}{2}\|\mathbf{z}-\mathbf{x}\|_2^2 \\ &= \min_{\mathbf{z}}\sum_{i=1}^{n}\|(z_i)_g\|_2 + \frac{\beta}{2s}s\|\mathbf{z}-\mathbf{x}\|_2^2 \\ &= \min_{\mathbf{z}}\sum_{i=1}^{n}\|(z_i)_g\|_2 + \frac{\beta}{2s}\sum_{i=1}^{n}\|(z_i)_g - (x_i)_g\|_2^2 \\ &= \min_{\mathbf{z}}\sum_{i=1}^{n}\left(\|(z_i)_g\|_2 + \frac{\beta}{2s}\|(z_i)_g - (x_i)_g\|_2^2\right), \end{aligned} \tag{14}$$

where $(x_i)_g$ is similarly defined as $(z_i)_g$ in Section 1.

For example, we set s=3 and define $(z_i)_g=(z_i, z_{i+1}, z_{i+2})$. The generalized $\ell_{2,1}$ norm $\|\mathbf{z}\|_{2,1}$ can be treated as the generalized ℓ_1 norm of generalized points, whose entry $(z_i)_g$ is also a vector, and the absolute value of each entry is treated as the ℓ_2 norm of $(z_i)_g$. See Fig. 1 a intuitively, where the top line is the vector **z**, the rectangles with dashed line are original $(z_i)_g$, and the rectangles with solid line are the generalized points. In the case of PBC, we know that each line of Fig. 1 a is translated equal. We operate the vector **x** similarly as the vector **z**. Putting these generalized points (rectangles with solid line in the figure) as the columns of a matrix, we can regard $\|\mathbf{z}-\mathbf{x}\|_2^2$ as the matrix Frobenius norm $\|[(z_i)_g] - [(x_i)_g]\|_F^2$, where $[(z_i)_g]$ is a matrix as in Fig. 1 a with every line being its row ($[(x_i)_g]$ is similar). This is why the second equality in (14) holds.

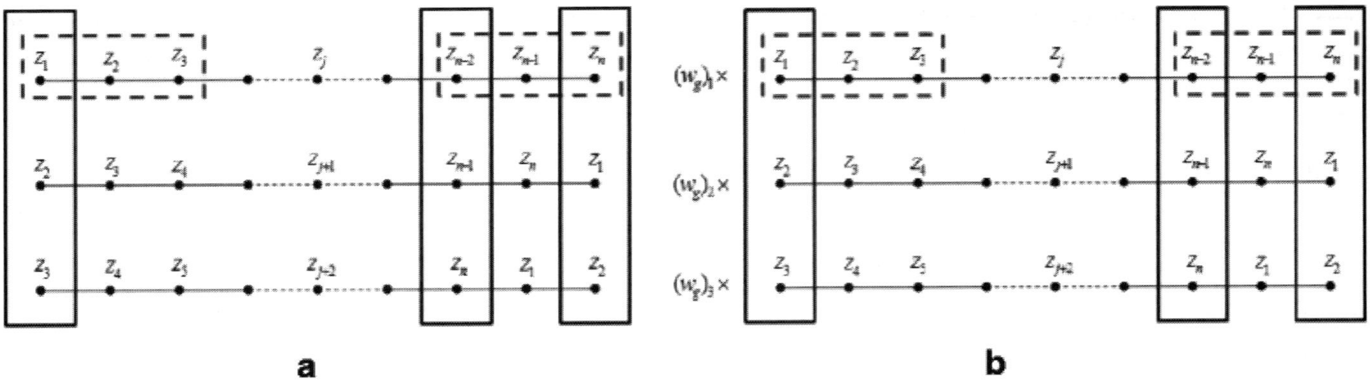

Figure. 1: Vector case. The top line is the vector **z**, the rectangles with dashed line are original group vector $(z_i)_g$, and therectangles with solid line are the generalized points, which changes the overlapping to non-overlapping. **a** Vector. **b**Weighted vector

Therefore, generally, for each i of the last line of (14), from the Eqs. (7) and (11), we can obtain

$$\arg\min_{(z_i)_g} \|(z_i)_g\|_2 + \frac{\beta}{2s}\|(z_i)_g - (x_i)_g\|_2^2 = \mathrm{Sh}_2\left((x_i)_g, \frac{\beta}{s}\right),$$

$$(z_i)_g = \max\left\{\|(x_i)_g\|_2 - \frac{s}{\beta}, 0\right\}\frac{(x_i)_g}{\|(x_i)_g\|_2},$$

$$(z_i)_g = \left([(z_i)_g]_1, [(z_i)_g]_2, \ldots, [(z_i)_g]_s\right). \tag{15}$$

Similarly as Fig. 1 a, for each i, the ith entry x_i (or z_i) of the vector **x** (or **z**) may appear s times, so we need to compute each z_i for s times in s different groups.

However, the results from (15) are not able to be satisfied simultaneously, because the results z_i in s different groups are different from (15). Moreover, for each i of the last line in (14), the result (15) is given by that the vector $(z_i)_g$ is parallel to the vector $(x_i)_g$ due to Section 2.1.

Notice this point and ignore that $(z_i)_g$=**0** or $(x_i)_g$=**0**, particularly for s=4 and $(z_i)_g=(z_{i-1},z_i,z_{i+1},z_{i+2})$, the vector **z**can be split as follows,

$$\begin{aligned}
\mathbf{z} = \;& (z_1,\ z_2,\ z_3,\ z_4,\ z_5,\ \cdots,\ z_{n-2},\ z_{n-1},\ z_n\ )\\
= & +\tfrac{1}{4}\ (z_1,\ z_2,\ z_3,\ 0,\ 0,\ \cdots,\ 0,\ 0,\ z_n\ )\\
& +\tfrac{1}{4}\ (z_1,\ z_2,\ z_3,\ z_4,\ 0,\ \cdots,\ 0,\ 0,\ 0\ )\\
& +\tfrac{1}{4}\ (0,\ z_2,\ z_3,\ z_4,\ z_5,\ \cdots,\ 0,\ 0,\ 0\ )\\
& +\ \cdots\\
& +\tfrac{1}{4}\ (z_1,\ 0,\ 0,\ 0,\ 0,\ \cdots,\ z_{n-2},\ z_{n-1},\ z_n\ )\\
& +\tfrac{1}{4}\ (z_1,\ z_2,\ 0,\ 0,\ 0,\ \cdots,\ 0,\ z_{n-1},\ z_n\ ).
\end{aligned} \tag{16}$$

Let $(z_i)g'=(0,\cdots,0,z_{i-1},z_i,z_{i+1},z_{i+2},0,\cdots,0)$ be the expansion of $(z_i)_g$, with $(z_1)g'=(z_1,z_2,z_3,0,\cdots,0,z_n)$,$(z_{n-1})g'=(z_1,0,0,\cdots,0,z_{n-2},z_{n-1},z_n)$,$(z_n)g'=(z_1,z_2,0,\cdots,0,z_{n-1},z_n)$. Let $(x_i)g'=(0,\cdots,0,x_{i-1},x_i,x_{i+1},x_{i+2},0,\cdots,0)$ be the expansion of $(x_i)_g$ similarly as $(z_i)g'$. Then, we have z $\mathbf{z} = \frac{1}{4}\sum_{i=1}^{n}(z_i)'_g$, and $\mathbf{x} = \frac{1}{4}\sum_{i=1}^{n}(x_i)'_g$. Moreover, we can easily obtain that $\|(z_i)'_g\|_2 = \|(z_i)_g\|_2$ and $\|(x_i)'_g\|_2 = \|(x_i)_g\|_2$ for every i.

On one hand, the Euler equation of $f_m(\mathbf{z})$ (with s=4) is given by

$$\beta(\mathbf{z}-\mathbf{x}) + \frac{(z_1)'_g}{\|(z_1)'_g\|_2} + \cdots + \frac{(z'_n)_g}{\|(z'_n)_g\|_2} \ni \mathbf{0}, \tag{17}$$

$$\frac{\beta}{4}\sum_{i=1}^{n}\left((z_i)'_g - (x_i)'_g\right) + \frac{(z_1)'_g}{\|(z_1)'_g\|_2} + \cdots + \frac{(z_i)'_g}{\|(z_i)'_g\|_2} + \cdots + \frac{(z_n)'_g}{\|(z_n)'_g\|_2} \ni \mathbf{0}. \tag{18}$$

From the deduction of the two-dimensional shrinkage formula (7) in Section 2.1, we know that the necessary condition of minimizing the ith term of the last line in (14) is that $(z_i)_g$ is parallel to $(x_i)_g$. That is, $(z_i)g'$ is parallel to $(x_i)g'$ for every i. For example,

$$(z_2)'_g = (z_1, z_2, z_3, z_4, 0, \cdots, 0, 0)//(x_2)'_g = (x_1, x_2, x_3, x_4, 0, \cdots, 0, 0). \tag{19}$$

Then, we obtain $\frac{(z_i)'_g}{\|(z_i)'_g\|_2} = \frac{(x_i)'_g}{\|(x_i)'_g\|_2}$.. Therefore, (18) can be changed to

$$\frac{\beta}{4}\sum_{i=1}^{n}\left((z_i)'_g - (x_i)'_g\right) + \frac{(x_1)'_g}{\|(x_1)'_g\|_2} + \cdots + \frac{(x_i)'_g}{\|(x_i)'_g\|_2} + \cdots + \frac{(x_n)'_g}{\|(x_n)'_g\|_2} \ni \mathbf{0}, \tag{20}$$

$$\beta(\mathbf{z}-\mathbf{x}) + \frac{(x_1)'_g}{\|(x_1)'_g\|_2} + \cdots + \frac{(x_i)'_g}{\|(x_i)'_g\|_2} + \cdots + \frac{(x_n)'_g}{\|(x_n)'_g\|_2} \ni \mathbf{0}, \tag{21}$$

$$\mathbf{z} \ni \mathbf{x} - \frac{1}{\beta}\left(\frac{(x_1)'_g}{\|(x_1)'_g\|_2} + \cdots + \frac{(x_i)'_g}{\|(x_i)'_g\|_2} + \cdots + \frac{(x_n)'_g}{\|(x_n)'_g\|_2}\right), \tag{22}$$

for each component, we obtain

$$z_i \ni x_i - \frac{1}{\beta}\left(\frac{x_i}{\|(x_{i-2})'_g\|_2} + \frac{x_i}{\|(x_{i-1})'_g\|_2} + \frac{x_i}{\|(x_i)'_g\|_2} + \frac{x_i}{\|(x_{i+1})'_g\|_2}\right). \tag{23}$$

Therefore, when $(z_i)_g \neq \mathbf{0}$ and $(x_i)_g \neq \mathbf{0}$, we find a minimizer of (3) on the direction that all the vectors $(z_i)_g$ are parallel to the vectors $(x_i)_g$. In addition, when $\frac{\beta}{4}\left((z_i)'_g - (x_i)'_g\right) + \frac{(z_i)'_g}{\|(z_i)'_g\|_2} = 0$, (18) holds, then $\left(\frac{\beta}{4} + \frac{1}{\|(z_i)'_g\|_2}\right)(z_i)'_g = (x_i)'_g$; therefore, $\frac{(z_i)'_g}{\|(z_i)'_g\|_2} = \frac{(x_i)'_g}{\|(x_i)'_g\|_2}$ holds. Moreover, because of the strict convexity of $f_m(\mathbf{z})$, we know that the minimizer is unique.

On the other hand, when $(z_i)_g$=**0** or $(x_i)_g$=**0**, our method may not obtain good results. When $(x_i)_g$=**0**, we know that the minimizer of the subproblem $\min_{(z_i)_g} \|(z_i)_g\|_2 + \frac{\beta}{2s}\|(z_i)_g - (x_i)_g\|_2^2$ is exactly that $(z_i)_g = \mathbf{0}$. When $(x_i)_g \neq$ 0is exactly that $(z_i)_g=\mathbf{0}$. When $(x_i)_g\neq\mathbf{0}$ and the minimizer of the subproblem $\min_{(z_i)_g}\|(z_i)_g\|_2+\frac{\beta}{2s}\|(z_i)_g-(x_i)_g\|_2^2$ is that $(z_i)_g=\mathbf{0}$ is that $(z_i)_g=\mathbf{0}$ (since the parameter β/s is two small), our method is not able to obtain good results. For example, that $(z_i)_g=\mathbf{0}$ while $(z_{i+1})_g\neq\mathbf{0}$ makes the element z_i in $\mathbf{z}$ take different values in different subproblems. However, we can obtain an approximate minimizer in this case, which is that the element z_i is a simple summation of corresponding subproblems containing z_i. We will show that in the experiments of Section 5, the approximate minimizer is also good. Moreover, when we take this problem as a subproblem of the image processing problem, we can set the parameter β to be large enough to avoid this drawback.

In addition, from (23), we can know that the element z_i of the minimizer can be treated as in s subproblems independently and then we combine them. In conclusion, after some manipulations, we can get the following two general formulas. (I)

$$\arg\min_{\mathbf{z}} \quad \|\mathbf{z}\|_{2,1}+\frac{\beta}{2}\|\mathbf{z}-\mathbf{x}\|_2^2=\mathrm{Sh_{OGS}}(\mathbf{x},\beta), \tag{24}$$

with

$$\mathrm{Sh_{OGS}}(\mathbf{x},\beta)_i=z_i=\max\left\{1-\frac{1}{\beta}F(x_i),0\right\}x_i. \tag{25}$$

Here, for instance, when group size s=4, $(z_i)_g=(z_{i-1},z_i,z_{i+1},z_{i+2})$ in $\|\mathbf{z}\|_{2,1}$ and $(x_i)_g$ is defined similarly as $(z_i)_g$, we have $F(x_i)=\left(\frac{1}{\|(x_{i-2})_g\|_2}+\frac{1}{\|(x_{i-1})_g\|_2}+\frac{1}{\|(x_i)_g\|_2}+\frac{1}{\|(x_{i+1})_g\|_2}\right)$ is contained in $F(x_i)$ if and only if $\|(x_j)_g\|_2$ has the component x_i, and we follow the convention (1/0)=1 in $F(x_i)$ because $\|(x_i)_g\|_2=0$ implies $x_i=0$ and the value of $F(x_i)$ is insignificant in (25). (II)

$$\arg\min_{\mathbf{z}} \quad \|\mathbf{z}\|_{2,1}+\frac{\beta}{2}\|\mathbf{z}-\mathbf{x}\|_2^2=\mathrm{Sh_{OGS}}(\mathbf{x},\beta), \tag{26}$$

with

$$\mathrm{Sh_{OGS}}(\mathbf{x},\beta)_i=z_i=G(x_i)\cdot x_i. \tag{27}$$

Here, symbols are the same as (I), and $G(x_i)=\max\left(\frac{1}{s}-\frac{1}{\beta\|(x_{i-2})_g\|_2},0\right)+\max\left(\frac{1}{s}-\frac{1}{\beta\|(x_{i-1})_g\|_2},0\right)+\max\left(\frac{1}{s}-\frac{1}{\beta\|(x_i)_g\|_2},0\right)+\max\left(\frac{1}{s}-\frac{1}{\beta\|(x_{i+1})_g\|_2},0\right)$.

We call them Formula (I) and Formula (II) respectively throughout this paper. When β is sufficiently large or sufficiently small, the former two formulas are the same. For the other values of β, from the experiments, we find that Formula (II) is better approximate to the MM iteration method than Formula (I), so we list the algorithm for Formula (II) as follows for finding the minimizer of (3). If we would use Formula (I), we can only change the final three steps, which does not change the computation cost.

We can see that Algorithm 1 only needs 2 times of convolution computations with time complexity n∗s, which is just the same time complexity as one-step iteration in the MM method of [10]. Therefore, our method is much more efficient than the MM method or other variable duplication methods. In Section 5, we will give the numerical experiments for comparison between our method and the MM method. Moreover, if s=1, our Algorithm 1 degenerates to the classic soft thresholding as our formula (27) degenerates to (6). Moreover, when β is sufficiently large or sufficiently small, the results of our new explicit algorithm are almost the same as the MM method (after 20 or more iteration steps) in [10].

Algorithm 1 Direct shrinkage algorithm for the minimization problem (3)

Input:

Given vector $\mathbf{x}=(x_1,x_2,\cdots,x_n)$, group size s, parameter β.

Compute:

1. Definition of $(x_i)_g$, for example $(x_i)_g=\left(x_{i-s_l},\cdots,x_i,\cdots,x_{i+s_r}\right)$, with $s_l=[\frac{s-1}{2}]$ and $s_r=[\frac{s}{2}]$ $(s_l+s_r+1=s)$.
   $\mathbf{w}$ = ones(1,s) = [1,1,···,1].
   If s is even, then $\mathbf{w}=[\mathbf{w},0]$ and $s=s+1$.
   Let $\mathbf{w}_r$ = fliplr(w) be fliping w over 180 degrees.
2. Compute $X_n=\left(\|(x_1)_g\|_2,\cdots,\|(x_i)_g\|_2,\cdots,\|(x_n)_g\|_2\right)$, by convolution of w and x.
3. Compute $X_n'=\max\left(\frac{1}{s}-1./(X_n\cdot\beta),0\right)$ point-wise.
4. Compute $G(x_i)$ by correlation of w and X_n', or by convolution of $\mathbf{w}_r$ and X_n'.
5. Compute z by multiplying x with $G(x_i)$ point-wise.

Remarks 2.

Our new explicit algorithm can be treated as an average estimation algorithm for solving all the overlapping group subproblem independently. In Section 5, our numerical experiments show that the results of our two formulas are almost the same as the MM method in [10] when β is sufficiently large or sufficiently small and is approximate to the MM method for other β. Moreover, this point illuminates that the OGS regularizers coincide to maintain the property of the classical sparse regularizers and smoothen the local regions. For instance, TV with OGS can preserve edges and simultaneously overcome staircases in our experiments and [11, 12].

For the problems (4) (See Fig. 1 b intuitively) and (5), we can get similar algorithms. We omit the evolution details here and give them as in Appendix. Here, we directly give the algorithm for solving the problem (5) as follows under Formula (II).

We can also see that Algorithm 2 only needs 2 times of convolution computations with time complexity n∗s, which is just the same time complexity as one-step iteration in the MM method of [10]. Therefore, our method is much more efficient than the MM method.

Algorithm 2 Direct shrinkage algorithm for the minimization problem (4)

Input:
Given matrix A, group size $s = K_1 \times K_2$, parameter β, weight vector W_g.
Compute:
1. Pad W_g to be an odd-by-odd matrix.
If K_1 is even, then $W_g =$[zeros$(1, K_2)$; W_g] and $K_1 = K_1 + 1$.
If K_2 is even, then $W_g =$[zeros$(K_1, 1)$, W_g] and $K_2 = K_2 + 1$.
Let W_g^r = fliplr(W_g) be fliping W_g over 180 degrees.
2. Compute
$A_{temp} = (\|(A_{1,1})_g\|_2, \cdots, \|(A_{1,K_2})_g\|_2; \cdots; \|(A_{K_1,1})_g\|_2, \cdots, \|(A_{K_1,K_2})_g\|_2)$, by correlation of $W_g.^2$ (pointwise square) and A, or by convolution of $W_g^r.^2$ and A.
3. Compute $A'_{temp} = \max\left(\frac{1}{\|W_g\|_F^2} - 1./(A_{temp}\cdot\beta), 0\right)$ pointwise.
4. Compute $G(A_{i,j})$ by convolution of $W_g.^2$ and A'_{temp}, or by correlation of $W_g^r.^2$ and A'_{temp}.
5. Compute X by multiplying A with $G(A_{i,j})$ point-wise.

SEVERAL EXTENSIONS

Other Boundary Conditions

In Section 2, we gave the explicit shrinkage formulas for one class of OGS problems (3), (4), and (5). In order to achieve a simply deduction, we assume that PBC is used. One may confuse that whether PBC is always good for regularization problems such as signal processing or image processing, since natural signals or images are often asymmetric. However, in these problems, assuming a kind of boundary condition is necessary for simplifying the problem and making the computation possible [24]. There are kinds of BCs, such as zero BC (ZBC) and reflective BC. PBC is often used in optimization because it can be computed fast as above or other methods, for example, computation of matrix of block circulant with circulant blocks (BCCB) by fast Fourier transforms [20, 21, 24].

In this section, we consider other BC such as ZBC and reflective BC. For simplification, we only consider the vector case, while the results can be easily expanded to the matrix case as that in Section 2. When ZBC is used, we can expand the original vectors (or signal) $\mathbf{z}$ (= $(z_i)_{i=1}^n$) and $\mathbf{x}$ by two s-length vectors on the both hands of the original vectors, which are $\tilde{\mathbf{z}}$ (= $[\tilde{z}_{-s}, \cdots, \tilde{z}_{-1}, (\tilde{z}_i)_{i=1}^n, \tilde{z}_{n+1}, \cdots, \tilde{z}_{n+s}] = [\mathbf{0}_s, (z_i)_{i=1}^n, \mathbf{0}_s]$) and $\tilde{\mathbf{x}}$, respectively then, the results and algorithms in Section 2 are similar as the case of PBC on $\tilde{\mathbf{z}}$ and $\tilde{\mathbf{x}}$.. See Fig. 2 a intuitively.

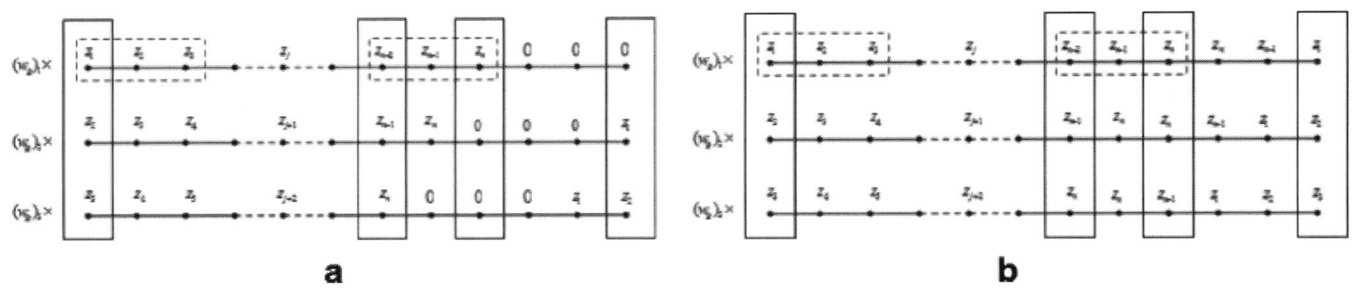

Figure. 2: The case of expanding vectors with other BCs to use the formulas under PBC. **a** ZBC. **b** Reflective BC

Moreover, our numerical results will show that the results from a different BC are almost the same in practice (see Section 5.1); and according to the definition of weighted generalized $\ell_{2,1}$ norm, ZBC seems better than PBC. Therefore, we will choose the ZBC to solve the problems (3), (4), and (5) in the following sections.

When reflective BC is assumed, the results are also the same. We only need to extend the original vector **x** to x^x^with reflective BC. See Fig. 2 b intuitively.

Nonpositive Weights and Different Weights in Groups

In this section, we show that the weight vector w_g and matrix W_g in the former sections can contain arbitrary sentries with arbitrary real numbers. On one hand, the zero value can be the arbitrary entries of the weight vector or matrix. For example, the original ℓ_1 regularization problem (1) can be seen as a special form of weighted generalized $\ell_{2,1}$ norm regularization problems (4) with s=1 or s=3 and w_g=(1,0,0) for the example of Fig. 1 b.

On the other hand, the norm is with the property of positive homogeneity, $\|(-1)\cdot w_g\|_{any} = \|w_g\|_{any}$, where "$\|\cdot\|_{any}$" can be arbitrary norm such as $\|\cdot\|_1$ and $\|\cdot\|_2$. Therefore, for the regularization problems (4) (or (5)), the weight vectorw $_g$ (or matrix W_g) is the same as $|w_g|$ (or $|W_g|$), where $|\cdot|$ the point-wise absolute value. Therefore, in general, our results are true, whatever the number of entries included in the weight vector w_g and matrix W_g and whatever the real number value of these entries.

However, if the weight w_g is dependent on the group index i, for example, $w_{i,g} \circ (z_i)_g = ((w_{i,g})_1 z_i, (w_{i,g})_2 z_{i+1}, (w_{i,g})_3 z_{i+2})$, we cannot solve the relevant problems easily since our method fails. As we mentioned above, we only focus on that the weight w_g is independent on the group index i, that is, $w_{i,g} = w_g$ for all i, which means that it is with translation invariant overlapping groups.

Applications in TV regularization problems with OGS

The TV regularizer was firstly introduced by Rudin et al. [25] (ROF), which is widely used in many fields, i.e., denoising and deblurring problems [20, 21, 26, 27, 28, 29, 30, 31]. Several fast algorithms such as Chambelle [26] and fast TV deconvolution (FTVd) [20, 21] have been proposed. Its corresponding minimization task is

$$\min_f \|f\|_{TV} + \frac{\mu}{2}\|Hf - g\|_2^2, \tag{28}$$

where $\|f\|_{TV} := \sum_{1\le i,j\le n} \|(\nabla f)_{i,j}\|_2 = \sum_{1\le i,j\le n} \sqrt{|(\nabla_x f)_{i,j}|^2 + |(\nabla_y f)_{i,j}|^2}$ (called isotropic TV, ITV), or $\|f\|_{TV} := \sum_{1\le i,j\le n} \|(\nabla f)_{i,j}\|_1 = \sum_{1\le i,j\le n} |(\nabla_x f)_{i,j}| + |(\nabla_y f)_{i,j}|$ (called anisotropic TV, ATV),

alled anisotropic TV, ATV), H denotes the blur matrix, and g denotes the given observed image with blur and noise. Operator ∇:Rn2→R2×n2∇:Rn2→R2×n2 denotes the discrete gradient operator (under PBC) which is defined by $(\nabla f)_{i,j} = ((\nabla_x f)_{i,j}, (\nabla_y f)_{i,j})$, with

$$(\nabla_x f)_{i,j} = \begin{cases} f_{i+1,j} - f_{i,j} & \text{if } i < n, \\ f_{1,j} - f_{n,j} & \text{if } i = n, \end{cases} \quad (\nabla_y f)_{i,j} = \begin{cases} f_{i,j+1} - f_{i,j} & \text{if } j < n, \\ f_{i,1} - f_{i,n} & \text{if } j = n, \end{cases}$$

for i,j=1,2,⋯,n, where $f_{i,j}$ refers to the ((j−1)n+i)th entry of the vector f (it is the (i,j)th pixel location of the n×n image, and this notation remains valid throughout the paper unless otherwise specified). Notice that H is a matrix of BCCB structure when PBC are applied [24].

Recently, Selesnick et al. [13] proposed an OGS TV regularizer to one-dimensional signal denoising. They applied the MM method to solve their model. Their numerical experiments showed that their method can overcome staircase effects effectively and get better results. However, their method has the disadvantages of the low speed of computation and the difficulty to be directly extended to the two-dimensional image case. More recently, Liu et al. [11] proposed an OGS TV regularizer for two-dimensional image denoising and deblurring under Gaussian noise, and Liu et al. [12] proposed an OGS TV regularizer for image deblurring under impulse noise. Both of them used a variable substitution method and the ADMM framelet with an inner MM iteration for solving the subproblems similar as (3). Therefore, their methods may spend more time than our methods. Moreover, when the MM method is used in the inner iterations in [11, 12], they can only solve the ATV case but not the ITV case with OGS while our methods can solve both ATV and ITV cases.

Firstly, we define the ATV case with OGS under Gaussian noise and impulse noise respectively which is similar as [11, 12]. For the Gaussian noise case,

$$\min_f \|(\nabla_x f)\|_{W,2,1} + \|(\nabla_y f)\|_{W,2,1} + \frac{\mu}{2}\|Hf - g\|_2^2. \tag{29}$$

For the impulse case,

$$\min_f \|(\nabla_x f)\|_{W,2,1} + \|(\nabla_y f)\|_{W,2,1} + \mu\|Hf - g\|_1. \tag{30}$$

We call the former model as TV OGS L_2 model and the latter model as TV OGS L_1 model, respectively.

Then, we defined the ITV case with OGS. For Gaussian noise and impulse noise, we only change the former two terms of (29) and (30) respectively by $\|A\|_{W,2,1}$. Here, A is a high-dimensional matrix (or tensor) with each entry $A_{i,j} = ((\nabla_x f)_{i,j}; (\nabla_y f)_{i,j})$.

Remarks 3.

In particular in this work, A can be treated as $((\nabla_x f); (\nabla_y f))$ for simplicity in vector and matrix computation, although the computation of $\|A\|_{W,2,1}$ is should be treated as pointwise with $A_{i,j} = ((\nabla_x f)_{i,j}; (\nabla_y f)_{i,j})$ since the computation on Ais almost all pointwise.

Moreover, we consider constrained model as listing in [11, 12, 32]. For any true digital image, its pixel value can attain only a finite number of values. Hence, it is natural to require all pixel values of the restored image to lie in a certain interval [a,b], see [32] for more details. In general, with the easy computation and the certified results in [32], we only consider all the images located on the standard range [0,1]. Therefore,we define a projection operator PΩPΩon the set $\Omega = \{f \in \mathbb{R}^{n\times n} | 0 \le f \le 1\}$,

$$\mathcal{P}_\Omega(f)_{i,j} = \begin{cases} 0, & f_{i,j} < 0, \\ f_{i,j}, & f_{i,j} \in [0,1], \\ 1, & f_{i,j} > 1. \end{cases} \tag{31}$$

Constrained TV OGS L_2 Model

For the constrained model (the ATV case) (called CATVOGSL2), we have

$$\min_{u\in\Omega, v_x, v_y, f} \left\{ \|(v_x)\|_{W,2,1} + \|(v_y)\|_{W,2,1} + \frac{\mu}{2}\|Hf - g\|_2^2 : v_x = \nabla_x f, v_y = \nabla_y f, u = f \right\}. \tag{32}$$

The augmented Lagrangian function [33] of (32) is

$$\begin{aligned} \mathcal{L}(v_x, v_y, u, f; \lambda_1, \lambda_2, \lambda_3) &= \|v_x\|_{W,2,1} - \lambda_1^T(v_x - \nabla_x f) + \frac{\beta_1}{2}\|v_x - \nabla_x f\|_2^2 \\ &+ \|v_y\|_{W,2,1} - \lambda_2^T(v_y - \nabla_y f) + \frac{\beta_1}{2}\|v_y - \nabla_y f\|_2^2 \\ &- \lambda_3^T(u - f) + \frac{\beta_2}{2}\|u - f\|_2^2 + \frac{\mu}{2}\|Hf - g\|_2^2, \end{aligned} \tag{33}$$

where $\beta_1, \beta_2 > 0$ are penalty parameters and $\lambda_1, \lambda_2, \lambda_3 \in$ ∈Rn2 are the Lagrange multipliers.

The solving algorithm is according to the scheme of ADMM in Gabay [34], and we refer to some applications in image processing which can be solved by ADMM, e.g., [35,36,37,38,39,40,41,42,43].Foragiven $(v_x^k, v_y^k, u^k, f^k; \lambda_1^k, \lambda_2^k, \lambda_3^k)$, the next iteration $(v_x^{k+1}, v_y^{k+1}, u^{k+1}, f^{k+1}; \lambda_1^{k+1}, \lambda_2^{k+1}, \lambda_3^{k+1})$ is generatedasfollows: 1.Fix $f = f^k, \lambda_1 = \lambda_1^k, \lambda_2 = \lambda_2^k, \lambda_3 = \lambda_3^k$ k and minimize (33) with respect to v_x, v_y, and u. With respect to v_x and v_y,

$$\begin{aligned} v_x^{k+1} &= \arg\min \|v_x\|_{W,2,1} - \lambda_1^{k^T}(v_x - \nabla_x f^k) + \frac{\beta_1}{2}\|v_x - \nabla_x f^k\|_2^2 \\ &= \arg\min \|v_x\|_{W,2,1} + \frac{\beta_1}{2}\|v_x - \nabla_x f^k - \frac{\lambda_1^k}{\beta_1}\|_2^2, \end{aligned} \tag{34}$$

$$\begin{aligned} v_y^{k+1} &= \arg\min \|v_y\|_{W,2,1} - \lambda_2^{k^T}(v_y - \nabla_y f^k) + \frac{\beta_1}{2}\|v_y - \nabla_y f^k\|_2^2 \\ &= \arg\min \|v_y\|_{W,2,1} + \frac{\beta_1}{2}\|v_y - \nabla_y f^k - \frac{\lambda_2^k}{\beta_1}\|_2^2. \end{aligned} \tag{35}$$

It is obvious that problems (34) and (35) match the framework of the problem (5); thus, the minimizers of (34) and (35) can be obtained by using the formulas in Section 2.2.

With respect to u,

$$u^{k+1} = \arg\min -\lambda_3^{k^T}(u - f^k) + \frac{\beta_2}{2}\|u - f^k\|_2^2$$
$$= \arg\min \frac{\beta_2}{2}\|u - f^k - \frac{\lambda_3^k}{\beta_2}\|_2^2.$$

The minimizer is given explicitly by

$$u^{k+1} = \mathcal{P}_\Omega\left[f^k + \frac{\lambda_3^k}{\beta_2}\right] \tag{36}$$

Compute f^{k+1} by solving the normal equation

$$(\beta_1(\nabla_x^*\nabla_x + \nabla_y^*\nabla_y) + \mu H^*H + \beta_2 I)f^{k+1}$$
$$= \nabla_x^*(\beta_1 v_x^{k+1} - \lambda_1^k) + \nabla_y^*(\beta_1 v_y^{k+1} - \lambda_2^k) + \mu H^*g$$
$$+\beta_2(u^{k+1} - \frac{\lambda_3^k}{\beta_2}), \tag{37}$$

where " *" denotes the conjugate transpose, see [44] for more details. Since all the parameters are positive, the coefficient matrix in (37) are always invertible and symmetric positive-definite. In addition, H, ∇_x, ∇_y and their conjugate transpose have BCCB structure under PBC. We know that the computations with BCCB matrix can be very efficient by using fast Fourier transforms. 3. Update the multipliers via

$$\begin{cases} \lambda_1^{k+1} = \lambda_1^k - \gamma\beta_1(v_x^{k+1} - \nabla_x f^{k+1}), \\ \lambda_2^{k+1} = \lambda_2^k - \gamma\beta_1(v_x^{k+1} - \nabla_x f^{k+1}), \\ \lambda_3^{k+1} = \lambda_3^k - \gamma\beta_2(u^{k+1} - f^{k+1}). \end{cases} \tag{38}$$

Based on the discussions above, we present the ADMM algorithm for solving the convex CATVOGSL2 model (32), which is shown as Algorithm 3.

Algorithm 3 CATVOGSL2 for the minimization problem (32)

initialization:

Starting point $f^0 = g$, $k = 0$, β_1, β_2, γ, μ, group size $K_1 \times K_2$, weighted matrix W_g, $\lambda_i^0 = 0, i = 1,2,3$.

iteration:

1. Compute v_x^{k+1} and v_y^{k+1} according to (34) and (35).

2. Compute u^{k+1} according to (36).

3. Compute f^{k+1} by solving (37).

4. update $\lambda_i^0 = 0, i = 1,2,3$ according to (38).

5. $k = k + 1$.

until a stopping criterion is satisfied.

Algorithm 3 is an application of ADMM for the case with two blocks of variables (v_x,v_y,u) and f. Thus, if the step (1) of Algorithm 3 can be solved exactly, its convergence is guaranteed by the theory of ADMM [34, 35, 37, 43]. Although step (1) of Algorithm 3 cannot be solved exactly, we can find a convergent series to ensure the convergence as [35]. Besides, our numerical experiments verify the convergence of Algorithm 3.

Remarks 4.

For the image f, we use PBC for the fast computation. However, for v_x, v_y, we use ZBC. Because v_x and v_y denote the gradient of the image, ZBC seems better for the definition of the generalized norm $\ell_{2,1}$ on v_x and v_y as mentioned in Section 3.1. Therefore, the two kinds of BCs for the image and its gradient are different and independent. These remain valid throughout the paper unless otherwise specified.

For the ITV case, the constrained model (CITVOGSL2) is

$$\min_{u\in\Omega,A,f}\left\{\|A\|_{W,2,1} + \frac{\mu}{2}\|Hf - g\|_2^2 : A = (\nabla_x f; \nabla_y f), u = f\right\}. \tag{39}$$

The detail will be presented in the next section for the constrained TV OGS L_1 model. We call this relevant algorithm CITVOGSL2.

Constrained TV OGS L_1 Model

For the constrained model (the ITV case) (called CITVOGSL1), we have

$$\min_{u\in\Omega,A,f}\{\|A\|_{W,2,1} + \mu\|z\|_1 : z = Hf - g, A = (\nabla_x f; \nabla_y f), u = f\}. \tag{40}$$

The augmented Lagrangian function of (40) is

$$\mathcal{L}(v_x, v_y, z, w, f; \lambda_1, \lambda_2, \lambda_3, \lambda_4) = \|A\|_{W,2,1} - (\lambda_1^T, \lambda_2^T)\left(A - \begin{bmatrix}\nabla_x f\\ \nabla_y f\end{bmatrix}\right)$$
$$+\frac{\beta_1}{2}\left\|A - \begin{bmatrix}\nabla_x f\\ \nabla_y f\end{bmatrix}\right\|_2^2$$
$$+\mu\|z\|_1 - \lambda_3^T(z - (Hf - g))$$
$$+\frac{\beta_2}{2}\|z - (Hf - g)\|_2^2$$
$$-\lambda_4^T(w - f) + \frac{\beta_3}{2}\|w - f\|_2^2, \tag{41}$$

where β_1,β_2,β_3>0 are penalty parameters and $\lambda_1, \lambda_2, \lambda_3, \lambda_4 \in \mathbb{R}^{n^2}$ are the Lagrange multipliers.

For a given $(A^k, z^k, u^k, f^k; \lambda_1^k, \lambda_2^k, \lambda_3^k, \lambda_4^k)$, the next iteration $(A^{k+1}, z^{k+1}, u^{k+1}, f^{k+1}; (A^{k+1}, z^{k+1}, u^{k+1}, f^{k+1}; \lambda_1^{k+1}, \lambda_2^{k+1}, \lambda_3^{k+1}, \lambda_4^{k+1})$ is generated as follows:

1. Fix $f = f^k, \lambda_1 = \lambda_1^k, \lambda_2 = \lambda_2^k, \lambda_3 = \lambda_3^k, \lambda_4 = \lambda_4^k$ and minimize (41) with respect to A, z, and u. With respect to A,

$$A^{k+1} = \arg\min \|A\|_{W,2,1} - (\lambda_1^T, \lambda_2^T)(A - \begin{bmatrix}\nabla_x f^k\\ \nabla_y f^k\end{bmatrix}) + \frac{\beta_1}{2}\left\|A - \begin{bmatrix}\nabla_x f^k\\ \nabla_y f^k\end{bmatrix}\right\|_2^2,$$
$$\begin{bmatrix}A_1^{k+1}\\ A_2^{k+1}\end{bmatrix} = \arg\min \|A\|_{W,2,1} + \frac{\beta_1}{2}\left\|\begin{bmatrix}A_1\\ A_2\end{bmatrix} - \begin{bmatrix}\nabla_x f^k\\ \nabla_y f^k\end{bmatrix} - \begin{bmatrix}\lambda_1^k/\beta_1\\ \lambda_2^k/\beta_1\end{bmatrix}\right\|_2^2. \tag{42}$$

It is obvious that problem (42) match the framework of the problem (5), thus the solution of (42) can be obtained by using the formulas in Section 2.2.

With respect to z,

$$z^{k+1} = \arg\min \mu\|z\|_1 - \lambda_3^{k^T}(z - (Hf^k - g)) + \frac{\beta_2}{2}\|z - (Hf^k - g)\|_2^2$$
$$= \arg\min \mu\|z\|_1 + \frac{\beta_2}{2}\|z - (Hf^k - g) - \frac{\lambda_3^k}{\beta_2}\|_2^2.$$

The minimization with respect to z can be given by (6) and (9) explicitly, that is,

$$z^{k+1} = \mathrm{sgn}\left\{Hf^k - g + \frac{\lambda_3^k}{\beta_2}\right\} \circ \max\left\{|Hf^k - g + \frac{\lambda_3^k}{\beta_2}| - \frac{\mu}{\beta_2}, 0\right\}. \quad (43)$$

With respect to u,

$$\begin{aligned} u^{k+1} &= \arg\min -\lambda_4^{k^T}(u - f^k) + \tfrac{\beta_3}{2}\|u - f^k\|_2^2 \\ &= \arg\min \tfrac{\beta_3}{2}\|u - f^k - \tfrac{\lambda_4^k}{\beta_3}\|_2^2. \end{aligned}$$

The minimizer is given explicitly by

$$u^{k+1} = \mathcal{P}_\Omega\left[f^k + \frac{\lambda_4^k}{\beta_3}\right] \quad (44)$$

2. Compute f^{k+1} by solving the following normal equation similarly as the last section.

$$\begin{aligned} &(\beta_1(\nabla_x^*\nabla_x + \nabla_y^*\nabla_y) + \beta_2 H^*H + \beta_3 I)f^{k+1} \\ &= \nabla_x^*(\beta_1 A_1^{k+1} - \lambda_1^k) + \nabla_y{}^*(\beta_1 A_2^{k+1} - \lambda_2^k) + \\ &H^*(\beta_2 z^{k+1} - \lambda_3^k) + \beta_2 H^* g + \beta_3(u^{k+1} - \tfrac{\lambda_4^k}{\beta_3}). \end{aligned} \quad (45)$$

3. Update the multipliers via

$$\begin{cases} \begin{bmatrix}\lambda_1^{k+1}\\ \lambda_2^{k+1}\end{bmatrix} = \begin{bmatrix}\lambda_1^{k}\\ \lambda_2^{k}\end{bmatrix} - \gamma\beta_1\left(\begin{bmatrix}A_1^{k+1}\\ A_2^{k+1}\end{bmatrix} - \begin{bmatrix}\nabla_x f^{k+1}\\ \nabla_y f^{k+1}\end{bmatrix}\right), \\ \lambda_3^{k+1} = \lambda_3^k - \gamma\beta_2(z^{k+1} - (Hf^{k+1} - g)), \\ \lambda_4^{k+1} = \lambda_4^k - \gamma\beta_3(u^{k+1} - f^{k+1}). \end{cases} \quad (46)$$

Based on the discussions above, we present the ADMM algorithm for solving the convex CITVOGSL1 model (40), which is shown as Algorithm 4.

Algorithm 4 CITVOGSL1 for the minimization problem (40)

initialization:
Starting point $f^0 = g$, $k = 0$, $\beta_1, \beta_2, \beta_3, \gamma, \mu$, group size $K_1 \times K_2$, weighted matrix W_g, $\lambda_i^0 = 0, i = 1,2,3,4$.
iteration:
1. Compute A^{k+1} according to (42),
and compute z^{k+1} according to (43).
2. Compute u^{k+1} according to (44).
3. Compute f^{k+1} by solving (45).
4. update $\lambda_i^0 = 0, i = 1,2,3,4$ according to (46).
5. $k = k + 1$.
until a stopping criterion is satisfied.

Algorithm 4 is an application of ADMM for the case with two blocks of variables (A,z,u) and f. Thus, if step (1) of Algorithm 4 can be solved exactly, its convergence is guaranteed by the theory of ADMM [34, 35, 37, 43]. Although step (1) of Algorithm 4 cannot be solved exactly, we can find a convergent series to ensure the convergence as [35]. Besides, our numerical experiments verify the convergence of Algorithm 4.

For the ATV case, the constrained model (CATVOGSL1) is

$$\min_{u\in\Omega, v_x, v_y, f} \left\{ \begin{array}{l} \|(v_x)\|_{W,2,1} + \|(v_y)\|_{W,2,1} + \mu\|z\|_1 : \\ z = Hf - g, v_x = \nabla_x f, v_y = \nabla_y f, u = f \end{array} \right\}. \quad (47)$$

The detail has been presented in the last section for the constrained TV OGS L_2 model. We call this relevant algorithm CATVOGSL1.

Numerical results

In this section, we present several numerical results to illustrate the performance of the proposed method. All experiments are carried out on a desktop computer using MATLAB 2010a. Our computer is equipped with an Intel Core i3-2130 CPU (3.4 GHz), 3.4 GB of RAM, and a 32-bit Windows 7 operation system.

Comparison with the MM Method for one-Dimensional Signal Denoising

As an illustrative example, we apply our direct formulas in one-dimensional signal denoising. We only compare the results of our explicit shrinkage formulas with the most recent MM iteration method proposed in [10] as a simple example. The one-dimensional group sparse signal **z** is shown in the top left of Fig. 3. The noisy signal **x** in the top right of Fig. 3 is obtained by adding independent white Gaussian noise with standard deviation σ=0.5 same as [10]. The denoising model is as follows

$$\arg\min_{\mathbf{z}} \quad \mathrm{Fun}(\mathbf{z}) = \|\mathbf{z}\|_{2,1} + \frac{\beta}{2}\|\mathbf{z} - \mathbf{x}\|_2^2. \quad (48)$$

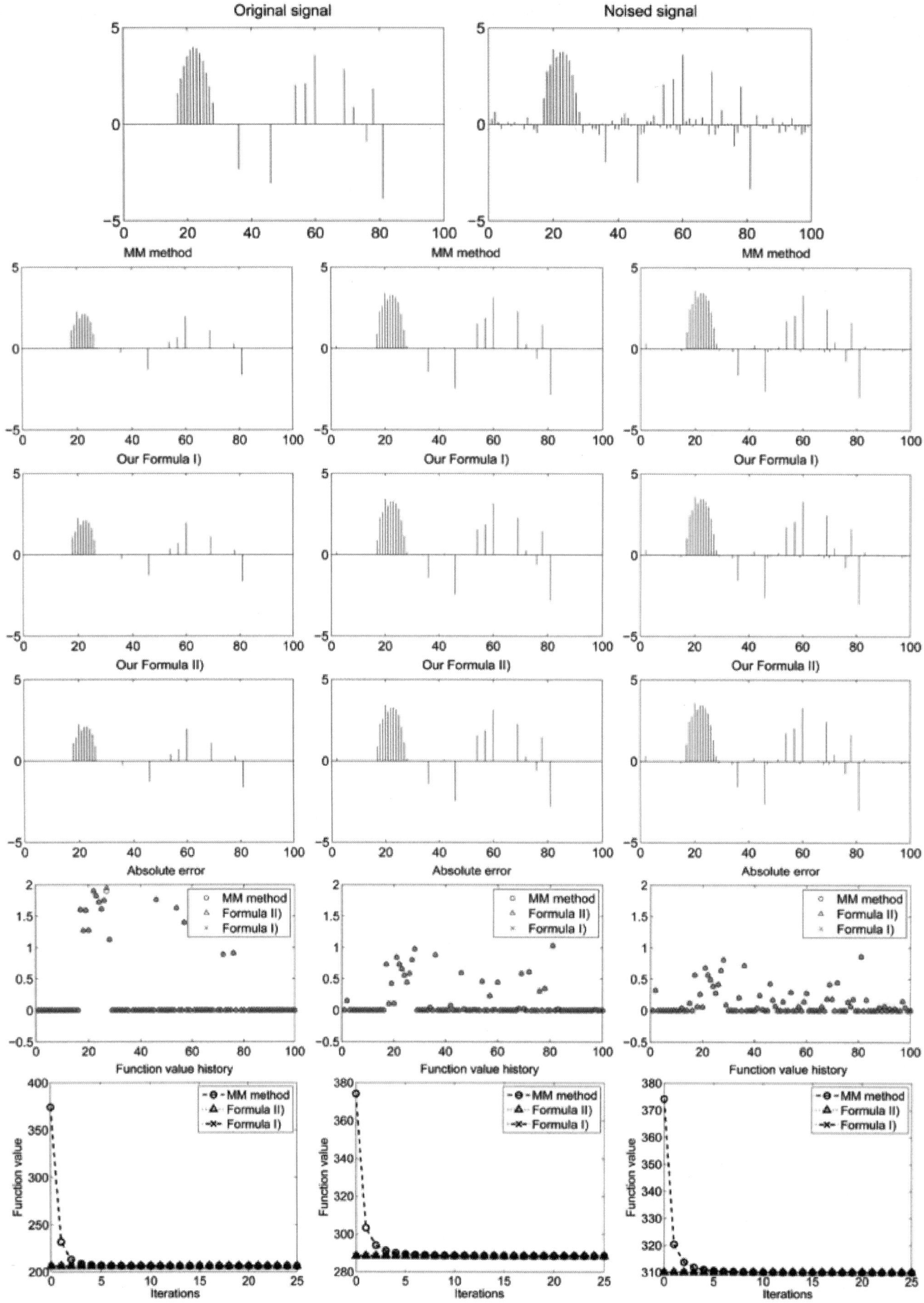

FigURE. 3: Comparison between our method and the MM method for one-dimensional signal denoising. Second row to bottom row, from top to bottom, results of the MM method, our Formula (I), our Formula (II) respectively, absolute error, and function value history. From left to right, β=3,10,15 respectively

We test several different parameters β (any integer from 1 to 50) for the comparison. The results of our formulas are almost the same as the MM method for almost these βs. Due to the limit space, we only illustrate parts of the results (β=3,10,15) in Fig. 3. We take both our Formulas (I) and (II) for the comparison with the MM method. From the figure, we can visually see that our results by our two kinds of formulas are almost the same as the MM method for different β. Moreover, for the practical problems, β = 10, can be better than others from the figure. This shows that our formulas are feasible, useful, and effective, because our method can only need the same computation cost as one-step iteration in the MM method while the MM method needs 25 iterations as [10]. That means our method is 25 times faster than the MM method. From the bottom line of Fig. 3, we can see that the MM method may take 5 iterations to be convergent, which is the reason why Liu et al. in [11] and Liu et al. in [12] only choose 5 inner iterations. However, in this case, our method is still 5 times faster than the MM method.

Remarks 5.

Although this model is not the best model for signal denoising, it shows the superiority of our method. Moreover, the authors in [10] also take similar experiments.

Comparison with the MM Method on Problem (5)

In this section, we give another example by solving the matrix case problem (5) for comparison with the MM method. Without loss of generality, we let A=rand(100,100) and A(45:55,45:55)=0 in the example. In particular, we set weighted matrix $\mathbb{R}^{3\times 3}((W_g)_{i,j} \equiv 1)$ and list (5) again

$$f_A(X) = \min_X \text{Fun}(X) = \|X\|_{2,1} + \frac{\beta}{2}\|X - A\|_F^2. \quad (49)$$

We use different parameters β and use the number of MM iteration by 20 steps. For comparison, we expand our result by explicit shrinkage formulas to length 20 (the same as the MM iteration steps). The values of the object function Fun(X) against iterations are illustrated in the top line in Fig. 4. The cross-sectional elements of the minimizer A are shown in the bottom line in Fig. 4. We choose both ZBC and PBC for our method and the MM method. It can be seen that there is only a little difference between the two kinds of BCs.

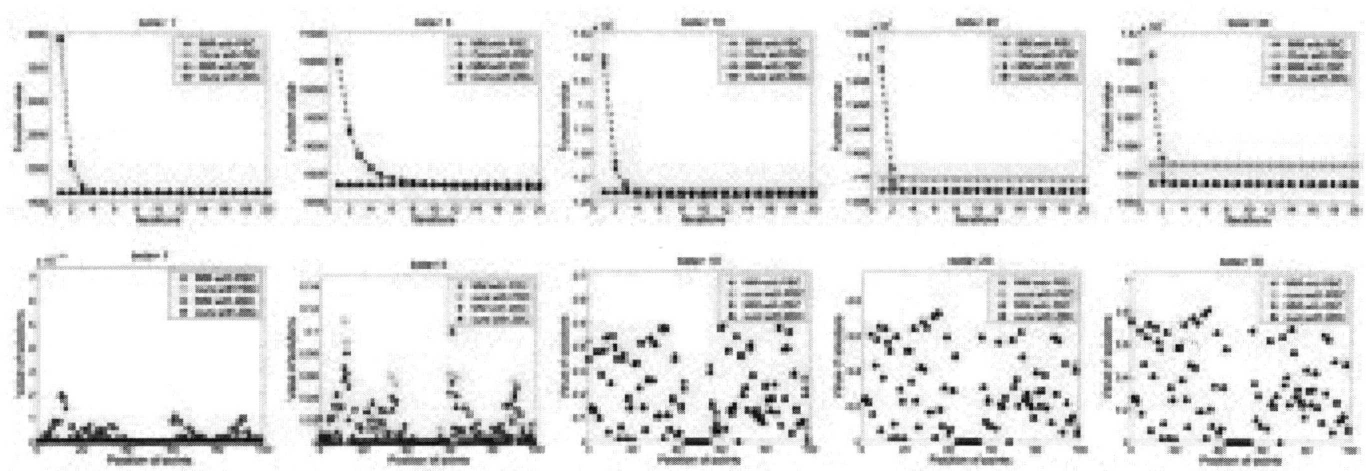

Figure. 4: Comparison between our method and the MM method on two kinds of BCs, ZPC, and PBC. Top row, the function value history. Bottom row, the cross line of the final minimizer—matrix X. From left to right, β=1,5,10,20,30

It is obvious that our method is much more efficient than the MM method since our time complexity is just the same as one-step iteration in the MM method due to the explicit shrinkage formula. From the top line in Fig. 4, we observe that the MM method is also fast because it only needs less than 20 steps (sometimes 5) to converge. The related error of the function value of two methods is much less than 1 % for different parameters β. From the bottom line in Fig. 4, we can see that when β≤1 which is sufficiently small, and when β≥30 which is sufficiently large, our result is almost the same as the final results by the MM iteration method. This shows that the results by our method are almost the same as the MM method. When 1<β<30, the minimizer computed by our method is approximate to the MM method both on the error of the function value and on the minimizer X.

In Table 1, we show the numerical comparison of our method and the MM method on three parts, the related error of function value (ReEfAReEfA), related error of minimizer X (ReE_X), and the mean absolute error of X (MAE_X). The three terms are defined by $ReE_{f_A} = \frac{|f_{A_{MM}} - f_{A_{ours}}|}{f_{A_{ours}}}$, $ReE_X = \frac{\|X_{MM} - X_{ours}\|_F}{\|X_{ours}\|_F}$, and $MAE_X =$ mean absolute error of $(X_{MM} - X_{ours})$, where X_{MM} , and MAE_X =mean absolute error of $(X_{MM} - X_{ours})$, where X_{MM} and X_{ours} are the final solution X of the MM method and our method, respectively. $f_{A\,MM}$ and $f_{A\,ours}$ are the final function value of MM and our method, respectively.

Table 1: Numerical comparison of our method and the MM method for two kinds of BCs, ZBC, and PBC

BC	β=	1	5	7	10	15	20	30	50
ZBC	ReEfAReEfA	5.9e-14	0.0055	0.0028	6.5e-4	1.4e-4	5.4e-5	1.4e-5	2.6e-6
	ReE_X	—	—	0.0207	0.0017	1.9e-4	4.7e-5	7.5e-6	7.8e-7
	MAE_X	3.4e-15	0.0094	0.0130	0.0067	0.0029	0.0016	6.9e-4	2.4e-4
PBC	ReEfAReEfA	6.5e-14	0.0053	0.0025	4.1e-4	5.5e-5	1.1e-5	6.3e-7	1.3e-6
	ReE_X	—	—	0.0193	0.0014	1.3e-4	2.9e-5	4.3e-6	4.5e-7
	MAE_X	3.8e-15	0.0097	0.0130	0.0063	0.0026	0.0014	5.9e-4	2.1e-4

ReE of $f_A(X)$ denotes the related error of final function value fun (X). ReE of X denotes related error of minimizer X. MAE of Xdenotes the mean absolute error of X

From the table and the figure, we can see that our formula can almost get the same results as the MM method whenβ is sufficiently large and approximate results similar as the MM method for other β. This is another proof for the feasibility of our formula.

Similarly as weighted matrix $W_g \in \mathbb{R}^{3\times3}((W_g)_{i,j} \equiv 1)$, we also tested other weighted matrix for more than 1000 times. For example, when A=rand(100,100) and A(A>=0.5)=0 (or A(A<=0.5)=1) (every element is in [0,1]), we find that, generally $\beta \leq \frac{\|W_g\|_2}{\sqrt{s}}$ is sufficiently small and $\beta \geq 30 \cdot \frac{\|W_g\|_2}{\sqrt{s}}$ is sufficiently large. These results illustrate that the former when $W_g \in \mathbb{R}^{3\times3}((W_g)_{i,j} \equiv 1)$ β≥30 which is sufficiently large again. However, in practice, if β is too large, then A=X, and the minimization problem (49) is meaningless. In our experiments after more than 1000 tests, we find that, when every element of A is in [0,1], $30 \frac{\|W_g\|_2}{\sqrt{s}} \leq \beta \leq 300 \cdot \frac{\|W_g\|_2}{\sqrt{s}}$ is sufficiently large but not too large to make the minimization problem meaningless. In this case, that means we may tune the parameter to be better in practice. On the other hand, when some elements of A are not in [0,1], we can first project or stretch A into the region [0,1] and then choose the better parameter β. Moreover, although the parameter is not chosen to be in this region, the result can also be treated as an approximate solution with small error. That means our formulas are feasible, useful, and effective.

Comparison with TV Methods and TV with OGS with Inner Iteration MM Methods for image Deblurring and Denoising

In this section, we compare all our algorithms with other methods. All the test images are shown in Fig. 5, one 1024×1024 image as (a) Man and three 512×512 images as (b) Car, (c) Parlor, and (d) Housepole.

Figure. 5: Original images. The versions of the images in this paper are special formats which are converted by Photoshop from the sources downloaded from http://decsai.ugr.es/cvg/dbimagenes/ . **a** Man. **b** Car. **c** Parlor. **d** Housepole

The quality of the restoration results is measured quantitatively by using the peak signal-to-noise ratio (PSNR) in decibel (dB) and the relative error (ReE):

$$\text{PSNR} = 10\log_{10}\frac{n^2\text{Max}_I^2}{\|f-\bar{f}\|_2^2}, \quad \text{ReE} = \frac{\|f-\bar{f}\|_2}{\|\bar{f}\|_2},$$

where $\bar{f}$ and f denote the original and restored images, respectively, and Max_I represents the maximum possible pixel value of the image. In our experiments, Max_I=1.

The stopping criterion used in our work is set to be as other methods

$$\frac{|\mathcal{F}^{k+1} - \mathcal{F}^k|}{|\mathcal{F}^k|} < 10^{-5}, \quad (50)$$

where FkFk is the objective function value of the respective model in the kth iteration.

We compare our methods with some other methods, such as Chan et al.'s TV method proposed in [32], Liu et al.'s method proposed in [11], and Liu et al.'s method proposed in [12]. Both the latter two methods [11, 12] are with inner iteration MM methods for OGS TV problems, where the number of the inner iterations is set 5 by them. In particular, for a fair comparison, we set weighted matrix Wg∈R3×3((Wg)i,j≡1) Wg∈R3×3((Wg)i,j≡1) in all the experiments of our methods as in [11, 12].

Experiments for the Constrained TV OGS L_2 Model

In this section, we compare our methods (CATVOGSL2

and CITVOGSL2) with some other methods, such as Chan et al.'s method proposed in [32] (Algorithm 1 in [32] for the constrained TV-L2 model) and Liu et al.'s method proposed in [11].

We set the penalty parameters $\beta_1=35$, $\beta_2=20$, for the ATV case, $\beta_1=100$, $\beta_2=20$ for the ITV case and relax parameter $\gamma=1.618$. The blur kernels are generated by MATLAB built-in function (i)fspecial('average',9) for 9×9 average blur. We generate all blurring effects using the MATLAB built-in function imfilter(I,psf,'circular','conv') under PBC with "I" as the original image and "psf" as the blur kernel. We first generated the blurred images operating on images (a)–(c) by the former Gaussian blurs and further corrupted by zero mean Gaussian noise with BSNR =40. The BSNR is given by

BSNR=10log10\textmdvarianceofg\textmdvarianceofη.

BSNR=10log10\textmdvarianceofg\textmdvarianceofη.

where g and η are the observed image and the noise, respectively.

The numerical results of the three methods are shown in Table 2. We have tuned the parameters for all the methods as in Table 2. From Table 2, we can see that PSNR and ReE of our methods (both ATV and ITV cases) are almost same as [11], which applied MM inner iterations to solve the subproblems (34) and (35) (only for the ATV case). However, each outer iteration of our methods is nearly twice faster than [11] from the experiments. The time of each outer iteration of our methods is almost the same as the traditional TV method in [32]. In Fig. 6, we display the restored "Parlor" images from different algorithms. We can see that OGS TV regularizers can get clearer edges on the "desk" of the image than TV regularizer.

Figure. 6: Top row: blurred and noisy image (left), restoration images of Chan [32] (middle), Liu [11] (right). Bottom row: original image (left), restoration images of CATVOGSL2 (middle), CITVOGSL2 (right)

Table 2: Numerical comparison of Chan [32], Liu [11], CATVOGSL2, and CITVOGSL2 for images (a)–(c) in Fig. 5

Images		**(a) Man**				**(b) Car**				**(c) Parlor**			
Method	μ	**Itr**	**PSNR**	**Time**	**ReE**	**Itr**	**PSNR**	**Time**	**ReE**	**Itr**	**PSNR**	**Time**	**ReE**
Chan [32]	0.5	15	30.34	6.02	0.0730	11	31.13	2.00	0.0426	12	31.70	2.01	0.0511
Liu [11]	1	12	30.60	9.06	0.0708	12	31.68	3.93	0.0386	13	32.46	4.21	0.0464
CAOL2	1	7	30.62	3.37	0.0711	9	31.68	1.76	0.0399	7	32.40	1.34	0.0472
CIOL2	1	8	30.59	3.70	0.0716	11	31.60	2.04	0.0403	8	32.03	1.61	0.0492

PSNR dB, Time s, Itr iterations, μ regularization parameter ($\times 10^5$), CAOL2 CATVOGSL2, CIOL2 CITVOGSL2

Now, we compute the complexity of each step of our methods and Liu et al.'s method [11]. Firstly, we know that the complexity of all the methods is 512×512×18×4 (4 times n log2n) except the OGS subproblems. Then, for the OGS subproblems, the MM method in [11] with five-step inner iteration, the complexity is 512×512×90×2 (2 subproblems). The complexity of our methods is 512×512×18×2 in CATVOGSL2 and 512×512×18 in CITVOGSL2. Therefore, the total complexity of [11] is $512^2 \times 252$, the total complexity of CATVOGSL2 is $512^2 \times 108$, and the total complexity of CITVOGSL2 is $512^2 \times 90$. That means each step of our methods is more than double faster than the inner iteration method in [11]. In the next section, the common computation parts of our methods and the inner iteration method are much more, and then our methods are only nearly double faster.

Remarks 6.

We do not list the results of image (d) because they are almost the same as images (b) and (c). Moreover, when $\beta_1<30$ which is not sufficiently large, the numerical results are also good while we did not list them. This shows that the approximate part of our shrinkage formula is also good, and that when the inner step in [11, 12] is chosen to be 5, the numerical experiments are convergent although they did not find a convergence control sequence. In addition, in our experiments, we find that the results of the ATV case are better than the ITV case, which is a little different from the classic TV methods. Moreover, we find that if the parameters of [11] are chosen to be the same as ours, the solutions of our method and their method are always the same on PSNR and visual presentation while our method may save more time (it remains valid for [12]).

Experiments for the Constrained TV OGS L_1 Model

In this section, we compare our methods (CATVOGSL1 and CITVOGSL1) with some other methods, such as Chan et al.'s method proposed in [32] (Algorithm 2 in [32] for the constrained TV-L1 model) and Liu et al.'s method proposed in [12].

Similarly as the last section, we set the penalty parameters $\beta_1=80$, $\beta_2=2000$, $\beta_3=1$, for the ATV case, $\beta_1=80$, $\beta_2=2000$, $\beta_3=1$, for the ITV case and relax parameter $\gamma=1.618$. The blur kernel is generated by MATLAB built-in function fspecial('gaussian',7,5) for 7×7 Gaussian blur with standard deviation 5. We first generated the blurred images operating on images (b)–(d) by the former Gaussian blur and further corrupted them by salt-and-pepper noise from 30 to 40 %. We generate all noise effects by MATLAB built-in function imnoise(Bl,'salt & pepper',level) with "Bl" the blurred image and fix the same random matrix for different methods.

The numerical results by the three methods are shown in Table 3. We have tuned the parameters manually to give the best PSNR improvement for Chan [32] as in Table 3 for different images. For Liu [12], we choose the given parameters μ default as 100, 80 for 30 and 40 %, respectively. For our method CATVOGSL1, we set μ as 180, 140 for 30 and 40 %, respectively. For our method CITVOGSL1, we set μ as 140, 100 for 30 and 40 %, respectively. In the experiments, we find that the parameters of our methods are robust and have wide range to choose. Therefore, we set the same μ for different images.

Table 3: Numerical comparison of Chan [32], Liu [12], CATVOGSL1, and CITVOGSL1 for images (a)–(d) in Fig. 5

Is	NL	Chan [32] μ/Itr	PSNR	Time	ReE	Liu [12] Itr	PSNR	Time	ReE
(a)	30	25/130	29.59	61.57	0.0796	35	30.92	30.23	0.0683
	40	18/ 99	28.85	46.41	0.0866	37	30.11	31.54	0.0749
(b)	30	25/128	29.71	22.90	0.0501	35	31.81	12.93	0.0393
	40	20/ 98	28.59	17.80	0.0570	40	30.70	14.99	0.0447
(c)	30	26/138	30.21	25.65	0.0607	35	32.15	12.84	0.0485
	40	22/105	29.10	19.27	0.0689	37	31.04	13.67	0.0551
(d)	30	26/127	30.41	23.17	0.0975	36	32.47	13.15	0.0769
	40	20/ 95	29.44	17.43	0.1091	39	31.43	14.27	0.0867
		CATVOGSL1				CITVOGSL1			
(a)	30	32	31.17	17.97	0.0663	41	31.34	21.56	0.0651
	40	32	30.10	18.03	0.0751	37	30.06	19.95	0.0754
(b)	30	26	31.77	5.97	0.0395	29	31.75	6.26	0.0396
	40	26	30.47	6.12	0.0459	29	30.22	6.38	0.0472
(c)	30	25	32.30	5.82	0.0477	30	32.08	6.75	0.0489
	40	26	31.01	5.76	0.0553	27	30.50	6.19	0.0587
(d)	30	25	32.54	5.68	0.0764	29	32.51	6.75	0.0766

Is	N	Chan [32]				Liu [12]			
	L	μ/Itr	PSNR	Time	ReE	Itr	PSNR	Time	ReE
	40	28	31.35	6.40	0.0875	32	31.06	7.33	0.0905

PSNR dB, Time s, Is images, NL noise level (%)

From Table 3, we can also see that PSNR and ReE of our methods (both ATV and ITV cases) are almost the same as Liu [12], which applied MM inner iterations to solve the subproblems (34) and (35) (only for the ATV case). However, each outer iteration of our methods is nearly twice faster than Liu [12] from the experiments. The time of each outer iteration of our methods is almost the same as the traditional TV method in Chan [32]. Moreover, we can also see that sometimes ATV is better than ITV and sometimes on the contrary for OGS TV. Finally, in Fig. 7, we display the degraded image, the original image, and the restored images for 30 % level of noise on image (d) by the four methods. From the figure, we can see that both our methods and Liu [12] can get better edges (handrail and window) than Chan [32].

Figure. 7: Top row: blurred and noisy image (left), restoration images of Chan [32] (middle), Liu [12] (right). Bottom row: original image (left), restoration images of CATVOGSL1 (middle), CITVOGSL1 (right)

CONCLUSIONS

In this paper, we propose the explicit shrinkage formulas for one class of OGS regularization problems with translation invariant overlapping groups. These formulas can be extended to several other regularization OGS problems as a subproblem in many fields. In this work, we apply our results in OGS TV regularization problems—deblurring and denoising problems. Furthermore, we also extend the image deblurring problems with OGS ATV in [11, 12] to both ATV and ITV cases. Since the formulas are very simple, these results can be easily extended to many other applications such as multichannel deconvolution and compress sensing, which we will consider in the future. In addition, in the experiments, we only choose all the entries of the weight matrix W_g equal to 1. We will test for other weights in the future on more experiments in order to choose the better or best weights for some other applications.

REFERENCES

1. W Deng, W Yin, Y Zhang, in Proc. SPIE. Group sparse optimization by alternating direction method (SPIESan Diego, US, 2013), pp. 88580R–15.
2. J Huang, T Zhang, The benefit of group sparsity. Ann. Stat.**38**(4), 1978–2004 (2010). doi:http://dx.doi.org/10.1214/09-AOS778.
3. R Chartrand, B Wohlberg, in IEEE International Conference on Acoustics, Speech and Signal Processing (ICASSP). A nonconvex ADMM algorithm for group sparsity with sparse groups (IEEEVancouver, BC, 2013), pp. 6009–6013.
4. M Stojnic, F Parvaresh, B Hassibi, On the reconstruction of block-sparse signals with an optimal number of measurements. IEEE Trans. Signal Process. **57**(8), 3075–3085 (2009).

5. L Jacob, G Obozinski, J-P Vert, in Proceedings of the 26th Annual International Conference on Machine Learning. ICML '09. Group lasso with overlap and graph lasso (ACMNew York, 2009), pp. 433–440.

6. A Majumdar, RK Ward, in IEEE International Conference on Acoustics, Speech and Signal Processing (ICASSP). Classification via group sparsity promoting regularization (IEEETaipei, 2009), pp. 861–864.

7. E Elhamifar, R Vidal, in IEEE Conference on Computer Vision and Pattern Recognition (CVPR). Robust Classification using structured sparse representation (IEEEProvidence, 2011), pp. 1873–1879.

8. J Gao, Q Shi, TS Caetano, Dimensionality reduction via compressive sensing. Pattern Recognit. Lett. **33**(9), 1163–1170 (2012).

9. YN Liu, F Wu, YT Zhuang, Group sparse representation for image categorization and semantic video retrieval. Sci. China Inform. Sci. **54**(10), 2051–2063 (2011).

10. P-Y Chen, IW Selesnick, Translation-invariant shrinkage/thresholding of group sparse signals. Signal Process. **94**(0), 476–489 (2014).

11. J Liu, T-Z Huang, IW Selesnick, X-G Lv, P-Y Chen, Image restoration using total variation with overlapping group sparsity. Inform. Sci. **295**(20), 232–246 (2015).

12. G Liu, T-Z Huang, J Liu, X-G Lv, Total variation with overlapping group sparsity for image deblurring under impulse noise. Plos One. **10**(4), 0122562 (2015).

13. IW Selesnick, P-Y Chen, in IEEE International Conference on Acoustics, Speech and Signal Processing (ICASSP). Total variation denoising with overlapping group sparsity (IEEEVancouver, BC, 2013), pp. 5696–5700.

14. M Figueiredo, J Bioucas-Dias, in Signal Processing with Adaptive Sparse Structured Representations - SPARS11. An alternating direction algorithm for (overlapping) group regularization (Edinburgh, Scotland, UK, 2011).

15. R Jenatton, J-Y Audibert, F Bach, Structured variable selection with sparsity-inducing norms. J. Mach. Learn. Res.**12:**, 2777–2824 (2011).

16. M Kowalski, Sparse regression using mixed norms. Appl. Comput. Harmonic Anal. **27**(3), 303–324 (2009).

17. G Peyré, JM Fadili, in EUSIPCO. Group sparsity with overlapping partition functions (Barcelona, Spain, 2011), pp. 303–307.

18. F Bach, R Jenatton, J Mairal, G Obozinski, Structured sparsity through convex optimization. Stat. Sci. **27**(4), 450–468 (2012).

19. I Bayram, in IEEE International Conference on Acoustics, Speech and Signal Processing (ICASSP). Mixed norms with overlapping groups as signal priors (IEEEPrague, 2011), pp. 4036–4039.

20. Y Wang, J Yang, W Yin, Y Zhang, A new alternating minimization algorithm for total variation image reconstruction. SIAM J. Imaging Sci. **1**(3), 248–272 (2008).

21. J Yang, W Yin, Y Zhang, Y Wang, A fast algorithm for edge-preserving variational multichannel image restoration. SIAM J. Imaging Sci.**2**(2), 569–592 (2009). doi:http://dx.doi.org/10.1137/080730421.

22. B Wohlberg, R Chartrand, J Theiler, in IEEE International Conference on Acoustics, Speech and Signal Processing (ICASSP). Local principal component pursuit for nonlinear datasets (IEEEKyoto, 2012), pp. 3925–3928.

23. P Sprechmann, I Ramirez, G Sapiro, YC Eldar, C-HiLasso: a collaborative hierarchical sparse modeling framework. IEEE Trans. Signal Process. **59**(9), 4183–4198 (2011).

24. PC Hansen, JG Nagy, DP O'Leary, Deblurring images: matrices, spectra, and filtering, 1st edn. (Society for Industrial and Applied Mathematics, Philadelphia, United States, 2006).**MATH**

25. LI Rudin, S Osher, E Fatemi, Nonlinear total variation based noise removal algorithms. Physica D: Nonlinear Phenomena.**60**(1–4), 259–268 (1992).

26. A Chambolle, An algorithm for total variation minimization and applications. J. Math. Imaging Vis.**20**(1–2), 89–97 (2004).**MathSciNet**

27. G Liu, T-Z Huang, J Liu, High-order TVL1-based images restoration and spatially adapted regularization parameter selection. Comput. Math. Appl. **67**(10), 2015–2026 (2014).

28. X-L Zhao, W Wang, T-Y Zeng, T-Z Huang, MK Ng, Total variation structured total least squares method for image restoration. SIAM J. Sci. Comput. **35**(6), 1304–1320 (2013).

29. X-L Zhao, F Wang, MK Ng, A new convex optimization model for multiplicative noise and blur removal. SIAM J. Imaging Sci.**7**(1), 456–475 (2014).

30. C Chen, MK Ng, X-L Zhao, Alternating direction method of multipliers for nonlinear image restoration problems. IEEE Trans. Image Process. **24**(1), 33–43 (2015).

31. L-J Deng, H Guo, T-Z Huang, A fast image recovery algorithm based on splitting deblurring and denoising. J. Comput. Appl. Math. **287:**, 88–97 (2015).

32. R Chan, M Tao, X Yuan, Constrained total variation deblurring models and fast algorithms based on alternating direction method of multipliers. SIAM J. Imaging Sci. **6**(1), 680–697 (2013).

33. J Nocedal, SJ Wright, Numerical optimization. 1431–8598 (Springer, New York, 2006).

34. D Gabay, B Mercier, A dual algorithm for the solution

of nonlinear variational problems via finite element approximation. Comput. Math. Appl.**2**(1), 17–40 (1976). **MATH**

35. J Eckstein, D Bertsekas, On the Douglas-Rachford splitting method and the proximal point algorithm for maximal monotone operators. Math. Programm. **55**(1–3), 293–318 (1992). doi:http://dx.doi.org/10.1007/BF01581204.

36. R Glowinski, Numerical methods for nonlinear variational problems. 1434–8322 (Springer, Berlin Heidelberg, 1984).

37. B He, H Yang, Some convergence properties of a method of multipliers for linearly constrained monotone variational inequalities. Oper. Res. Lett.**23**(3–5), 151–161 (1998).

38. E Esser, Applications of Lagrangian-based alternating direction methods and connections to split Bregman. Los Angeles: UCLA CAM report, 9–31 (2009). ftp://arachne.math.ucla.edu/pub/camreport/cam09-31.pdf.

39. M Ng, P Weiss, X Yuan, Solving constrained total-variation image restoration and reconstruction problems via alternating direction methods. SIAM J. Sci. Comput. **32**(5), 2710–2736 (2010).

40. X Zhang, M Burger, X Bresson, S Osher, Bregmanized nonlocal regularization for deconvolution and sparse reconstruction. SIAM J. Imaging Sci. **3**(3), 253–276 (2010).

41. J Yang, Y Zhang, W Yin, A fast alternating direction method for TVL1-L2 signal reconstruction from partial fourier data. IEEE J. Selected Topics Signal Process. **4**(2), 288–297 (2010).

42. X Zhang, M Burger, S Osher, A unified primal-dual algorithm framework based on Bregman iteration. J. Sci. Comput. **46**(1), 20–46 (2011).

43. S Boyd, N Parikh, E Chu, B Peleato, J Eckstein, Distributed optimization and statistical learning via the alternating direction method of multipliers. Foundations Trends Mach. Learn. **3**(1), 1–122 (2011).**MATH**

44. C Wu, X Tai, Augmented Lagrangian method, dual methods, and split Bregman iteration for ROF, vectorial TV, and high order models. SIAM J. Imaging Sci. **3**(3), 300–339 (2010).

CHAPTER 10

MULTIVIEW VIDEO PLUS DEPTH TRANSMISSION VIA VIRTUAL-VIEW-ASSISTED COMPLEMENTARY DOWN/UPSAMPLING

Zhi Jin[1,2], Tammam Tillo[1], Jimin Xiao[1] and Yao Zhao[3]

[1]Department of Electrical and Electronic Engineering, Xi'an Jiaotong-Liverpool University, Ren Ai Road 111, 215123 Suzhou, People's Republic of China

[2]Department of Electrical Engineering and Electronics, University of Liverpool, Merseyside L69

[3]BX, Liverpool, UK

ABSTRACT

Multiview video plus depth is a popular 3D video format which can provide viewers a vivid 3D feeling. However, its requirements in terms of computational complexity and transmission bandwidth are more than that of conventional 2D video. To mitigate these limitations, some works have proposed to reduce the amount of transmitted data by adopting different resolutions for different views, and consequently, the transmitted video is called mixed resolution video. In order to further reduce the transmitted data and maintain good quality at the decoder side; in this paper, we propose a down/upsampling algorithm for 3D multiview video which systematically takes into account the video encoder and decoder. At the encoder side, the rows of the two adjacent views are downsampled following an interlacing and complementary fashion, whereas, at the decoder side, the discarded pixels are recovered by fusing the virtual view pixels with the directional interpolated pixels from the complementary downsampled views. Moreover, the patterns of the texture surrounding the discarded pixels are used to aid the data fusion, so as to enhance edges recovery. Meanwhile, with the assistance of virtual views, at the decoder side, the proposed approach can effectively recover the discarded high-frequency details. The experimental results demonstrate the superior performance of the proposed framework.

INTRODUCTION

The development of 3D technologies and communication networks makes 3D video applications increasingly popular. An example of this technology is 3D multiview television which allows to cover a wide view angle of the scene. However, delivering a large number of high-quality views to end users is a challenging task due to the limitations of data transmission and storage capacity [1]. Therefore, some advanced video coding standards have been proposed to compress video data, such as H.264/MVC [2] and HEVC [3, 4, 5, 6]. Besides that, some data representation formats for the 3D multiview data have been adopted for efficient representation. One widely accepted format is the multiview video plus depth (MVD) format [7], which consists of textures and the associated per-pixel depth data (the latter describes the geometric relationship between objects in the scene and the capturing cameras [8]). Since this

Citation: Zhi Jin, Tammam Tillo, Jimin Xiao and Yao Zhao, Multiview video plus depth transmission via virtual-view-assisted complementary down/upsampling, DOI 10.1186/s13640-016-0119-4.

format allows any intermediate view within a certain range to be generated, with the assistance of the depth-image-based rendering (DIBR) technique [9]. Therefore, it can, to a large extent, reduce the number of transmitted views. However, the required data of 3D multiview video is still very large.

Because of the above reasons, many works have focused on reducing the amount of transmitted data at the encoder side and recovering it at the decoder side for low bit rate transmission [10, 11, 12]. Additionally, given that depth maps consist of large homogeneous areas, they require less transmission bit rate compared with texture [13]. Therefore, the reduction and recovery of texture data have drawn more attention than that of the depth data. In order to speed up the encoding process and reduce the overall bit rate, in [14], Garcia et. al adopted a low-resolution (LR) and full-resolution (FR) frames mixed video sequence. Based on this, a mixed resolution coding approach was proposed where the first M frames in the sequence were encoded at FR and the rest frames were coded at LR. Mixed resolution (MR) view frameworks have been proposed in [15, 16] for multiview video coding, where at least one view is coded at LR, while the others are coded at FR. These frameworks can reduce the amount of transmitted and stored data in comparison with the full FR framework. In [17, 18], a MR-MVD framework was adopted at the encoder side and 3D warping-generated virtual views were utilized to recover FR frames from the LR frames at the decoder side. With proper reconstruction algorithms at the decoder side, MR-MVD framework can well enhance the overall transmission efficiency.

In terms of reducing the transmission and storage data size, for stereoscopic video, the MR format can save 3/8 transmitted texture data compared with FR format, while downsampling both views by half can reduce the amount of transmitted texture data to 1/2 of the original format. Furthermore, the neighboring views have the same frame size which is suitable for the MVC coding approach. Two frame-compatible coding frameworks for stereoscopic video were proposed in [19, 20]. In [19], two views were decimated to half of their original size by downsampling filters, which were designed based on frame contents and the targeted interpolation coefficients. At the decoder side, the reconstructed frames were demultiplexed and interpolated into the full resolution. Since the frame downsampling pattern at the encoder side was evaluated as a function of the interpolation method, it can, to a large extent, reduce the interpolation errors. Unfortunately, this paradigm is not suitable for video applications because the downsampling pattern is frame dependent which means that the temporal redundancy cannot be efficiently removed by the video encoder. Thus, for video applications, there is a need to use temporally static downsampling patterns. In [20], an opposite parity packing arrangement for stereoscopic video was proposed to multiplex the two views. At the decoder side, the optimal disparity vector for each block was obtained from calculating the smallest differences between the matched blocks in coarsely interpolated FR left view and right view. However, in this work, some theoretical supports for the proposed view packing arrangement were missing. Moreover, although both [19, 20] had shown superior performance than other state-of-the-art frame-compatible coding frameworks, they mainly focus on stereoscopic video which may not be optimal for multiview or MVD video.

In this paper, a systematical down/upsampling framework for MVD video is proposed to enhance the coding performance at low bit rate (Fig. 1). In the proposed downsampling approach, the rows of two adjacent texture views are downsampled following an interlacing and complementary pattern, before compression. The aim of this downsampling approach is to facilitate the upsampling at the decoder side while exploiting the feature of 1D DIBR. The LR views will be upsampled by fusing the virtual view pixels with directional interpolated pixels with the aid of the pattern of the texture surrounding the discarded pixels. This approach has two benefits. Firstly, the high frequency information contained in one LR view can be properly utilized to upsample the other LR view through the generated virtual views. Secondly, the proposed directional interpolation approach can overcome the limitation of the virtual-view-based upsampling mechanism, which suffers in the areas corresponding to depth map discontinuity. Hence, by taking advantage of these two strategies, the discarded pixels can be recovered effectively. Experimental results have shown that the proposed algorithm achieves superior performance with respect to the filter-based interpolation algorithms and other state-of-the-art algorithms. The proposed upsampling approach will be named directional data fusion upsampling (DDFU) throughout this paper.

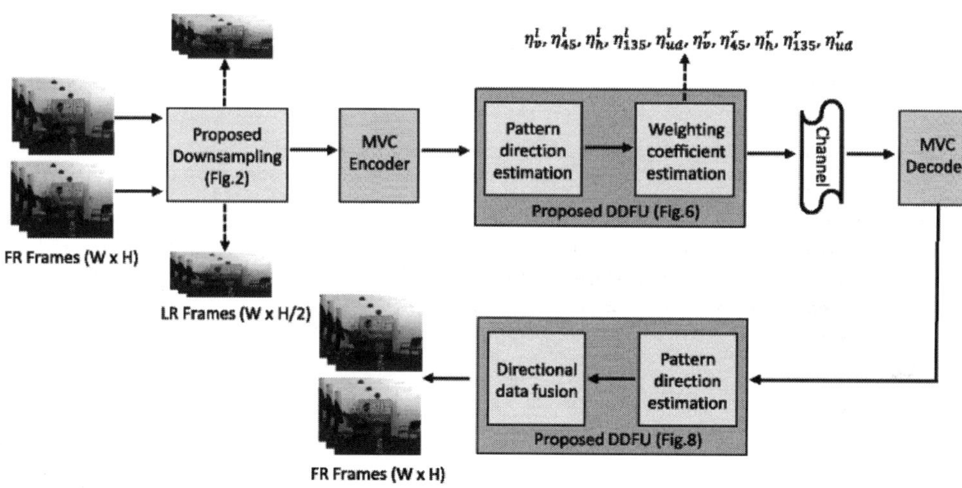

Figure. 1: Framework of the proposed down/upsampling method. Framework of the proposed down/upsampling method for a stereo video

The rest of this paper is organized as follows. Section 2 describes the details of the proposed downsampling algorithm, and the upsampling algorithm is introduced in Section 3. The generalization of the proposed method is presented in Section 4, and experimental results are presented in Section 5. Finally, the conclusions are in Section6.

PROPOSED INTERLACING-AND-COMPLEMENTARY-ROW-DOWNSAMPLING

A proper downsampling approach for multiview video needs to take into account the fact that different views cover almost the same scene. Hence, between neighboring views, there is considerable amount of inter-view redundancy. In this work, an interlacing-and-complementary-row-downsampling method is proposed by taking the features of proposed upsampling and multiview video into account, as shown in Fig. 2.

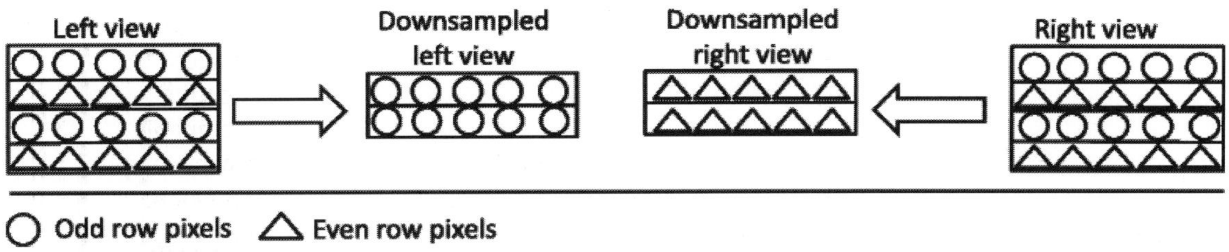

Figure. 2: The proposed downsampling method. The proposed interlacing-and-complementary-row-downsampling process for a stereo video

In the view multiplexing approaches, the generated sequence is from the mapping of two downsampled views. The process of multiplexing is carried out before the video encoding stage. There are variety of options for both the downsampling and view combination [21]. Due to inter-view redundancy, interlacing-and-complementary downsampling approaches could maintain more information than the non interlacing and complementary ones. In the following parts and aided with a graphical example, three downsampling approaches will be compared. In these three scenarios, we assume that two calibrated cameras in a parallel configuration setting and the same image plane (the most common camera configuration) are used to record an uneven bar structure (similar to the artistic gymnastics apparatus), as shown in Fig. 3 d. Figure 3 a–c shows the front, side, and top view of the stereoscopic orthographic projection of the scene, respectively. The viewed scene of the first and second cameras is shown in Fig. 4 a. The output of the vertical interlacing and complementary downsampling approach (i.e., column-wise downsampling) is shown in Fig. 4 b, where the gray areas indicate the "discarded areas" during downsampling process. It is possible to see that the left black bar of the uneven bar structure is missing in both views. Hence, neither intra-view or inter-view interpolation can help to recover this part. This is because the column-wise downsampling approach causes some "blind areas," where objects cannot be seen in any of the two views. The "blind areas" could be easily seen in Fig. 5, which shows the top view of the prospective projection of a scene with two pinhole cameras. The area enclosed by red lines can be viewed by both cameras, whereas, the yellow and blue bands indicate discarded areas in view 1 and view 2, respectively, due to the column-wise downsampling. Some areas (indicated by black) inevitably end up being discarded in both views; thus, any object falling in any of these black areas cannot be recovered by inter-view interpolation, and consequently, these areas are called "blind areas".

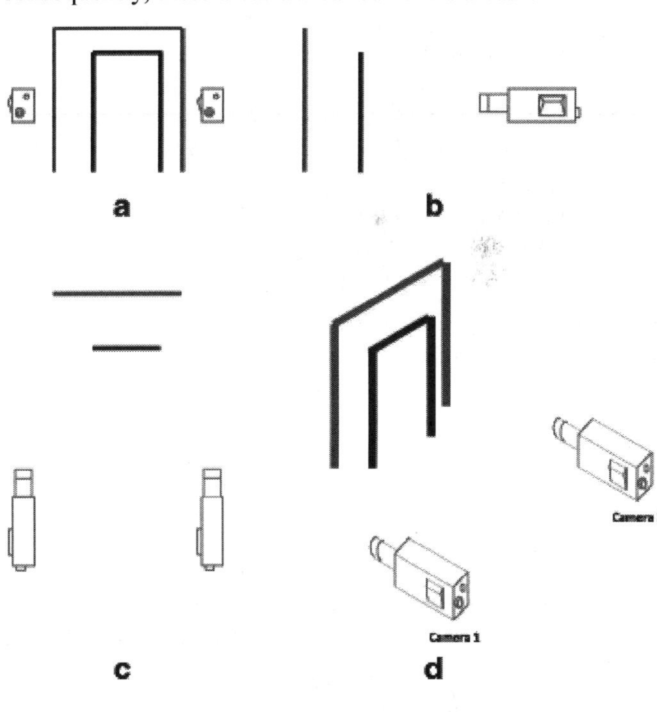

Figure. 3: Captured scene description. **a**, **b**, and **c** show the front, side, and top view of the stereoscopic orthographic projection of uneven bar structure viewed by two cameras in a parallel configuration setting as depicted in (**d**)

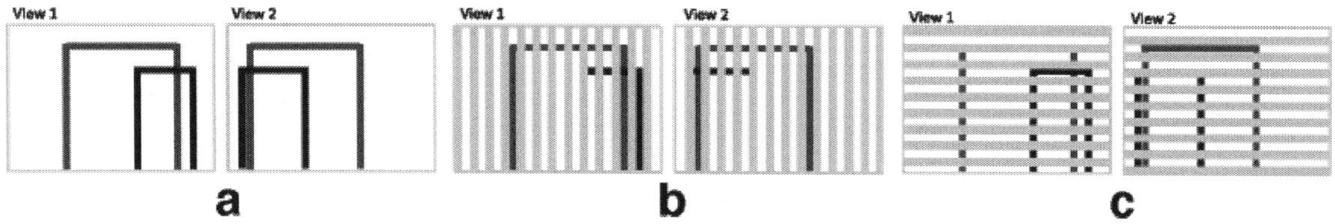

Figure. 4: The outputs of different downsampling methods. **a** the left side and right side of each frame shown the captured scene by the corresponding cameras without downsampling, respectively; **b** the output of the vertical downsampling method (i.e., column-wise downsampling); **c** the output of the interlacing and complementary row-wise downsampling method

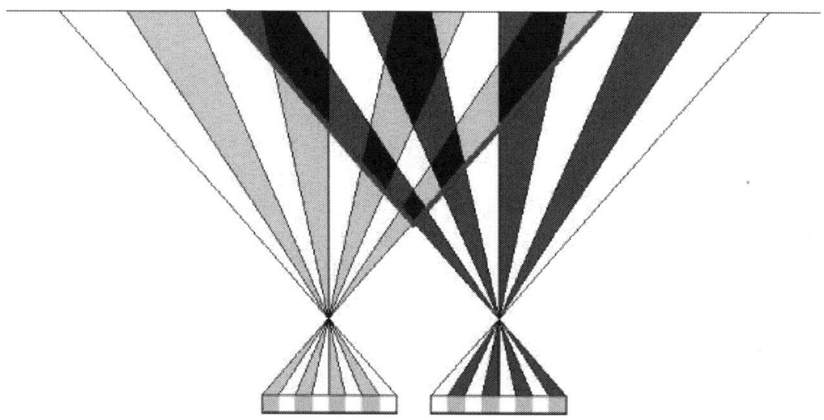

Figure. 5: The top viewof the prospective projection. The top view of the prospective projection of a scene using a pinhole camera model for the column-wise downsampling approach; area which could be viewed by both cameras (before the downsampling process) is enclosed in red; yellow bands indicate areas that cannot be seen in view 1 due to the column-wise downsampling approach; unviewed areas in view 2 due to the downsampling are indicated by blue. Areas depicted in black cannot be viewed in both views, those are called "blind areas"

Compared with column-wise downsampling, the output of row-wise one is shown in Fig. 4 c. It indicates that the proposed interlacing-and-complementary-row-downsampling will almost always guarantee that the object can be seen in the rows of one of the two views, except for some small objects with a one-pixel-width projection size in the camera plane. Nevertheless, the probability of this situation is low, and it also happens for the column-wise downsampling approach. Consequently, this row-wise downsampling can better exploit the warping feature of the DIBR technique and, as a result, can enhance the upsampling performance.

The chessboard downsampling approach can be regarded as the combination of row- and column-wise ones. It is able to achieve highest intra-view upsampling performance, since each to-be-filled pixel has four adjacent pixels in both horizontal and vertical directions which provide more information during interpolation. However, the chessboard pattern usually requires a comparatively higher bit rate due to low spatial and temporal correlations [19]. Furthermore, for each row of the chessboard downsampled views, it is possible to notice that the top view of the prospective projection of a scene is similar to the one shown in Fig. 5. Therefore, it could be conjectured that the chessboard approach also suffers from some "blind areas"; thus, its performance is better than the column-wise approach while being worse than the row-wise approach.

VIRTUAL VIEW-ASSISTED DIRECTIONAL DATA FUSION UPSAMPLING

In order to reduce the required resources and the amount of transmitted data, downsampling of the texture sequences is performed before the compression stage. In this paper, motivated by the findings in Section 2, the downsampled texture frames are generated by discarding the even rows in the left view and the odd rows in the right view of the stereo video, respectively.

Let the left and right FR frames be defined as $\mathbf{V}_f^l$ and $\mathbf{V}_f^r$, respectively, with size W×H, and the downsampled left and right LR frames as $\mathbf{V}_l^l$ and $\mathbf{V}_l^r$, respectively, with size W×H/2. Figure 6 shows the main stages of the proposed FR recovery mechanism. The downsampled views are expanded to their original size with the positions of the discarded pixels left empty (this stage is indicated by ① in Fig. 6). The expanded left view is represented by $\mathbf{V}_e^l$ where $V_e^l(2n, m) = 0, 1 \leq n \leq H/2, 1 \leq m \leq W,$ whereas, the expanded right view is represented by $\mathbf{V}_e^r$ where $V_e^r(2n-1, m) = 0, 1 \leq n \leq H/2, 1 \leq m \leq W$. Then, in the second stage indicated by ② based on the direction estimation results, a directional interpolation method is used to generate the corresponding interpolated frames, and these are denoted by $\mathbf{V}_i^l$ and $\mathbf{V}_i^r$ for the left and right views, respectively. Meanwhile, in the third stage③, the DIBR technique is applied on the expanded frames using the corresponding depth maps in order to generate the virtual views at the counterpart viewpoints, i.e., the left side virtual view $\mathbf{V}_v^l$ is generated by the right side expanded view. As a consequence, all the even rows in the left virtual view are warped from the even rows in the right view. Similarly, for the right virtual view, all the odd rows are warped from the odd rows in the left view. Therefore, based on the above design which aims to make the recovery of discarded pixels work in synergy with the downsampling stage, the virtual view becomes a potential source of information to efficiently recover the discarded pixels. Virtual views and directional interpolated views as the outputs of the two parallel stages,② and ③, are fused to generate the final FR frames at stage ④. This fusion process is driven by the pattern direction of the texture around each of the discarded pixels, so as to exploit the potential of stages ② and ③.

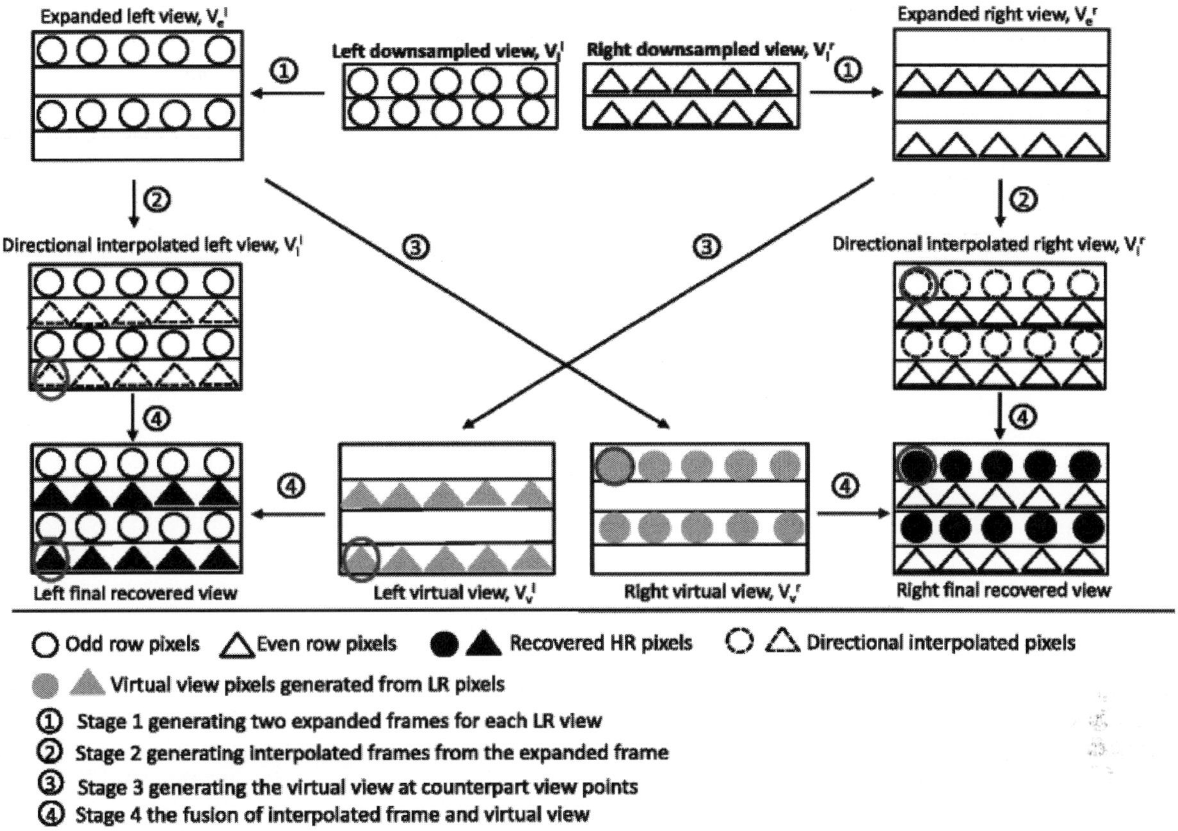

Figure. 6: The proposed discarded pixels recovery process

PCA-Based Pattern Direction Estimation

Knowing the dominant direction of the texture surrounding each discarded pixel allows better exploitation of the virtual and interpolated frames in recovering the discarded pixels. For example, texture patterns with horizontal edges usually cannot be accurately estimated from their upper and lower neighbors. Hence, exploiting the virtual view pixels can greatly help to recover such pixels.

To get the pattern direction, in this work, a principal components analysis (PCA)[22]-based method is used. This approach evaluates the gradients of the surrounding pixels for each discarded pixel, and then, the dominant direction is determined [23], where PCA can be implemented by evaluating the singular value decomposition (SVD) [24] of the data.

In general, the gradient at V(x,y) can be obtained by $\nabla V(x,y)=[\partial V(x,y)/\partial x, \partial V(x,y)/\partial y]^T$, and this could be approximated for discrete applications as

$$\nabla V(x,y) \approx \begin{pmatrix} \frac{1}{2}(V(x+\Delta,y) - V(x-\Delta,y)) \\ \frac{1}{2}(V(x,y+\Delta) - V(x,y-\Delta)) \end{pmatrix} \tag{1}$$

$\Delta=1$ offers the best approximation, however, taking into account that half of the rows are discarded, then Δ needs to be 2 while evaluating the gradients of the surrounding pixels of a discarded pixel. This ensures that $V(x+\Delta,y)$, $V(x-\Delta,y)$, $V(x,y+\Delta)$, and $V(x,y-\Delta)$ are available[1].

It is worth noticing that the horizontal neighbors (i.e., left and right neighbors) of the discarded pixels are unavailable; therefore, the dominant direction for each discarded pixel will be inferred from the four corner pixels of a 3×3 overlapping window centered at the discarded pixel. For example, the discarded pixel p_5, in Fig. 7, has two discarded neighbors, namely p_4 and p_6, so in order to maintain an equivalent number of neighbors and symmetric structure around p_5, the two pixels p_2 and p_8 will not be taken into account while evaluating the dominant pattern direction. In other words, only the gradients of the corner pixels p_1, p_3, p_7, and p_9 will be evaluated[2]. The gradients of the surrounding pixels of the discarded pixel at position (x,y) will be then arranged into a 4×2 matrix **G**, as follows:

$$\mathbf{G} = \begin{bmatrix} \nabla V(x-1,y-1)^T \\ \nabla V(x-1,y+1)^T \\ \nabla V(x+1,y-1)^T \\ \nabla V(x+1,y+1)^T \end{bmatrix} \tag{2}$$

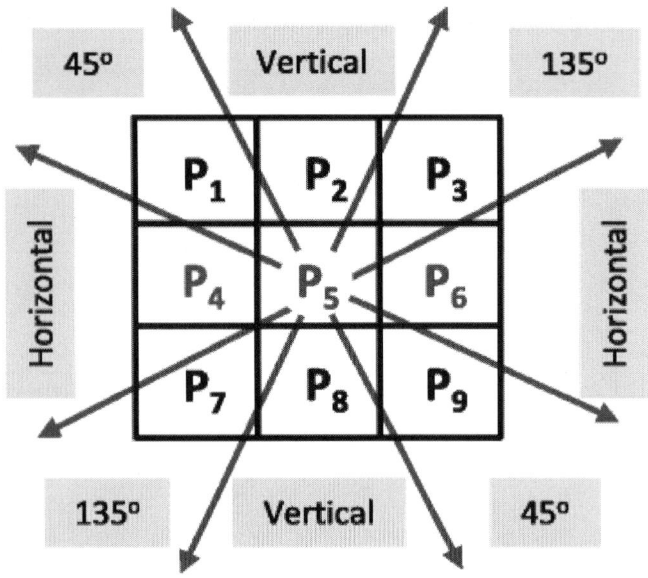

Figure. 7: Missing pixels classification based on texture pattern directions. The overlapping window centered at the discarded pixel p_5. The dominant pattern direction will be categorized into five groups. In this figure, only the remarkably dominant patterns are shown which are horizontal, 45° diagonal, vertical, and 135° diagonal directions.

The SVD of the matrix **G** will be computed as $\mathbf{G}=\mathbf{U}\,\mathbf{S}\,\mathbf{V}^T$, where **S** is a 4×2 diagonal matrix and the ratio of the diagonal elements in **S** (i.e., S_{11}/S_{22}) represents the energy of the dominant gradient. **U** and **V** are orthogonal matrices with size 4×4 and 2×2, respectively, and the angle of dominant gradient is $\theta=\arctan(v_{21}/v_{11})$. For the remarkably dominant gradient (i.e., $S_{11}/S_{22}\geq Th$ where Th is a threshold to define the remarkably dominant gradient), this angle will be used to determine the pattern directions of the discarded pixel, which are horizontal, 45° diagonal, vertical, and 135° diagonal directions as shown in Fig. 7. For the pixels from texture uniform areas whose energy in all four directions is almost equal, there is no remarkably dominant directional pattern (i.e., $S_{11}/S_{22}<Th$), will be classified into the "undefined" direction category. This process will be carried forward at both the encoder and decoder sides.

Weighting Coefficient Estimation of Directional Data Fusion

Since all of the discarded pixels are classified into five categories: horizontal, 45° diagonal, vertical, 135° diagonal, and undefined direction, the directional interpolated frames $\mathbf{V}_i^l$ and $\mathbf{V}_i^r$ are generated based on this classification. The horizontal pixels are the average of corresponding four nearest corner pixels, and the undefined directional pixels are recovered by vertical interpolation, since the vertical neighbors are the closest to the discarded pixels. For the other three directions' pixels, they are recovered by linearly interpolation along the pattern direction. In the fourth stage of the proposed upsampling algorithm, the discarded pixels are recovered by fusing the interpolated pixels with the virtual view pixels in order to exploit the advantages of both types of approach and to compensate the compression distortion.

To recover the discarded information, at the fusion stage, each discarded pixel is filled by a weighted average of the counterpart pixels in V_v and V_i as shown:

$$\hat{V}^l(2n,m)=\eta^l V_i^l(2n,m)+(1-\eta^l)V_v^l(2n,m) \tag{3}$$

The value of the weighting coefficient, η^l, is in the range [0,1]. This value, in theory, should be evaluated for each missing pixel, and it determines the relative contribution of the directional interpolated pixel with respect to the virtual view pixel. The fusing coefficients could be obtained by minimizing the L2 distance between the recovered pixels and their counterpart original pixels, as follows.

$$\sum_{m=1}^{W}\sum_{n=1}^{H/2}\left(\hat{V}^l(2n,m)-V_f^l(2n,m)\right)^2 \tag{4}$$

Holes and disoccluded areas in the virtual views are excluded during the fusion process and in these areas; the discarded pixels are directly recovered by directional interpolation. Since the original FR frame is only available at the encoder side, this means that all the fusing coefficients need to be transmitted for each frame to the decoder side. Obviously, this makes the pixel-by-pixel estimation of the weighting coefficient impractical. Hence, in this paper, a direction-mask-based weighting coefficient estimation is proposed, which can hugely reduce the transmitted side information. At the encoder and decoder sides, all the discarded pixels will be classified into five categories and respectively represented by five binary masks M_h, M_{45}, M_v, M_{135}, and M_{ud}. In this way, the binary value "1" in M_h indicates that the discarded pixel in that position has a horizontal texture pattern; in this case, the same position in M_v, M_{45}, M_{135}, and M_{ud} will have "0" binary value. For each direction, one weighting coefficient will be estimated by (4). Therefore, Eq. (3) could be rewritten in matrix format, while taking into account the five pattern categories, as follows:

$$\begin{aligned}\hat{\mathbf{V}}^l &= \eta_h^l\mathbf{M}_\mathbf{h}^\mathbf{l}.*\mathbf{V}_i^l+(1-\eta_h^l)\mathbf{M}_\mathbf{h}^\mathbf{l}.*\mathbf{V}_v^l\\ &+\eta_{45}^l\mathbf{M}_\mathbf{45}^\mathbf{l}.*\mathbf{V}_i^l+(1-\eta_{45}^l)\mathbf{M}_\mathbf{45}^\mathbf{l}.*\mathbf{V}_v^l\\ &+\eta_v^l\mathbf{M}_\mathbf{v}^\mathbf{l}.*\mathbf{V}_i^l+(1-\eta_v^l)\mathbf{M}_\mathbf{v}^\mathbf{l}.*\mathbf{V}_v^l\\ &+\eta_{135}^l\mathbf{M}_\mathbf{135}^\mathbf{l}.*\mathbf{V}_i^l+(1-\eta_{135}^l)\mathbf{M}_\mathbf{135}^\mathbf{l}.*\mathbf{V}_v^l\\ &+\eta_{ud}^l\mathbf{M}_\mathbf{ud}^\mathbf{l}.*\mathbf{V}_i^l+(1-\eta_{ud}^l)\mathbf{M}_\mathbf{ud}^\mathbf{l}.*\mathbf{V}_v^l\end{aligned} \tag{5}$$

where $\hat{\mathbf{V}}^l$ denotes the recovered frame. The operation.* represents the element-by-element multiplication of two matrixes. The optimal weighting coefficient for each direction can be obtained applying (4) on each direction.

Given that the encoder and decoder work on the same set of data to estimate the pattern direction, there is no need to transmit the five direction masks and only the directional weighting coefficients for the left view (i.e., η_h^l, η_{45}^l, η_v^l, η_{135}^l, and η_{ud}^l) and the right view, need to be estimated at the encoder side and transmitted to the decoder side. Obviously, the overhead bit cost of transmitting the weighting coefficients is negligible in comparison to the bit cost of texture and depth map. In the experimental results section, the term "DDFU" will be used to refer to this proposed full version scheme. In addition, DDFU can be simplified to only transmit the weighting coefficients of the first frame, which will be used for the fusion of all the remaining frames as well. This simplification is possible as the content of each frame does not change significantly, especially for the sequences with slow motion. Based on this observation, the simplified approach can further reduce the amount of transmitted side information with little quality degradation. In the experimental section, the term "DDFU (first frame η)" will be used to refer to this simplified scheme.

Although directional data fusion process happens at both encoder and decoder sides, it has different targets. At the encoder side, it involves the original FR frames to estimate the optimal weighting coefficients for all the directions. Then, as the outputs, these coefficients are sent to the decoder side and used in the fusion process to reconstruct the FR frames. A graphic representation of the proposed directional data fusion process at the decoder side is shown in Fig. 8. By receiving the weighting coefficients and using the same pattern direction estimation process, virtual view pixels and directional interpolated pixels can be fused to generate the recovered frame using Eq. (5).

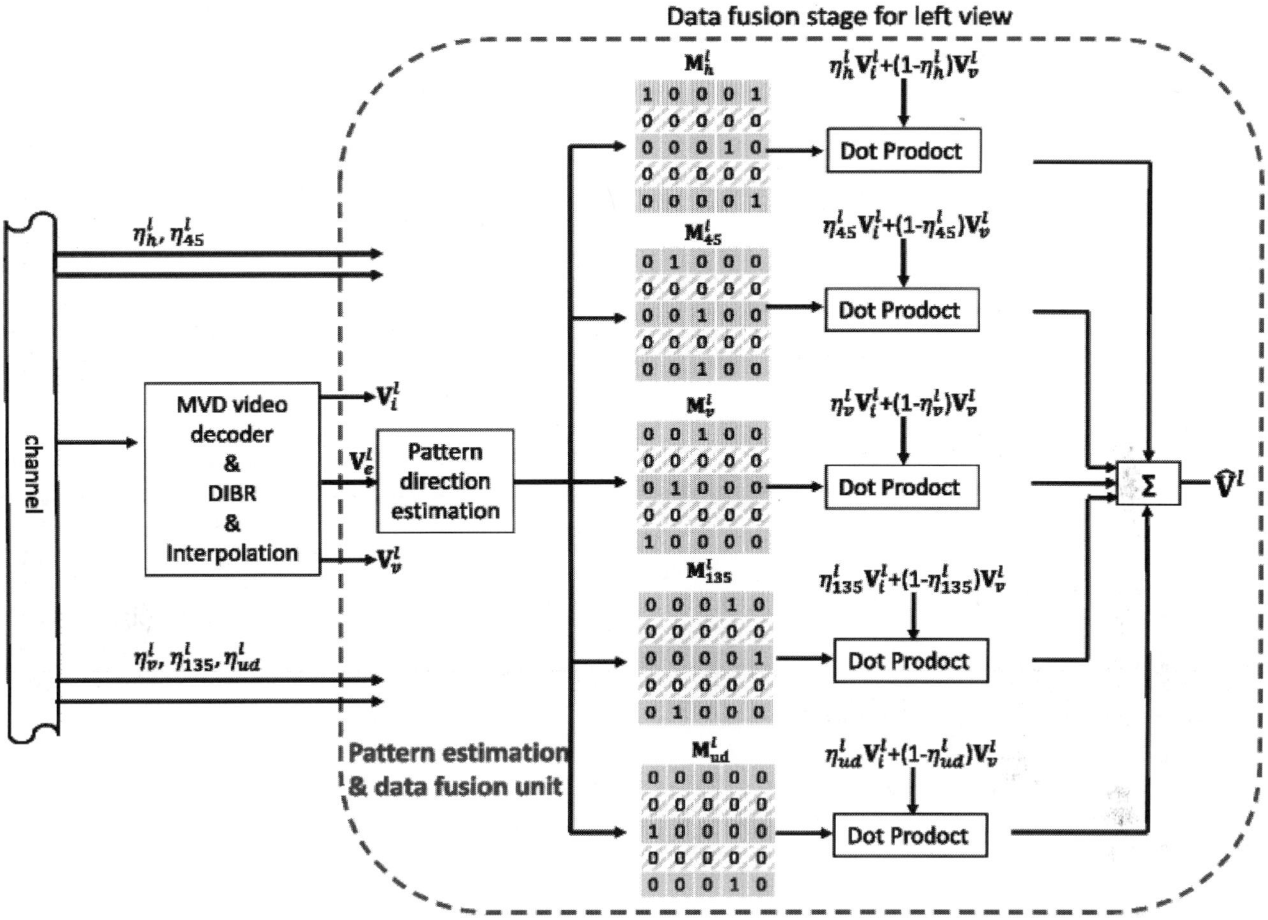

Figure. 8: Directional data fusion process. The process of data fusion by directional weighting coefficients and corresponding directional binary masks

GENERALIZATION OF PROPOSED DOWN/UPSAMPLING METHOD

The proposed down/upsampling with virtual view-assisted data fusion algorithm can also be applied to MVD video in addition to stereo video. Since in this kind of video more neighboring views and the corresponding depth maps are available, at a given viewpoint, more virtual view versions can be utilized. With the aid of these virtual views, the quality of the final recovered FR views can be considerably improved. As depicted in Fig. 9 in a N views multiview video, V_e^1 is the expanded view at viewpoint 1 and similarly for V_e^2. V_i^1 is the directional interpolated view at viewpoint 1, and V_v^{21}, V_v^{31}, and V_v^{N1} are the virtual views generated from the adjacent views 2,3,N at viewpoint 1. In this case, the discarded pixels are recovered by fusing interpolated pixels with one of the available virtual views that gives the minimum differences when compared with original FR pixels. Subsequently, the fusion coefficients are transmitted. In this way, the proposed algorithm can also effectively recover the FR frames.

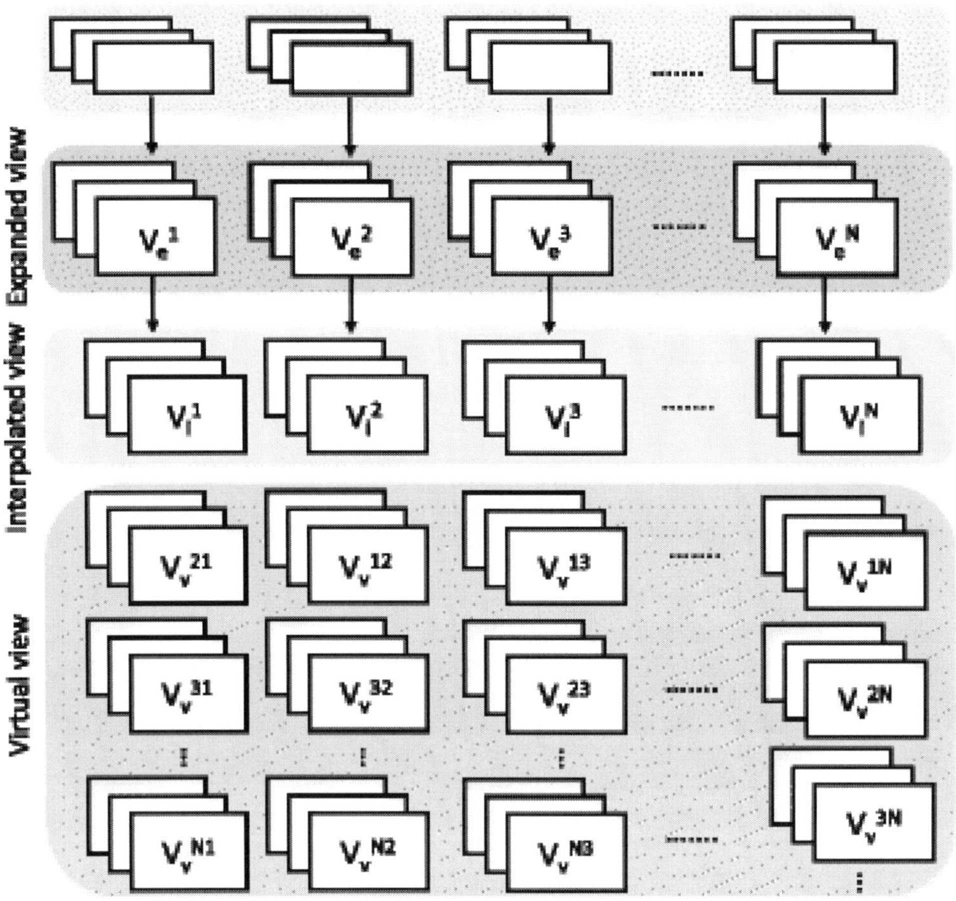

Figure. 9: The proposed algorithm for multiview video

EXPERIMENTAL RESULTS

To objectively evaluate the performance of the proposed method, several experiments were conducted on public 3D video datasets [25, 26] "Doorflower," "Kendo," "Dog," "Balloons," "Newspaper," and "Undo-Dancer". Some parameters and content characteristics of the testing sequences are listed in Table 1 for reference. For each sequence, both the left and right views had been interlacing and complementary row downsampled with a factor 2 before encoding. JMVC 5.0 [27] was used for compression, and six different QPs, namely 34, 37, 40, 43, 46, and 49, were used to code the texture and depth map sequences. The temporal GOP size and the total number of encoded frames was 8 and 80, respectively, while the delta QP and the differential QP between the base layer and sublayer in hierarchical-B picture structure was set to zero in all layers. The virtual views at the decoder side were rendered using a 1D DIBR technique from one reference view to another view without any postprocessing (i.e., no hole filling).

Table 1: The parameters and characteristics of each used sequence

	Size	Camera	Left	Right	Content's motion
Doorflower	1024×768	Fixed	View10	View08	Moderate
Kendo	1024×768	Moving	View03	View05	Complex
Dog	1280×960	Fixed	View38	View39	Medium
Balloons	1024×768	Moving	View03	View05	Complex
Newspaper	1024×768	Fixed	View04	View06	Simple
Undo-Dancer	1920×1088	Moving	View02	View05	Complex

Performance Evaluation on Stereoscopic Video

The first set of simulations aims to evaluate the effectiveness of the proposed approach by comparing the rate distortion performance with FR video coding approach, filter-based approach, and the state-of-the-art approaches [19, 28]. Figure 10 shows the coding performance of these methods on all of the testing sequences. In the comparison, a 6-tap Lanczos filter has been used at encoder and decoder sides for down/upsampling, respectively, for the filter-based approach. From Fig. 10, the effectiveness of the proposed approach over the matched-filter approach and FR coding at low bit rate can be appreciated for all the testing sequences. The proposed method has higher PSNR results than the matched-filter method, and the maximum PSNR gains can be 0.81 and 0.76 dB on sequences "Undo-Dancer" and "Doorflower", respectively. This is due to the proposed fusion mechanism that can well preserve the edges. In addition, the better quality of the depth maps is the more contributions the virtual view pixels can make. Generally, the matched-filter approach has good coding performance on the smooth areas; therefore, for the sequences (e.g. "Dog") containing more smooth areas, it is comparable to the proposed method. However, if more texture is contained in the sequences, the proposed method can be more advantageous over the matched filter method. In summary, the average PSNR gains across different bit rates for all the sequences range from 0.18 to 0.45 dB.

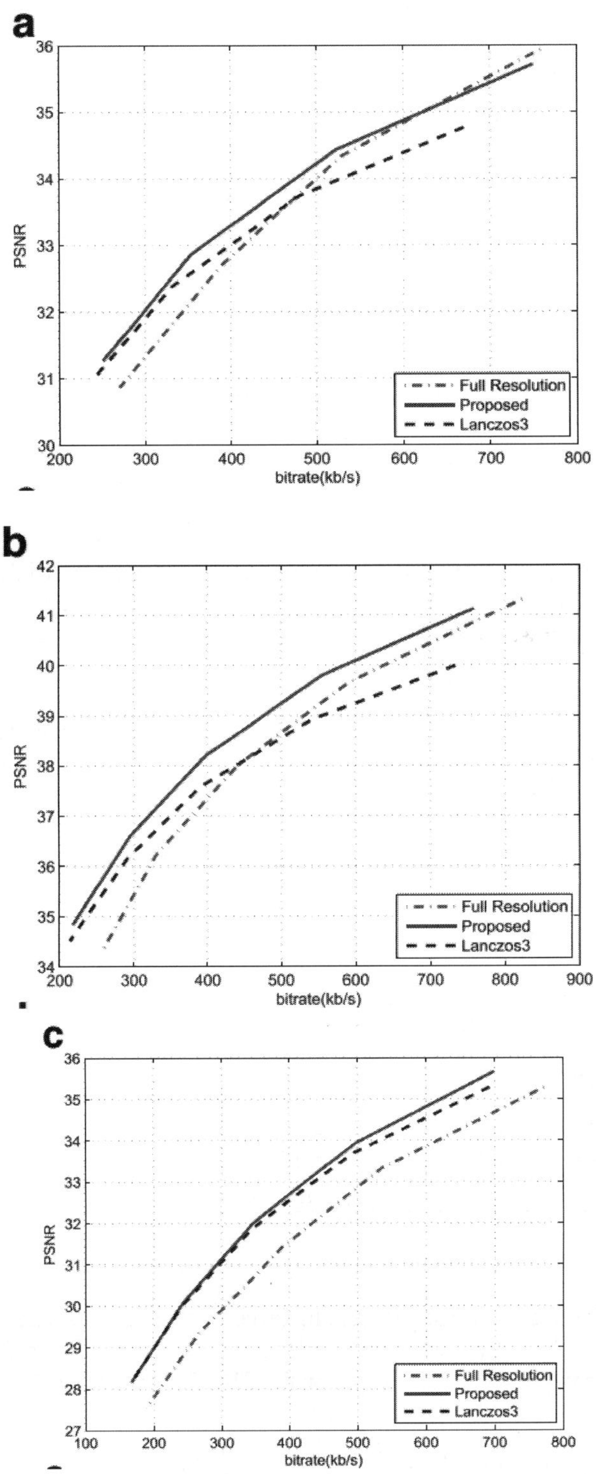

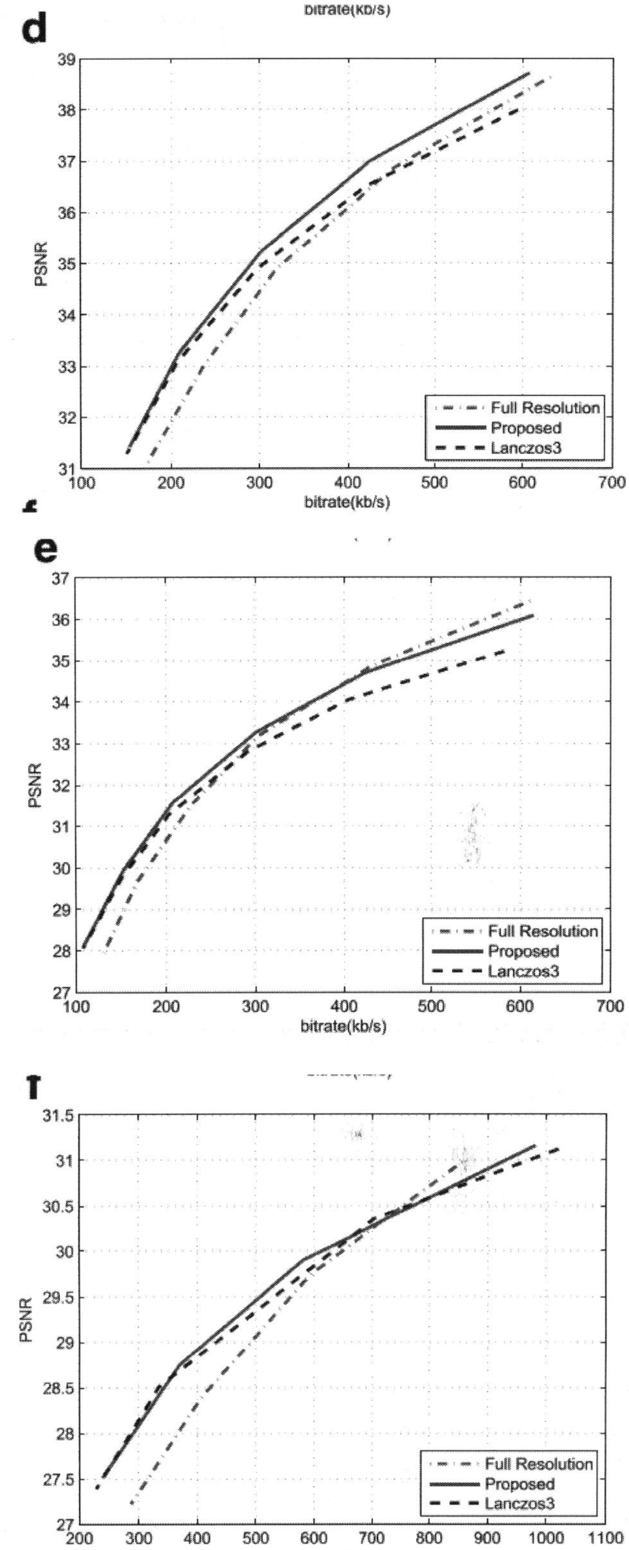

Figure. 10: The rate distortion curves. The rate distortion curves for the testing sequences. **a** Doorflower. **b** Kendo. **c** Dog. **d**Balloons. **e** Newspaper. **f** Dancer

To further evaluate the effectiveness of the proposed method, two fair comparisons with the frame-compatible coding methods in [19, 28][3] are carried out by adopting the same testing sequences with the same resolution used in their work, the same coding standard, and the same coding parameters as [19, 28]. The comparison results of [19] are shown in Fig. 11, where LF1 represents the direct downsampling (i.e., even rows in both left and right views are discarded without low pass filtering), CAIS represents the proposed method in [19], and IF1 and IF2 are two interpolation filters with coefficients {1,−5,20,20,−5,1}/32 and {−3,28,8,−1}/32, respectively, as proposed in [19]. The depth sequences used in the proposed DDFU are generated by the method presented in [29], and depth map bit rates have been included in the total bit rate. Indicated by these results, the gain of the proposed method is larger than that of [19]. Figure 12 shows the comparison results of the proposed approach with respect to [28]. In [28], two views are asymmetrically downsampled in frame-compatible coding and the left view has higher quality. Separately comparing the left view and right view, it is obvious that even the left view in the proposed work contains less information (downsampled by factor of 2) than the left view in [28] (downsampled by factor of 8/5); the left view recovery performance of the proposed work is comparable to [28]. Meanwhile, the recovery performance of the right view of the proposed approach is much higher than that in [28]. From the comparison with frame-compatible coding methods, the superiority of the proposed method is due to jointly take into account the features of down/upsampling, the inter-view redundancy, and virtual views.

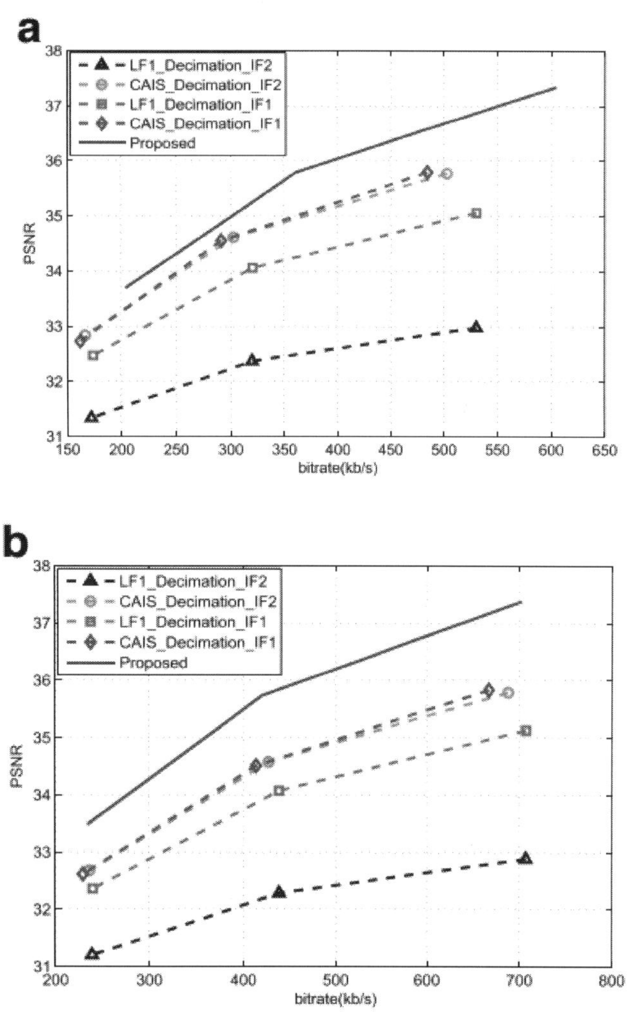

Figure. 11: Comparison with [19]. The rate distortion curves for the testing sequences. **a** Doorflower. **b** Laptop, for the proposed approach and [19]

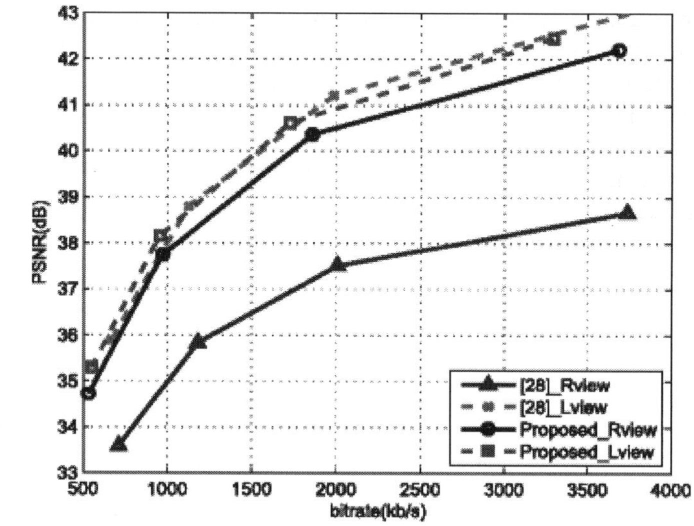

Figure. 12: Comparison with [28]. The rate distortion curves for the testing sequence balloons for the proposed approach and [28]

The visual results of zoomed-in parts of sequences "Doorflower" and "Undo-dancer" are shown in Fig. 13. It is possible to note that the edges recovered by DDFU are sharper than those recovered by the benchmark method. Although the proposed DDFU recovered frame also has some blurred areas, nevertheless, it still achieves a better visual quality than matched-filter-interpolated frame. Figure 13 d shows a portion of the original left view of "Undo-Dancer', and its recovered versions using the matched-filter-based approach and the proposed approach in Fig. 13e, f, respectively. Since the one-pixel-wide edge is difficult to recover properly by using only the surrounding pixels, the advantage of the DDFU method is more obvious in the highlighted areas by a red ellipse in Fig. 13 e, f. From this comparison, it can be seen that the proposed approach can recover the one-pixel-wide edge without blurring.

Figure. 13: The visual results. Comparison between proposed DDFU method and benchmark method. **a**, **b**, and **c** are the results of Original, Benchmark, and DDFU, respectively, on zoomed-in part of the sequence Doorflower; **d**, **e**, and **f** are the results of Original, Benchmark, and DDFU, respectively, on zoomed-in part of the sequence Undo-Dancer

Performance of Each Stage of the Proposed Method

In this subsection, several experiments have been conducted to validate the necessity and effectiveness of direction estimation and data fusion step in the proposed algorithm. Since the category of the to-be-filled pixels is determined by the estimated texture pattern, accurate pattern direction estimation plays an important role in the fusion process. Therefore, to verify its effectiveness, Fig. 14 b shows the pattern estimation result on the uncompressed frame, whereas, Fig. 14 c, d shows the estimation results on the compressed frame with Q P=34 and Q P=40, respectively. For reference, Fig. 14 a shows the original uncompressed texture frame from the "Doorflower" sequence with three highlighted parts containing clear patterns. Different colors are used to distinguish the five directions, so the colors dark red, red, orange, yellow, and white are used to represent vertical, 135° diagonal, horizontal, and 45° diagonal edges and the undefined pattern areas, respectively. In this paper, areas are regarded as undefined pattern areas when $S_{11}/S_{22} \leq Th$ where $Th=4$. The accuracy of the adopted pattern detection algorithm could be appreciated from Fig. 14 b, c. By comparing these two figures, the direction estimation results of the three highlighted parts are almost the same. It demonstrates that the accuracy of the pattern estimation is barely affected by the compression distortion.

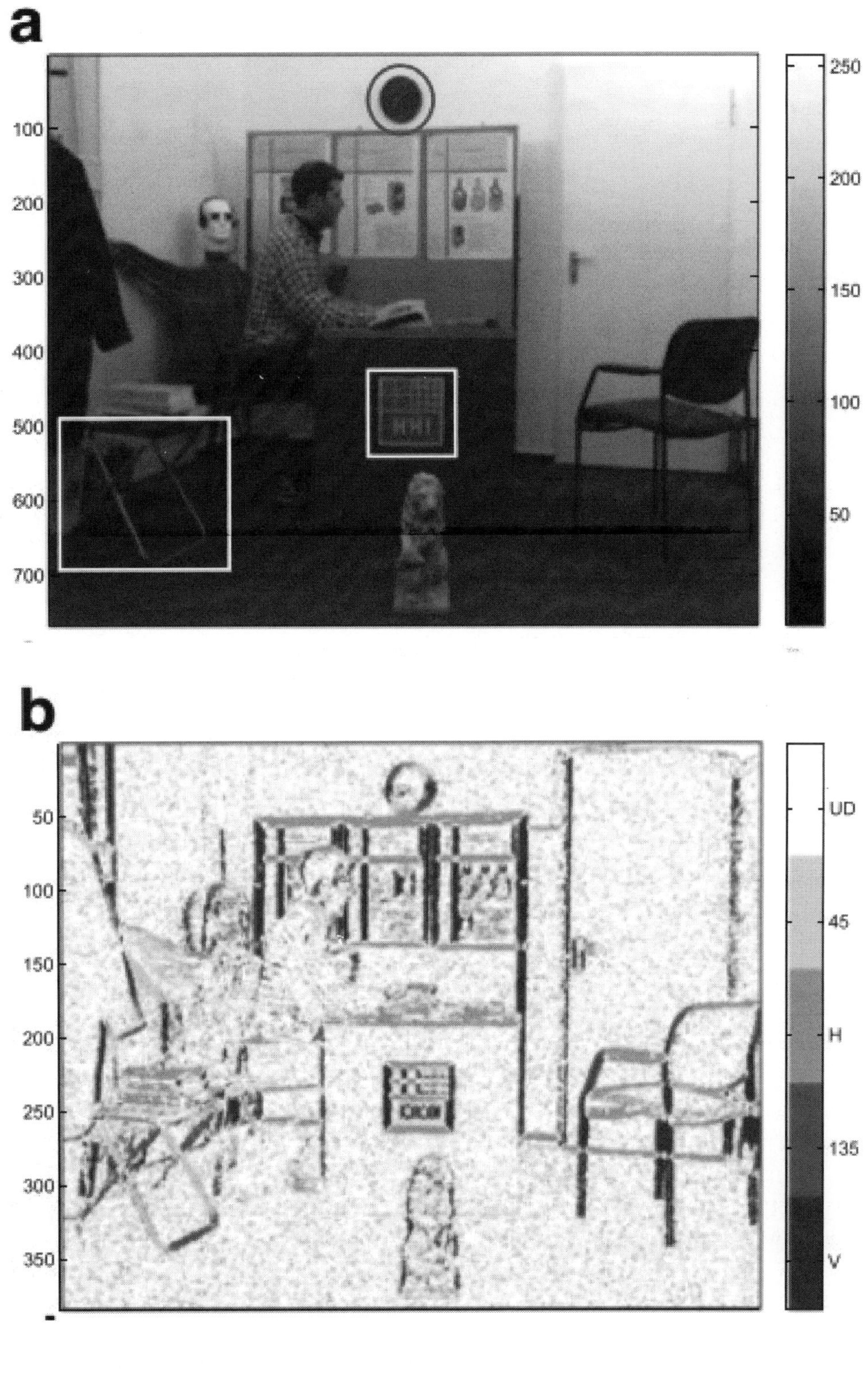

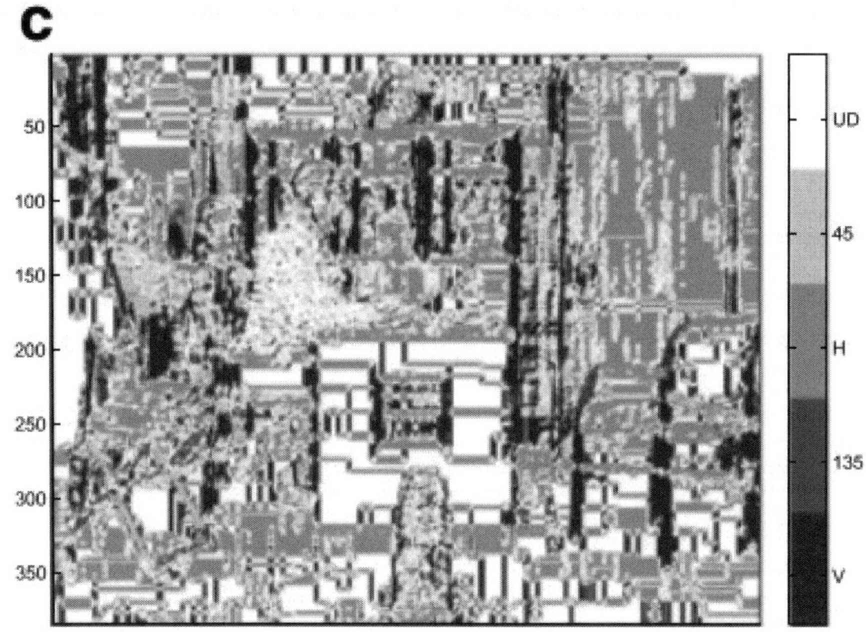

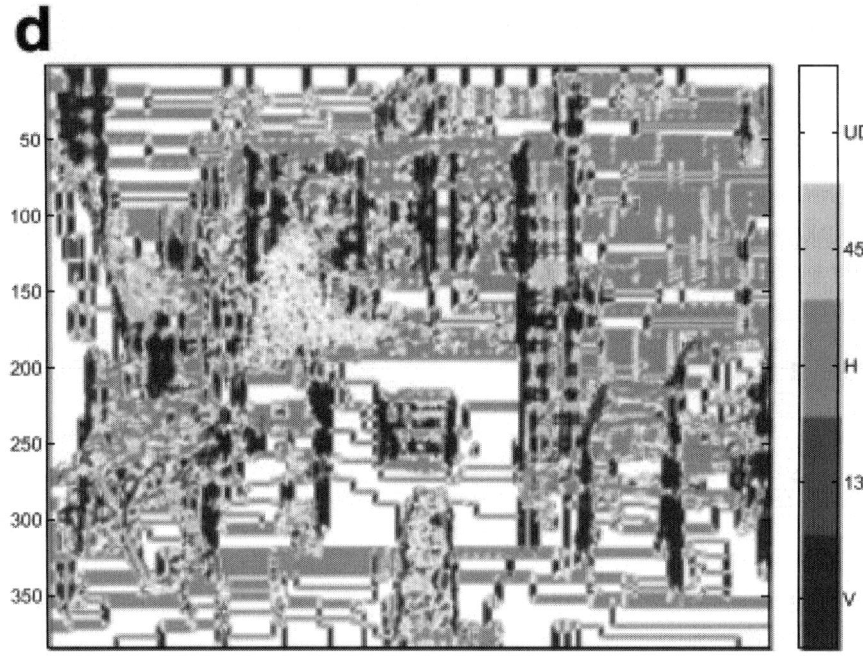

Figure. 14: The pattern direction estimation results. **a** Original texture; the pattern direction estimation results on (**b**) original uncompressed texture, **c** compressed texture with QP = 34, and (**d**) compressed texture with QP = 40; the colors dark red, red, orange, yellow, and white represent vertical, 135° diagonal, horizontal, 45° diagonal, and undefined direction pixels, respectively. (For clearness, the directional estimation results on the discarded pixels are scaled up to the same size as the original texture; their real size is shown on the y-axis of each figure)

To show the level of contribution of the virtual views in the fusion stage and how the texture pattern direction influences the fusing process, the average fusion coefficients η_h, η_v, η_{45}, η_{135}, and η_{ud} of both views are reported in Table 2 for the six testing sequences and different QPs. The smaller the value of η is, the more important the virtual view pixels are for the recovery of discarded pixels. In the fusion stage, the contribution of virtual view depends on several factors, such as the adopted DIBR technique and depth map quality. It is worth noticing that even with advanced rendering techniques, the generated virtual view may still face a problem in generating high-quality and aligned texture around depth discontinuous areas, where the adopted directional interpolation can compensate well. From this table, it can be seen that virtual view pixels are more important to recover the pixels with horizontal pattern than other directions. On the other hand, the directional interpolated frame is more important to recover the pixels with vertical pattern. For example, the average η_h for the "Undo-Dancer" sequence at Q P=34 is 0.11 versus η_v=0.97. The η_{45}, η_{135}, and η_{ud} values for the two diagonal patterns and undefined pattern lay in between the horizontal and vertical cases, which are 0.63,0.62,0.19 for Q P=34, respectively. Moreover, it should be noted that for the sequence "Undo-Dancer" which is a computer graphic sequence and consequently has an accurate depth map, the virtual view pixels provide a greater contribution to the final recovered FR frames, in all five directions, with respect to other sequences. As expected, this contribution is remarkably higher for the horizontal pattern.

Table 2: The two-views-all-frames average values of η_h, η_v, η_{45}, η_{135}, and η_{ud} for each sequence and for different QPs

Doorflower							
QP		34	37	40	43	46	49
Average η	η_h	0.23	0.24	0.30	0.37	0.47	0.55
	η_v	0.97	0.95	0.91	0.89	0.86	0.82
	η_{45}	0.77	0.76	0.78	0.77	0.75	0.75
	η_{135}	0.80	0.80	0.79	0.80	0.82	0.79
	η_{ud}	0.58	0.61	0.62	0.65	0.64	0.63
Undo-Dancer							
QP		34	37	40	43	46	49
Average η	η_h	0.11	0.16	0.23	0.36	0.50	0.58
	η_v	0.97	0.98	0.98	0.96	0.95	0.93
	η_{45}	0.63	0.72	0.79	0.85	0.83	0.80
	η_{135}	0.62	0.72	0.79	0.82	0.81	0.78
	η_{ud}	0.19	0.28	0.40	0.53	0.62	0.66
Kendo							
QP		34	37	40	43	46	49
Average η	η_h	0.83	0.81	0.77	0.73	0.68	0.64
	η_v	0.95	0.89	0.86	0.81	0.78	0.74
	η_{45}	0.94	0.91	0.89	0.84	0.77	0.71
	η_{135}	0.93	0.90	0.87	0.83	0.80	0.75
	η_{ud}	0.86	0.83	0.79	0.75	0.69	0.66
>Balloons							
QP		34	37	40	43	46	49
Average η	η_h	0.86	0.85	0.82	0.77	0.72	0.65
	η_v	0.95	0.95	0.95	0.94	0.91	0.87
	η_{45}	0.95	0.95	0.93	0.90	0.87	0.82
	η_{135}	0.95	0.95	0.94	0.91	0.89	0.87
	η_{ud}	0.93	0.93	0.91	0.88	0.84	0.78
Dog							
QP		34	37	40	43	46	49
Average η	η_h	0.95	0.95	0.92	0.89	0.85	0.77
	η_v	1.00	0.98	0.95	0.94	0.89	0.80
	η_{45}	0.99	0.96	0.94	0.90	0.84	0.78
	η_{135}	0.99	0.97	0.95	0.93	0.86	0.78
	η_{ud}	0.99	0.97	0.95	0.91	0.90	0.82
Newspaper							

QP		34	37	40	43	46	49
Average η	η_h	0.89	0.89	0.86	0.85	0.81	0.77
	η_v	1.00	1.00	1.00	1.00	0.99	0.96
	η_{45}	1.00	1.00	1.00	0.98	0.94	0.93
	η_{135}	1.00	1.00	1.00	0.98	0.96	0.94
	η_{ud}	0.98	0.99	0.98	0.93	0.92	0.89

Performance Evaluation on Multiview Video

When testing on multiview video, the View1, View3, and View5 of sequences "Kendo" and "Balloons" and the View2, View4, and View6 of sequence "Newspaper" are adopted. For multiview testing, the same codec setting up is used as the two-view testing and after proposed downsampling method, each LR view is in half size of its corresponding FR view. The performance comparison at the decoder side is shown in Table 3. From the reported results, it is possible to note that the proposed down/upsampling method can also work properly in multiview video system. In this case, the highest PSNR gain can be up to 1.19dB for the "Kendo" sequence. Compared with the two-view video case, the PSNR gains of multiview video become higher and the average gain for all the sequences and all QPs is 0.4 db. These gains are obtained due to the availability of multiple virtual view candidates. This ensures that the more suitable virtual view pixels are merged with the interpolated view. Moreover, compared with the average ηvalue for two-view testing, in multiview testing, the virtual view has more impacts at low bit rate (Table 4).

Table 3: The upsampling performance comparison on multiview video

Kendo							
QP		34	37	40	43	46	49
Bit rate (kb/s)		1073	808	608	476	361	283
PSNR	Lanc	37.12	35.88	34.38	32.79	30.94	29.10
	Pro	38.31	36.71	34.95	33.19	31.26	29.39
ΔPSNR		1.19	0.83	0.57	0.40	0.32	0.29
Balloons							
QP		34	37	40	43	46	49
Bit rate(kb/s)		1150	824	577	426	309	245
PSNR	Lanc	36.54	34.97	33.12	31.37	29.54	27.86
	Pro	37.07	35.32	33.36	31.54	29.67	28.00
ΔPSNR		0.53	0.36	0.23	0.17	0.14	0.14
Newspaper							
QP		34	37	40	43	46	49
Bit rate(kb/s)		1134	809	572	426	317	253
PSNR	Lanc	34.12	32.90	31.44	29.94	28.23	26.55
	Pro	34.85	33.41	31.77	30.16	28.39	26.67
Δ PSNR		0.73	0.51	0.33	0.22	0.15	0.12

Table 4: The two-view all-frame average values of η_h, η_v, η_{45}, η_{135}, and η_{ud} for three multiview sequences and different QPs

Kendo							
QP		34	37	40	43	46	49
Average η	η_h	0.91	0.89	0.85	0.79	0.71	0.62
	η_v	0.99	0.96	0.91	0.87	0.82	0.75
	η_{45}	0.98	0.94	0.91	0.86	0.81	0.71
	η_{135}	0.96	0.91	0.89	0.84	0.77	0.72
	η_{ud}	0.95	0.91	0.85	0.77	0.70	0.62
Balloons							
QP		34	37	40	43	46	49
Average η	η_h	0.94	0.91	0.86	0.82	0.74	0.63
	η_v	1.00	0.99	0.95	0.91	0.87	0.80
	η_{45}	0.97	0.96	0.94	0.89	0.84	0.78
	η_{135}	1.00	0.98	0.93	0.90	0.84	0.78
	η_{ud}	0.99	0.95	0.92	0.88	0.82	0.74

Newspaper							
QP		34	37	40	43	46	49
Average η	η_h	0.87	0.86	0.84	0.83	0.78	0.73
	η_v	1.00	1.00	1.00	1.00	0.99	0.95
	η_{45}	1.00	1.00	1.00	0.99	0.96	0.92
	η_{135}	1.00	1.00	1.00	0.99	0.95	0.92
	η_{ud}	0.99	0.99	0.98	0.96	0.93	0.87

Performance of Simplified Method

In the basic implementation of the DDFU algorithm, the encoder needs to transmit the five weighting coefficients, η, for each frame and each view. Clearly, it needs to evaluate them by minimizing (4). However, given that in most cases, there are no major changes in the scene content, it is reasonable to assume that those coefficients do not change very much from frame to frame; hence it is not necessary to evaluate them for each frame. This assumption could be verified by Fig. 15 which shows the trend of the weighting coefficients versus frame number. Thus, one way to reduce the overhead transmission of the proposed approach is using directional weighting coefficients of the first frame for the whole sequence (called DDFU (first frame η)). In this approach, the weighting coefficients are only estimated for the first frame and then used for the whole sequence. To verify the effectiveness of this simplified approach, its performance has been compared with the proposed DDFU approach and another simplified approach called in the following DDFU (user defined η), the latter adopts user defined coefficients at the decoder side for the whole sequence. The pre-set values for the DDFU (user defined η) used are $\eta_v=1$, $\eta_h=0$, $\eta_{45}=0.5$, $\eta_{135}=0.5$, and$\eta_{ud}=1$ for the left and right views which means that all vertical edges and undefined pattern areas are recovered by the directional interpolation algorithm. All recovered horizontal edges are obtained from the virtual view pixels, and the two diagonal direction pixels are obtained by equally fusing the directional interpolated pixels with the virtual view pixels. The results of this comparison are listed in Fig. 16.

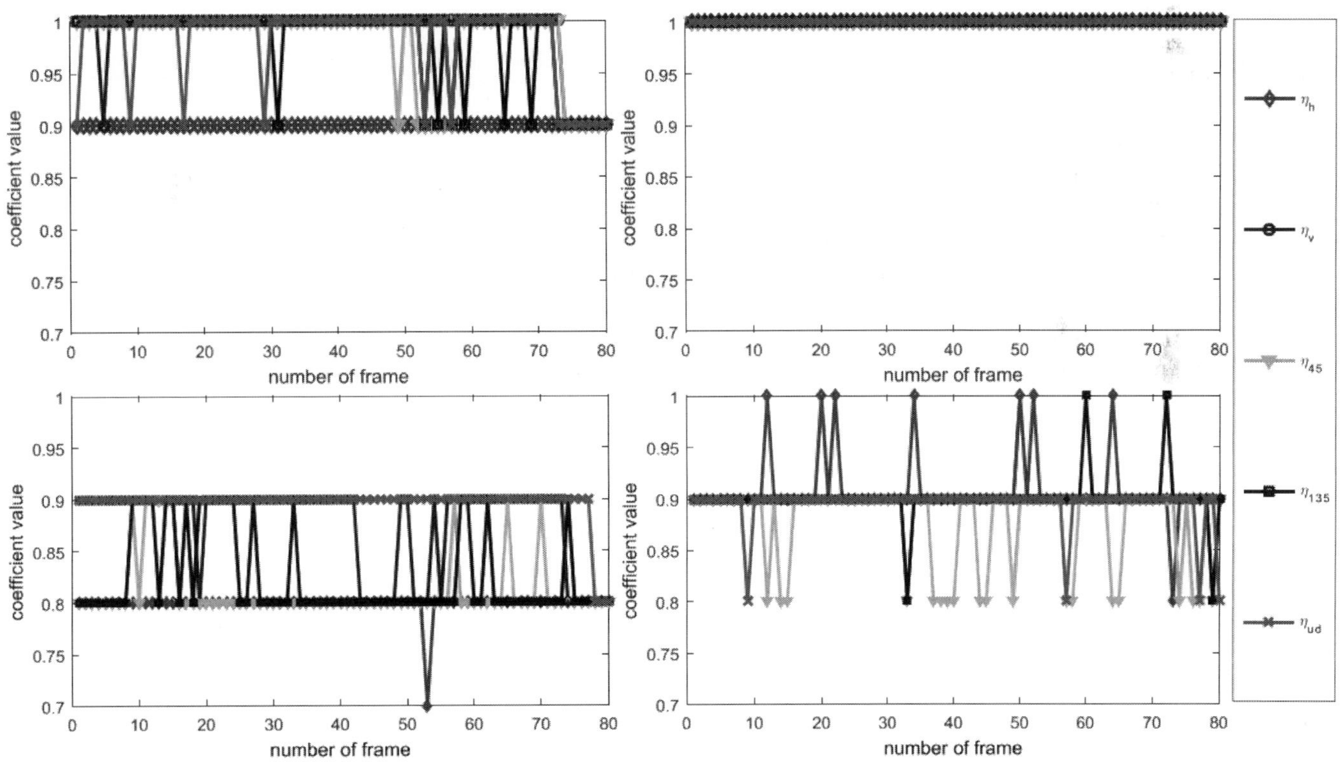

Figure. 15: The variance of five η coefficients. The pane shows the five coefficients for the sequence "Dog". The top left and rightfigures of the pane are the weighting coefficients of the left and right views, respectively, when QP = 34; the bottom leftand right figures of the pane are the weighting coefficients of the left and right views, respectively, when QP = s46

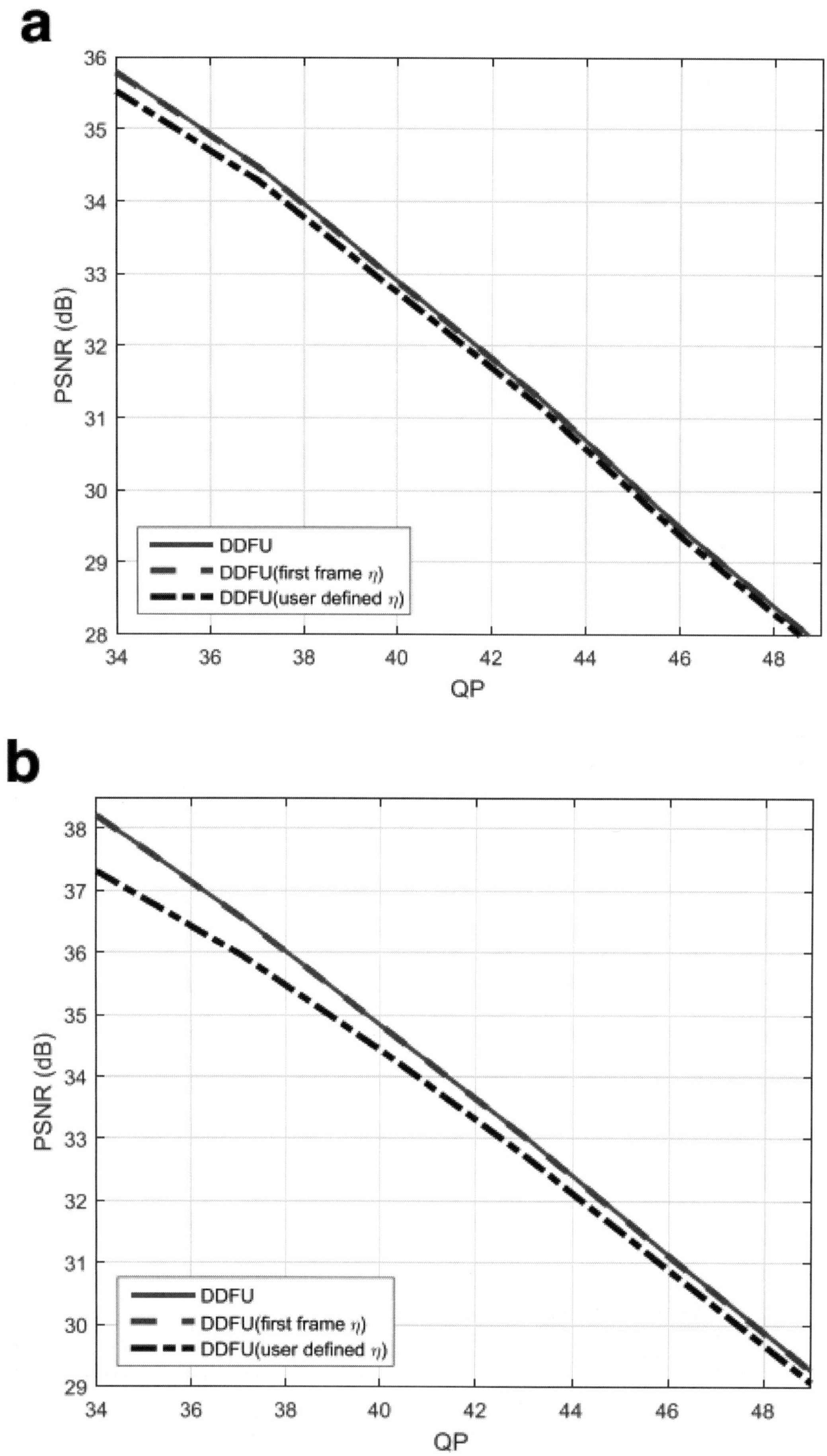
a
36
35
34
33
32
31
30
29
28
PSNR (dB)
34
36
38
40
42
44
46
48
QP
DDFU
DDFU(first frame η)
DDFU(user defined η)
b
38
37
36
35
34
33
32
31
30
29
PSNR (dB)
34
36
38
40
42
44
46
48
QP
DDFU
DDFU(first frame η)
DDFU(user defined η)

c

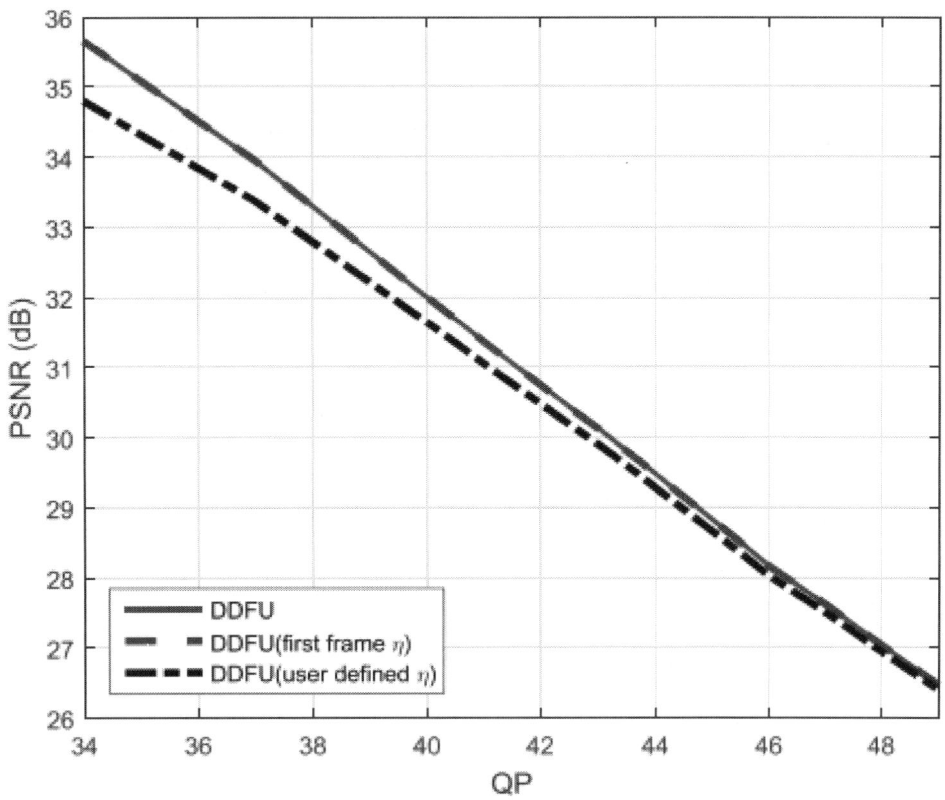
PSNR (dB)
QP
DDFU
DDFU(first frame η)
DDFU(user defined η)

d

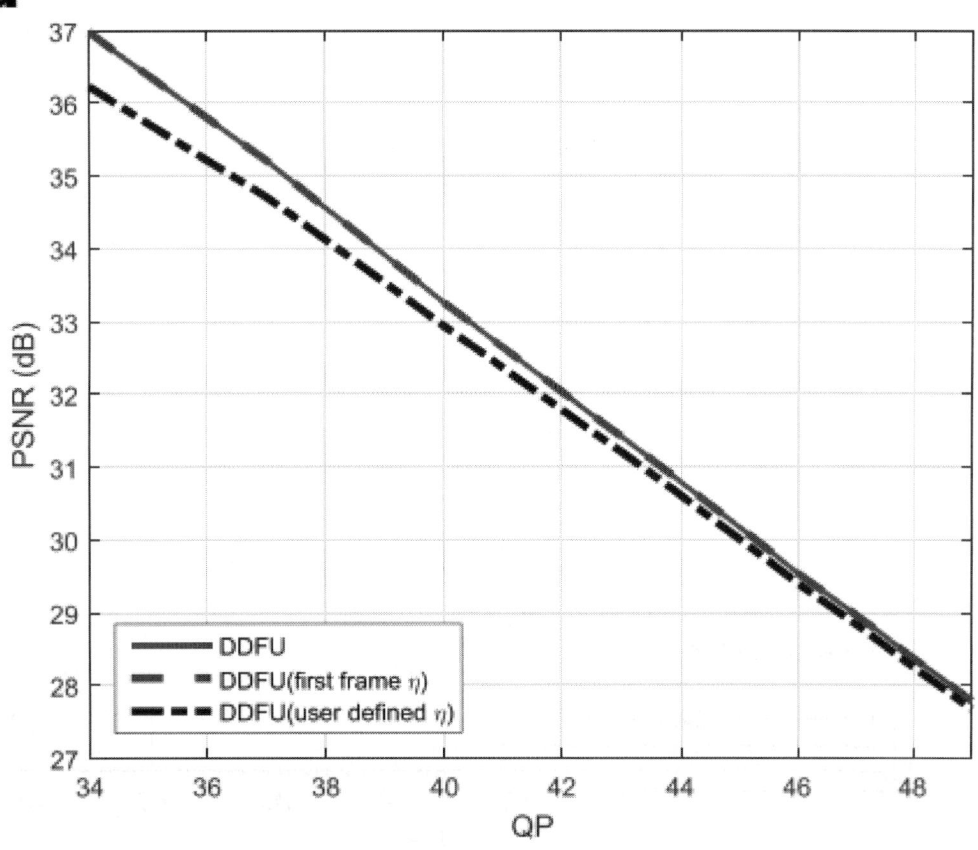
PSNR (dB)
QP
DDFU
DDFU(first frame η)
DDFU(user defined η)

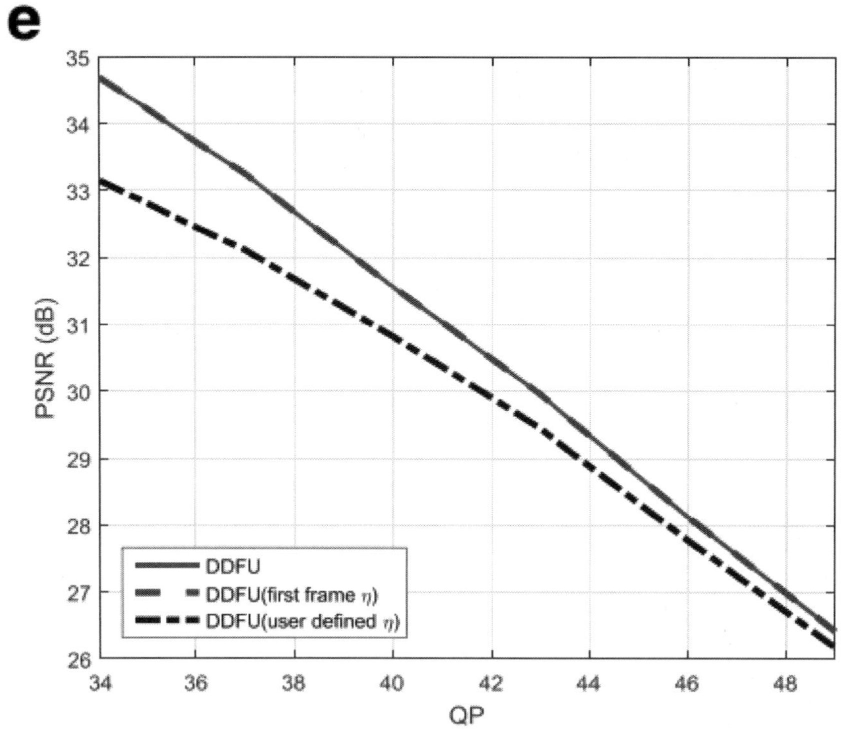

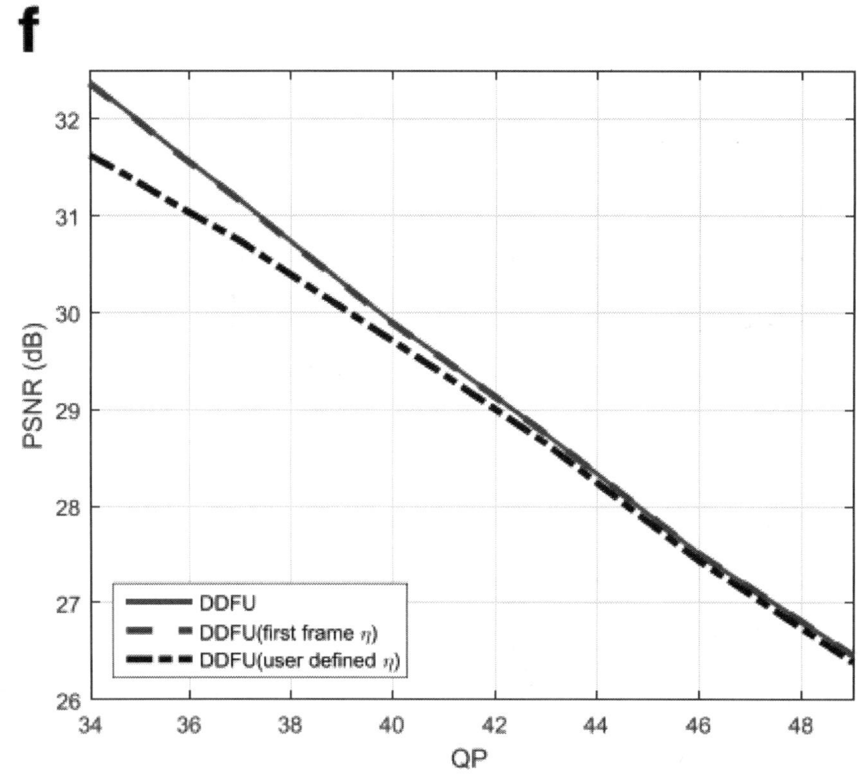

Figure. 16: The comparisons of proposed simplified approach. The comparisons of proposed simplified approach with full version approach and user-defined coefficient approach on the testing sequences (**a**) Doorflower, **b** Kendo, **c** Dog, **d** Balloons,**e** Newspaper, and **f** Dancer

From Fig. 16, it can be seen that DDFU and DDFU (first frame η) have almost the same performance for all sequences, which demonstrates the validity and effectiveness of the simplified approach. By comparing the results of DDFU and DDFU (user defined η), the importance of adapting the coefficients to the scene content can be appreciated. The results in Fig. 16 show that the performance of DDFU (first frame η) are better than that of DDFU (user defined η). This is due to the η values for the DDFU (first frame η) are based on the content of the testing sequence, if the content of the sequence does not vary hugely frame by frame, neither does the value of η, while the values of the predetermined η are user defined values, which means they do not take the content of the sequence into account. The performance of DDFU (user defined η) highly depends on how close the predetermined values are to the frame-by-frame evaluated coefficients.

CONCLUSIONS

In this paper, an interlacing-and-complementary-row-downsampling method is employed on the two adjacent views of a multiview video at the encoder side to reduce the transmitted data. This downsampling method allows the proposed directional data fusion upsampling (DDFU) algorithm to recover the discarded pixels by exploiting the information of the downsampled views and the corresponding virtual views. In the proposed upsampling approach, edge directions around the discarded pixels are estimated by principal components analysis. This information is subsequently used to steer the fusion of the virtual view with the directional interpolated pixels. The aim behind this is to exploit the inter-view redundancy to minimize the overall system distortion, which is a combination of the compression distortion and the distortion introduced by the downsampling process. Therefore, different from filter-based interpolation algorithms, the advantages of virtual views have been exploited by the proposed method. Moreover, it has been shown that the proposed algorithm achieves superior performance in comparison with filter-based interpolation algorithms and the state-of-the-art algorithms. The future work will be devoted to exploiting the temporal correlation in video sequences to control the fusion process.

REFERENCES

1. A Vetro, T Wiegand, G Sullivan, Overview of the stereo and multiview video coding extensions of the H.264/MPEG-4 AVC standard. Proc. IEEE. **99:**, 626–642 (2011).
2. M Flierl, A Mavlankar, B Girod, Motion and disparity compensated coding for multiview video. IEEE Trans. Circ. Syst. Video Technol. **17:**, 1474–1484 (2007).
3. C Yan, Y Zhang, J Xu, F Dai, J Zhang, Q Dai, F Wu, Efficient parallel framework for HEVC motion estimation on many-core processors. IEEE Trans. Circ. Syst. Video Technol. **24:**, 2077–2089 (2014).
4. C Yan, Y Zhang, J Xu, F Dai, L Li, Q Dai, F Wu, A highly parallel framework for HEVC coding unit partitioning tree decision on many-core processors. IEEE Signal Process. Lett. **21:**, 573–576 (2014).
5. C Yan, Y Zhang, F Dai, X Wang, L Li, Q Dai, Parallel deblocking filter for HEVC on many-core processor. Electron. Lett. **50:**, 367–368 (2014).
6. C Yan, Y Zhang, F Dai, J Zhang, L Li, Q Dai, Efficient parallel HEVC intra-prediction on many-core processor. Electron. Lett.**50:**, 805–806 (2014).
7. P Merkle, A Smolic, K Muller, T Wiegand, Efficient prediction structures for multiview video coding. IEEE Trans. Circ. Syst. Video Technol. **17:**, 1461–1473 (2007).
8. C Fehn, in Conference Record of the Thirty-Seventh Asilomar Conference on Signals, Systems and Computers, 2. A 3D-TV system based on video plus depth information (IEEE, 2003), pp. 1529–1533.
9. C Fehn, in Electronic Imaging, International Society for Optics and Photonics. Depth-image-based-rendering (DIBR), compression, and transmission for a new approach on 3d-tv (SPIEUSA, 2004), pp. 93–104.
10. J Wu, Y Xing, G Shi, L Jiao, in 16th IEEE International Conference on Image Processing (ICIP). Image compression with downsampling and overlapped transform at low bit rates (IEEECairo, Egypt, 2009), pp. 29–32.
11. V Nguyen, Y Tan, W Lin, in IEEE International Symposium on Circuits and Systems (ISCAS). Adaptive downsampling/upsampling for better video compression at low bit rate (IEEESeattle, USA, 2008), pp. 1624–1627.
12. X Wu, X Zhang, X Wang, Low bit-rate image compression via adaptive down-sampling and constrained least squares upconversion. IEEE Trans. Image Process. **18:**, 552–561 (2009).
13. C Fehn, A 3D-TV approach using depth-image-based rendering (DIBR), VIIP 03 (ACTA, Calgary, Canada, 2003).
14. D Garcia, T Fonseca, R Queiroz, in 18th IEEE International Conference on Image Processing (ICIP). Video compression complexity reduction with adaptive down-sampling (IEEEBrussels, Belguim, 2011), pp. 745–748.
15. H Sawhney, Y Guo, K Hanna, R Kumar, in SIGGRAPH. Hybrid stereo camera: an IBR approach for synthesis of very high resolution stereoscopic image sequences (ACMNY, USA, 2001), pp. 451–460.
16. P Aflaki, M Hannuksela, J Hakkinen, P Lindroos, M Gabbouj, in 17th IEEE International Conference on Image Processing (ICIP). Subjective study on compressed asymmetric stereoscopic video (IEEEHong Kong, 2010), pp. 4021–4024.
17. D Garcia, C Dorea, R Queiroz, Super resolution for multiview images using depth information. IEEE Trans. Circ. Syst. Video Technol. **22:**, 1249–1256 (2012).
18. Z Jin, T Tillo, C Yao, J Xiao, Y Zhao, Virtual view assisted video super-resolution and enhancement. IEEE Trans. Circ. Syst. Video Technol. **26**(3), 467–478 (2016).
19. Y Zhang, X Ji, H Wang, Q Dai, Stereo interleaving video coding with content adaptive image subsampling. IEEE Trans.Circ. Syst. Video Technol. **23:**, 1097–1108 (2013).
20. X Ji, Y Zhang, Q Dai, Opposite parity packing arrangement for stereoscopic video coding.Electron. Lett. **47:**, 1180–1181 (2011).
21. A Vetro, in 17th IEEE International Conference on Image Processing (ICIP). Frame compatible formats for 3D video distribution (IEEEHong Kong, 2010), pp. 2405–2408.
22. I Jolliffe, Principal Component Analysis (Springer, New York, 2002).
23. B Yang, Z Gao, X Zhang, in IEEE International Conference on Multimedia and Expo (ICME). Principal components analysis-based edge-directed image interpolation (IEEEMelbourne, Australia, 2012), pp. 580–585.
24. G Stewart, On the early history of the singular value decomposition. SIAM Rev. **35:**, 551–566 (1993).
25. Fraunhofer Heinrich Hertz Institute, MPEG 3D video Sequences. http://ftp.hhi.de/HHIMPEG3DV/. Accessed 17 June 2012.

26. Nagoya University, FTV Test Sequences (2008). http://www.tanimoto.nuee.nagoya-u.ac.jp/mpeg/mpeg_ftv.html. Accessed 17 June 2012.

27. Y Chen, P Pandit, S Yea, C Lim, in Joint Video Team (JVT) of ISO/IEC/MPEG and ITU-T/VCEG, Doc. JVT-AE207. Draft reference software for MVC (Joint Video TeamLondon, UK).

28. J Chiang, S Chen, in 2014 IEEE Global Conference on Signal and Information Processing (GlobalSIP). Frame-compatible asymmetric stereo video coding considering human perception(IEEEAtlanta, USA, 2014), pp. 956–959.

29. Z Wang, Z Zheng, in IEEE Conference on Computer Vision and Pattern Recognition (CVPR). A region based stereo matching algorithm using cooperative optimization (IEEEAnchorage, USA), pp. 1–8.

CHAPTER 11

ROBUST MULTIMODAL FACE AND FINGERPRINT FUSION IN THE PRESENCE OF SPOOFING ATTACKS

Peter Wild·Petru Radu ,Lulu Chen ,James Ferryman

University of Reading, School of Systems Engineering, Whiteknights, Reading RG6 6AY, United Kingdom

ABSTRACT

Anti-spoofing is attracting growing interest in biometrics, considering the variety of fake materials and new means to attack biometric recognition systems. New unseen materials continuously challenge state-of-the-art spoofing detectors, suggesting for additional systematic approaches to target anti-spoofing. By incorporating liveness scores into the biometric fusion process, recognition accuracy can be enhanced, but traditional sum-rule based fusion algorithms are known to be highly sensitive to single spoofed instances. This paper investigates 1-median filtering as a spoofing-resistant generalised alternative to the sum-rule targeting the problem of partial multibiometric spoofing where m out of $n$biometric sources to be combined are attacked. Augmenting previous work, this paper investigates the dynamic detection and rejection of liveness-recognition pair outliers for spoofed samples in true multi-modal configuration with its inherent challenge of normalisation. As a further contribution, bootstrap aggregating (bagging) classifiers for fingerprint spoof-detection algorithm is presented. Experiments on the latest face video databases (Idiap Replay-Attack Database and CASIA Face Anti-Spoofing Database) and fingerprint spoofing database (Fingerprint Liveness Detection Competition 2013) illustrate the efficiency of proposed techniques.

INTRODUCTION

Fingerprint and face biometrics as most widely adopted traits are being exposed to an increasing threat of presentation attacks. Consequently, there are numerous studies[1] and [2] and open challenges [3] and [4] on anti-spoofing techniques assessing the spoofing detector's ability to distinguish between genuine and fake attempts for especially these two traits. Recently, the integration of anti-spoofing scores with recognition scores has received considerable attention [5], [6] and [7]. The standard approach, as outlined in [5], has been to reject spoofed samples before comparing them against the gallery template. However, recognition scores can be helpful in the probe-attack spoofing detection problem and liveness scores can impact on the recognition task. Considering imposters with access to fake fingers or face photographs reveals an impact on overall accuracy (shifted imposter score distribution for non-zero-effort attempts [7]) and assuming a correlation between successful spoofs achieving a higher score and their corresponding liveness score is likely (and shown) to help in the final judgment of the decision task, especially in an ensemble of classifiers where this paper looks for outliers. It is therefore

Citation: Peter Wild, Petru Radu, Lulu Chen, James Ferryman, Robust multimodal face and fingerprint fusion in the presence of spoofing attacks, Pattern Recognition, Volume 50, February 2016, Pages 17-25, ISSN 0031-3203, http://dx.doi.org/10.1016/j.patcog.2015.08.007.

useful to investigate the benefits of dealing with a holistic (liveness and verification) multi-class problem rather than two separate classification problems (live vs. fake and genuine vs. impostor). If a system involves multiple modalities there is an even larger variety of different ways to treat the problem of combining liveness and recognition scores. Multibiometrics using face and fingerprint biometrics comes with many benefits including expected increased accuracy, higher universality (absence of single characteristics), efficiency (fast indexing), but its robustness to spoofing attempts has been shown to be compromised [8] and [9]. Furthermore, with the inclusion of multiple modalities the attacker has an even more extended choice to select the easiest modality to be attacked. It is therefore desirable to find new techniques coping with spoofing attacks, which are subject to investigation in this paper. The paper focuses on three objectives: (1) investigation of spoofing robustness in multibiometrics; (2) development of novel methods towards anomaly detection for increased systematic anti-spoofing; and (3) proposition of a novel bootstrap aggregating (bagging) of classifiers method combining features in fingerprint counter-spoofing.

With regard to the first topic on spoofing robustness in multibiometrics, the paper tests degradation in accuracy for the "partial multibiometric spoofing" scenario, where *m* out of*n* samples are spoofed, highlighting the tradeoff between accuracy and security for different fusion methods. Fig. 1 illustrates this concept. The sensitivity of a recognition and liveness fusion method with regard to spoofing is especially interesting in multimodal configuration, where scores originate from different underlying distributions and multiple traits facilitate a selection of the modality to be attacked. The paper analyses the impact of the number of spoofed fingers or spoofed face on accuracy using the latest biometric datasets. The relative robustness of several score-level fusion rules can be used to choose the most robust fusion rule [9].

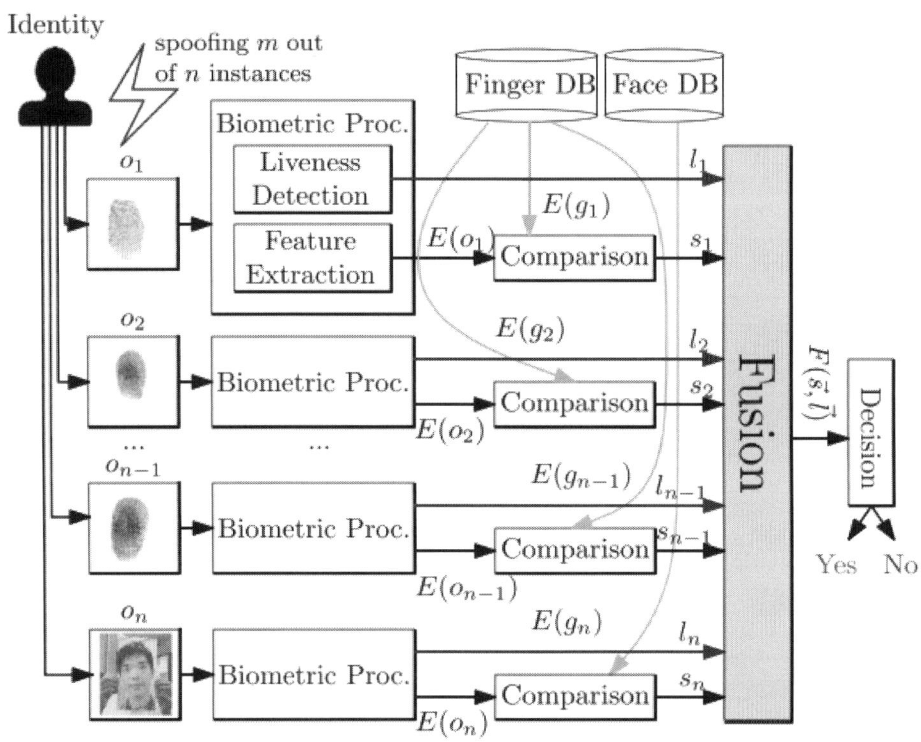

Figure. 1: Partial multibiometric spoofing of observations *oi* given templates *gi* fusing scores *si* and liveness values *li*.

As a second outlined contribution, this paper presents a novel multibiometric spoofing-aware fusion method following the idea of anomaly detection and extending research in[10] to multiple modalities. This paper investigates 1-median-based fusion using outlier detection applied in a multibiometric setup. Note that the extension to multiple modalities raises further questions with regard to normalisation. For different modalities, scores generally follow different distributions. Therefore, counter-spoofing is much more challenging than for single-modality approaches, including multi-instance or multi-algorithm approaches. Further, this work presents further theoretical considerations and discusses parameter choice in detail. Recognition scores and liveness scores are likely to be dependent, as spoofing tries to achieve a high recognition score in order to successfully claim the alien (spoofed) identity. In partial multibiometric spoofing this information can be used to further discriminate between genuines and impostors. Despite spoofing sensitivity of traditional fusion techniques, it is a reasonable assumption to claim a higher difficulty for attackers to spoof multiple modalities at the same time or even to obtain the necessary samples to produce a fake fingerprint or face mask. On the other hand, special spoofing-robust fusion schemes might exhibit a reduced level of accuracy. This trade-off between cost and security to limit drawbacks [5] and [11] is investigated.

Third and last, as a by-product of evaluations the paper further presents a novel spoofing detector again employing a fusion principle: bootstrap aggregating (bagging) of classifiers. This technique is employed in combining the decision outcome of multiple different classifiers. Using also multiple features to be more robust vs. changes in materials (see [12]), the paper aims at investigating this technique in the employed system as an anti-fingerprint spoofing technique towards integral fusion concepts in robust anti-spoofing. Bagging is shown to outperform state-of-the-art detectors on the most challenging LivDet 2013 Crossmatch subset database.

The remainder of this paper is organised as follows: Section 2 introduces the problems of anti-spoofing and spoofing-aware fusion in biometrics. The proposed methods of bagging for spoof-detection and 1-median filtering for spoofing-resistant multibiometric fusion are outlined in Section 3. Section 4 highlights experimental results with regard to the proposed and investigated techniques. This includes a discussion of methods towards anomaly detection in multibiometrics, highlighting parameter choice and optimisation for the proposed 1-median filtering. Section 5 concludes this paper with an outlook on future work.

RELATED WORK

There are several anti-spoofing or liveness detection algorithms extracting features (usually trained for modality, sensor, material, etc.), in order to determine whether a biometric sample is either *live* or *fake*. For evaluation purposes, *ferrlive* (rate of misclassified live samples) and *ferrfake* (rate of misclassified fake samples) are employed. Whereas for individual modalities the anti-spoofing problem is well defined and evaluated separately from biometric system performance, research on fusion between match scores and liveness factors is still in its infancy [13]. Recently, [14]suggested a framework for verification systems under spoofing attacks. Within the framework [8] adopted in this paper, liveness and recognition scores are combined considering the scenario of probe-spoofing only (i.e. no gallery-spoofing, enforced by e.g., attended enrolment). Formally, given a vector of biometric observation (units, e.g. fingers, eyes) $\vec{o} = (o_1, \dots o_n)$ from one or more modalities, and corresponding claimed identity template $\vec{g} = (g_1, \dots, g_n)$, the task of the fusion module F is to compute a unified decision score, using comparison scores $\vec{s} = (s_1, \dots, s_n)$ and (probe) liveness values $\vec{l} = (l_1, \dots, l_n)$, so that the verification task V (authentication based on threshold η) can be formulated as follows:

$$V(\vec{o}, \vec{g}) := \begin{cases} \text{accept} & \text{if } F(\vec{s}, \vec{l}) \geq \eta; \\ \text{reject} & \text{else.} \end{cases} \qquad (1)$$

Let i be the current index and $E(o_i), E(g_i)$ refer to extracted (modality-specific) features of samples, then $s_i = C(E(o_i), E(g_i)) \in [0,1]$ is used to denote the normalised comparison result of o_i, g_i and $l_i = L(o_i) \in [0,1]$ denote the likeliness of a genuine (live) sample. Clearly, it is desirable to find a method F unaffected in performance if m out of the n elements of $\vec{o}$ are spoofed. This testing setup is referred to as "partial multibiometric spoofing", introduced in [10] and extended in this work towards multiple modalities. Note that this notion of live or spoofed probes vs. always-live enrolled gallery samples (assuming attended enrollment) leads to a simpler modelling (2 classes distinguishing live probe from spoof or live, but different sources) than in the general asymmetric case (8 classes based on live/spoof probe, live/spoof gallery sample, and same/different source) or symmetric case (6 classes) [7], fully concentrating on a dichotomous authentication task, which can be evaluated in the traditional way using receiver operating characteristics.

On Combining Anti-Spoofing and Recognition

Marasco et al. [5] are among the first considering fusion of liveness with recognition scores separately for each modality, using simple rejection of spoofed samples. If a spoofing attempt is indicated, the current modality matching score is ignored. This initial study is extended in [15] evaluating sequential fusion, classifier fusion, and Bayesian Belief Networks for combining match scores and liveness measures, highlighting the superiority of the latter method for the LivDet2009 dataset but also that accuracy is decreased when taking liveness detection into account. Chingovska et al. [6] evaluate binary decision rules and Logistic Regression (LR) as decision and score-level fusion techniques combining face recognition and liveness scores addressing the integration (but neglecting the partial spoofing problem) of liveness. They report higher resistance to spoofing attacks (91.54% vs. 10%) but are outperformed by LR approaches achieving both, high verification accuracy and good spoofing detection. Recently, Poh et al. [7]have targeted the problem of integrating spoofing and matching scores in a probe and gallery-spoofing scenario, investigating Gaussian Copula-based Bayesian classifiers and a mixture of linear classifiers for this task. While their method outperforms classical Support Vector Machine (SVM) based techniques, the approach needs training with regards to the full range of attacks.

The assessment of traditional fusion rules (this work is using Kittler et al.'s classical framework [16]) in the presence of spoofing attacks is a further relevant sub-problem and addressed in this work. Rodrigues et al. [8] and [17] first addressed this security issue of spoofing attacks against a multimodal biometric system. They presented two methods, one using likelihood ratio and another employing fuzzy logic, both exceeding the accuracy of traditional fusion rules. Also Akhtar et al. [18] studied the impact of spoofing on parallel and serial fusion rules for face and fingerprint reporting that score-level fusion methods from the literature are not robust to spoofing attacks and that serial fusion gave better results for an overall assessment of performance, verification time, user acceptability and robustness.

Anti-Spoofing in Fingerprint and Face Recognition

In fingerprint recognition, there are two general ways to address the spoofing problem: either by actively assessing the liveness (e.g. by measuring pulse, perspiration patterns, or blood pressure), or by passively analysing patterns of spoofed materials (e.g. lack of detail, pattern differences). The latter type, which is the subject of interest in this paper, reveals high risk of material and sensor-dependence [12]. An excellent recent survey of spoofing methods in fingerprint recognition can be found in [2]. Among the most common techniques for static (extracted from single image) texture-based anti-spoofing methods are statistical features [19], Power Spectrum Fourier analysis [20], Ridge Frequency Analysis [19], Local Binary Patterns (LBP) [21] and Local Phase Quantisation [22].

However, recent developments towards material-independent static anti-spoofing suggest to combine multiple features and probably even detectors. Fumera et al. [13]give a good introduction into the problem of combining multiple liveness detectors for a single modality, fusion of liveness detector and matcher for a single modality, and anti-spoofing capabilities of ad hoc fusion rules combining multiple comparison scores.

Face spoofing counter-measures can broadly be classified into texture-based and motion-based counter-measures. A good overview on face counter-spoofing may be found in [23]. The first category assessing textural properties is the more widespread group with approaches like LBP [24], or statistical features [25] exploiting the observation that images/ videos with spoofed faces (printed or replayed) do not exhibit the same noise-level like genuine samples. The second type of motion-based approaches targets the reproduction of (flat) printed photographs or re-display of faces on tablets exploiting the difference in 3D appearance of spoofed approaches. For fusion purposes this paper focuses on the first type and employs an existing anti-spoofing system [26].

PROPOSED METHODS

In order to solve the problem of robust face and fingerprint fusion in the presence of spoofing attacks, this work proposes 1-median filtering for enhanced tolerance with regards to a number of attack-outliers in the ensemble of score-liveness tuples, and bagging of classifiers for enhanced (fused) spoofing resistance. Both methods are described in detail in the next subsections.

Score and Liveness Fusion Using 1-Median Filtering

Aiming to overcome the limitations of traditional sum-rule based techniques, which are known to be very susceptive towards outliers and thus easy to be attacked in multibiometric configuration, where an attacker can target the weakest link in the chain of combined biometric units to be attacked (e.g. using a particular available latent fingerprint), 1-median filtering [10] is investigated as a method for joint face and fingerprint score-and-liveness fusion. The motivations in the definition of 1-median filtering are (1) an extension towards a hybrid between sum rule and median rule as in Kittler et al.'s classical fusion methods [16] to find an optimal compromise between (0-spoof) accuracy and (m -spoof) robustness performance; and (2) an incorporation of high-dimensional information to be combined (score $\vec{s}$ and liveness $\vec{l}$ pairs as introduced in Section 2).

Based on the median rule's property to be less affected by outliers (which is very beneficial for spoofing resistance), this fusion method can be formulated as follows:

$$F_{mf}(\vec{s}) := \frac{1}{\sum_{i=1}^{n} M(\vec{s}, s_i)} \sum_{i=1}^{n} M(\vec{s}, s_i) s_i. \tag{2}$$

$$M(\vec{s}, s_i) := \begin{cases} 1 & \text{if } \left| s_i - \operatorname*{med}_{j=1}^{n} s_j \right| < \phi; \\ 0 & \text{else.} \end{cases} \tag{3}$$

Note that parameter ϕ limiting the zone of influence allows for an arbitrary tradeoff between the sum rule (ϕ=∞ results in $F_{mf}(\vec{s}) = \frac{1}{n}\sum_{i=1}^{n} s_i$, the classical sum rule) and the median rule (for ϕ sufficiently small, the definition becomes $F_{mf}(\vec{s}) =$, $med_{j=1}^{n} s_j$, the median rule). One of the important tasks is to find a suitable (trained) parameter ϕ, which can be either fixed or a function of the scores. The choice of ϕ is not straightforward and should rely on the underlying distribution's properties. Parameter selection is discussed in Section 4.7.

As a motivation for the selection of the median, consider the following theoretical considerations: as observed in [6] unimodal non-zero-effort imposter score distributions (comparing a spoofed sample with a genuine reference) are shifted towards the genuine distribution (comparing two live genuine samples) compared to zero-effort imposters (comparing two live samples from different identities). In an equally weighted mixture-model for random variables Xi, we have $E(\frac{1}{n}\sum_{i=1}^{n} X_i) = \frac{1}{n}\sum_{i=1}^{n} E(X_i) = \mu$ assuming independent, normalised (same mean μ and variance σ^2) distributions. However, for the variance, we get $Var(\frac{1}{n}\sum_{i=1}^{n} X_i) =$. $\frac{1}{n^2} Var(\sum_{i=1}^{n} X_i) = \frac{1}{n}\sigma^2$. While this illustrates the positive effect of fusion on imposter scores (narrowing the variance), it also clearly illustrates that if one of the random variables follows a degraded spoof-imposter distribution with lower mean, this is likely leading to a bimodal distribution (especially if n is large). Assuming the distributions can be modelled by Gaussians, a mixture of two normal distributions with highly unequal means has a positive kurtosis, as the smaller distribution lengthens the tail of the more dominant one. While there are exceptions to the rule [27], as a rule of thumb, it is generally suggested, that in skewed distributions, the mean is farther out the longer tail than the median [28], therefore a better representative in the filtering process (which is likely to succeed as can be seen from theoretical considerations if the number of spoofed modalities is low compared to n). However, note that the crucial pre-assumption is a proper normalisation, which ideally should leave genuine score distribution almost unaffected. Further, note that median filtering is not to be mixed up with image- or kernel-based combination and works on score (and liveness-) values to be combined.

Algorithm 1.

Median filtering

Data: comparison pairs' scores $\vec{s} = \{s_1, s_2, \dots s_n\}$ and probe-liveness
$\vec{l} = \{l_1, l_2, \dots l_n\}$, trained parameters ϕ, Ψ

Result: verification decision $p \in [0, 1]$ indicating live and genuine match

m ← $\operatorname*{med}_{j=1}^{n} \begin{bmatrix} s_j \\ l_j \end{bmatrix}$;

for *each score pair* (s_i, l_i) **do**
 if $\left\| \begin{bmatrix} s_i \\ l_i \end{bmatrix} - m \right\| < \phi$ **then**
  reject sample from $\vec{s}, \vec{l}$ and update $n \leftarrow n - 1$;
 end
end

f ← $\frac{1}{n}\sum_{i=1}^{n} \begin{bmatrix} s_i \\ l_i \end{bmatrix}$; p ← logreg$(f, \Psi)$;

The outlined technique can easily be extended to 2D for combining points (s_i, l_i) of recognition and liveness scores, using the geometric-median (1-median). This is the point minimising the sum of distances to the sample points using score *si* and liveness *li*as coordinate values.

$$F^2_{mf}(\vec{s},\vec{l}) := \frac{1}{\sum_{i=1}^{n} M\left(\begin{bmatrix}\vec{s}\\ \vec{l}\end{bmatrix},\begin{bmatrix}s_i\\ l_i\end{bmatrix}\right)} \sum_{i=1}^{n} M\left(\begin{bmatrix}\vec{s}\\ \vec{l}\end{bmatrix},\begin{bmatrix}s_i\\ l_i\end{bmatrix}\right)\begin{bmatrix}s_i\\ l_i\end{bmatrix}. \tag{4}$$

$$M\left(\begin{bmatrix}\vec{s}\\ \vec{l}\end{bmatrix},\begin{bmatrix}s_i\\ l_i\end{bmatrix}\right) := \begin{cases} 1 & \text{if } \left\| \begin{bmatrix}s_i\\ l_i\end{bmatrix} - \underset{j=1}{\overset{n}{med}} \begin{bmatrix}s_j\\ l_j\end{bmatrix} \right\| < \phi; \\ 0 & \text{else.} \end{cases} \tag{5}$$

Fig. 2 illustrates how median filtering uses the median as a seed point to select all points in a local neighbourhood, computing the centroid of the set of filtered points as an even better local representative. As the median is less affected by outliers (left example) it is beneficial in the presence of outliers, whereas in case samples are less scattered (right example), no samples are rejected. Algorithm 1 illustrates all the steps. Note that the additional processing time needed for the comparison should have a negligible impact, as *n* is traditionally rather small.

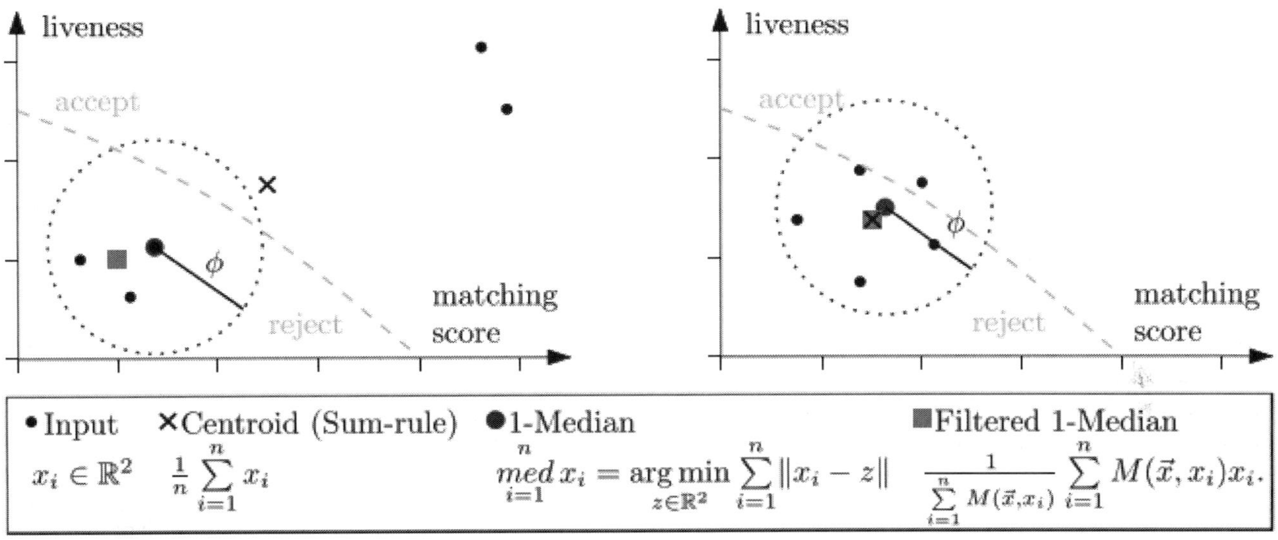

Figure. 2: 1-Median filtering vs. sum-rule in 2D with outliers (left) and without (right).

Note that the 1-median is not necessarily an input point and for performance reasons an approximation (e.g., coordinate-wise median) might be sufficient. As the task of the fusion module is to come to a final single decision score, a further mapping to a single scalar is necessary. For this task, LR or SVMs can be employed to find the hyperplane $\Psi : \vec{w} \cdot \vec{x} - \vec{b} = 0$ optimally separating the sets of genuine and zero-/*m*-spoof impostors, where *m* is the number of spoofed samples in the joint fusion scheme

$$F_{mf}(\vec{s},\vec{l}) := dist(F^2_{mf}(\vec{s},\vec{l}),\Psi). \tag{6}$$

Separability becomes more difficult for larger values of *m* (spoofed samples), the presented implementation uses $m=\lfloor n/2 \rfloor$. Threshold variation is equal to moving the hyperplane separating the two (genuine and impostor) joint score- and liveness-distributions. Training of Ψ is discussed in Section 4.5.

Bagging-Based Fingerprint Liveness Detection

For fingerprint anti-spoofing relatively poor performance compared to other test sets is reported for the Crossmatch subset of LivDet 2013 fingerprint database [4]. In order to improve those results, this paper proposes the following novel spoof detection algorithm. The employed setup follows a three-stage architecture with preprocessing, feature extraction, and classifier fusion. Fig. 3 illustrates the processing chain. In the preprocessing stage, the fingerprint image is segmented and aligned. The background of the image is removed using Otsu's thresholding [29] and the region of interest is automatically cropped at a dimension of 248 by 256 pixels. Feature extraction extracts global properties and local texture details using three methods selected as representative methods (wavelet-based, statistical and frequency-based) to make maximal use of the fusion technique:

1.2 *D Gabor filters* [30]: These filters as a product of a Gaussian and a sinusoid capturing local details are parameterised by Gaussian space constants δx and δy, frequency *f* of the modulating sinusoid and orientation θ

$$G(x,y,f) = \frac{1}{2\pi\delta_x\delta_y} e^{-1/2(x'^2/\delta_x^2 + y'^2/\delta_y^2)} \cos(2\pi f x')$$
$$x' = x\sin\theta + y\cos\theta \quad y' = x\cos\theta - y\sin\theta \tag{7}$$

Similar to [31]θ is set to $0°, 45°, 90°$ and $135°$ at frequency f=0.1, corresponding to 10 pixels (typical inter ridge distance). The filtered region of interest is divided in blocks of 20×20 pixels. For each block, mean and standard deviation are computed leading to a total of 880 features.

2. *Gray level co-occurrence matrix* (*GLCM*): 6 features each were extracted from 20 by 20 sized pixel-blocks (leading to a total feature vector size of

1100 components) computing local characteristics following [31]: maximum probability, entropy, contrast, energy, homogeneity and inverse difference moment of order *k*.

3. *Fourier Transform (FT) based features*: As global features on the Fourier-transformed image, the sum of absolute differences between pairs of concentric circles (at distance 1–3 pixels, evaluated at 25 locations) are computed, similar to[32], yielding 180 components.

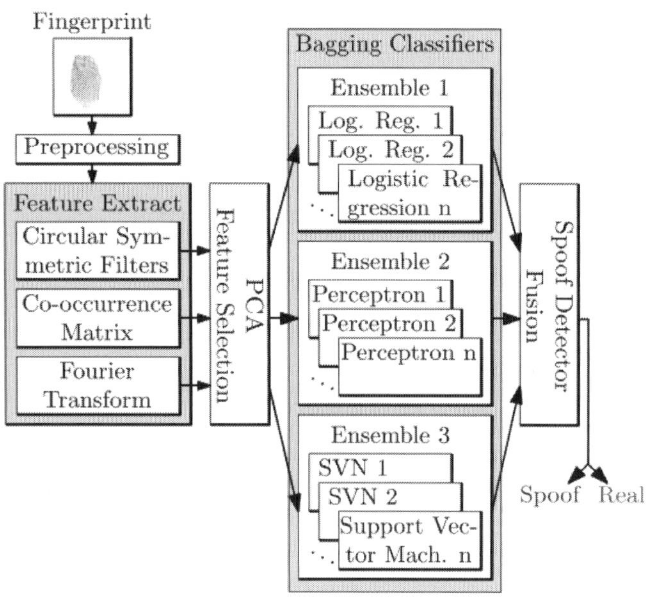

Figure. 3: Proposed fingerprint counter-spoofing based on bagging classifiers.

The final feature of 2160 components is obtained by concatenating the 3 individual feature vectors described above and Principal Component Analysis (PCA) is applied as a feature selection procedure retaining 99% of the variance of the data with 80 PCA components. Besides a fusion of features, the suggested anti-spoofing algorithm employs a multiple classifier framework, which distinguishes itself from a standard multiple classifier system by applying a bagging [33] technique for its component (base) classifiers.

The Bootstrap AGGregatING (bagging) method [33] is used to add base classifiers to the base ensemble using bootstrap replicates on the training set. The bootstrap method facilitates determining the probability distribution of the data without using the Central Limit Theorem [34]. The idea behind bootstrap sampling is to create an artificial random list of the labelled training set by picking some labels more than once. One classifier is trained on this random list and is added to the base ensemble. In the operational phase, the base classifiers are applied to the input feature and their outputs are combined at the decision level by using majority vote. To benefit from the variations of the training set, it is better if the base classifiers are unstable (e.g. neural networks and tree classifiers). In the present work, three base classifiers are employed: (1) regularised LR; (2) single layer perceptron, and; (3) SVM. The three base classifiers are trained n=100 times on different bootstrap replicates of the training data. In the operational phase, the 300 classifiers decisions are recorded as 0 or 1, where 1 indicates that the classifier believes that the image is spoofed. The final spoofing score *sf* for one test fingerprint image is given as

$$s_f = \frac{\sum_{i=1}^{m} \sum_{j=1}^{n} D_j^i}{mn} \tag{8}$$

where *m* is the number of base classifiers types, *n* is the number of bootstrap replicates and *Dji* is the decision of base classifier of type *I* trained on the bootstrap replicate number *j* . In the proposed implementation, m=3, n=100. The value of *sf* is in the range [0,1]. By setting and adjusting a threshold t∈[0,1], different operating points of the anti-spoofing approach are configurable.

EXPERIMENTS AND DISCUSSION

In order to evaluate the suggested 1-median filtering and bagging approaches, a modularised setup is employed, using state-of-the-art feature extraction, recognition and spoofing algorithms described in the following sections. After an introduction into database, metrics and employed reference setup, this section concentrates on questions related to baseline performance of spoofing detectors evaluating the bagging classifier approach and inter-relation of recognition and anti-spoofing. Then, partial multibiometric spoofing in multibiometric face and fingerprint context is assessed, considering recognition-only and joint recognition and liveness fusion techniques.

Setup: Database and Metrics

As in many other approaches assessing face and fingerprint fusion [18] and [35], also this work builds on a chimeric dataset pairing face and fingerprints originally originating from different people. This approach is justifiable, since fingerprints and faces as biometric modalities can be assumed to be independent. While also true multi-biometric databases exist with multiple traits collected from the same person, this does not extend to spoofing datasets. Further, spoofing datasets are created for liveness detection purposes and therefore usually do not have to come with a large number of genuine samples, which are needed for the intended recognition-based assessment. In contrast to spoofing evaluations assessing *ferrfake* and *ferrlive* (see Section 2) measures, this evaluation refers to Receiver Operating Characteristics (ROC). Note that for m>0 spoofed samples these refer to pairs of Spoof False Acceptance Rate (SFAR) and Genuine Acceptance Rate (GAR, the percentage of genuine users being accepted), rather than False Acceptance Rate (FAR) and GAR pairs, see [36]. For comparing recognition performance (S)EER is employed as the (Spoof) Equal Error Rate where GAR=(S)FAR and decidability index (*d* -Prime) as $d' = |\mu_1 - \mu_2| / \sqrt{(\sigma_1^2 + \sigma_2^2)/2}$ measuring the separation of distributions with mean μi and standard deviation *i*.

Paired data originate from the following datasets (Table 1):

- *LivDet 2013 CrossMatch* [4]: The 4500 images of 99 users support up to 3 genuine samples per finger and a varying number of spoofed samples made from BodyDouble, Latex, Playdoh and WoodGlue.
- *Idiap ReplayAttack* [37]: The counter-spoofing video database of 1300 clips of 50 clients with 320 × 240

pixels resolution provides 8 genuine and 40 attack samples per user offering a *controlled* (homogeneous background) and challenging *adverse*recording setup. There are 4 mobile attacks using iPhones, 4 high-resolution iPad replays, and 2 hard-copy prints.

- *CASIA AntispoofingFace* [38]: This database of 600 clips of 50 clients with 640 × 480 pixels resolution comes with 3 genuine and 9 fake samples per user, offering low, medium and high quality setups and 3 fake attacks.

Table 1: Employed test databases

Mod.	Database/set	Images	Users	Training	Testing
FP	LivDet2013 Crossmatch	4500	99	Left hand	Right hand
Face	ReplayAttack	1300	50	*train* set	*devel, enroll, test*
Face	AntispoofingFace	600	50	–	*train, test*

A chimeric dataset is compiled, combining faces and fingers from the databases above, forming a new set of 85 classes. Note the number of classes, 85, is due to the restriction of LivDet to right-hands only and guaranteeing a minimum number of genuine and spoof fingers to simulate the selection of *n* out of *m* spoofed samples of different fingers in a random way. Testing uses right hands only to allow for a training of the employed counter-spoofing detector and learning-based parameters of median filtering. Spoofing attempts are simulated by randomly replacing *m* out of *n* samples (4 fingers and 1 face) with spoofs.

Setup: Baseline System

For experiments, the following baseline system is employed:

- *NIST Biometric Image Software* [39]: For fingerprint feature extraction (using minutiae detection *mindtct*) and comparison (using *bozorth3* in 1:1 verify mode). While the final score is not normalised, (capped) min–max normalisation is used to map scores to the unit interval $[0,1]$.
- *Neurotechnology VeriLook 5.5* [40]: For processing video/still image face samples. The off-the-shelf software extracts a 4–35 kB template via facial reference points and is able to account for off-axis registration (using 15 degrees roll, pitch and yaw parameters). Note that quality assurance (ISO/IEC 19794-5:2005) was deactivated to account for low-quality samples in the dataset. The setup employed in this work used the *low matching speed* setting with switched-off threshold (such that final scores could be obtained).
- *LBP-TOP Face-Liveness* [24]: Using the open-source implementation in [26] for face liveness-detection. The LBP-TOP operator calculates LBP features at three orthogonal planes that intersect in the centre pixel. The features are extracted from each separate plane and then concatenated together. A multi-resolution description is then generated such that the histograms along the time domains (*XT* and *YT*) are concatenated for different values of time *t*. Fig. 4 illustrates the process. Compared with traditional 2D LBP features, LBP-TOP can capture spatio-temporal features combining information from both image and time domains. SVMs are then applied for classification.

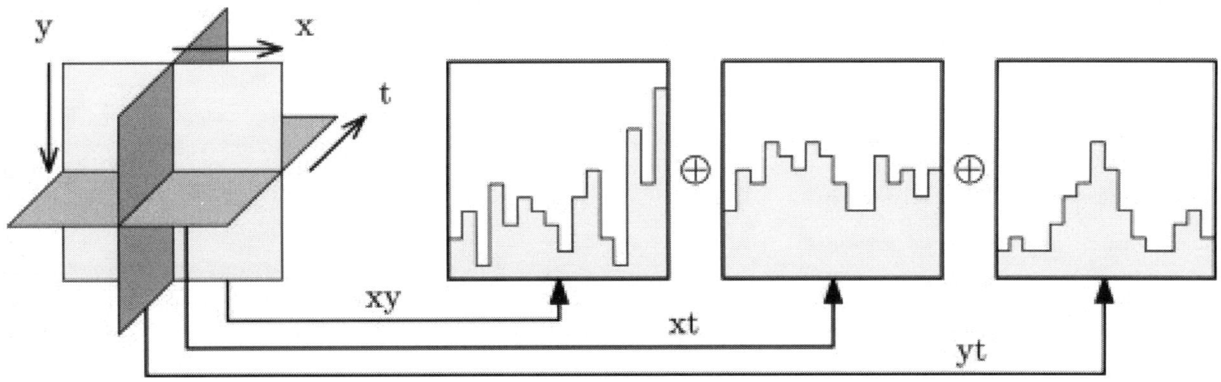

Figure. 4: LBP-TOP process [24]: for each of three planes intersecting at one pixel, LBP histograms are computed and concatenated.

- *Bagging Fingerprint-Liveness*: As introduced in Section 3.

Baseline Performance of Spoofing Detectors

In a first experiment, the detection performance of employed spoofing detectors is investigated. For anti-spoofing performance assessment, *ferrfake* and *ferrlive* rates are computed, using the underlying LivDet 2013 Crossmatch dataset for the fingerprint modality, and the ReplayAttack database for face. The implementation of the face spoofing-detection algorithm based on the open source package provided from the original work [26] yielded an accuracy of 85% on ReplayAttack. A similar accuracy is obtained for the presented fingerprint detection scheme, however the rate is much more remarkable given the high quality of the underlying spoofing dataset. Counter-spoofing using bagging classifiers yielded an accuracy of 84% on the Crossmatch set (the method is trained using a distinct subset of the test database using fingers from left hands), a value which significantly outperforms the best accuracy reported (68.8%) in the results of the LivDet liveness detection competition [4] for this subset. Given that integrated feature-level and classifier fusion in spoofing detectors is not a common practice, recognition rates are very promising and suggest to explore this topic even further in the future.

On the Mutual Impact of Liveness and Recognition

Spoofing systems are usually evaluated on their own without taking recognition into consideration [6]. However, the joint operation of liveness and recognition systems in practice raises a series of questions, most notably how to combine recognition and liveness information. This section highlights that not only liveness values are useful in justifying the authenticity of an identification claim, but also vice versa: the recognition score of a template is actually helpful to judge the presence of a spoofing attack. In LivDet, counter-spoofing performance is measured in terms of *ferrfake* and *ferrlive* rates referring to finger images as inputs, not comparisons. If recognition scores are to be considered in the evaluation of spoofing detection, it is important not to assume specific properties of the comparison. In real-world applications, however, a fake fingerprint will be employed to fake the originating identity causing its corresponding score to be distributed according to the spoof-imposter score distribution, whereas a live fingerprint is to originate from either genuine or zero-impostor distributions. The impostor score distribution is likely to shift towards the genuine score distribution in evaluations considering non-zero-effort impostors [6], whereas the genuine score distribution remains unaffected. Since spoofs are unlikely to be perfect, scores are typically degraded, which can be used to judge the presence of a fake sample using, e.g. fuzzy logic as employed in [8]. Simple fuzzy-rule based (inverting the recognition score contribution after threshold t=0.6) product fusion of recognition and liveness scores using randomly determined identity claims in experiments increased the spoofing detection capability from 18.36% *ferrfake=ferrlive* to 13.86% for the LivDet CrossMatch set. The integration of spoofing scores into recognition accuracy can further increase recognition accuracy when considering 1-spoof impostors in evaluations, e.g. by simply rejecting the sample [15]. Therefore, the real challenge is to find a suitable tradeoff between recognition accuracy and spoofing robustness, subject to investigation in the next sections.

Combining Liveness and Recognition Scores with Logistic Regression

With the positive impact of recognition on spoofing detection and vice versa, it is reasonable to proceed towards a holistic framework integrating both evidence as discussed in 2 and 3 and illustrated in Fig. 1. When combining both sets of scores individually using sum rule, logistic regression can be used to learn a more robust boundary. This boundary is trained (and then outliers eliminated using the presented 1-median filtering approach) in already combined (fused) space, which is able to clearly distinguish between different zero-effort and spoof detection. The 1-spoof and 2-spoof examples are used to account for the median tolerating a number of outliers up to half of the samples. The trained decision hyperplane in Fig. 5 has the following form:

$$\Psi : y = -13.52x + 12.4257. \tag{9}$$

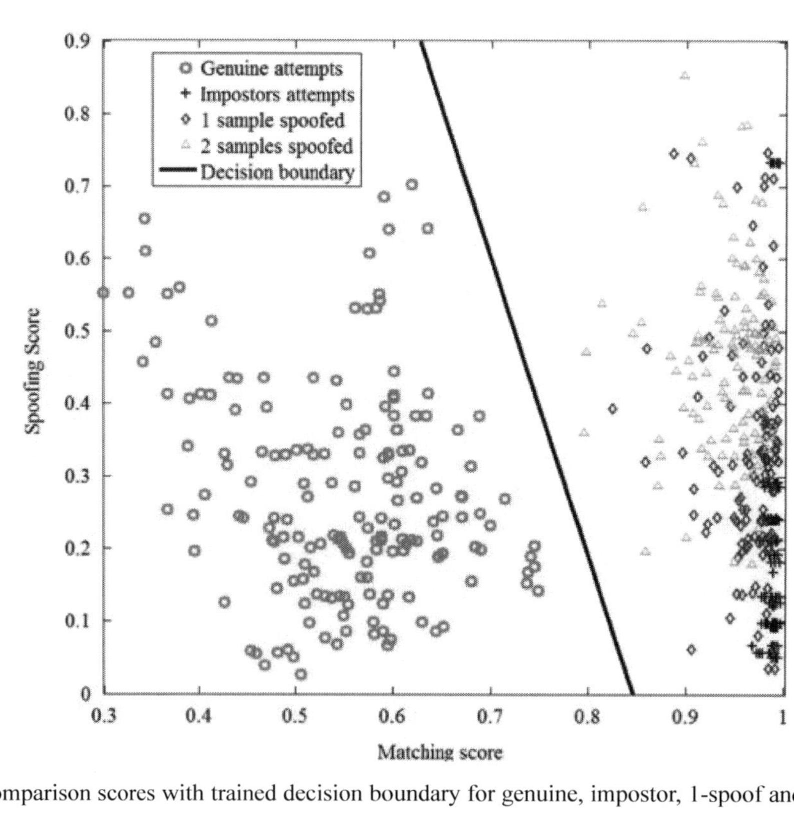

Figure. 5: Fused liveness-comparison scores with trained decision boundary for genuine, impostor, 1-spoof and 2-spoof pairs for finger-and-face fusion.

The experiment also clearly illustrates how *m*-spoof distributions are shifted towards the genuine distribution with increased *m*.

Classical Fusion in Partial Multibiometric Spoofing

In partial multibiometric spoofing, *m* out of *n* biometrics samples of an identity are spoofed (*n* =5 with 4 fingerprints and 1 face in experiments, using equal probabilities). That is, an attacker is assumed to have access to m=0,1,...,n latent fingerprints or face masks/print-outs for the attempt to spoof the system, referred to as an *m-spoof*attack. Recognition rates for the common fusion rules sum, product and median are evaluated in this scenario and results listed in Table 3. Confirming assumed behaviour, experiments in this work show that spoofing clearly impacts on recognition. For simple sum rule fusion it is evident that recognition is affected if even a single sample is spoofed, EER is degraded from 0% to 2.32% (*d*-Prime from 2.91 to 2.62). While perfect separation in the first case is certainly also attributable to the size of the dataset, the degradation is clearly visible, taking also the large number of 5 combinations of 2 modalities in a single authentication attempt in our particular setup into account. Further *m*-spoofing of faces and fingerprints suggest an increase of abs. 2–3% using sum rule for every additional spoofed sample (EERs of 2.32–12.24% for 1–5 spoofs), which is even more pronounced compared to previous fingerprint-only experiments [10] and spoofing of all samples did not always lead to acceptance (due to degraded accuracy of spoofed samples). The ROCs in Fig. 6 and Fig. 7 illustrate the degradation for sum and median rules, respectively.

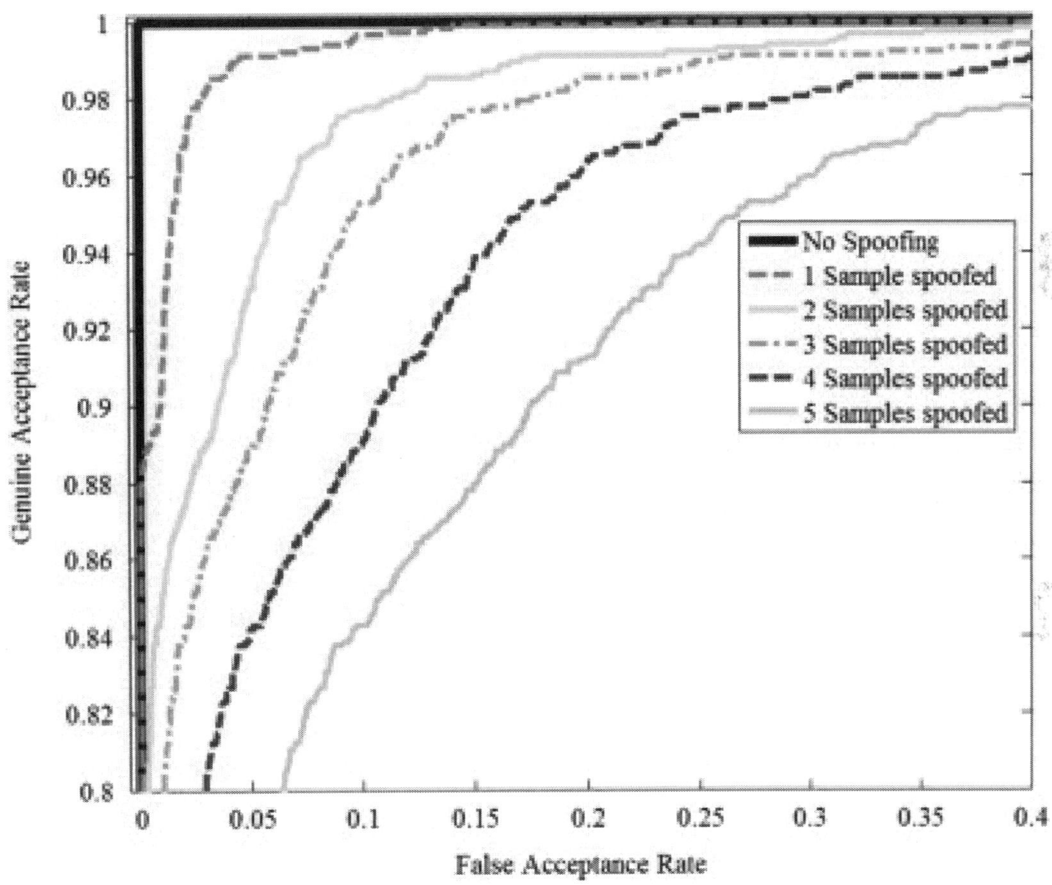

Figure. 6: ROC for partial multibiometric spoofing using sum rule.In contrast to previous experiments on fingerprints only, median rule has shown to be even more successful in combining results. There is little difference between 0-spoof, 1-spoof, and 2-spoof samples (EERs of 0.42%, 0.87%, and 1.20%), suggesting better tolerance vs. spoofing attempts clearly outperforming the sum-rule. However, this comes at a price of clearly degraded initial performance. Further, results indicated that the underlying distributions have a huge impact on the performance of the median rule (reported results refer to min–max-normalised scores). It is therefore important to consider learning distributions beforehand and employ a proper normalisation method. If median rule is extended to median filtering, the low 0% EER of 0-spoofs can be retained and still a better spoofing resistance than sum rule is observed. The product rule performed slightly worse than sum rule with 1–5 spoof EERs of 3.35–15.72%.

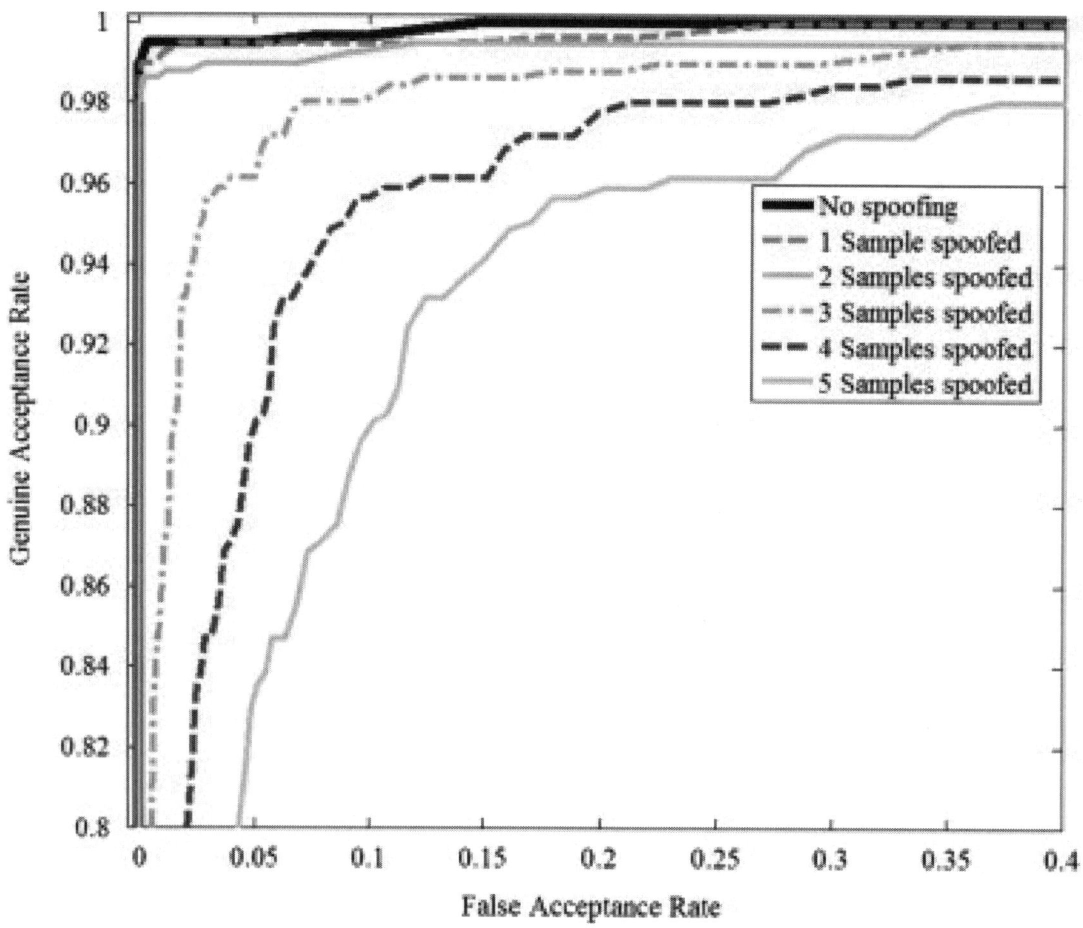

Figure. 7: ROC for partial multibiometric spoofing using median rule.

As an interesting side-aspect of the evaluation conducted in this paper, the effect of face-only vs. fingerprint-only spoofing is investigated. If always a specific modality is spoofed, a clear discrepancy of spoofing success can be observed: for 1-spoof spoofing restricted to spoofing the face image only (without spoofing detection) 7.33% EER is obtained (using sum rule), whereas fingerprint-only spoofing results in 1.19% EER. This underlines the difficulty of finding suitable counter-spoofing fusion methods being able to tackle these different shifts in distributions. Note that however the quality of the employed face database was also very challenging and thus impacting on this result. Experiments continue with the setup using random selection of spoofing samples (i.e. 20% probability to spoof face and 4×20% probability to spoof fingerprint).

Median Filtering

In order to evaluate the suggested 1-median filtering an experiment using LR on the joint liveness-and-score pairs is conducted, incorporating liveness information into decision. Obtained EER for the method show much more stable results and also ROC curves are much flatter, see Fig. 8 plotted using log-scale. Median-filtering on probabilities of already combined scores and liveness measures is able to retain EERs below 2% over all spoofing attempts, within a narrow band (0.81–1.81%), and at the same time retains a very high 0-spoof accuracy (0.47%). Results refer to using a filter radius of $\phi=3\sigma$, i.e. relative to the standard deviation of fused samples. Also corresponding *d*-Prime clearly illustrate the much better separation of genuine and *m*-impostor distribution. However, the tradeoff is a slightly degraded initial 0-spoof performance of 0.47% (which however, is much better than previously published results in [10] due to better spoofing detection).

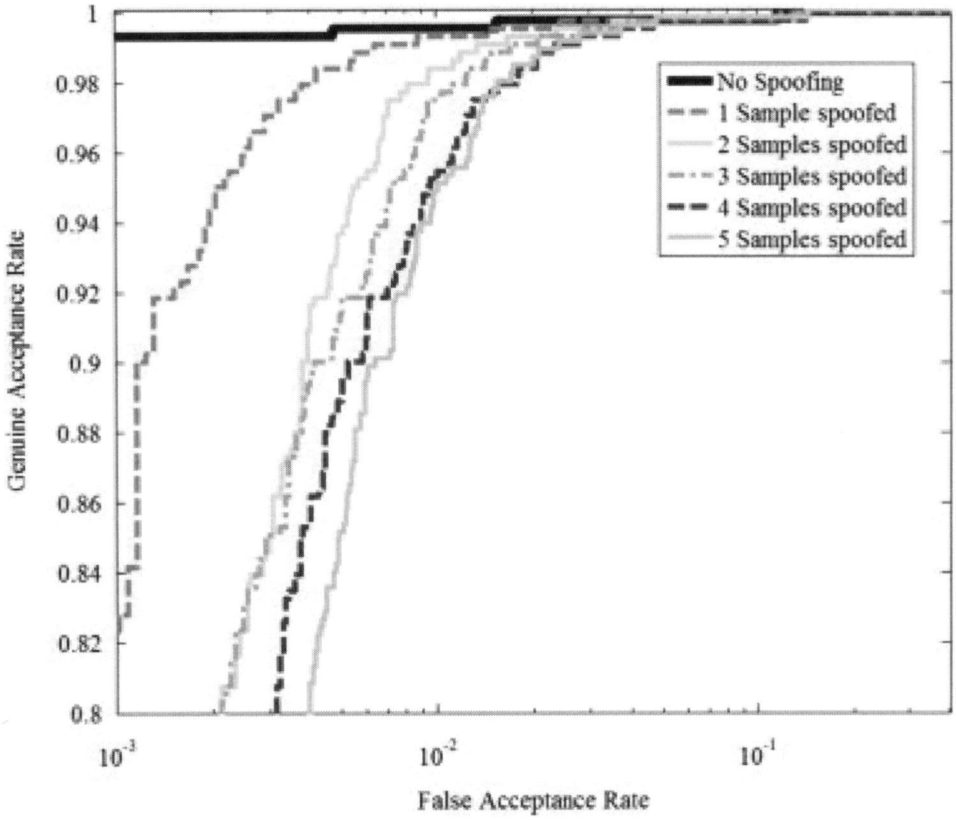

Figure. 8: ROC for partial multibiometric spoofing using 1-median filtering + LR.

In a further experiment we verified the superiority of the median compared to the mean when looking at simple recognition-score *m*-spoof imposter distributions. From Table 2we can see that (1) mean recognition scores decrease with the number of spoofed samples indicating an effect of the spoofing effort; and (2) median consequently delivered better performance suppressing the negative impact of spoofed samples (note the small changes especially for up to 2-spoof in contrast to the mean) confirming theoretical considerations. Finally, we also tested the effect on variance and found that variance is increasing for a larger number of spoofed samples in multibiometric configuration (e.g., $\sigma^2=0.00012$ for 0-spoof, 0.0222 for 3-spoofed and 0.0293 for the 5-spoof case).

Table 2: Average mean vs. median of *m*-spoof imposter distribution

Unit	0-spoof	1-spoof	2-spoof	3-spoof	4-spoof	5-spoof
Median	0.991	0.989	0.984	0.965	0.935	0.921
Mean	0.989	0.964	0.940	0.914	0.889	0.873

Finally, a critical task in setting up parameters for the median filtering is a suitable choice for the filter radius ϕ. The introduction of filter parameter ϕ results in a tradeoff permitting for better choices between the median rule ($\phi\approx 0$), which is better for suppression of spoofing attempts, and the sum rule ($\phi=\infty$), which delivers the best zero-spoof performance. While one method is to employ static values of ϕ, in order to avoid over-fitting with regards to the training set, dynamic selection methods of ϕ are investigated, based on the scattering of input scores (e.g., as a factor of σ being the standard deviation of the *n* scores/tuples to be combined). As can be seen from filter evaluations on score-only combinations in Table 3, small values of ϕ (0.5) deliver a closer performance to median filtering, whereas larger values of ϕ are closer to the performance of the sum rule. This way an arbitrary compromise between classical accuracy using the sum rule and potentially slightly degraded 0-spoof performance but higher spoofing resistance, as for the median rule, can be obtained. While the paper does not aim to provide an assessment of computational cost, note that the overhead introduced by median filtering is minimal, as the number *n* of employed features is typically a fixed and low number.

Table 3: EER/SEER (in %) and *d*-Prime results of face-and-finger fusion on the test set varying the number *m* of spoofed samples

Method	(S)EER						d-Prime					
	m=0	*m*=1	*m*=2	*m*=3	*m*=4	*m*=5	*m*=0	*m*=1	*m*=2	*m*=3	*m*=4	*m*=5
Sum rule	**0**	2.32	5.71	7.52	10.38	12.24	2.91	2.62	2.27	2.11	1.89	1.67
Product rule	**0**	3.35	7.61	9.48	12.64	15.72	6.44	3.40	2.31	2.03	1.69	1.42
Median rule	0.42	0.87	1.20	3.91	6.74	10.03	2.38	2.37	2.31	2.14	1.89	1.68
Median filtering (ϕ=1σ)	0	1.71	3.65	5.22	8.02	10.60	2.66	2.45	2.19	2.06	1.87	1.69
Median filtering (ϕ=0.5σ)	0	1.03	2.04	3.50	6.23	9.53	2.29	2.17	2.01	1.90	1.75	1.69
1-Median filter	0.47	**0.81**	**1.16**	**1.39**	**1.71**	**1.81**	3.18	3.16	3.12	3.11	3.10	3.08

Rejection of Spoofing Samples

Finally, the classical alternative in many implementations implementing counter-spoofing functionality is to reject samples during quality assurance, if they appear to originate from a spoofed source. Unfortunately spoofing detector errors in multibiometric configuration add up and cause a high number of falsely rejected genuine attempts. In order to virtually compare this method with the presented integrated approach, the final score is set to 1, if and only if one or more of the spoofing scores were greater than a threshold (t=0.44 is used). The resulting (virtual, as normally this would result in a Failure to Acquire error) EER of approx. 22% clearly illustrates the superiority of integrated recognition-and-spoofing fusion. Further, especially for commercial applications falsely rejected users are considered critical, whereas any threshold is typically set at a very conservative level which limits the use of employed techniques. Compared to techniques integrating liveness results, an advantage is that no information is lost and the overall scores can be taken into account.

CONCLUSION

Experiments in this paper show that 1-spoofing in face and fingerprint fusion can successfully be targeted by employing median instead of sum rule for combinations using its property to be less affected by a certain number of outliers. However, this comes at the cost of a reduced 0-spoof performance. Its extension to 1-median filtering is able to find arbitrary trade-off points between sum and median rule, allowing for better flexibility in choosing the right tradeoff between accuracy and security. The paper investigated how spoofing detection and recognition can mutually benefit from each other and evaluated 1-median filtering as a novel multibiometric fusion method integrating liveness and recognition scores. Results yielded more stable results for this method in partial multibiometric spoofing configuration, where m out of n samples of an identity are spoofed (EERs 0.47–1.81% vs. 0–12.24% for the sum rule). The paper presented an analysis of the filter radius in median filtering and investigated the impact of face vs. fingerprint spoofing. Finally, bootstrap aggregating (bagging) classifiers were proposed for anti-spoofing and shown to deliver highly accurate results (84% accuracy) on the challenging LivDet2013 crossmatch dataset. We believe the following remaining questions should be further investigated in future work: a closer investigation of score normalisation issues for median filtering; an extension towards user-adaptive anti-spoofing and recognition fusion, and; integration of extrinsic factors (e.g. acquisition conditions) and/or quality-related measurements into the fusion scheme.

REFERENCES

1. A. Anjos, J. Komulainen, S. Marcel, A. Hadid, M. Pietikainen, Face anti-spoofing: visual approach, in: S. Marcel, et al., (Eds.), Handbook of Biometric AntiSpoofing, Springer, London, 2014, pp. 65–82.
2. E. Marasco, A. Ross, A survey on antispoofing schemes for fingerprint recognition systems, ACM Comput. Surv. 47 (2) (2014) 28:1–28:36.
3. M. Chakka, et al., Competition on counter measures to 2-d facial spoofing attacks, in: Proceedings of the International Joint Conference on Biometrics, 2011, http://dx.doi.org/10.1109/IJCB.2011.6117509.
4. L. Ghiani, et al., Livdet 2013 fingerprint liveness detection competition, in: Proceedings of the International Conference on Biometrics, 2013, http://dx.doi. org/10.1109/ICB.2013.6613027.
5. E. Marasco, P. Johnson, C. Sansone, S. Schuckers, Increase the security of multibiometric systems by incorporating a spoofing detection algorithm in the fusion mechanism, in: C. Sansone, et al. (Eds.), Proceedings of the International Workshop on MCS, Lecture Notes in Computer Science, vol. 6713, Springer, Berlin, 2011, pp. 309–318, http://dx.doi.org/10.1007/978-3-642-21557-5_33.

6. I. Chingovska, A. Anjos, S. Marcel, Anti-spoofing in action: Joint operation with a verification system, in: Proceedings of the International Conference on Computer Vision and Pattern Recognition, Workshop, 2013, pp. 98–104, http:// dx.doi.org/10.1109/CVPRW.2013.22.
7. N. Poh, R. Wong, G.-L. Marcialis, Toward an attack-sensitive tamper-resistant biometric recognition with a symmetric matcher: a fingerprint case study, in: Proceedings of the Symposium on Computational Intelligence in Biometrics and Identity Management, 2014, pp. 1–6.
8. R.N. Rodrigues, L.L. Ling, V. Govindaraju, Robustness of multimodal biometric fusion methods against spoof attacks, J. Vis. Lang. Comput. 20 (3) (2009) 169–179.
9. Z. Akhtar, G. Fumera, G. Marcialis, F. Roli, Evaluation of multimodal biometric score fusion rules under spoof attacks, in: Proceedings of the International Conference on Biometrics, 2012, pp. 402–407, http://dx.doi.org/10.1109/ICB. 2012.6199784.
10. P. Wild, P. Radu, L. Chen, J. Ferryman, Towards anomaly detection for increased security in multibiometric systems: spoofing-resistant 1-median fusion eliminating outliers, in: Proceedings of the International Joint Conference on Biometrics, 2014, http://dx.doi.org/10.1109/BTAS.2014.6996293.
11. A. Rattani, N. Poh, A. Ross, A Bayesian approach for modeling sensor influence on quality, liveness and match score values in fingerprint verification, in: Proceedings of the International Workshop on Information Forensics and Security, 2013, pp. 37–42, http://dx.doi.org/10.1109/WIFS.2013.6707791.
12. A. Rattani, A. Ross, Automatic adaptation of fingerprint liveness detector to new spoof materials, in: Proceedings of the International Joint Conference on Biometrics, 2014, http://dx.doi.org/10.1109/BTAS.2014.6996254.
13. G. Fumera, G. Marcialis, B. Biggio, F. Roli, S. Schuckers, Multimodal antispoofing in biometric recognition systems, in: S. Marcel, et al., (Eds.), Handbook of Biometric Anti-Spoofing, Springer, London, 2014, pp. 165–184.
14. I. Chingovska, A. Rabello dos Anjos, S. Marcel, Biometrics evaluation under spoofing attacks, IEEE Trans. Inf. Forensics Secur. 9 (12) (2014) 2264–2276. http://dx.doi.org/10.1109/TIFS.2014.2349158.
15. E. Marasco, Y. Ding, A. Ross, Combining match scores with liveness values in a fingerprint verification system, in: Proceedings of the International Conference on Biometrics: Theory, Applications and Systems, 2012, pp. 418–425, http://dx.doi.org/10.1109/BTAS.2012.6374609.
16. J. Kittler, M. Hatef, R.P. Duin, J. Matas, On combining classifiers, IEEE Trans. Pattern Anal. Mach. Int. 20 (3) (1998) 226–239. http://dx.doi.org/10.1109/ 34.667881.
17. R. Rodrigues, N. Kamat, V. Govindaraju, Evaluation of biometric spoofing in a multimodal system, in: Proceedings of the International Conference on Biometrics: Theory, Applications and Systems (BTAS), 2010, pp. 1–5, http:// dx.doi.org/10.1109/BTAS.2010.5634531.
18. Z. Akhtar, G. Fumera, G. Marcialis, F. Roli, Evaluation of serial and parallel multibiometric systems under spoofing attacks, in: Proceedings of the International Conference on Biometrics: Theory, Applications and Systems, 2012, pp. 283–288, http://dx.doi.org/10.1109/BTAS.2012.6374590.
19. A.Abhyankar, S. Schuckers, Fingerprint liveness detection using local ridge frequencies and multiresolution texture analysis techniques, in: Proceedings of the International Conference on Image Processing, 2006, pp. 321–324, http://dx.doi.org/10.1109/ICIP.2006.313158.
20. P. Coli, G. Marcialis, F. Roli, Power spectrum-based fingerprint vitality detection, in: Proceedings of the Workshop on Automatic Identification Advanced Technologies, 2007, pp. 169–173. http://dx.doi.org/10.1109/AUTOID.2007. 380614.
21. S. Nikam, S. Agarwal, Fingerprint liveness detection using curvelet energy and co-occurrence signatures, in: Proceedings of the International Conference on Computer Graphics, Imaging and Visualisation, 2008, pp. 217–222, http://dx. doi.org/10.1109/CGIV.2008.9.
22. L. Ghiani, G. Marcialis, F. Roli, Fingerprint liveness detection by local phase quantization, in: Proceedings of the International Conference on Pattern Recognition, 2012, pp. 537–540.
23. A. Anjos, M.M. Chakka, S. Marcel, Motion-based counter-measures to photo attacks in face recognition, IET Biometrics 3 (3) (2013) 147–158.
24. G. Zhao, M. Pietikainen, Dynamic texture recognition using local binary patterns with an application to facial expressions, IEEE Trans. Pattern Anal. Mach. Int. 29 (6) (2007) 915–928. http://dx.doi.org/10.1109/TPAMI.2007.1110.
25. X. Tan, Y. Li, J. Liu, L. Jiang, Face liveness detection from a single image with sparse low rank bilinear discriminative model, in: K. Daniilidis, et al. (Eds.), ECCV 2010, Lecture Notes in Computer Science, vol. 6316, 2010, pp. 504–517.
26. T. de Freitas Pereira, A. Anjos, J.M. de Martino, S. Marcel, LBP-TOP based countermeasure against face spoofing attacks, in: Proceedings of the Workshop on Computer Vision, Springer, 2013, pp. 121–132, http://dx.doi.org/10. 1007/978-3-642-37410-4_11.
27. P.T. von Hippel, Mean, median, and skew: correcting a textbook rule, J. Stat. Educ. 13 (2) (2005) 1–13.
28. D. Moore, G. McCabe, Introduction to the Practice of Statistics, Freeman, New York, 2003.
29. N. Otsu, A threshold selection method from gray-level histograms, IEEE Trans. Syst. Man Cybern. 9 (1) (1979) 62–66.
30. J. Daugman, Complete discrete 2-d Gabor transforms by neural networks for image analysis and compression, IEEE Trans. Acoust. Speech Signal Process. 36 (7) (1988) 1169–1179. http://dx.doi.org/10.1109/29.1644.
31. S.B. Nikam, S. Agarwal, Gabor filter-based fingerprint anti-spoofing, in: J. Blanc-Talon, et al. (Eds.), Proceedings of the International Conference on ACIVS, Lecture Notes in Computer Science, vol. 5259, Springer, Berlin,

2008, pp. 1103–1114, http://dx.doi.org/10.1007/978-3-540-88458-3_100.

32. H. Choi, R. Kang, K. Choi, A. Jin, J. Kim, Fake-fingerprint detection using multiple static features, Opt. Eng. 48 (2009) 047202–047202–13.

33. L.Breiman,Baggingpredictors,Mach.Learn.24(2)(1996) 123–140. http://dx. doi.org/10.1023/A:1018054314350.

34. B. Efron, R.J. Tibshirani, An Introduction to the Bootstrap, Chapman & Hall, New York, 1994.

35. N. Poh, A. Merati, J. Kittler, Heterogeneous information fusion: a novel fusion paradigm for biometric systems, in: Proceedings of the International Joint Conference on Biometrics, 2011, pp. 1–8, http://dx.doi.org/10.1109/IJCB.2011. 6117494.

36. P. Johnson, B. Tan, S. Schuckers, Multimodal fusion vulnerability to non-zero effort (spoof) imposters, in: Proceedings of the International Workshop on Information Forensics and Security, 2010, pp. 1–5, http://dx.doi.org/10.1109/ WIFS.2010.5711469.

37. I. Chingovska, A. Anjos, S. Marcel, On the effectiveness of local binary patterns in face anti-spoofing, in: Proceedings of the International Conference of the Biometrics Special Interest Group (BIOSIG), 2012, pp. 1–7.

38. Z. Zhang, J. Yan, S. Liu, Z. Lei, D. Yi, S. Li, A face antispoofing database with diverse attacks, in: Proceedings of the International Conference on Biometrics, 2012, pp. 26–31, http://dx.doi.org/10.1109/ ICB.2012.6199754.

39. NIST, Biometric Image Software V4.2, 2013, ⟨http:// www.nist.gov/itl/iad/ig/ nbis.cfm⟩.

40. Neurotechnology Inc, VeriLook SDK, 2014, ⟨http:// www.neurotechnology.com/ verilook.html⟩.

CHAPTER 12 FALL DETECTION IN DUSKY ENVIRONMENT

Ying-Nong Chen[1], Chi-Hung Chuang[2], Hsin-Min Lee[1], Chih-Chang Yu[3] and Kuo-Chin Fan[1]

[1] Department of Computer Science and Information Engineering, National Central University

[2] Department of Applied Informatics, Fo Guang University

[3] Digital Education Institute, Institute for Information Industry

ABSTRACT

Accidental fall is the most prominent factor that causes the accidental death of elder people due to their slow body reaction. Automatic fall detection technology integrated in a health-care system can assist human monitoring the occurrence of fall, especially in dusky environments. In this paper, a novel fall detection system focusing mainly on dusky environments is proposed. In dusky environments, the silhouette images of human bodies extracted from conventional CCD cameras are usually imperfect due to the abrupt change of illumination. Thus, our work adopts a thermal imager to detect human bodies. The proposed approach adopts a coarse-to-fine strategy. Firstly, the downward optical flow features are extracted from the thermal images to identify fall-like actions in the coarse stage. The horizontal projection of motion history images (MHI) extracted from fall-like actions are then designed to verify the incident by the proposed nearest neighbor feature line embedding (NNFLE) in the fine stage. Experimental results demonstrate that the proposed method can distinguish the fall incidents with high accuracy even in dusky environments and overlapping situations.

INTRODUCTION

Accidental fall is the most prominent factor that causes the accidental death of elder people due to their slow body reaction. Fall accidents usually occur at night with nobody except the elder people if they live alone. It is usually too late to remedy the tragedy when the body is discovered hours or days after with the occurrence of accidental fall. In the occurrence of fall incident, humans usually lie flat on the ground. However, we cannot merely use the images to perceive whether this person is lying on the ground. Hence, we have to detect and avoid the risk caused by fall action. According to the survey, a sudden fainting or body imbalance is the main reason to cause a fall. No matter what reasons, fall is a warning that the subject may be in danger. Moreover, the silhouette images of human bodies are hard to be extracted from conventional CCD cameras in dusky environments due to the illumination constraint. If the incidents occur in a dusky and unattended environment, people usually miss the prime time for rescue. To remedy this problem, a fall detection system using a thermal imager (see Fig. 1) to capture the images of human bodies is proposed in this paper. By using the thermal imager, the human bodies can be accurately located even in a dusky environment. For comparison, Fig. 2a shows the images obtained by a CCD camera in a dusky environment, whereas Fig. 2b shows the images obtained by a thermal imager in the same environment.

Citation: Ying-Nong Chen, Chi-Hung Chuang, Hsin-Min Lee, Chih-Chang Yu and Kuo-Chin Fan, Fall detection in dusky environment, DOI 10.1186/s13640-016-0115-8.

It is obvious that the thermal imagers can extract more clear and intact human bodies in the dusky environments than CCD cameras.

Figure. 1: The thermal imager

(a)

(b)

Figure. 2: Image extraction results captured by (**a**) CCD camera and (**b**) thermal imager

Moylan [1] illustrated the gravity of falls as a health risk with abundant statistics. Larson [2] described the importance of falls in elderly. The National Center for Health Statistics showed that more than one third of ages 65 or older fall each year. Moreover, 60 % of lethal falls occur at home, 30 % occur in public region, and 10 % happen in health-care institutions for ages 65 or older [3]. In the literatures of fall detection, Tao et al. [4] applied the aspect ratio of the foreground object to detect fall incidents. Their system firstly tracks the foreground objects and then analyzes the sequences of features for fall incident detection. Anderson et al. [5] also applied the aspect ratio of the silhouette to detect fall incidents. The rationale based mainly on the fact that the aspect ratio of the silhouette is usually very large when the fall incidents occur. On the contrary, the aspect ratio is much smaller when the fall incidents do not occur. Juang [6] proposed a neural fuzzy network method to classify the human body postures, such as standing, bending, sitting, and lying down. In [7], Foroughi et al. proposed a fall detection method using an approximated eclipse of human body silhouette and head pose as features for multi-class support vector machine (SVM). Rougier et al. [8] applied the motion history image (MHI) and variations of human body shape to detect falls. In [9], Foroughi et al. proposed a modified MHI integrating the time motion image (ITMI) as the motion feature. Then, the eigenspace technique was used for motion feature reduction and fed into individual neural network for each activity. Liu et al. [10] proposed a nearest neighbor classification method to classify the ratio of human body silhouette of fall incidents. In order to differentiate between the fall and lying, the time difference between fall and lying was used as a key feature. Liao et al. [11] proposed a slip and fall detection system based on Bayesian Belief Network

(BBN). They used the integrated spatiotemporal energy (ISTE) map to obtain the motion measure. Then, the BBN model of the causality of the slip and fall was constructed for fall prevention. Olivieri et al. [12] proposed a spatiotemporal motion feature to represent activities termed motion vector flow instance (MVFI) templates. Then, a canonical eigenspace technique was used for MVFI template reduction and template matching.

In this paper, a novel fall detection mechanism based on coarse-to-fine strategy which is workable in dusky environments is proposed. In the coarse stage, the downward optical flow features are extracted from the thermal images to identify fall-like actions. Then, the horizontal projected motion history image (MHI) features of fall-like actions are used in the fine stage to verify the fall by the nearest neighbor feature line embedding. The contributions of this work are listed as follows: (1) using the thermal imager instead of CCD camera to capture intact human body silhouettes; (2) proposing a coarse-to-fine strategy to detect fall incidents; (3) proposing a nearest neighbor feature line embedding method for fall detection which improves the original nearest feature line embedding method; (4) proposing a scheme to detect fall incidents even though occlusion occurs.

The rest of this paper is organized as follows. In Section 2, the concept of nearest feature line embedding (NFLE) algorithm presented in our previous work [13] will be briefly reviewed. Then, the fall detection based on coarse-to-fine strategy and the nearest neighbor feature line embedding (NNFLE) algorithm are presented in Section 3. Experimental results are illustrated in Section 4 to demonstrate the soundness and effectiveness of the proposed fall detection method. Finally, conclusions are given in Section 5.

NEAREST FEATURE LINE EMBEDDING (NFLE)

The NFLE transformation is a linear transformation method based on a nearest feature space (NFS) strategy [13] originating from an nearest linear combination (NLC) methodology [14]. Since the points on the nearest feature line (NFL) are linearly interpolated or extrapolated from each pair of feature points, the performance is better than those of point-based methods. In addition, the NFL metric is embedded into the transformation through the discriminant analysis phase instead of in the matching phase.

Consider Nd-dimensional samples $X=[x_1, x_2, \ldots x_N]$ constituting N_c classes, the corresponding class label of x_i is denoted as $l_{x_i} \in \{1,2,3,\ldots N_c\} l_{x_i} \in \{1,2,3,\ldots N_c\}$ and a specified point $y_i = w^T x_i$ in the transformed space. The distance from point y_i to the feature line is defined as $\|y_i - f^{(2)}(y_i)\|$, in which $f^{(2)}$ is a function generated by two points, and $f^{(2)}(y_i)$ is the projected point of the line. A number of C_2^{N-1} possible lines for point y_i will be generated. The scatter computation of feature points to feature lines can be obtained and embedded in the discriminant analysis. In consequence, this approach is termed as NFLE.

In NFLE, the objective function in the equation

$$F = \sum_i \left(\sum \left\| y_i - f^{(2)}(y_i) \right\|^2 w^{(2)}(y_i) \right) \quad (1)$$

is minimized.

The weight values $w^{(2)}(y_i)$ (being 1 or 0) constitute a connected relationship matrix of size $N \times C_2^{N-1}$ for Nfeature points to their corresponding projection points $f^{(2)}(y_i)$. Consider the distance $\left\|y_i - f^{(2)}_{m,n}(y_i)\right\|$ for point y_i to a feature line $L_{m,n}$ that passes through two points y_m and y_n; the projection point f $f^{(2)}_{m,n}(y_i)$ can be represented as a linear combination of points y_m and y_n by $f^{(2)}_{m,n}(y_i) = y_m + t_{m,n}(y_n - y_m)$, in which $t_{m,n} = (y_i - y_m)^T(y_m - y_n)/(y_m - y_n)^T(y_m - y_n)$. The mean square distance for all training samples to their corresponding NFLs is minimized and its representation is given by the following lemma.

Lemma 2.1: The mean square distance from the training points to the NFLs can be represented in the form of a Laplacian matrix.

See Fig. 3 for illustration. For a specified point y_i, the vector from point y_i to the projection point $nf^{(2)}_{(m,n)}(y_i)$ of the NFL $NL_{m,n}$ which passes through points y_m and y_n can be obtained as follows:

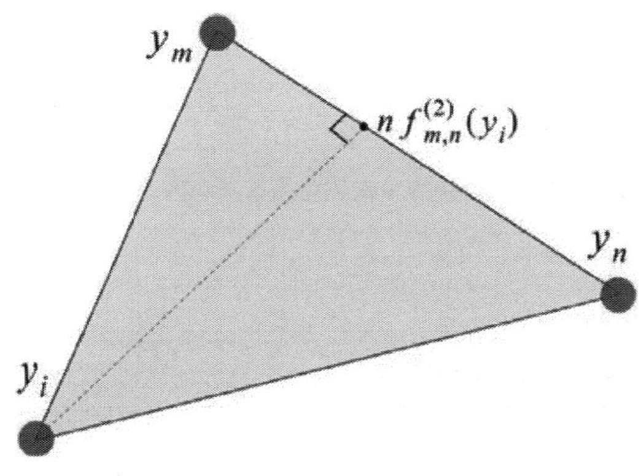

Figure. 3: Projection of NFL

$$\begin{aligned} y_i - nf^{(2)}_{(m,n)}(y_i) &= y_i - y_m + t_{m,n}(y_m - y_n) \\ &= y_i - (1 - t_{m,n})y_m - t_{m,n}y_n \\ &= y_i - t_{n,m}y_m - t_{m,n}y_n \\ &= y_i - \sum_j M_{i,j}y_j \end{aligned} \quad (2)$$

Here, $t_{n,m} = 1 - t_{m,n}$ and $i \neq m \neq n$. Two values in the ith row in matrix M are set as $M_{i,m} = t_{n,m}$ and $M_{i,n} = t_{m,n}$. The other values in the ith row are set as $M_{i,j} = 0$ if $j \neq m \neq n$. In general, the mean square distance for all training points to their NFLs is obtained as follows:

$$\begin{aligned} \left(\sum \left\| y_i - nf^{(2)}(y_i) \right\|^2 \right) &= \sum_i \left(y_i - \sum_j M_{i,j} y_j \right)^2 \\ &= tr\left(Y^T (I-M)^T (I-M) Y \right) \\ &= tr\left(Y^T (D-W) Y \right) \\ &= tr\left(w^T X L X^T w \right) \end{aligned} \quad (3)$$

in which $\sum_j M_{i,j} = 1$ and $L = D - W$. From the conclusions of [15], matrix W is defined as $W_{i,j} = (M + M^T - M^T M)_{i,j}$ when $i \neq j$ and zero otherwise. The function in (3) can thus be represented by a Laplacian matrix.

Moreover, when the K NFLs are chosen from C_2^{N-1} possible combinations, the objective function in (1) is also represented as a Laplacian matrix as stated in the following theorem.

Theorem 2.1: The objective function in (1) can be represented as a Laplacian matrix that preserves the locality among samples.

The objective function F in (1) is first decomposed into K components. Each component denotes the mean square distances for point y_i to the kth NFL. The first component matrix $M_{i,j}(1)$ denotes the connectivity relationship matrix between point x_i and the NFL $L_{m,n}$ for $i,m,n=1,\dots,N$ and $i\neq m\neq n$. Two non-zero terms, $M_{i,n}(1)=t_{m,n}$ and $M_{i,m}(1)=t_{n,m}$, exist at each row of matrix $M_{i,j}(1)$ and satisfy $\sum_j M_{i,j}(1)=1$. According to Lemma 2.1, it is represented as a Laplacian matrix $w^TXL(1)X^Tw$. In general, $M_{i,n}(k)=t_{m,n}$ and $M_{i,m}(k)=t_{n,m}$ for $i\neq m\neq n$ if line $L_{m,n}$ is the kth NFL of point x_i and zero otherwise. All components are derived in a Laplacian matrix representation, $w^TXL(k)X^Tw$, for $k=1,2,\dots,K$. Therefore, function F in (1) becomes

$$
\begin{aligned}
F &= \sum_i \sum_{m\neq n} \left\| y_i - f^{(2)}_{m,n}(y_i) \right\|^2 w^{(2)}_{m,n}(y_i) \\
&= \sum_i \sum_{m\neq n} \left(y_i - t_{n,m}y_m - t_{m,n}y_n \right)^2 w^{(2)}_{m,n}(y_i) \\
&= \sum_i \left(y_i - \sum_j M_{i,j}(1)y_j \right)^2 + \sum_i \left(y_i - \sum_j M_{i,j}(2)y_j \right)^2 \\
&\quad + \dots + \sum_i \left(y_i - \sum_j M_{i,j}(K)y_j \right)^2 \\
&= tr(Y^T(I-M(1))^T(I-M(1))Y) \\
&\quad + Y^T(I-M(2))^T(I-M(2))Y \\
&\quad + \dots + Y^T(I-M(K))^T(I-M(K))Y) \\
&= tr(Y^T(D-W(1))Y) + Y^T(D-W(2))Y \\
&\quad + \dots + Y^T(D-W(K))Y) \\
&= tr(Y^T(L(1))Y + Y^T(L(2))Y + \dots Y^T(L(L))Y) \\
&= tr(Y^TLY) = tr(w^TXLX^Tw)
\end{aligned}
\tag{4}
$$

where $W_{i,j}(k)=(M(k)+M(k)^T - M(k)^T M(k))_{i,j}$ and $L(k)=D(k)-W(k)$, for $k=1,2,\dots,K$, and $L=L(1)+L(2)+\dots+L(K)$. Since the objective function in (4) can be represented as a Laplacian matrix, the locality of the samples is also preserved in the low-dimensional space. More details are given in [13].

Consider the class labels in supervised classification, the two parameters, K_1 and K_2 are manually determined for the computation of the within-class scatter $\mathbf{S}_w$ and the between-class scatter $\mathbf{S}_b$, respectively

$$
\mathbf{S}_w = \sum_{p=1}^{N_c} \left(\sum_{\substack{x_i\in C_p \\ f^{(2)}\in F^{(2)}_{K_1}(x_i,C_p)}} \left(x_i - f^{(2)}(x_i)\right)\left(x_i - f^{(2)}(x_i)\right)^T \right)^2 \tag{5}
$$

$$
\mathbf{S}_b = \sum_{p=1}^{N_c} \left(\sum_{\substack{l=1 \\ l\neq p}}^{N_c} \sum_{\substack{x_i\in C_p \\ f^{(2)}\in F^{(2)}_{K_2}(x_i,C_l)}} \left(x_i - f^{(2)}(x_i)\right)\left(x_i - f^{(2)}(x_i)\right)^T \right) \tag{6}
$$

in which $F^{(2)}_{K_1}(x_i,C_p)$ indicates the K_1 NFLs within the same class C_p of point x_i and $F^{(2)}_{K_2}(x_i,C_l)$ is a set of the K_2 NFLs belonging to the different classes of point x_i. The Fisher criterion $tr(\mathbf{S}_B/\mathbf{S}_W)$ is then maximized to find the projection matrix w which is composed of the eigenvectors with the corresponding largest eigenvalues. A new sample in the low-dimensional space can be obtained by the linear projection $y=w^Tx$. After that, the NN (one-NN) matching rule is applied to classify the samples. The training algorithm for the NFLE transformation proposed in our previous work [13] is described in Fig. 4.

Training algorithm for the NFLE transformation	
Input:	A training set $X=[x_1,x_2,\dots,x_N]\in R^{d\times N}$ and two parameters K_1 and K_2.
Output:	The projection transformation W.
Step 1.	PCA projection: Sample data are transformed from space R^d into a low-dimensional subspace R^n. Let w_{PCA} denote the transformation matrix of PCA.
Step 2.	Computation of the within-class scatter: The possible feature lines are first generated from the samples within the same class of a specified point X. Find the projection points and calculate the corresponding distances. Select K_1 vectors with the smallest distances and compute the within-class scatter $\mathbf{S}_w$ by Equation (5).
Step 3.	Computation of the between-class scatter: Generate the possible feature lines from the samples that belong to the different class of point X. Find the projection points and calculate the corresponding distances. Select K_2 discriminant vectors with the smallest distances from point X to the feature lines. The between-class scatter $\mathbf{S}_b$ is obtained from Equation (6).
Step 4.	Fisher criterion maximization: The Fisher criterion $w^* = \arg\max_{\mathbf{w}} \mathbf{S}_b/\mathbf{S}_w$ is maximized to obtain the best transformation matrix, which is composed of r eigenvectors with the largest eigenvalues.
Step 5.	Output the final transformation matrix: $W=w_{PCA}w^*$.

Figure. 4: Training algorithm for the NFLE transformation

Although the point-to-line strategy is successfully adopted in the training phase instead of the classification phase for the nearest feature line-based transformation, some drawbacks still remained and limited its performance. The problems are as follows: (1) extrapolation/interpolation inaccuracy: NFLE may not preserve the locality precisely when prototypes are far away from the probes (the probes are the training samples that would be projected on the NFL, and the prototypes are the training samples that generate the NFL); (2) high computation complexity: a large number of feature lines are generated when there are too many training samples; and (3) singular problem: the NFLE needs the inverse procedure to find the final transformation matrix w, which is troubled by the problem of singularity especially when the sample size is small. Motivated from the three problems of NFLE, we propose a modified NFLE algorithm to avoid the above three problems. Meanwhile, the algorithm is optimized for detecting fall incidents. The reason why we apply the modified NFLE is stated as follows: NFLE generates virtual training samples by linearly interpolating or extrapolating each pair of feature points. By doing so, the generalization and data diversity are increased. However, three drawbacks are shown as well. For the completeness and no repetition, the details of the three problems of NFLE and the proposed modified NFLE (NNFLE) algorithm are elaborated in Section 3.4.

THE PROPOSED FALL DETECTION MECHANISM

The proposed fall detection mechanism consists of two modules including human body extraction and fall detection. In human body extraction module, temperature frames obtained from the thermal imager are processed with image processing techniques to obtain intact human body contours and silhouettes. In fall detection module, a coarse-to-fine strategy is devised to verify fall incidents. In the coarse stage, the downward optical flow features are extracted from the temperature images to identify possible fall down actions. Then, the 50-dimensional temporal-based motion history image (MHI) feature vectors are projected into the nearest neighbor feature line space to verify the fall down incident in the fine stage. Figure 5 depicts the proposed system flow diagram. The details associated with each step including the human body extraction, the analysis of optical flows in the coarse stage, the extraction of MHIs in the fine stage, and the nearest neighbor feature line embedding for fall verification are described in the following contexts.

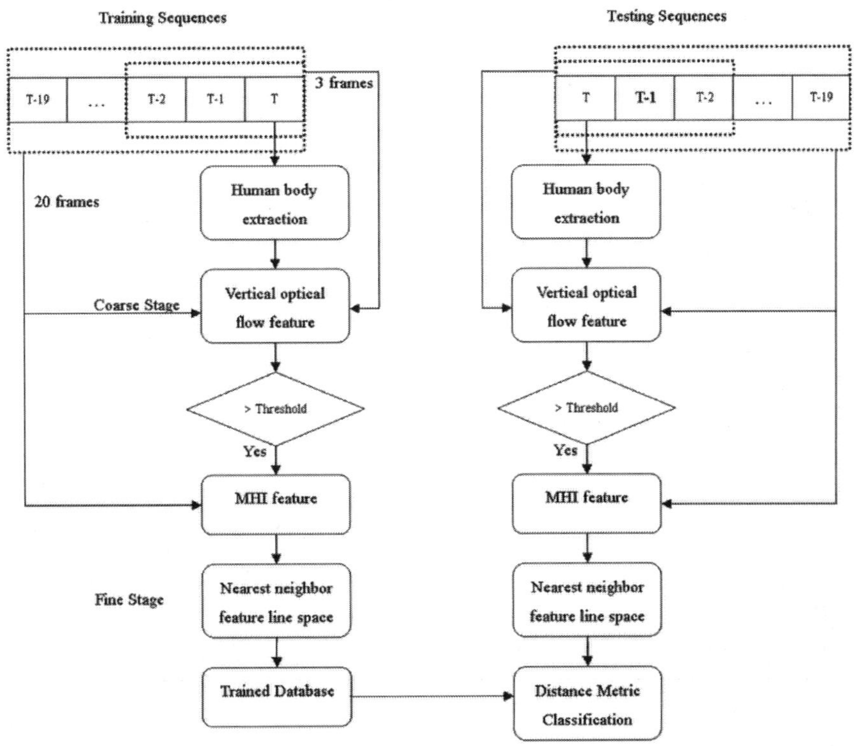

Figure. 5: Flow diagram in training and testing the fall detector

Human Body Extraction

To improve fall detection accuracy, complete silhouettes of human body must be extracted to obtain accurate bounding box of human body. To this end, the temperature images captured from a thermal imager are binarized by Otsu's method firstly. Then, the morphological closing operation is employed to obtain a complete human silhouette. Finally, a labeling process is performed to locate each human body in the image and filter out background noises. The process of human body extraction is depicted in Fig. 6. Figure 6a shows the temperature images captured from the thermal imager, Fig. 6b shows the Otsu's binarization results, and Fig. 6c shows the results of morphological closing operation. The bounding box of the human silhouette can be successfully generated after the morphological closing operations.

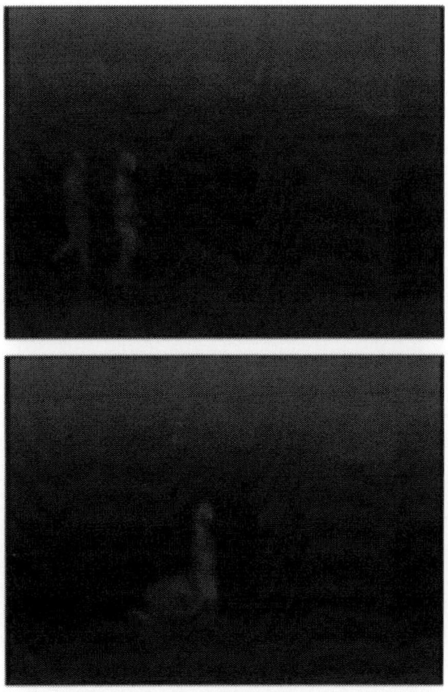

(a)

(b)

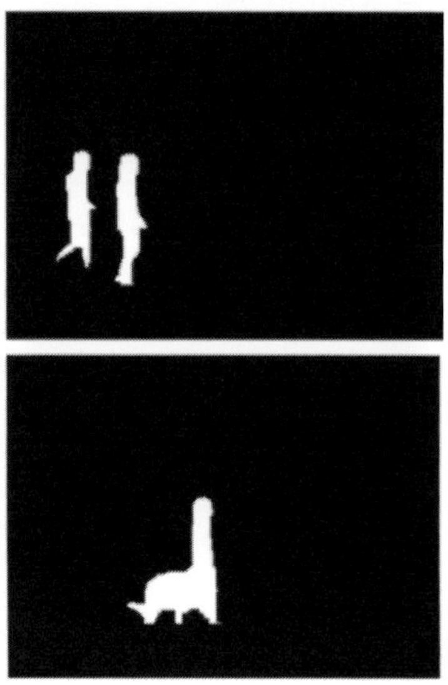

(c)

Figure. 6: Human body extraction. **a** Temperature gray level images. **b** Binarization results. **c** Morphological closing operation results

Optical Flow in the Coarse stage

After the bounding box of human body has been determined, a coarse-to-fine strategy is utilized to verify fall incidents. The purpose of the coarse stage is to identify possible fall actions. Wu [16] had shown that a fall could be described by the increase in horizontal and vertical velocities. Moreover, this work observes that the histogram of vertical optical flows has also demonstrated the significant difference between walking and falling (see Fig. 7). In our work, a multi-frame optical flow method proposed by Wang [17] is adopted to extract the downward optical flow features inside the extracted bounding box (see Fig. 8) in this stage. A possible fall action can be identified by two heuristic rules:

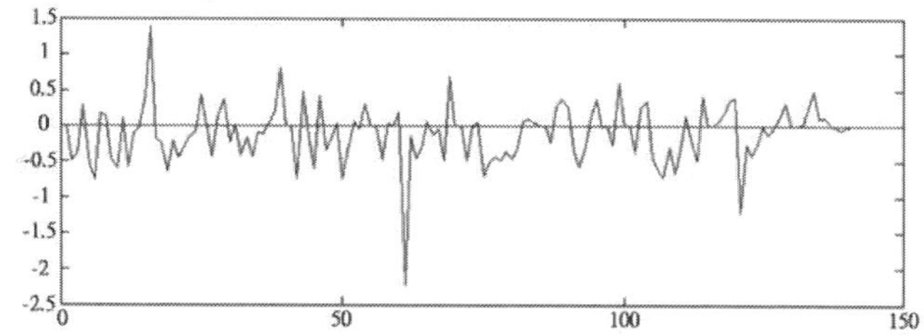

(a)

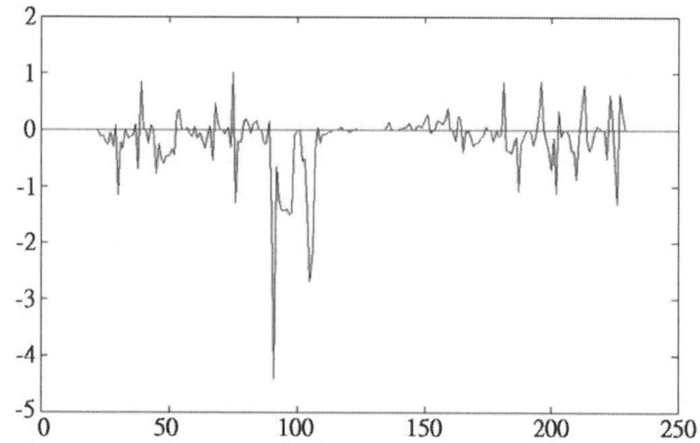

(b)

Figure. 7: The histogram of vertical optical flow of (**a**) walking and (**b**) falling down

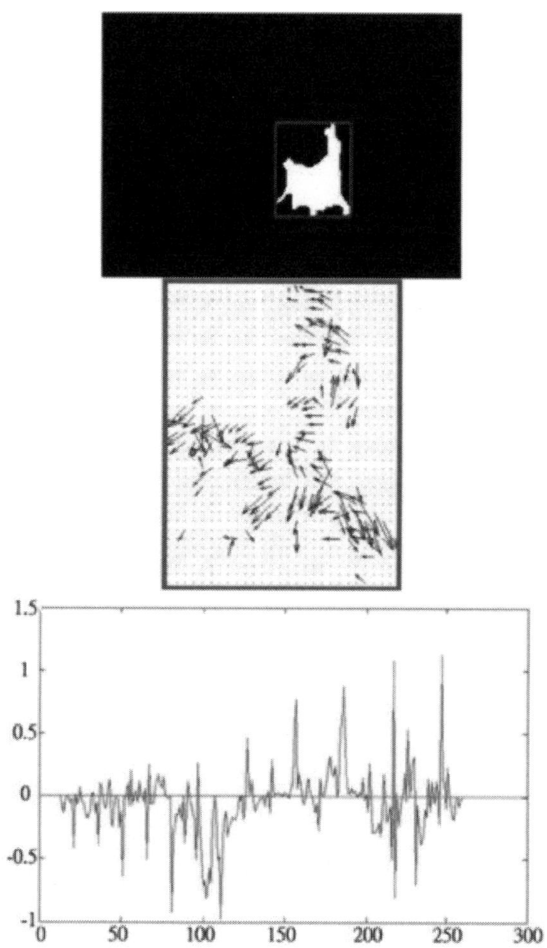

(a)

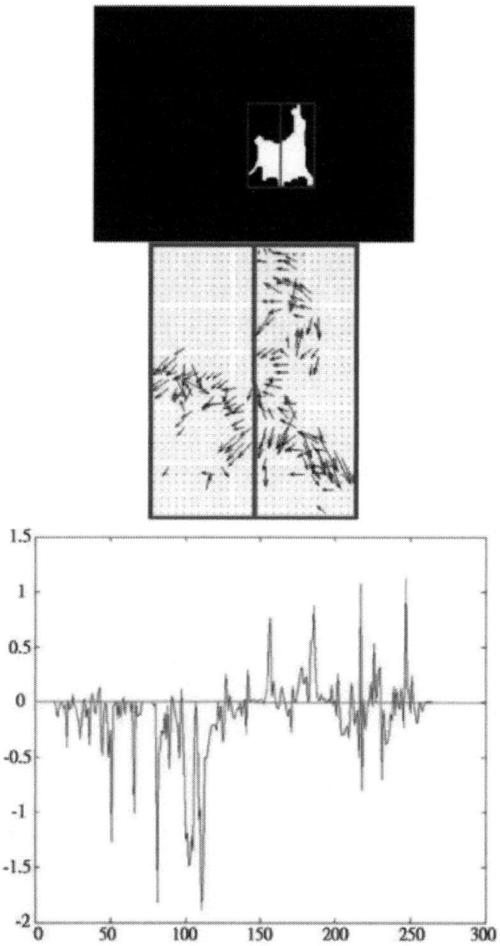

(b)

Figure. 8: Fall incident in overlapping situation. The *first row* is the silhouettes, the *second row* is the corresponding optical flow results, and the *third row* is the histograms of vertical optical flows. **a** The results generated by original method. **b** The results generated by using dividing method

(1) Rule 1: Given 20 consecutive frames, the average vertical optical flows exhibit downward more than 75 % of frames.

(2) Rule 2: The sum of the average vertical optical flows in 20 consecutive frames is larger than a threshold, say 10 in this study.

As shown in Fig. 8a, a fall incident may not be identified if the subject is overlapped by the other. To solve this problem, the bounding box is divided into two equal boxes if overlapping occurs. The width of the silhouette is used to identify whether the overlapping occurs or not. The optical flow features are then extracted in each divided box. The one which has larger average downward optical flow is used to identify possible fall action. As a result, the fall incidents can be extracted correctly as shown in Fig. 8b, and Fig. 8a demonstrates the result without using the bounding box division strategy.

Motion History Image in the Fine Stage

In the coarse stage, most non-fall actions can be filtered out via the downward optical flow features. However, some fall-like actions are identified as fall incidents due to the swing of arms. To solve this problem, we devise a feature vectors which are formed by projecting the MHI horizontally to verify fall incidents in the fine stage. MHI proposed by Bobick [18] is a template which condenses a determined number of silhouette sequences into a gray scale image (as shown in Fig. 9a) which is capable of preserving dominant motion information. Since the main difference between fall and other actions is the vertical component changes, our work projects the MHI horizontally to obtain a 50-dimensional feature vectors using equation (7):

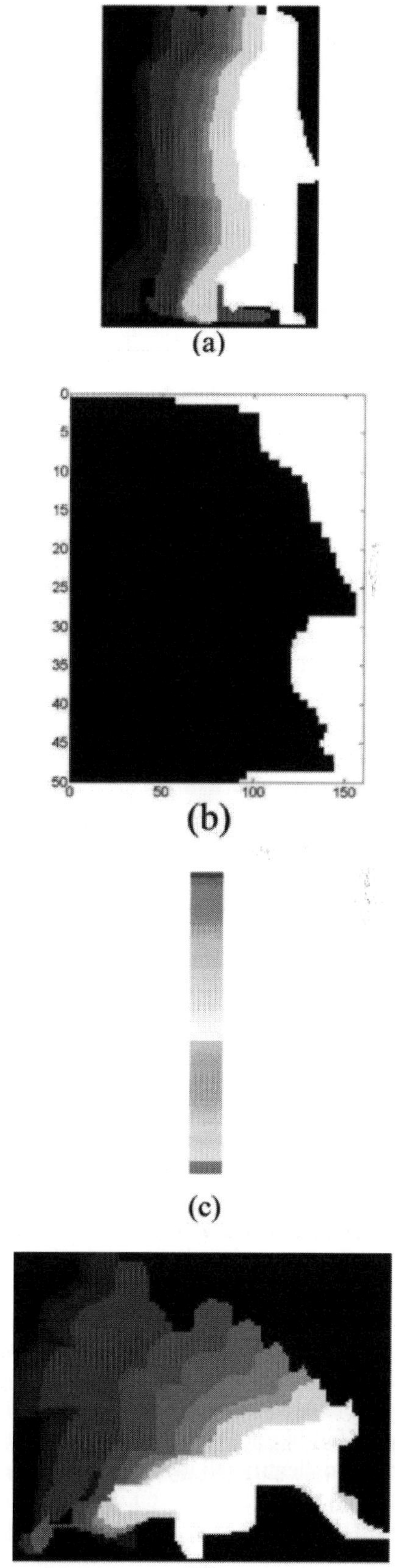

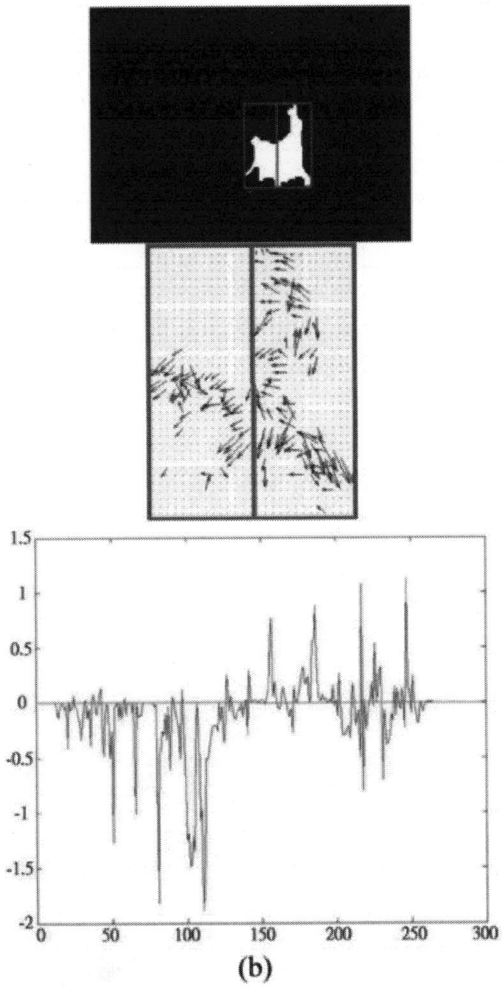

Figure. 8: Fall incident in overlapping situation. The *first row* is the silhouettes, the *second row* is the corresponding optical flow results, and the *third row* is the histograms of vertical optical flows. **a** The results generated by original method. **b** The results generated by using dividing method

(1) Rule 1: Given 20 consecutive frames, the average vertical optical flows exhibit downward more than 75 % of frames.

(2) Rule 2: The sum of the average vertical optical flows in 20 consecutive frames is larger than a threshold, say 10 in this study.

As shown in Fig. 8a, a fall incident may not be identified if the subject is overlapped by the other. To solve this problem, the bounding box is divided into two equal boxes if overlapping occurs. The width of the silhouette is used to identify whether the overlapping occurs or not. The optical flow features are then extracted in each divided box. The one which has larger average downward optical flow is used to identify possible fall action. As a result, the fall incidents can be extracted correctly as shown in Fig. 8b, and Fig. 8a demonstrates the result without using the bounding box division strategy.

Motion History Image in the Fine Stage

In the coarse stage, most non-fall actions can be filtered out via the downward optical flow features. However, some fall-like actions are identified as fall incidents due to the swing of arms. To solve this problem, we devise a feature vectors which are formed by projecting the MHI horizontally to verify fall incidents in the fine stage. MHI proposed by Bobick [18] is a template which condenses a determined number of silhouette sequences into a gray scale image (as shown in Fig. 9a) which is capable of preserving dominant motion information. Since the main difference between fall and other actions is the vertical component changes, our work projects the MHI horizontally to obtain a 50-dimensional feature vectors using equation (7):

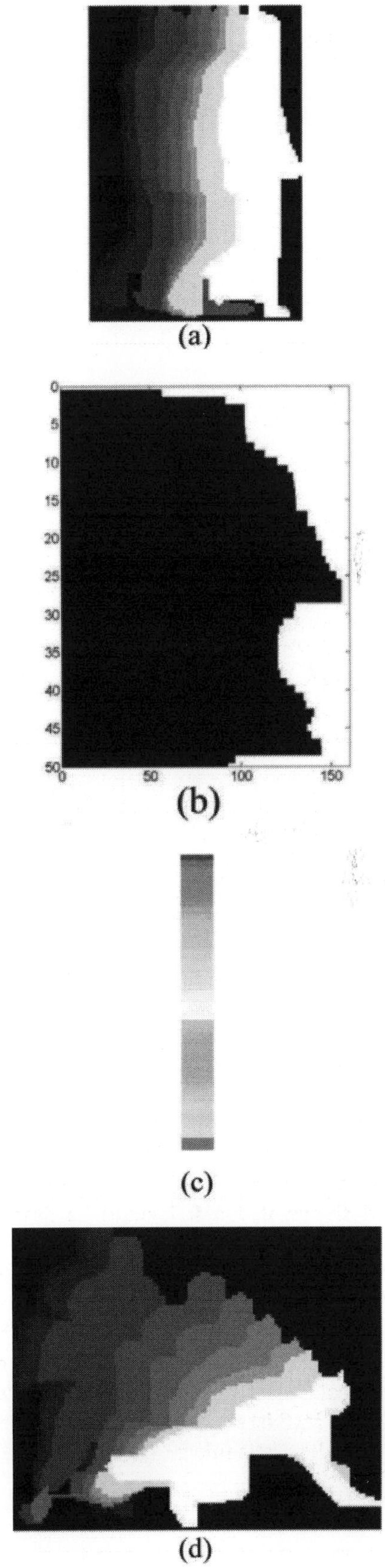

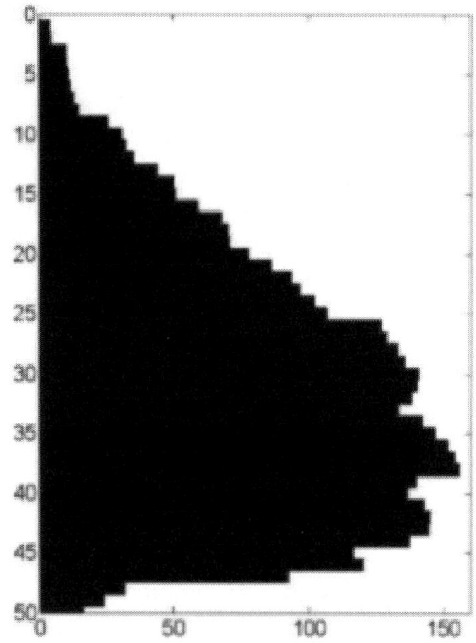

(e)

(f)

Figure. 9: Fine stage feature vector extraction. **a** MHI of walk. **b** Horizontal projection of walk MHI. **c** The obtained fine stage feature vector from walk MHI. **d** MHI of fall. **e** Horizontal projection of fall MHI. **f** The obtained fine stage feature vector from fall MHI

$$Q(i) = \frac{1}{U_w}\sum_{j=1}^{n} g\left(\left\lfloor \frac{U_h}{50} \times i \right\rfloor, j\right), \quad i = 1, 2, ..., 50 \tag{7}$$

where U_h, U_w, and $g(i,j)$ are the height, the width, and the pixel value of the motion energy in row i and column j, respectively. $Q(i)$ is the obtained 50-dimensional feature vectors. Figure 9c, f illustrates the comparison between the feature vectors of walk and fall in this study. The distributions of the walk action and the fall action are significantly different. As can be seen, the vertical motion information of the fall action is encoded directly with the horizontal projections, which can be viewed as extracting from MHI but not the silhouette. Therefore, the MHI features of fall-like actions will be fed into the constructed NNFLE verifier to identify fall incidents after the coarse stage.

Nearest Neighbor Feature Line Embedding (NNFLE)

Because the projection of MHI is a high-dimensional feature vector, a dimensional reduction scheme is employed to extract more salient features for fall detection. In our previous work [13], NFLE has demonstrated its effectiveness in pattern recognition. However, three problems of the NFLE have also been indicated in Section 2. To mitigate the three problems of NFLE, a modified NFLE termed Nearest Neighbor Feature Line Embedding (NNFLE) is proposed as a fall verifier in the fine stage. Here, given a feature vector x_i, which is extracted from MHI, the proposed NNFLE method is formulated as the following optimization problem:

$$\max_{w} J(w) = \sum_{i=1}^{N} \|w^T x_i - w^T x_i^{\text{between}}\|^2 - \sum_{i=1}^{N} \|w^T x_i - w^T x_i^{\text{within}}\|^2 \tag{8}$$

where xwithinixiwithin indicates the projected point of x_i on the nearest neighbor feature lines (NNFLs) formed by the samples with the same labels, and xbetweenixibetween indicates the projected point of x_i on NNFLs formed by the samples with different labels from x_i. Here, it has to be mentioned that in the NFLE, each NFL is formed by the samples with the same class. However, in the proposed NNFLE, the NNFLs on which the projected point xbetweenixibetween of the x_i could be formed by the samples with different labels from each other. In other words, all the other classes are treated as one class while calculating the projected point xbetweenixibetween.

With some algebraic operation, the $J(w)$ can be simplified to the following form:

$$\begin{aligned} J(w) &= \sum_{i=1}^{N} \|w^T x_i - w^T x_i^{\text{between}}\|^2 - \sum_{i=1}^{N} \|w^T x_i - w^T x_i^{\text{within}}\|^2 \\ &= \sum_{i=1}^{N} tr\left[w^T \left(x_i - x_i^{\text{between}}\right)\left(x_i - x_i^{\text{between}}\right)^T\right] \\ &\quad - \sum_{i=1}^{N} tr\left[w^T \left(x_i - x_i^{\text{within}}\right)\left(x_i - x_i^{\text{within}}\right)^T\right] \\ &= tr[w^T(\mathbf{S}_B - \mathbf{S}_W)w] \end{aligned} \tag{9}$$

Then, we impose a constraint $w^T w = 1$ on the proposed NNFLE. The transformation matrix w can thereby be obtained by solving the eigenvalue problem:

$$(\mathbf{S}_B - \mathbf{S}_W)w = \lambda w \tag{10}$$

Since the proposed NNFLE method does not need the inverse of any matrix, it can solve the singular problem of NFLE. However, the extrapolation and interpolation errors existing in NFLE may decrease the performance of locality preserving as shown in Fig. 10. Let us consider two feature line points $L_{2,3}$ and $L_{4,5}$ generated from two prototype pairs (x_2, x_3) and (x_4, x_5), respectively. Points $f_{2,3}(x_1)$ and $f_{4,5}(x_1)$ are two projection points $L_{2,3}$ of lines $L_{2,3}$ and $L_{4,5}$ for a query point x_1. From Fig. 11, it is clear that point x_1 is close to points x_2 and x_3 but far away from points x_4 and x_5. However, the distance $\|x_1 - f_{4,5}(x_1)\|$ for line $L_{4,5}$ is smaller than that for line $L_{2,3}$, i.e., $\|x_1 - f_{2,3}(x_1)\|$. The discriminant vector for line $L_{4,5}$ to point x_1 is hence selected instead of the other one. In addition, a great deal of computational time is needed due to the vast number of feature lines in the classification phase, e.g., C_2^{N-1} possible lines.

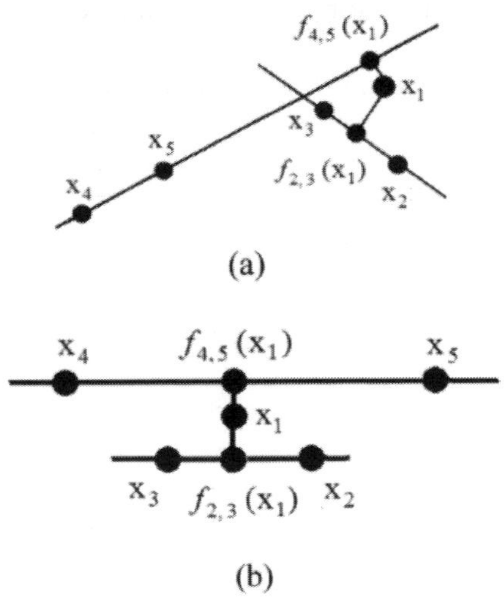

Figure. 10: a An extrapolation error. **b** An interpolation error

Training algorithm for the NNFLE transformation

Input: A training set $X = [x_1, x_2, \ldots x_N] \in R^{d \times N}$ and four parameters k_1, k_2, K_1, and K_2.

Output: The projection transformation W.

Step 1. PCA projection: Sample data are transformed from space R^d into a low-dimensional subspace R^n. Let W_{PCA} denote the transformation matrix of PCA.

Step 2. Computation of the within-class scatter: The possible NNFLs are generated from the k_1 nearest neighbor samples within the same class of a specified point x. Find the projection points and calculate the corresponding distances. Select K_1 vectors with the smallest distances and compute the within-class scatter S_W by Equation (11).

Step 3. Computation of the between-class scatter: Select k_2 nearest neighbor samples in different classes from a specified point x. Generate the possible NNFLs, find the projection points and calculate the corresponding distances. Select K_2 discriminant vectors with the smallest distances from point x to the feature lines. The between-class scatter S_B is obtained from Equation (12).

Step 4. Objective function maximization: $w^* = \arg\max_{W}(S_B - S_W)$ is maximized to obtain the best transformation matrix, which is composed of r eigenvectors with the largest eigenvalues.

Step 5. Output the final transformation matrix: $W = W_{PCA} W^*$.

Figure. 11: Training algorithm for the NNFLE transformation

To overcome the inaccuracy problem resulted from extrapolation and interpolation, feature lines for a query point are generated from the k nearest neighborhood prototypes. More specifically, when two points x_m and x_n belong to the nearest neighbors of a query point x_i, a straight line passing through points x_m and x_n is NNFL. The discriminant vector $x_i - f_{m,n}(x_i)$ is chosen for the scatter computation. The selection strategy for discriminant vectors in NNFLE is designed as follows:

(1) The within-class scatter $\mathbf{S}_W$: The NNFLs are generated from the k_1 nearest neighbor samples within the same class for the computation of the within-class scatter matrix, i.e., a set $F^+_{k_1}(x_i)$.

(2) The between-class scatter $\mathbf{S}_B$: Select k_2 nearest neighbor samples in different classes from a specified point x_i, i.e., a set $F^-_{k_2}(x_i)$, to generate the NNFLs and calculate the between-class scatter matrix.

$$\mathbf{S}_W = \sum_{p=1}^{N_c} \left(\sum_{\substack{\mathbf{x}_i \in C_p \\ f^{(2)} \in F^+_{k_1}(x_i)}} (x_i - f(x_i))(x_i - f(x_i))^T \right) \tag{11}$$

$$\mathbf{S}_B = \sum_{p=1}^{N_c} \left(\sum_{\substack{l=1 \\ l \neq p}}^{N_p} \sum_{\substack{x_i \in C_p \\ f \in F^-_{k_2}(\mathbf{x}_i)}} (x_i - f(x_i))(x_i - f(x_i))^T \right) \tag{12}$$

The training algorithm for the NNFLE transformation proposed in this study is described in Fig. 11.

The proposed NNFLE method is a simple and effective method to alleviate the extrapolation and interpolation errors. In addition, the scatter matrices are also generated based on the Fisher's criterion and represented as a Laplacian matrix. Moreover, the complexity of NNFLE is more efficient than that of NFLE. Consider N training samples, C_2^{N-1} possible feature lines will be generated and C_2^{N-1} distances have to be calculated for a specified point. The K_1 nearest feature lines are chosen from all possible lines to calculate the class scatter. The time complexity is $O(N^2)$ for line generation and $O(2N^2 \log N)$ for distance sorting. At the same time, the time complexity for selecting the K_1 nearest feature lines is $O(k^2)+O(2k^2 \log k)$ when nearest prototypes are chosen for line generation. Extra overhead $O(N \log N)$ is needed for finding the k nearest prototypes. When N is large, traditional method needs longer time to calculate the class scatter.

EXPERIMENTAL RESULTS

In this section, experimental results conducted on fall incident detection are illustrated to demonstrate the effectiveness of the proposed method. This work compares the proposed method with two state-of-the-art methods. Results are evaluated by using the simulated video data set captured from outdoor scenes. The data set is formed by 320 videos. In each video, the environment is in the dusky environments as shown in Fig. 2. Only the thermal imager can effectively capture the human silhouette under the environments. Table 1 tabulates the data sets used in the experiments. In this study, videos used for training are different from that used for testing. More specifically, training videos and testing videos were captured under different conditions (different places at different time). Among these data sets, video sequences which contain only one subject are utilized to compare the performance of the proposed method with other state-of-art fall detection methods and the results will be illustrated in Section 4.1. The identification capability of coarse-to-fine verifier is evaluated and illustrated in Section 4.2. In Section 4.3, the performance of the proposed method is evaluated by using video sequences which contain multiple subjects. Different from the other researches, the experimental results in Section 4.3 demonstrate that the proposed method can effectively detect fall incidents even when multiple persons overlap.

Table 1: The data sets used in the experiments

Actions	Number of training videos	Number of testing videos
Walk (one person)	30 (5135 frames)	50 (17,125 frames)
Fall (one person)	30 (545 frames)	50 (1822 frames)
Walk (multiple persons)	30 (5069 frames)	50 (16,130 frames)
Fall (multiple persons)	30 (460 frames)	50 (1810 frames)

Performance Comparisons of Various Fall Detection Algorithms

The data sets used in this subsection contain only one subject in each video sequence. Two state-of-the-art methods, BBN [11] and CPL [12], are implemented for comparison. The CPL takes a sequence as a sample, whereas the BBN and our proposed method take a frame as a sample. Therefore, the performance comparison of these three methods is based on each video sequence. In the experiments, 60 video sequences of one person are used as training sets and 100 video sequences of one person are used for testing. In addition, the projection matrix$w$ of the proposed NNFLE is constructed from the eigenvectors of $S_b - S_w$ with the largest corresponding eigenvalues when the objective function J is maximized. In our work, the dimensionality of feature vectors is reduced by the PCA transformation to remove noises. More than 99 % of the feature information is kept in the PCA process. After the PCA transformation, the optimal projection transformations are obtained for the proposed NNFLE method. All of the testing frames are matched with the trained prototypes using the NN matching rule. The performance comparisons of these three methods are tabulated in Table 2. From Table 2, we can notice that the proposed coarse-to-fine strategy of fall detection outperforms the other two methods. It implies that the proposed method is much more effective than the other two methods.

Table 2: The fall detection performance on the data set (%)

Method	Classification action (videos)	Reference action (videos)	
		Fall	Walk
CPL	Fall	92.00 (46/50)	8.00 (4/50)
	Walk	10.00 (5/50)	90.00 (45/50)
BBN	Fall	80.00 (40/50)	20.00 (10/50)
	Walk	12.00 (6/50)	88.00 (44/50)
NFLE	Fall	94.00 (47/50)	6.00 (3/50)
	Walk	6.00 (3/50)	94.00 (47/50)
Ours	Fall	98.00 (49/50)	2.00 (1/50)
	Walk	0.00 (0/50)	100.00 (50/50)

The Identification Capability of Coarse-to-Fine Verifier

In this subsection, the discriminability of the proposed coarse-to-fine strategy is analyzed as tabulated in Table 3. The identification capability is evaluated by the total number of frames of those video sequences which contains only one person. Among these 18,947 frames, there are 1822 and 17,125 frames of "fall" and "walk" actions predefined, respectively. As depicted in Table 3, the proposed method can identify most of the walk actions in the coarse stage. Only a small amount of fall-like actions are needed to be verified in the fine stage. In other words, almost all of the fall actions can pass through the coarse stage filter. Hence, the proposed coarse stage is very useful for pre-filtering non-fall actions so that the performance of the NNFLE classifier in the fine stage is less affected by the noisy data in both training and testing phases.

Table 3: The identification capability of coarse stage and fine stage of the proposed method

Classification actions	Reference actions in coarse stage (frames)		Reference actions in fine stage (frames)	
	Walk	Fall	Walk	Fall
Walk	17,079/17,125	46/17,125	46/46	0/46
Fall	68/1822	1754/1822	153/1754	1601/1754

Performance Evaluation of Fall Detection Under Overlapping Situations

In this subsection, the performance evaluation of fall detection in overlapping situations is illustrated. Video sequences which contain multiple persons are used for evaluation. Similar to the comparison described in Section 4.1, the NN matching rule is adopted to identify each testing frames in the fine stage. In the experiments, the performance evaluation of fall detection under overlapping situations is conducted based on each video sequence. Here, 30 video sequences are used for training and 100 video sequences are used for testing. The detection results are tabulated in Table 4. The proposed method utilizing coarse-to-fine strategy can effectively detect fall incidents while two persons are overlapping each other, and the performance is almost the same as that of the "one person fall" data sets described in Section 4.1.

Table 4: The performance evaluation of fall detection under overlapping situations (%)

Method	Classification action (videos)	Reference action (videos)	
		Fall	Walk
NFLE	Fall	92.00 (46/50)	10.00 (5/50)
	Walk	6.00 (3/50)	90.00 (45/50)
Ours	Fall	96.00 (48/50)	4.00 (2/50)
	Walk	0.00 (0/50)	100.00 (50/50)

CONCLUSIONS

In this paper, a novel fall detection mechanism based on a coarse-to-fine strategy in dusky environment is proposed. The human body in dusky environment can be successfully extracted using the thermal imager, and fragments inside the human body silhouette can also be significantly reduced as well. In the coarse stage, the optical flow algorithm is applied on thermal images. Most of walk actions are filtered out by analyzing the downward flow features. In the fine stage, the projected MHI is used as the features followed by the NNFLE method to verify fall incidents. The proposed NNFLE method, which adopts a nearest neighbor selection strategy, is capable of alleviating extrapolation/interpolation inaccuracies, singular problem, and high computation complexity. Experimental results demonstrate that the proposed method outperforms the other state-of-the-art methods and can effectively detect fall incidents even when multiple subjects are moving together.

REFERENCES

1. KC Moylan, EF Binder, Falls in older adults: risk assessment, management and prevention. Am. J. Med. 120(6), 493–497 (2007)
2. L Larson, TF Bergmann, Taking on the fall: the etiology and prevention of falls in the elderly. Clin. Chiropr. 11(3), 148–154 (2008)
3. S Gs, Falls among the elderly: epidemiology and prevention. Am. J. Prev. Med. 4(5), 282–288 (1988)
4. J Tao, M Turjo, M-F Wong, M Wang, Y-P Tan, Fall incidents detection for intelligent video surveillance, in *Proceedings of the 15th international conference on communications and signal processing*, 2005, pp. 1590–1594
5. D Anderson, JM Keller, M Skubic, X Chen, Z He, Recognizing falls from silhouettes, in *Proceedings of the 28th IEEE EMBS annual international conference*, 2006
6. CF Juang, CM Chang, Human body posture classification by neural fuzzy network and home care system applications. IEEE Trans. SMC, Part A 37(6), 984–994 (2007)
7. H Foroughi, N Aabed, A Saberi, HS Yazdi, An eigenspace-based approach for human fall detection using integrated time motion image and neural networks, in *Proceedings of the IEEE International Conference on Signal Processing (ICSP)*, 2008
8. C Rougier, J Meunier, AST Arnaud, J Rousseau, Fall detection from human shape and motion history using video surveillance, in *Proceedings of the 21st International Conference on Advanced Information Networking and Applications Workshops*, vol. 2, 2007, pp. 875–880
9. H Foroughi, A Rezvanian, A Paziraee, Robust fall detection using human shape and multi-class support vector machine, in*Proceedings of the Sixth Indian Conference on CVGIP*, 2008
10. CL Liu, CH Lee, P Lin, A fall detection system using k-nearest neighbor classifier. Expert Syst. Appl. 37(10), 7174–7181 (2010)
11. YT Liao, CL Huang, SC Hsu, Slip and fall event detection using Bayesian Belief Network. Pattern Recogn. 45, 24–32 (2012)
12. DN Olivieri, IG Conde, XAV Sobrino, Eigenspace-based fall detection and activity recognition from motion templates and machine learning. Expert Syst. Appl. 39(5), 5935–5945 (2012)
13. YN Chen, CC Han, CT Wang, KC Fan, Face recognition using nearest feature space embedding. IEEE Trans. Pattern Anal. Mach. Intell. 33(6), 1073–1086 (2012)
14. SZ Li, J Lu, Face recognition using the nearest feature line method. IEEE Trans. Neural Netw. 10(2), 439 (1999). -433, 1999
15. S Yan, D Xu, B Zhang, HJ Zhang, S Lin, Graph embedding and extensions: general framework for dimensionality reduction. IEEE Trans. Pattern Anal. Mach. Intell. 29(1), 40–51 (2007)
16. G Wu, Distinguishing fall activities from normal activities by velocity characteristics. J. Biomech. 33(11), 1497–1500 (2000)
17. CM Wang, KC Fan, CT Wang, Estimating optical flow by integrating multi-frame information. J. Inf. Sci. Eng. 24(6), 1719–1731 (2008)
18. AF Bobick, JW Davis, The recognition of human movement using temporal templates. IEEE Trans. Pattern Anal. Mach. Intell. 23(3), 257–267 (2001)

CHAPTER 13

AN OPTIMISATION APPROACH TO THE DESIGN OF A FAST, COMPACT AND DISTINCTIVE BINARY DESCRIPTOR

Mariusz Oszust

Department of Computer and Control EngineeringRzeszow University of Technology

ABSTRACT

The desired local feature descriptor should be distinctive, compact and fast to compute and match. Therefore, many computer vision applications use binary keypoint descriptors instead of floating-point, rich techniques. In this paper, an optimisation approach to the design of a binary descriptor is proposed, in which the detected keypoint is described using several, scale-dependent patches. Each such patch is divided into disjoint blocks of pixels, and then, binary tests between blocks' intensities, as well as their gradients, are used to obtain the binary string. Since the number of image patches and their relative sizes influence the descriptor creation pipeline, a simulated annealing algorithm is used to determine them, optimising recall and precision of keypoint matching. The simulated annealing is also used for dimensionality reduction in long binary strings. The proposed approach is extensively evaluated and compared with SIFT, SURF and BRIEF on public benchmarks. Obtained results show that the binary descriptor created using the resulted pipeline is faster to compute and yields comparable or better performance than the state-of-the-art descriptors under different image transformations.

INTRODUCTION

Recently, many computer vision applications for object recognition, tracking or scene reconstruction, have been using local feature descriptors. Such descriptors capture an image content that surrounds the detected keypoint and express it in the form of a vector. Feature descriptors with real-valued vectors produce the highest quality results, both in terms of robustness and performance, against a variety of image transformations. Scale-Invariant Feature Transform (SIFT) [1, 2] and Speeded Up Robust Features (SURF) [3] approaches are such floating-point descriptors. However, their computation time is long. Also, high dimensionality of compared vectors has negative impact on the image matching time. Therefore, binary descriptors have become more attractive in recent years, since they are compact and faster to compare using Hamming metric. In most cases, hand-crafted binary descriptors are obtained using pairwise tests between intensities of predefined parts of described image patch, i.e., pixels or regions according to a sampling pattern [4–8]. However, binary descriptors can be long, what requires an additional procedure for their reduction. The bit selection procedures use machine learning approaches [8–10] or optimisation [11, 12].

Citation: Mariusz Oszust, An optimisation approach to the design of a fast, compact and distinctive binary descriptor, DOI 10.1007/s11760-016-0907-4.

In this paper, an approach to the design of the binary descriptor is proposed. In the approach, 3.5 M SURF keypoints were detected on 4000 transformed images. Then, a simulated annealing (SA) algorithm [13, 14] determined a solution that maximised recall, as well as precision of keypoint matching. The solution consists of a set of image patches and their sizes. The patches are divided into disjoint pixel blocks. Finally, the binary string was created as a result of pairwise tests on intensities and gradients of blocks. Such binary string can be long; therefore, the SA was run in order to determine the most important 128 or 256 bits. The main novelty of the proposed approach is the use of the SA for optimisation of parameters of the binary descriptor creation pipeline. The dimensionality reduction approach of the binary descriptor with this optimisation technique is also novel. The details of the descriptor and experimental results showing its promising performance are presented in the paper. SURF, SIFT and BRIEF [5] descriptors are evaluated for comparison.

The remaining part of the paper is organised as follows. Section 2 provides a presentation of the state-of-the-art approaches. A detailed description of the proposed method can be found in Sect. 3. Extensive evaluation of the descriptor on widely used image datasets is reported in Sect. 4. Section 5 concludes the paper.

RELATED WORK

A growing number of computer vision applications, incorporating different approaches to image content description, reveal a tendency to construct faster, and still maintaining desirable properties, solutions. Many of them are focused on interest point (keypoint) description [15–18]. Here, a keypoint is a meaningful pixel group which is repetitively detected in spite of different image transformations (e.g., illumination, viewpoint change, or rotation). SIFT [2] is the keypoint detector and the descriptor. This technique produces high-quality results, both in terms of robustness and performance against common image transformations. In SIFT, keypoints are found using difference of Gaussians applied in scale space. However, the keypoint detection approach present in SIFT requires time-consuming rescaling and smoothing of the image. Furthermore, high dimensionality of the descriptor (128 values) slows keypoint matching. Another, high-dimensional, real-valued descriptor that in some cases outperforms SIFT is SURF [3]. The method is based on SIFT, but it incorporates many simplifications reducing its complexity. In SURF, the orientation of the keypoint is extracted from a circular region with scale-dependant radius. The descriptor is created by union of vectors resulted from sums of horizontal and vertical Haar wavelet responses and their absolute values calculated for a square region centred at the keypoint. Here, the descriptor consists of 64 floating-point values.

In comparison with newly developed approaches, the performance of SIFT (and SURF) has not been significantly outperformed [18]. However, in order to speed up the keypoint description and to reduce the dimensionality of the descriptor, binary descriptors have been introduced. Most of them apply pairwise tests between intensities of image regions or pixels according to some sampling pattern. Intensity comparison makes these solutions robust to photometric changes. Binary Robust Invariant Scalable Keypoints (BRISK) [4] uses AGAST [19] corner detector to find keypoints; then, a circular pattern with equally spaced points is used in binary tests. In this technique, the orientation of the keypoint is estimated using local gradients between the sampling pairs. In Binary Robust Independent Elementary Features (BRIEF) [5], the binary vector is obtained in tests between the intensities of point pairs along the same lines. Points are selected randomly from an isotropic Gaussian distribution. Since it is not rotation invariant, Oriented FAST and Rotated BRIEF (ORB) [6] was developed. The sampling pairs in ORB are determined in a learning process involving a greedy algorithm working on 300 K keypoints, maximising the amount of information carried by the descriptor. Here, FAST [20] detects keypoints. The dominant orientation in ORB is obtained using the intensity centroid approach [21]. In spite of introduced improvements, ORB performs better than BRIEF only in presence of large rotation or scale change [17]. In Fast Retina Keypoint (FREAK) [7], the sampling pattern was inspired by the human visual system. Here, the learning process is similar to ORB's with additional rejection of correlated pairs. The orientation of the keypoint is estimated using local gradients of 45 sampling pairs. FREAK seems to be outperformed by other binary descriptors according to [22, 23]. In recently developed Local Difference Binary (LDB) [8, 11] descriptor, in turn, the image patch of a predefined size (e.g., 45×45, 50×50 pixels [11]) is divided into 4, 9, 16 and 25 square cells. Then, binary tests between values representing cells are applied. In this binary descriptor, gradients are compared, as well as cells' mean intensities. AdaBoost-based important bit selection method, working on 400 K patches from [24], is applied to produce 256 bit binary strings. The orientation of the keypoint is estimated using the intensity moments. In works [9, 10], a supervised learning framework that finds low-dimensional descriptors is proposed. Here, the content of the image patch is modelled using local nonlinear filters that are selected with boosting. Boosting also helps in selection of the most important bits. In another approach [12], the descriptor design and its dimensionality reduction are formulated as convex optimisation problems involving separation of positive and negative patches. The solutions are found using support vector machine (SVM) solvers.

THE APPROACH

In the following subsections, the proposed binary descriptor is presented. At first, its design is formulated as an optimization problem. Then, a simulated annealing algorithm is used to determine its creation pipeline. Finally, the simulated annealing is used for dimensionality reduction in long binary strings.

Optimization Problem

The proposed approach is partially inspired by Multiscale Block Local Binary Pattern (MBLBP) [25], in which authors propose to compare average intensities of neighbouring blocks of pixels in order to create a binary string. In MBLBP, three patches with 3×3, 9×9, and 15×15 pixels are divided into nine blocks (cells). MBLBP is a dense descriptor typically applied to all pixels in the image. Here, the binary string is created by concatenation of strings obtained for three patches. The pairwise tests are performed only between some cells. It is worth mentioning that BRISK descriptor was also inspired by the dense approach (DAISY [26]). MBLBP uses only three image patches, and their sizes are given in pixels. The similar limitation can be seen in [11]. The approach proposed in this paper follows the idea of nine cells (pixel blocks) for the image patch. Moreover, the number of image patches and their sizes for the pipeline of creation of the descriptor are found in optimisation experiments. The proposed solution also uses intensity gradients inspired by SURF.

The variability in the pipeline of the creation of the descriptor caused by the number of used patches and their size expressed in terms of relation between the length of patch's side and the kyepoint's scale, has led to the formulation of the following optimisation problem. In the problem, precision and recall of keypoint matching were maximised. At the beginning, a keypoint detector generates MiMi keypoints for an input image *i*, i=1,... ,N_i=1,...,N, where *N* denotes the number of input images. The keypoint K_i^k $(K_i^k = 1, \ldots, K_i^{M_i})$ is described with the help of*P* image patches. Each patch *p* (p=1,...,Pp=1,...,P) has its keypoint's scale multiplier SpSp and nine square pixel blocks $B_p^j, j = 1, \ldots, 9$ j=1,...,9. The SpSp determines the size of the patch, i.e., the patch's side in pixels is equal to SpSp multiplied by the keypoint's scale. The B5pBp5-th block is centred at the keypoint's location (*x*, *y*). In order to create the binary descriptor, pairwise tests are performed between blocks' intensities. Such test returns *true* (1) if a difference of values is smaller than 0, and *false* (0), otherwise. Since blocks for the *p*-th patch have the same number of pixels, sums of their intensities B_p^5- can be compared. This solution is motivated by efficient calculation of such sums using the integral image technique [3, 27]. Also, the first-order intensity gradients in the *x* and *y* directions $(D_x(B_p^j)$ and $D_y(B_p^j))$ are computed using Haar-like box filters and the integral image, as in SURF [3]. Gradients for BjpBpj-th block are obtained for its central pixel. The results of pairwise tests performed on gradients, in a form of the binary vector, are added to binary strings obtained for intensity *I*. Finally, for the *p*-th patch 108 binary tests are performed, and for the K_i^k-th keypoint 108*P*-bit string is obtained. Exemplary image patches and their division are presented on Fig. 1.

The number of image patches *P* and multipliers *S* can be seen as decision variables in the optimisation problem. Since *precision* and *recall* measures [17] reflect the quality of performance of a given local feature descriptor in image matching tests, they can be used as an objective function. Precision counts correctly matched pairs out of all returned matches, while recall, in turn, counts such pairs out of corresponding pairs. Higher values of both metrics are considered better; therefore, their multiplication was used as the objective function in the proposed approach.

Figure. 1: An exemplary four image patches centred at the keypoint and their division

Descriptor Design using Simulated Annealing

The simulated annealing algorithm (SA) [13, 14] reflects phenomena observed while slow cooling of molten metals. The annealing process achieves the global minimum state using random fluctuations in energy. Such fluctuations help the process to escape local minima. The SA algorithm requires definitions of: (1) a vector of decision variables *x*, (2) an objective function $F_C(x)$, (3) a method *G*(*x*) for generation of a neighbouring solution x'x', based on information stored in *x*, (4) a temperature drop *T*, and (5) an acceptance criteria of a weaker solution $F_A(x,x',T)$. The temperature drop *T* can be seen as the number of iterations of the SA. The algorithm consists of the following steps:

1. Create random x
2. Calculate $F_C(x)$
3. While $T > T_{stop}$ then:
   (a) Create $x' = G(x)$
   (b) Calculate $F_C(x')$
   (c) If $F_C(x')$ is better than $F_C(x)$ or $F_A(x, x', T)$ accept x', $x = x'$
   (d) Decrease T.

FAFA is calculated as follows:

$$F_A(x, x', T) = e^{\frac{\Delta F_C}{T}}, \quad (1)$$

where $\frac{\Delta F_C}{T}$ expresses the difference between values of $F_C(x)$ and $F_C(x')$. T is iteratively decreased, at the beginning causing that weaker solutions to have a higher chance to be accepted, i.e., if F_A is larger than a randomly selected number in the range [0, 1), the solution x' is accepted.

The problem of determination of the descriptor creation pipeline is solved using the S_A, x in this case refers to finding P and S. For the important bit selection problem, or dimensionality reduction, the solution x takes a form of a binary vector indicating which bits are used while maximising $F_C(x)$ and reducing the descriptor's binary string to a given length (e.g., 128, 256 bits). F_C is given as

$$F_C(x) = \sum_{i=1}^{N} P_i R_i, \quad (2)$$

where P denotes precision, R recall, and i is the input image. The $G(x)$ function changes values in S_p and adds or rejects the patch p. In the dimensionality reduction problem, $G(x)$ selects a predefined number of the most important bits randomly changing information carried by x.

The choice of the SA was motivated by its simplicity, in terms of implementation, and the small number of parameters. Furthermore, the proposed objective function does not require a process of preparation of negative examples, as in approaches with SVM or boosting. Population-based optimisation algorithms (e.g., genetic, immune, or swarm) were also taken into consideration. They evaluate and reject many solutions in iterations, since each solution requires time (and memory)-consuming computation of the objective function; the SA seems to be a reasonable approach, providing acceptable results with fewer objective function calls.

Once P and S are obtained, K_i^k keypoints are generated by a keypoint detector on the input image i. It is assumed that the detector provides keypoints with scales. Then, the keypoints are described in the following manner:

1. For each keypoint K_i^k
   (a) Select $p = 1, \ldots, P$ patches with size $S_p\sigma \times S_p\sigma$, where σ denotes keypoint's scale (in pixels).
   (b) For each patch
      i. For each B_p^j block, $j = 1, \ldots, 9$, within the p-th patch: Calculate $I(B_p^j)$, $D_x(B_p^j)$, and $D_y(B_p^j)$ associated with the centre of the block.
      ii. Perform $9!/(2! \times 9!) = 36$ pairwise tests for $I(B_p^j)$, $D_x(B_p^j)$ and $D_y(B_p^j)$.
   (c) Concatenate $108P$ binary strings into the descriptor of the keypoint K_i^k.

The dominant orientation of the detected interest point is estimated using wavelet responses in horizontal and vertical direction. A similar step is present in SURF, but here only half-wavelet responses are used. The orientation helps to determine the position of the central pixel of each outer block $(B_p^j, j \neq 5)$.

EXPERIMENTS

In the following subsections, experiments on the design of the pipeline for the creation of the descriptor and the most important bit selection are presented. Then, the resulting descriptor family is evaluated on public benchmarks.

Optimization

For the design of the descriptor's pipeline, 1K images were randomly selected from *25 K MIRFLICKR* dataset [28]. The dataset consists of images downloaded from the social photography site *Flickr*. They were rotated (90∘∘ and 45∘∘) and scaled (1/2 and 2/3). Then, 225 images from *Phos* dataset [29] were added to the resulting dataset. *Phos* contains 15 scenes captured under different illumination conditions (i.e., uniform and nonuniform). Each scene has selected one base image. It can be seen that the number of images being under different illumination changes is significantly smaller than rotated or scaled images. Since constructed binary descriptor uses pairwise intensity tests, it is, like other binary descriptors, robust against illumination changes. Therefore, this part of the learning dataset was smaller. 3,519,420 keypoints were detected by SURF on 5225 images, prior to the optimisation experiments. Integral images for the learning dataset, as well as lists of keypoints have large memory requirements, thus only one solution, x, was evaluated at a time, i.e., in terms of calculation of the objective function FcFc on used CPU with 16 GB RAM. The SA maximises F_C (see Eq. (2)) which uses precision and recall metrics. For a given input image, precision was calculated as the number of correct matches, i.e., the number of corresponding pairs of keypoints between the base image and the distorted one, to the number of all returned pairs. The match was accepted if a distance ratio between the first and the second close described keypoint on the second image was below or equal a threshold of 0.8 [2, 17] and the location of the keypoint on the second image was within three pixels from the expected location. The expected location of the keypoint was inferred using homography [17]. 4 K distorted images from*MIRFLICKR* were matched with the original 1 K images, and 15 reference images were selected from *Phos* for the same purpose. Since the dataset consists of 4210 pairs of images and used metrics have maximum value equal 1, the best (maximal) possible objective function value is 4210. This would indicate the case in which each detected keypoint was successfully matched. Due to image transformations, such large F_c value was not likely to be found.

The SA was run 30 times with T_{start}=1. The T was decreased using T_{next}=0.9Told in 100 iterations of the algorithm. The maximal possible number of patches P was set to 15, values of scale multipliers S_p were selected in range 1 to 150 with step 1. Scale multipliers could be floating-point values but the use of integers significantly reduced the search space of the SA algorithm. Table 1 contains mean, maximal, minimal and standard deviation of the objective function of solutions obtained in experiments. The best binary descriptor was the input to the second optimisation problem of finding the most 128 or 256 important bits. The dimensionality of the binary vector was reduced since it affects the matching time and the storage cost. Parameters of the SA for the second problem remained unchanged. It can be seen, the obtained mean values are promising, and the results are characterised by small standard deviations. The best result $(F_C(x_{best}) = 2119.1)$ was obtained for the set of P=12 patches with the following multipliers S_p=[7,10,16,19,23,28,30,34,44,56,69,89]. This gives the descriptor with 1296 bits. For comparison, objective function calculated for SURF approach was equal to 1642.2; BRIEF obtained $7F_C$=136.7 and SIFT 2889.6. SIFT used its own detector to generate keypoints, thus about half of the number of SURF keypoints were detected. The proposed descriptor using SIFT keypoints yielded $3F_C$=2744.3. They seem to be more suitable, but in experiments previously detected SURF points remained, since their number was larger, making F_C optimisation more difficult.

The creation of the descriptor using twelve patches can be time-consuming. Therefore, a simplified version of the descriptor was used. It contains only four patches (432 bits) with S_p=[5,10,15,20], and obtained $9F_C$=1683.9 with SURF keypoints.

Table 1: Objective function values obtained in experiments

Optimisation problem	Maximum	Minimum	Mean	SD
Descriptor design	2119.1	2093.3	2106.3	5.6
Selection of 256 bits	2069.9	2060.3	2064.3	2.9
Selection of 128 bits	2004.1	1975.1	1987.5	7.4

Evaluation

Evaluation of the proposed descriptor was made using two publicly available, widely used, Mikolajczyk et al. [15] (*Oxford*) and Heinly et al. [17] datasets. They contain image sequences with known homography between the first image in the sequence of images and the others. There are six to nine images in each sequence. Images exhibit an increasing amount of transformations, such as rotation (sequences: *Ceiling* [17], *Rome* [17], *Semper* [17]), scaling (*Venice* [17]), rotation with scaling (*Bark* [15], *Boat* [15]), view point change (*Wall* [15],*Graffiti* [15]), blur (*Bikes* [15], *Trees* [15]), illumination (*Day and night* [17]), exposure (*Leuven* [15]) and JPEG compression (*Ubc* [15]).

SURF, SIFT and BRIEF were selected for comparison due to their high performance reported in many works. BRIEF (512 bit) is representative of binary descriptors, and its performance was confirmed in work [17], despite being not fully rotation and scale invariant. The proposed descriptor, since then called Optimised Binary Robust fAst Features (OBRAF), is represented by its three versions, with 1296, 256, 128-bit vectors (twelve patches) and simple 432-bit version with four patches (nonoptimised BRAF). For each image pair, about 1 K SURF keypoints were detected and described. SIFT generated similar number of keypoints and described them, since, in the spirit of fairness, SIFT descriptor should be coupled with its own detector. Keypoints were matched using the same protocol as in the experiments with the optimisation. For comparative analysis, precision and putative match ratio [17] were used. Precision of matching has a practical significance, since all returned pairs often undergo further processing, e.g., using RANSAC [30]. Putative match ratio (PMR) counts how many detected keypoints were returned as matched. This measure is influenced by the descriptor's ability to differentiate similar image regions, and could be understood as a measure of distinctiveness of the descriptor. In some cases, the keypoints can be detected close to each other.

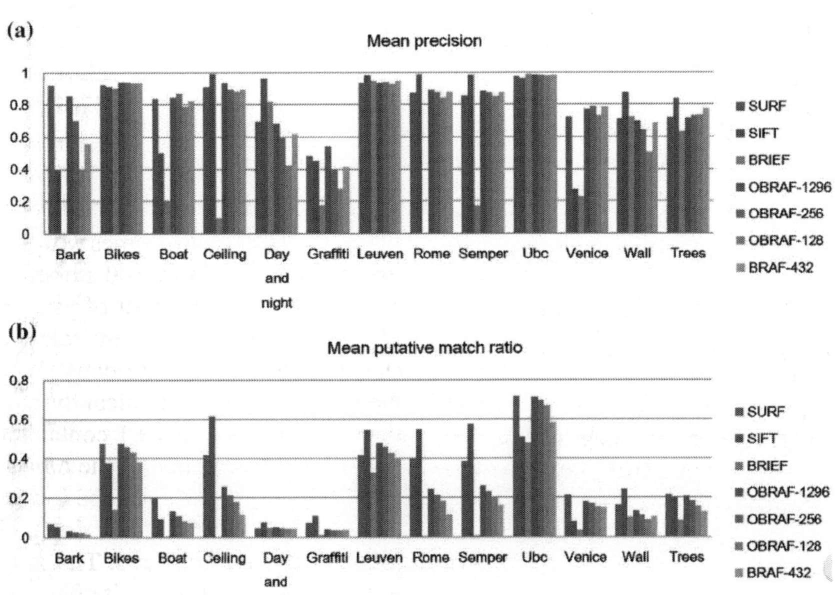

Figure. 2: Mean precision (**a**) and mean putative match ratio (**b**) for descriptors evaluated on Mikolajczyk et al. [15] and Heinly et al. [17] datasets

Figure 2 shows mean values of the precision and PMR results obtained for both datasets. For all 13 image sequences, OBRAF family of descriptors was better, in terms of precision values, than one of the compared floating-point descriptors. For the PMR metric, OBRAF descriptors were better than one of the floating-point counterparts for seven image sequences. BRIEF was outperformed by the other methods, even in tests without rotation and scale changes. The simplified version of the proposed descriptor, BRAF, in some cases performed better than SIFT or SURF. Since mean values were presented, the behaviour of the descriptors cannot be seen in the most difficult tests, i.e., when the metrics are evaluated for the last images in the sequences. Therefore, *Recall versus 1-Precision* curves obtained for the last images in sequences, considered as the most challenging, are shown on Fig. 3. Here, the proposed descriptor family achieved comparative or better performance than floating-point descriptors. In the *Night and day* sequence, BRIEF performed quite well, since it was designed for such transformation.

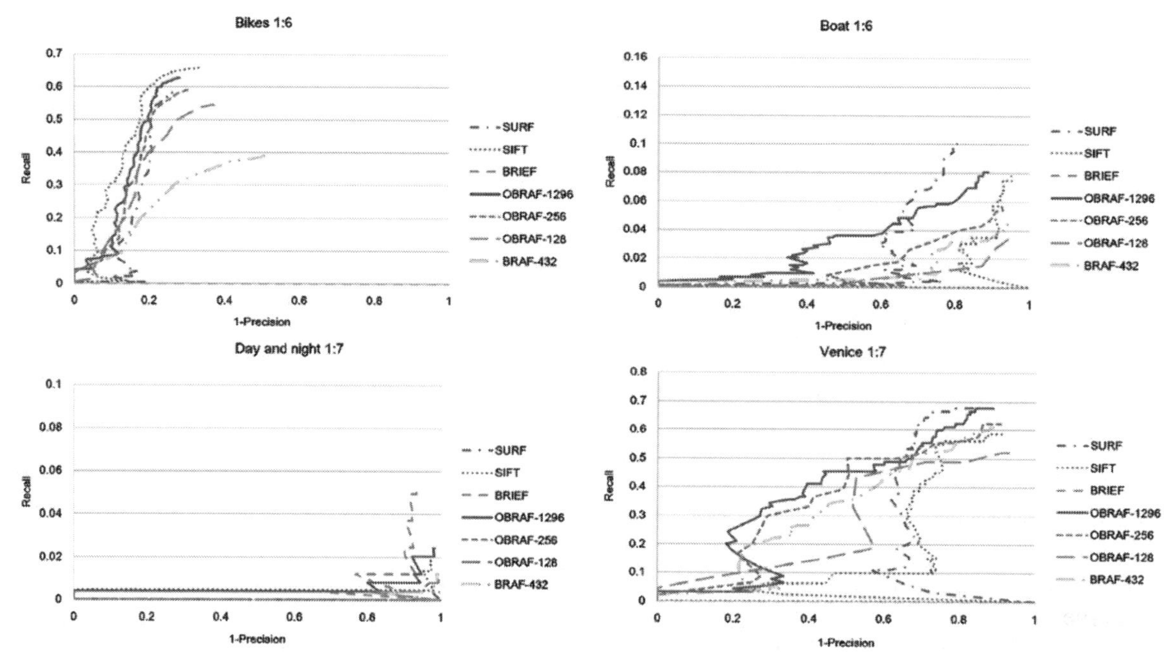

Figure. 3: *Recall versus 1-Precision* curves for compared descriptors evaluated on Mikolajczyk et al. [15] and Heinly et al. [17] datasets. *Curves* were obtained using the last image in a given image sequence. The name of the sequence is given above the curve, the pair of images is denoted as 1 : *t*, where *t* is the number of the last image, and *1* indicates the number of the reference image

Table 2 contains comparison of descriptor computation time. Since most descriptors in original works were computed on similar or faster machines using different numbers of keypoints, the table contains timings per keypoint. OBRAF is implemented in Java as a single-threaded application, and all presented experiments were run on Intel Core i7-2720QM 2.2 GHz, 16 GB of RAM, Microsoft Windows 7, and Java 8.0. Therefore, BoofCV [31] (Java) implementations of SIFT, SURF and BRIEF were used. SURF in BoofCV library is faster and produce better results than many widely used SURF implementations, e.g., *OpenSurf*, *ETH* or *OpenCV* [31]. It can be seen that OBRAF descriptor family obtained similar or better computation time than other descriptors. A simplified OBRAF version (BRAF), in which the dominant orientation is not estimated, is computed in 0.011 ms, 0.5 times faster than BRIEF that also does not incorporate this operation. It significantly outperforms other compared approaches. Cited descriptors were implemented in C++, and their computation time was mostly optimised. OBRAF's computation consists of many steps which are independent, e.g., each patch and each block within the patch can be processed independently what makes the descriptor easy to parallel. Therefore, some further improvements in shortening the computation time are expected.

The matching time is strongly affected by the descriptor length, and binary vectors can be efficiently compared (Hamming distance) on modern CPUs using binary *XOR* and *population count* instructions. The state-of-the-art sparse binary descriptors are seldom shorter than 128 bits. Since 128 and 256-bit OBRAF versions yielded promising performance results, they would also offer the state-of-the-art matching time.

Table 2: Descriptor computation time (per keypoint)

Name	Length (bits)	Computation time (ms)
OBRAF	1296, 256, 128	0.027⋆⊖⋆⊖, 0.044⋆⋆
BRAF	432	0.011⋆⊖⋆⊖, 0.026⋆⋆
SIFT	1024	6.156 [4], 0.407⋆⋆
SURF	512	0.43 [4], 0.66 [5], 0.67 [8], 0.15⋆⋆
BRISK	512	0.037 [4], 0.026 [8]
BRIEF	512	0.0187⊖⊖ [5], 0.053⋆⊖⋆⊖
ORB	256	0.056 [8]
LDB	256	0.048 [8]
FREAK	512	0.039 [8]

Results with ⋆⋆ were obtained in experiments reported in this paper (Java implementations), ⊖⊖denotes an approach, in which the orientation of the keypoint is not estimated

CONCLUSIONS

In this paper, a novel descriptor, OBRAF, is proposed as a result of an optimisation approach to the design of the descriptor creation pipeline. The SA algorithm was used to solve two problems. At first, it determined the number of patches and their sizes used for the description of the interest point. Then, the solution of the dimensionality reduction problem was found. In both cases, recall and precision of keypoint matching were used as the objective function. The obtained descriptor family was evaluated on two popular image datasets. Experimental results showed that OBRAF is faster than the state-of-the-art descriptors while maintaining comparable or better performance under different image transformations.

REFERENCES

1. Lowe, D.G.: Object recognition from local scale-invariant features. In: Proceedings of the 7th International Conference Computer Vision, vol. 2, pp. 1150–1157. Kerkyra, Sept 20–27 (1999). doi:10.1109/ICCV.1999.790410
2. Lowe, D.G.: Distinctive image features from scale-invariant keypoints. Int. J. Comput. Vision 60(2), 91–110 (2004). doi:10.1023/B:VISI.0000029664.99615.94 CrossRef
3. Bay, H., Ess, A., Tuytelaars, T., Gool, L.V.: SURF: speeded-up robust features. In: Proceedings of the European Conference Computer Vision, pp. 404–417. Graz, May 7–13 (2006). doi:10.1007/11744023_32
4. Leutengger, S., Chli, M., Siegwart, R.Y.: BRISK: binary robust invariant scalable keypoints. In: Proceedings of the International Conference on Computer Vision, pp. 2548–2555. Barcelona, Nov 6–13 (2011). doi:10.1109/ICCV.2011.6126542
5. Calonder, M., Lepetit, V., Strecha, C., Fua, P.: BRIEF: binary robust independent elementary features. In: Proceedings of the European Conference on Computer Vision, pp. 778–792. Heraklion, Sept 5–11 (2010). doi:10.1007/978-3-642-15561-1_56
6. Rublee, E., Rabaud, V., Konolige, K., Bradski, G.: ORB: an efficient alternative to SIFT or SURF. In: Proceedings of the International Conference on Computer Vision, pp. 2564–2571. Barcelona, Nov 6–13 (2011). doi:10.1109/ICCV.2011.6126544
7. Alahi, A., Ortiz, R., Vandergheynst, P.: Freak: fast retina keypoint. In: Proceedings of the International Conference on Computer Vision and Pattern Recognition, pp. 510–517. Providence, June 16–21 (2012). doi:10.1109/CVPR.2012.6247715
8. 8.Yang, X., Cheng, K.T.: Local difference binary for ultra-fast and distinctive feature description. IEEE Trans. Pattern Anal. 36(1), 188–194 (2014). doi:10.1109/TPAMI.2013.150CrossRef
9. Trzcinski, T., Christoudias, M., Lepetit, V., Fua, P.: Learning Image Descriptors with the Boosting-Trick. In: Pereira, F., Burges, C.J.C., Bottou, L., Weinberger, K.Q. (eds.) Advances in Neural Information Processing Systems 25. Proceedings of Neural Information Processing Systems Conference (NIPS), pp. 269–277. Curran Associates, Inc. (2012)
10. Trzcinski, T., Christoudias, M., Fua, P., Lepetit, V.: Boosting binary keypoint descriptors. In: Proceedings of the International Conference IEEE CVPR, pp. 2874–2881. Portland, June 23–28 (2013). doi:10.1109/CVPR.2013.370
11. Yang, X., Cheng, K.T.: LDB: an ultra-fast feature for scalable augmented reality on mobile devices. In: International Symposium IEEE Mixed and Augmented Reality, pp. 49–57. Atlanta, Nov 5–8 (2012). doi:10.1109/ISMAR.2012.6402537
12. Simonyan, K., Vedaldi, A., Zisserman, A.: Learning local feature descriptors using convex optimisation. IEEE Trans. Pattern Anal. 36(8), 1573–1585 (2014). doi:10.1109/TPAMI.2014.2301163CrossRef
13. Kirkpatrick, S., Vecchi, M.P.: Optimization by simulated

annealing. Science 220(4598), 671–680 (1983). doi:10.1126/science.220.4598.671MathSciNetCrossRef MATH

14. Goffe, W.L., Ferrier, G.D., Rogers, J.: Global optimization of statistical functions with simulated annealing. J. Econom. 60(1), 65–99 (1994). doi:10.1016/0304-4076(94)90038-8CrossRefMATH
15. Mikolajczyk, K., Tuytelaars, T., Schmid, C., Zisserman, A., Matas, J., Schaffalitzky, F., Kadir, T., Van Gool, L.: A comparison of affine region detectors. Int. J. Comput. Vis. 65, 43–72 (2005). doi:10.1007/s11263-005-3848-xCrossRef
16. Xie, X., Mirmehdi, M.: A Galaxy of Texture Features. Handbook of Texture Analysis. Imperial College Press, London (2008)
17. Heinly, J., Dunn, E., Frahm, J.M.: Comparative evaluation of binary features. In: Proceedings of the European Conference Computer Vision, pp. 759–773. Florence, Oct 7–13 (2012). doi:10.1007/978-3-642-33709-3_54
18. Miksik, O., Mikolajczyk, K.: Evaluation of local detectors and descriptors for fast feature matching. In: Proceedings of the 21st International Conference on IEEE Pattern Recognition, pp. 2681–2684. Tsukuba, Nov 11–15 (2012)
19. Mair, E., Hager, G.D., Burschka, D., Suppa, M., Hirzinger, G.: Adaptive and generic corner detection based on the accelerated segment test. In: Proceedings of the European Conference on Computer Vision, pp. 183–196. Crete, Sept 5–11 (2010). doi:10.1007/978-3-642-15552-9_14
20. Rosten, E., Porter, R., Drummond, T.: Faster and better: a machine learning approach to corner detection. IEEE Trans. Pattern Anal. 32, 105–119 (2010). doi:10.1109/TPAMI.2008.275CrossRef
21. Rosin, P.L.: Measuring corner properties. Comput. Vis. Image Underst. 73(2), 291–307 (1999). doi:10.1006/cviu.1998.0719CrossRef
22. Canclini, A., Cesana, M., Redondi, A., Tagliasacchi, M., Ascenso, J., Cilla, R.: Evaluation of low-complexity visual feature detectors and descriptors. In: Proceedings of the 18th International Conference on IEEE Digital Signal Processing, pp. 1–7. Fira, July 1–3 (2013). doi:10.1109/ICDSP.2013.6622757
23. Caetano, C., Avila, S., Guimaraes, S., Araujo, A.D.A.: Representing local binary descriptors with bossaNova for visual recognition. In: Proceedings of the 29th Annual ACM Symposium on Applied Computing, pp. 49–54. Gyeongju, March 24–28 (2014). doi:10.1145/2554850.2555058
24. Winder, S., Hua, G., Brown, M., Picking the best DAISY. In: Proceedings of the International Conference on IEEE CVPR, pp. 178–185. Miami, June 20–25 (2009). doi:10.1109/CVPR.2009.5206839
25. Liao, S., Zhu, X., Lei, Z., Zhang, L., Li, S.Z.: Learning Multi-scale Block Local Binary Patterns for Face Recognition LNCS, vol. 4642, pp. 828–837. Springer, New York (2007). doi:10.1007/978-3-540-74549-5_87
26. Tola, E., Lepetit, V., Fua, P.: Daisy: an efficient dense descriptor applied to wide-baseline stereo. IEEE Trans. Pattern Anal. 32(5), 815–830 (2010). doi:10.1109/TPAMI.2009.77CrossRef
27. Crow, F.C.: Summed-area tables for texture mapping. In: Proceedings of the 11th Conference on Computer Graphics and Interactive Techniques, pp. 207–212. ACM, Minneapolis, July 23–27 (1984). doi:10.1145/964965.808600
28. Huiskes, M.J., Lew, M.S.: The MIR flickr retrieval evaluation. In: Proceedings of the 1st International Conference on ACM Multimedia Information Retrieval, pp. 39–43. Vancouver, Oct 30–31 (2008). doi:10.1145/1460096.1460104
29. Vonikakis, V., Chrysostomou, D., Kouskouridas, R., Gasteratos, A.: A biologically inspired scale-space for illumination invariant feature detection. Meas. Sci. Technol. 24(7), 074024 (2013). doi:10.1088/0957-0233/24/7/074024CrossRef
30. Fischler, A.M., Bolles, R.C.: Random sample consensus: a paradigm for model fitting with applications to image analysis and automated cartography. Commun. ACM 24(6), 381–395 (1981). doi:10.1145/358669.358692MathSciNetCrossRef
31. Abeles, P.: Speeding up SURF. In: Proceedings of the 9th International Symposium Advances in Visual Computing, LNCS, vol. 8034, pp. 454–464. Springer, Rethymnon, July 29–31 (2013). doi:10.1007/978-3-642-41939-3_44

CHAPTER 14 ADAPTIVE KUWAHARA FILTER

Krzysztof Bartyzel

Department of Computer Image AnalysisJohn Paul II Catholic University of Lublin

ABSTRACT

A new filter was created by improving the standard Kuwahara filter. It allows more efficient noise reduction without blurring the edges and image preparation for segmentation and further analyses operations. One of the biggest and most common restrictions encountered in filter algorithms is the need for a declarative definition of the filter window size or the number of iterations that an operation should be repeated. In the case of the proposed solution, we are dealing with automatic adaptation of the algorithm to the local environment of each pixel in the processed image.

INTRODUCTION

Contextual transformations are operations in which the outcome depends on a modified pixel and its surroundings [4, 5]. Alternatively, the name "digital filters" is also used. The application of digital filters is very wide, and they are used in many areas of life including radiology, automatic control systems and machine vision. The main applications of digital filters are as follows:

- suppressing unwanted noise,
- improving image sharpness,
- removing specific defects ,
- visualizing certain image features,
- reconstructing partly destroyed images.

Citation: Krzysztof Bartyzel, Adaptive Kuwahara filter, DOI 10.1007/s11760-015-0791-3.

In this paper, I address the issues of reducing noise from digital images and present a new highly efficient algorithm. The issue of noise or other errors is extremely important because they can appear during image acquisition, transmission or compression and decompression. Therefore, image filtering is a very important aspect of image processing and preparation for further work such as image compression, edge detection or image segmentation. The application of noise reduction algorithms is also extensively used in medical imaging. A large number of articles discussing this subject have been published, e.g., [2, 3].

However, at the beginning I would like to present the presentation of the standard commonly used algorithms and to clarify the terms used.

It is highly likely that there exist filters that remove noise from digital images faster, gain a better rate of bug fixing or modify the edges of objects less than those used in this study. However, a comparison of all the existing solutions would require a large and difficult to estimate amount of work, at least because of the need to provide a similar level of implementation and other conditions of the experiment, and thus, it would exceed significantly the planned scope of the study. There is nothing, however, to prevent extending the scope of the comparison and to present its results in future publications.

The general concept of the context filters is to create the output image on the basis of the source image in such a way that the value of each pixel at coordinates (x,y) in the output image is determined based on a certain neighborhood of the pixel at coordinates (x,y) in the source image [4–9]. This neighborhood is defined as a filter window. Usually, it is assumed that the filter window is square and the size (width and height) is an odd integer number, e.g., 3×3,5×5,7×73×3,5×5,7×7. In this case, a pixel at coordinates (x,y) is exactly in the middle of the filter window. There are also solutions that collect statistical information about the entire image and then use them during the processing of each pixel.

CONTEXT FILTERS

Average Filter

The average filter is one of the simplest filters enabling removing noise and distortion. Unfortunately, it also introduces changes in the areas that have originally been noise and distortion free.

Let us assume that we construct a filter operating on the window N×N. In this case, the resulting value of the pixel is the average value of all the pixels contained within the current analysis window of the source image

$$m_\theta = \frac{1}{N \times N} \times \sum_{(x,y)\in\theta} \varphi(f(x,y)) \tag{1}$$

where:

- f is the source image function,
- $f(x,y)$ is the value of the pixel at coordinates (x, y),
- φ is a function calculating the value of a particular pixel. This function can take different values depending on the color space, the format or the depth of the image,
- N×N is the number of pixels in the current window,
- θrepresentsthecollectionofpixelsinthecurrentwindow.
- Someofthelargestdisadvantagesoftheaveragefilterare blurring noise and significant modification of correct pixels.

Median Filter

The concept of the median filter is very similar to that shown above. The only difference is that all elements in the current filter window have to be sorted and ordered in an ascending sequence. The resulting value is the value of the item in the exact middle point of the ordered sequence (position $\frac{N \times N}{2}$ for N×N window).

With the use of this filter, one can very effectively remove any local noise, without blurring the noise to larger areas, which was a major disadvantage of the average filter.

Adaptive Median Filter

The adaptive median filter, as its name suggests, is a modification of the median filter. This modification consists in allowing a situation where the window size is not constant, but changes dynamically according to the context. If you select too small a window, the median filter could not handle all the noises, while too large a window results in dissolving the smaller but desired image detail.

The adaptive median filter algorithm has the following form:

Step 1 For each pixel f(x,y) of the image f, we determine the list of neighbors in N×N window. Among these elements, we calculate the minimum, middle (median) and the maximum value and denote these values as fminfmin, fmedfmed and fmaxfmax, respectively.

Step 2 If $f_{\min} = f_{\text{med}}$ or $f_{\text{med}} = f_{\max}$, then the size of filter window is increasing and the actions done in the previous step for the new window have to be repeated. Otherwise, go to the next step.

For obvious reasons, there must exist the maximum allowable window size. If it is not possible to further increase the filter window size, then the output pixel is set to the neighbors list middle value determined in the previous step (fmedfmed).

Step 3 If $f_{\min} < f(x,y)$ and $f(x,y) < f_{\max}$, then the output pixel takes the value of $f(x, y)$, which means that item is not modified. Otherwise, the output pixel is set to the neighbors list middle value determined in the first step (f_{med}).

This filter is able to reduce even a large noise. Unfortunately, the high computational complexity of this filter is quite a serious disadvantage. In addition, if the maximum window size is set to too small a value, we can get unsatisfactory results. However, too large a window size may cause very unsatisfactory execution time.

Kuwahara Filter

Because the contour plays a very important role in the process of image analysis [4, 9], it is very important to ensure that the image smoothing does not affect the sharpness of contours. This could cause serious problems later, during segmentation. The Kuwahara filter [1] is an example of a filter which meets these requirements. This filter can be constructed for any window size. For readability, the algorithm will be described for the window of size of 3×3.

The filter window should be divided into four areas. Let us denote them as θ_k where $k \in \{0,1,2,3\}$. These four areas are highlighted in Fig. 1. The center pixel is marked with black color. If a square filter window consists of $(2 \times n + 1) \times (2 \times n + 1)$ elements, then each area will contain exactly (n+1)×(n+1) elements. In the described case, the filter window size is 3×3, so each area will consist of four elements. Then, for each of these areas, the values of average m_k and variance δ_k^2 are calculated. The average and variance are calculated according to the formula:

$$m_k = \frac{1}{(n+1) \times (n+1)} \times \sum_{(x,y)\in\theta_k} \varphi(f(x,y)) \tag{2}$$

$$\delta_k^2 = \frac{1}{(n+1)\times(n+1)} \times \sum_{(x,y)\in\theta_k} \left[\varphi\left(f(x,y)\right) - m_k\right]^2 \qquad (3)$$

where:

$k \in \{0, 1, 2, 3\}$,
f is the source image function,
$f(x, y)$ is the value of the pixel at coordinates (x, y),
φ is a function calculating the value of a particular pixel,
$\frac{1}{(n+1)\times(n+1)}$ is the number of pixels in the current area,
n is the value obtained directly from the filter window size.

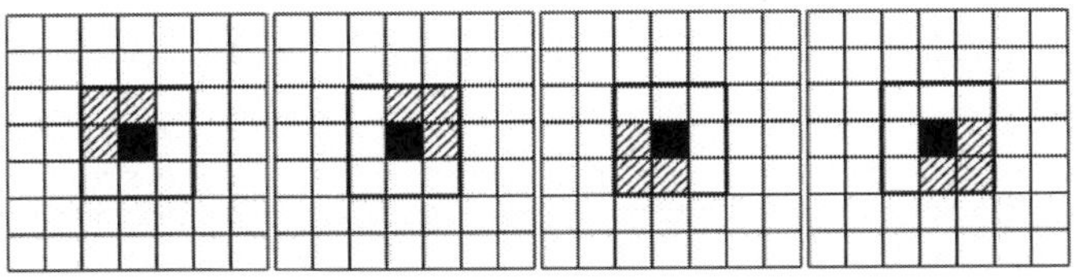

Figure. 1: Kuwahara filter window with highlighted four areas θkθk

Finally, we compare the variance of all four areas and look for an index of the area for which the variance is the smallest.

$$\delta_{\min}^2 = \min_{k\in\{1,2,3,4\}} \left(\delta_k^2\right) \qquad (4)$$

The resulting value of the center pixel is the average value of the area for which the variance was the smallest.

The result of increasing the size of the filter windows will be better noise reduction, but it can cause blurring of small details in the analyzed image.

ADAPTIVE KUWAHARA FILTER

The adaptive Kuwahara filter algorithm was created by combining two other filters: the adaptive median filter and the Kuwahara filter. The most important feature of the adaptive median filter is the possibility to adjust the window size of the filter to the results of a partial analysis obtained during the operation, whereas the major task of the Kuwahara filter is smoothing colors intensity and removing small noise while maintaining the edges. Unfortunately, neither of these filters are flaw free.

The first blurs the edges, making further analysis even more difficult. Furthermore, in connection with variable window size, computational complexity of the filter can be much higher than of other filters. On the other hand, the Kuwahara filter retains the edges, but it requires a strictly defined window size. An undesirable effect of the second filter is very significant pixelation, which is a mosaic of large, homogeneous, often rectangular areas.

The differences in the operation of the various algorithms can be seen in Figs. 2, 3, 4, 5, 6 and 7. Each figure consists of a sequence of five images; starting from the left side, they represent (a) original image, (b) after applying Kuwahara filter where window size is constant and equals 3×3, (c) after applying Kuwahara filter where window size is constant and equals 11×11, (d) after applying adaptive median filter and (e) after applying adaptive Kuwahara filter.

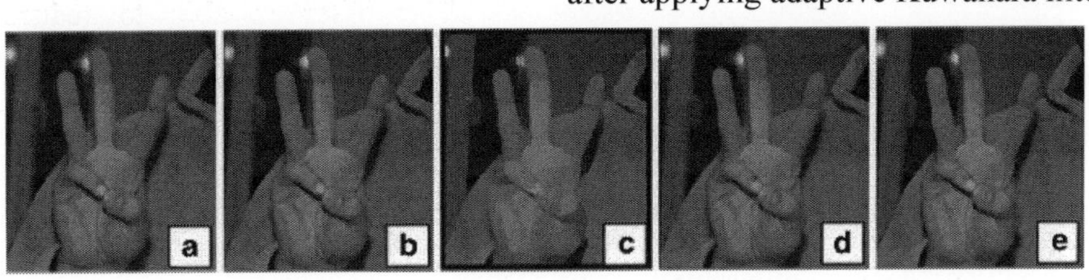

Figure. 2: Comparison of adaptive Kuwahara filter with other filters. Image without noise. Labels **a**,**b**, **c**, **d** and **e** correspond to **a** original image, **b** Kuwahara filter with constant window size3×3, **c** Kuwahara filter with constant window size 11×11, **d** adaptive median filter and **e** adaptive Kuwahara filter

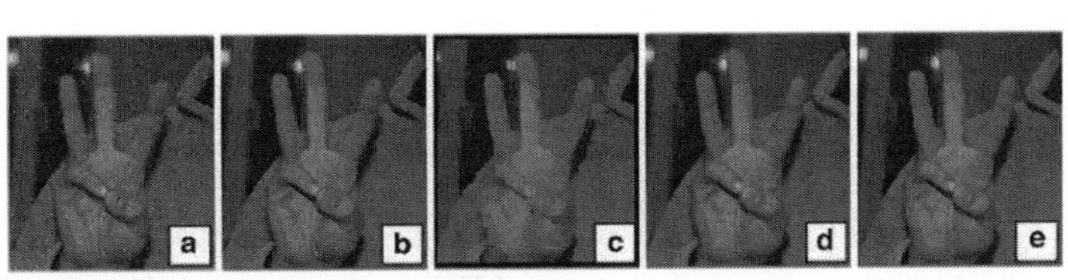

Figure. 3: Comparison of adaptive Kuwahara filter with other filters. Image has been modified by the addition of salt and pepper noise with uniform distribution on the surface of 1 %. Labels **a**, **b**, **c**, **d** and **e** correspond to **a** original image, **b** Kuwahara filter with constant window size 3×3, **c** Kuwahara filter with constant window size 11×11, **d** adaptive median filter and **e** adaptive Kuwahara filter

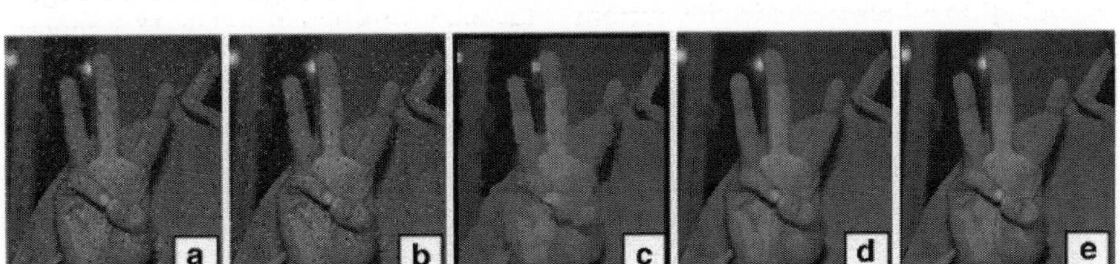

Figure. 4: Comparison of adaptive Kuwahara filter with other filters. Image has been modified by the addition of salt and pepper noise with uniform distribution on the surface of 2 %. Labels **a**, **b**, **c**, **d** and **e** correspond to **a** original image, **b** Kuwahara filter with constant window size 3×3, **c** Kuwahara filter with constant window size 11×11, **d** adaptive median filter and **e** adaptive Kuwahara filter

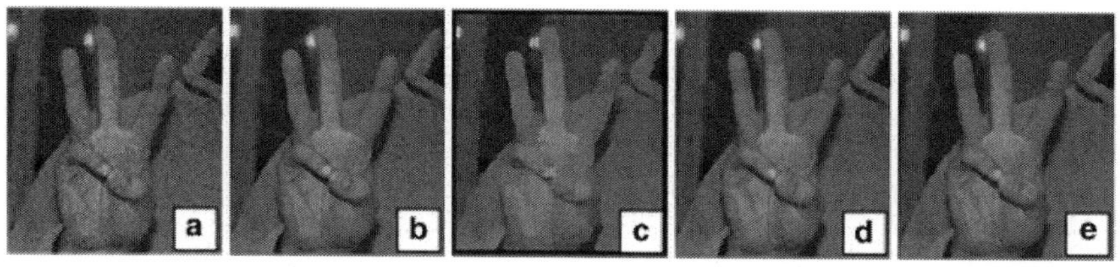

Figure. 5: Comparison of adaptive Kuwahara filter with other filters. Image has been modified by the addition of additive salt and pepper noise with uniform distribution on the surface of 5 %. Labels **a**, **b**, **c**, **d** and **e** correspond to **a** original image, **b** Kuwahara filter with constant window size 3×3, **c** Kuwahara filter with constant window size 11×11, **d**adaptive median filter and **e** adaptive Kuwahara filter

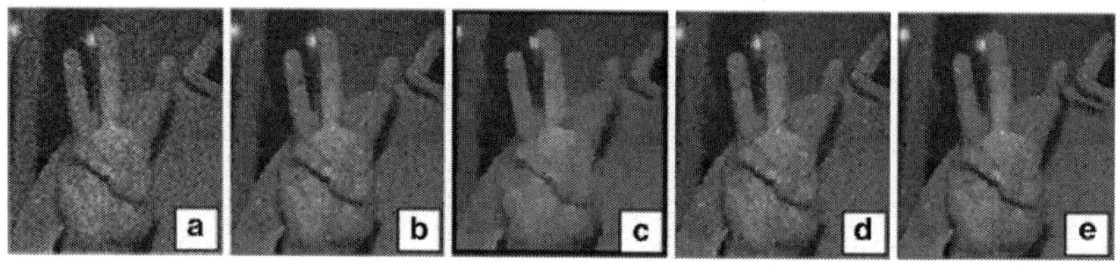

Figure. 6: Comparison of adaptive Kuwahara filter with other filters. Image has been modified by the addition of additive salt and pepper noise with uniform distribution on the surface of 25 %. Labels **a**, **b**, **c**, **d** and **e** correspond to **a** original image, **b** Kuwahara filter with constant window size 3×3, **c** Kuwahara filter with constant window size 11×11, **d**adaptive median filter and **e** adaptive Kuwahara filter

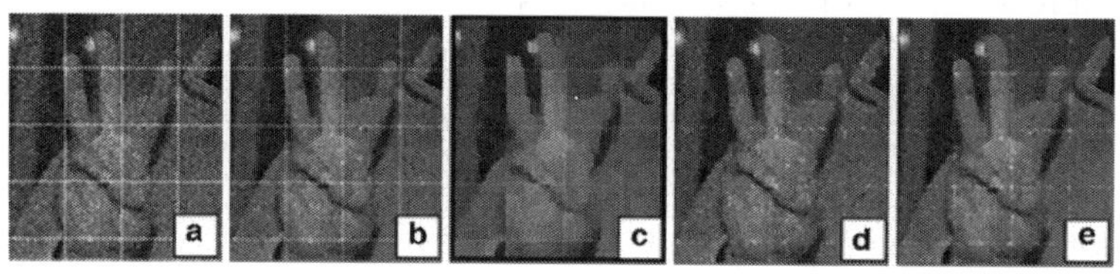

Figure. 7: Comparison of adaptive Kuwahara filter with other filters. Image has been modified by the addition of additive salt and pepper noise with uniform distribution on the surface of 25 % and white grid. Labels **a**, **b**, **c**, **d** and **e** correspond to **a** original image, **b** Kuwahara filter with constant window size 3×3, **c** Kuwahara filter with constant window size 11×11, **d** adaptive median filter and **e** adaptive Kuwahara filter

In addition, one can see how the presented filters behave in relation to images with different noise levels. Figure 2 is the original image without additional noise. Figure 3 has been modified by the addition of salt and pepper noise with uniform distribution on the surface of 1 %. Figure 4has been modified by the addition of salt and pepper noise with uniform distribution on the surface of 2 %. Figure 5 has been modified by the addition of additive salt and pepper noise with uniform distribution on the surface of 5 %. The brightness of the randomly selected pixel was changed (increased or decreased) by 25 % of the maximum value. Figure 6, like the previous one, has been modified by the addition of additive salt and pepper noise with uniform distribution on the surface of 25 %. The last picture (Fig. 7) is a modification of the previous one by adding a white grid.

ALGORITHM DESCRIPTION

The standard Kuwahara filter requires a strictly defined window size. In the case of the proposed modification of the filter, the window size is changing, just as it did in the adaptive median filter, depending on the local properties of the image. The initial window size is 3×3.

Step 1 The filter window should be divided into four areas by following the original algorithm. Let us denote them as θ_k where $k \in \{0, 1, 2, 3\}$. Initially, each of these areas will consist of four pixels. For the purposes of this algorithm, let us call these four areas the basic areas.

Step 2 For each of these areas, the values of mean m_k and variance $\delta^2_{\min}$ are calculated. As before, the value of a specific pixel can be the value of color intensity, brightness or any other calculated value. The mean and variance are calculated according to the formula:

$$m_k = \frac{1}{N_k} \times \sum_{(x,y)\in\theta_k} \varphi(f(x,y)) \quad (5)$$

$$\delta_k^2 = \frac{1}{N_k} \times \sum_{(x,y)\in\theta_k} \left[\varphi(f(x,y)) - m_k\right]^2 \quad (6)$$

where:

$k \in \{0,1,2,3\}$,

f is the source image function,

φ is a function calculating the value of a particular pixel,

N_k is the number of pixels in the current area, in the first cycle the value is 4.

Step 3 Each of the basic areas is considered separately. For the chosen area, the size of the window is increased by 1. 1.

Next, for the new window size, mean $\overline{m_k}$ and variance $\overline{\delta_k^2}$ have to be calculated according to the formulas presented in the previous step.

An example can be seen in Fig. 8. The filter window central element and the elements included in area θ_k as it increases are highlighted.

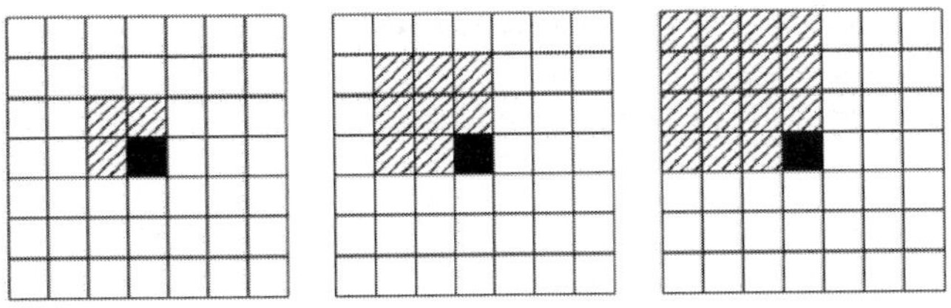

Figure. 8: Resizing the filter window and items included in the new area θ_k

If the variance of the new area $(\overline{\delta_k^2})$ is smaller than before the resizing of the filter window (δ_k^2), then the mean and variance of the basic area k take the newly calculated values:

$$m_k := \overline{m_k} \quad (7)$$

$$\delta_k^2 := \overline{\delta_k^2} \quad (8)$$

Then we continue to increase the size of the window for the basic area selected in the current step until its size reaches the maximum allowable size or until the variance of the newly enlarged area is greater than that calculated in the previous cycle of the algorithm. In this way, the minimum variance and the corresponding average value will be achieved for the basic area *k*. For further calculations, it is not necessary to know the size of the window for which the variance and the mean values were calculated.

The calculations shown in Step 3 must be repeated separately for each of the four basic areas.

Step 4 Finally, we compare the variance of all four areas. At this stage, each of the basic areas can be made with different quantity of items. We are looking for an index of the area for which the variance is the smallest.

$$\delta_{\min}^2 = \min_{k \in \{1,2,3,4\}} (\delta_k^2) \quad (9)$$

The resulting value of the output pixel is the average value of the basic area for which the variance was the smallest. Figure 9 presents an example of possible window size distribution for items included in the four basic areas and common parts of these areas. The center pixel is marked with black color.

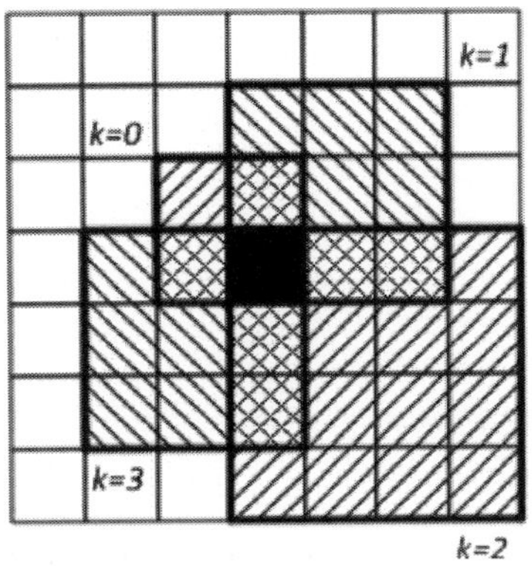

Figure. 9: An example of possible window size distribution for basic areas

Figure 10 presents a block diagram showing the operation of the developed algorithm.

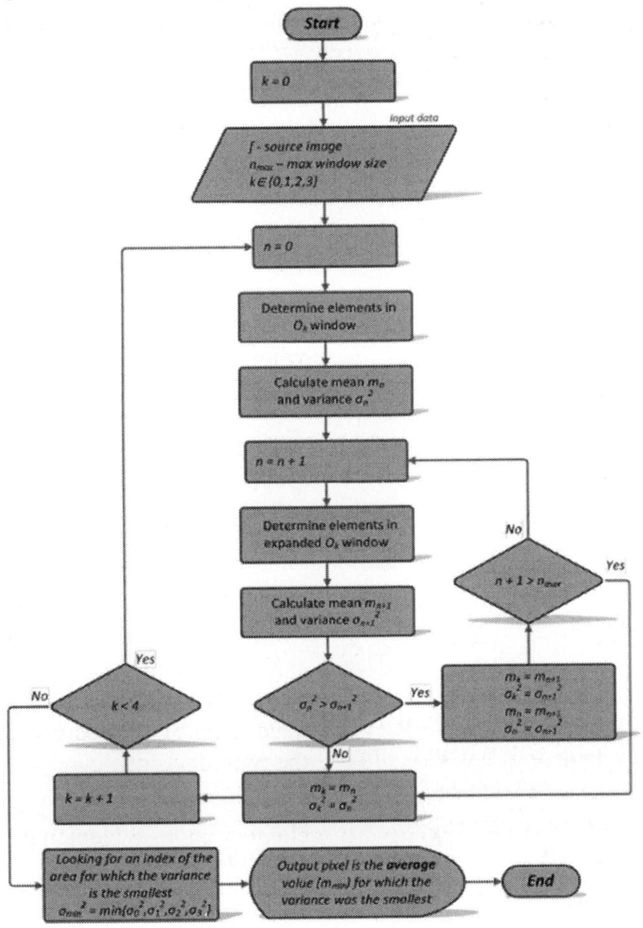

Figure. 10: Block diagram of the adaptive Kuwahara filter

COMPARISON OF ALGORITHMS

One of the most serious problems of the edge preserving smoothing filters is causing over blurring and removing details in low-contrast regions. Nonlinear edge preserving smoothing filters have been an actively researched continuously for decades. Worth mentioning is the articles [10–15] where authors present various approaches to the problem of edge preserving smoothing filters.

Very interesting approach of the advanced modification of the Kuwahara filter was introduced in the work [16] where authors used very sophisticated mathematical description on the theoretical considerations. The paper in any way does not cover the technical aspects of implementation or optimization. The only suggestion placed by the authors was information that definite integral can be approximate by discrete sums. It is a trivial remark and in any meaningful way does not clean the complexity of implementation.

The article authors suggest that processing time depends on the size of the filter kernel, the number of sectors and the size of input. However, based on the exemplary, implementation and the description of the algorithm can be concluded that the computational complexity is also influenced by the radius h parameter which is used for mapping Ω into a disk of h radius. There were not, however, shown any estimations considering the computational complexity of the presented algorithm. But, on the basis of the theoretical considerations, it can easily be concluded that the computational complexity of the adaptive Kuwahara filter is significantly smaller than anisotropic Kuwahara filter. This is, due to among other things, because the necessity of smoothing with a Gaussian filter and defining elliptic filter kernels. In addition, for each regions, in order to obtain weighted local averages and variances, complex calculations have to be performed.

The separate aspect is the fact that anisotropic Kuwahara filter was primarily designed for preserving shape boundaries and achieving the painterly effect look, without having to deal with individual brush strokes. One of the major similarities of the two filters (adaptive Kuwahara filter and anisotropic Kuwahara filter) is a robust against high-contrast noise, and the difference is the fact that the proposed new approach does not require from the user to provide any additional parameters.

Short tests were conducted to compare the quality of the noise reduction in the analyzed filters. Tests include previously discussed algorithms and anisotropic Kuwahara filter. Original images have been modified by the addition of additive salt and pepper noise with uniform distribution on the surface of 25 %. Finally, the original and filtered were compared with the use of the mean squared error algorithm. The results are shown in Table 1. In the MSE algorithm, smaller value indicates a better result. Figure 11 presents a sequence of seven images; starting from the left side, they represent original image, noised image, after applying average filter with where window size 7×7, after applying adaptive median filter, after applying Kuwahara filter where window size is constant and equals 7×7, after applying adaptive Kuwahara filter and finally after applying anisotropic Kuwahara filter.

Table 1: Comparation of the noise reduction filters (MSE algorithm)

SN	Average filter	adaptive median filter	Kuwahara filter	adaptive Kuwahara filter	anisotropic Kuwahara filter
1	16.62	15.64	16.40	**13.67**	15.50
2	17.88	17.45	17.58	**14.78**	15.17
3	21.37	**17.58**	21.42	18.09	19.08

Bold values indicate the smallest and the best value in row

Figure. 11: Comparation of the noise reduction filters on the Lena image

One can easily see proposed algorithm guarantees to achieve very good results. For the all examined examples, the effects of the adaptive Kuwahara filter were better than anisotropic Kuwahara filter. In addition, taking into account, lower computational complexity adaptive Kuwahara filter can be considered as a strong contender of the anisotropic Kuwahara filter.

SUMMARY

On the basis of the examples, it can be easily noticed that the modified Kuwahara filter is a promising improvement. The most important of its features includes the ability to preserving the edges of the objects. After applying the proposed filter, both the objects and the edges are changed much less than when using the standard Kuwahara filter. Additionally, the tendency to pixelation, compared to the standard Kuwahara filter, has been greatly reduced. Furthermore, it has the ability to remove noise from an image with comparable efficacy as that of the adaptive median filter as well as it retains a greater variety of color intensity in relation to the Kuwahara filter what can be confirmed by analyzing the results of MSE algorithm.

While creating presented solution, very important aspect was the optimization of computational complexity. The adaptive Kuwahara filter is very fast in operation, especially compared to the adaptive median filter. This suggests that the direction for further research should be a detailed analysis of the computational complexity of created algorithm and its comparison with other filters.

REFERENCES

1. Kuwahara, M., Hachimura, K., Eiho, S., Kinoshita, M.: Processing of RI-angiocardiographic images. In: Preston Jr, K., Onoe, M. (eds.) Digital Processing of Biomedical Images, pp. 187–202. Plenum, New York (1976)
2. Söderman, M., Holmin, S., Andersson, T., Palmgren, C., Babić, D., Hoornaert, B.: Image noise reduction algorithm for digital subtraction angiography: clinical results. Radiology269(2), 553–560 (2013)
3. Leipsic, J., LaBounty, T.M., Heilbron, B., Min, J.K., Mancini, G.B.J., Lin, F.Y., Taylor, C., Dunning, A., Earls, J.P.: Adaptive statistical iterative reconstruction: assessment of image noise and image quality in coronary CT angiography. Cardiopulm. Imaging, 195, September (2010)
4. Pratt, W.K.: Digital Image Processing. Wiley, New York (2001)CrossRefMATH
5. Rosenfeld, A.: Picture Processing by Computer. Academic Press, New York (1969)MATH
6. Bovik, A.C.: Handbook of Image and Video Processing. Academic Press, New York (2000)MATH
7. Hall, E.L.: Computer Image Processing and Recognition. Academic Press, New York (1979)MATH
8. Jahne, B.: Digital Image Processing. Springer, Berlin, Heidelberg (2005)MATH
9. Sonka, M., Hlavac, V., Boyle, R.: Image Processing Analysis and Machine Vision. Springer, US (1993) CrossRef
10. Tomita, F., Tsuji, S.: Extraction of multiple regions by smoothing in selected neighborhoods. IEEE Trans. Syst. Man Cybern. 7, 107–109 (1977)CrossRef
11. Nagao, M., Matsuyama, T.: Edge preserving smoothing. Comput. Gr. Image Process. 9, 394–407 (1979)CrossRef
12. van den Boomgaard, R.: Decomposition of the Kuwahara-Nagao operator in terms of linear smoothing and mor- phological sharpening. ISMM 2002 (2002)
13. Hu, H., de Haan, G.: Adding explicit content classification to nonlinear filters. Signal, Image Video Process. 5(3), 291–305 (2011)CrossRef
14. Nasimudeen, A., Nair, Madhu S., Tatavarti, Rao: Directional switching median filter using boundary discriminative noise detection by elimination. Signal, Image Video Process. 6(4), 613–624 (2012)CrossRef
15. Bhadouria, V.S., Ghoshal, D., Siddiqi, A.H.: A new approach for high density saturated impulse noise removal using decision-based coupled window median filter. Signal, Image Video Process. 8(1), 71–84 (2014) CrossRef
16. Kyprianidis, J.E., Kang, H., Döllner, J.: Image and video abstraction by anisotropic Kuwahara filtering. Comput. Gr. Forum, pp. 1955–1963 (2009)

CHAPTER 15

ROBUST LOCAL SIMILARITY FILTER FOR THE REDUCTION OF MIXED GAUSSIAN AND IMPULSIVE NOISE IN COLOR DIGITAL IMAGES

Bogdan Smolka,Damian Kusnik

Faculty of Automatic Control, Electronics and Computer ScienceSilesian University of Technology

ABSTRACT

In this paper, a novel technique designed for the suppression of mixed Gaussian and impulsive noise in color images is proposed. The new denoising scheme is based on a weighted averaging of pixels contained in a filtering block. The main novelty of the proposed solution lies in the new definition of the similarity between the samples of the processing block and a small window centered at the block's central pixel. Instead of direct comparison of pixels, a measure based on the similarity between a given pixel and the samples from the neighborhood of the central pixel is utilized. This measure is defined as the sum of distances in a given color space, between a pixel of the block and a certain number of most similar samples from the filtering window. The main advantage of the proposed scheme is that the new similarity measure is not influenced by the outliers injected into the image by the impulsive noise and the averaging process ensures the effectiveness of the new filter in the reduction of Gaussian noise. The experimental results prove that the novel filtering design is capable of suppressing mixed noise of high intensity and is competitive with respect to the state-of-the-art noise filtering methods.

INTRODUCTION

Noise reduction is one of the most frequently performed image processing operations, as the enhancement of degraded images or video streams is crucial for the success of the consecutive image processing steps.

Very often color images are corrupted by various types of noise, introduced by malfunctioning or damaged sensors in the image formation process, poor lighting conditions and aging of the storage material, electronic instabilities of the signal, transmission errors in noisy channels and atmospheric as well as electromagnetic interferences. As a result, images are often corrupted by a mixture of Gaussian and impulsive noise [2, 13, 19, 30, 36, 58].

The suppression of this kind of noise is a challenging task, as filters designed to cope with the Gaussian noise are not able to remove the impulses and the techniques designed for the impulsive noise removal are ineffective in the elimination of the noise modeled by the Gaussian distribution [6, 22, 23, 37, 49, 50].

Citation: Bogdan Smolka, Damian Kusnik, Robust local similarity filter for the reduction of mixed Gaussian and impulsive noise in color digital images, DOI: 10.1007/s11760-015-0830-0.

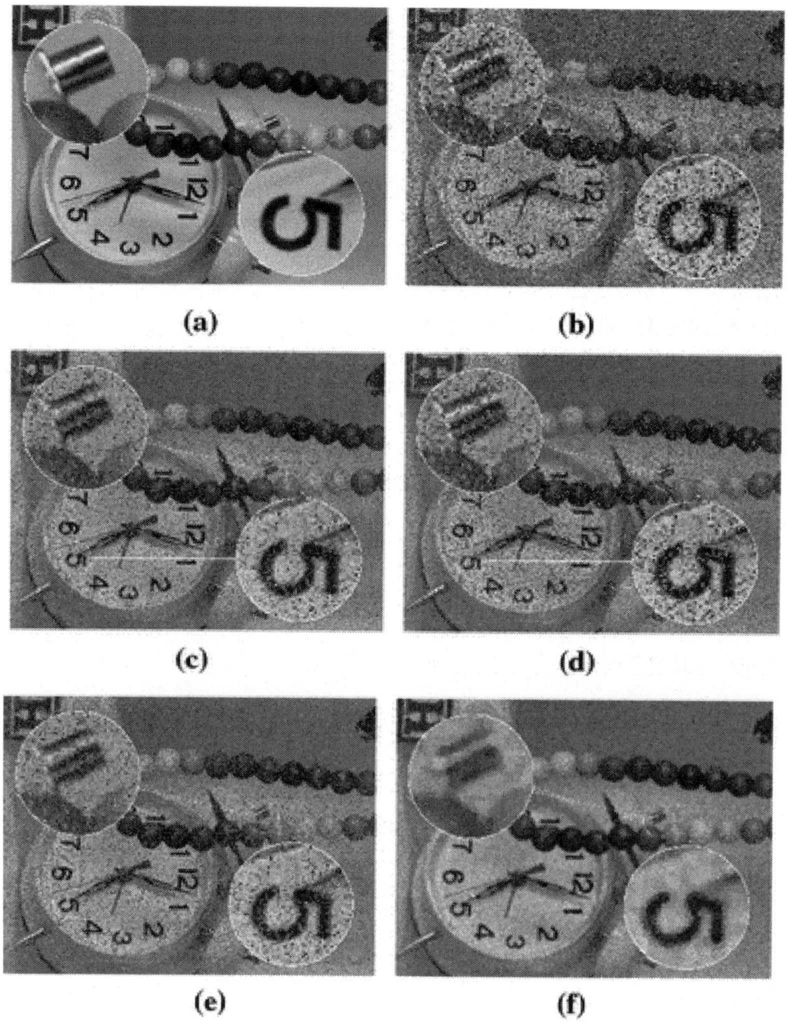

Figure. 1: Illustration of the inability of popular techniques to suppress mixed Gaussian and impulsive noise. **a** Test image. **b** Noisy image. **c** NLM filter [3]. **d** BM3D filter [9]. **e**Bilateral filter [52]. **f** Vector median filter [1]

This effect is depicted in Fig. 1, in which the restoration of a degraded image distorted by mixed noise is presented using filters intended for Gaussian or impulsive noise suppression. As can be observed, the popular, highly effective techniques like *Non-Local Means* (NLM) [3, 4], *Block-Matching and 3D Filtering* (BM3D) [9] or *Bilateral Filter* (BF) [52] are unable to suppress the impulses and the final filtering result is of unacceptable quality (see Fig. 1c–e). The same applies to the methods of impulsive noise removal, like vector median filter (VMF) [1], which are generally unable to reduce the Gaussian noise component as depicted in Fig. 1f.

The widely used VMF efficiently removes the impulses; however, as its output is one of the pixels from the filtering window, the VMF is not able to reduce the Gaussian noise. To alleviate this drawback, the average of the most reliable pixels, determined using the reduced ordering technique, is used to reduce the noise contamination [15, 21, 24, 29, 33, 34, 38, 47, 48].

Very often the mixed noise is being suppressed by applying first a filter suitable for the removal of impulsive and then the one intended for the reduction of Gaussian noise. However, such a scheme is not effective, as the consecutive application of two filters leads to significant image distortions and substantial increase in the computational burden [27].

Another approach to the problem of mixed noise reduction is based on the switching filtering [10, 18, 27], which detects the impulses and removes them using a suitable technique replacing the remaining samples with a filter designed for the Gaussian noise. Such procedure can be effective when the contamination is not severe; otherwise, the detection module fails to detect the impulses which are masked by the Gaussian noise.

An efficient family of filters is utilizing the concept of a peer group introduced in [5, 11, 41, 42] and its fuzzy extensions [27, 28], in which a combination of impulsive noise detection and a replacement scheme based on averaging is performed. Another group of filters relies on the concept of geodesic digital paths [8, 25, 26, 51], which determines the connection cost between pixels belonging to the processing window used as weights in the averaging process.

The authors of [14] described the *Rank of Ordered Absolute Differences* (ROAD) statistic for impulse detection and combined it with the bilateral denoising scheme [52], designing a trilateral filter able to efficiently reduce the mixed noise. This filter, although intended for gray-scale images, can be also applied to color images, replacing the difference in intensities by the distance in a given color space.

In [20], a method for the restoration of heavily damaged images, in which Bayesian classification of the input pixels is combined with the kernel regression framework, was proposed. In [46], similarly to the ROAD statistic, but for color images, a sum of distances to the *k*-nearest pixels is used to calculate the weights for each pixel that can be treated as a measure of pixel distortion and these coefficients are used in the weighted average of the pixels in the processing window.

In this paper, we propose a new technique which is able to cope simultaneously with the two kinds of noise. The paper is structured as follows. In Sect. 2, the new filtering design is described. In the next section, the results of experiments and a comparison with competitive filters are presented. Finally, some conclusions are drawn in Sect. 4.

PROPOSED FILTERING DESIGN

Various techniques of Gaussian noise reduction has been proposed in the recent years. A widely used family of filters is based on the BF [52] which takes into the account the similarity between pixels from a local processing window and their topographic distance. Assuming that the pixels of a local processing block BB of size (2r+1)×(2r+1) are denoted as $x_1,\ldots,x_N$, where x_1 is in the center and *N* is the number of pixels in BB, the output y_1 of BF which replaces the central pixel x_1 is

$$y_1 = \frac{\sum_{k=1}^{N} w_1(\boldsymbol{x}_1, \boldsymbol{x}_k) \cdot w_2(\boldsymbol{x}_1, \boldsymbol{x}_k) \cdot \boldsymbol{x}_k}{\sum_{k=1}^{N} w_1(\boldsymbol{x}_1, \boldsymbol{x}_k) \cdot w_2(\boldsymbol{x}_1, \boldsymbol{x}_k)}, \quad (1)$$

where the weights w_1 and w_2 are usually defined as

$$w_1(\boldsymbol{x}_1, \boldsymbol{x}_k) = \exp\left\{-\frac{\|\boldsymbol{x}_1 - \boldsymbol{x}_k\|}{2\sigma_1^2}\right\}, \quad (2)$$

where ||·|| denotes the Euclidean distance in the RGB color space, and

$$w_2(\boldsymbol{x}_1, \boldsymbol{x}_k) = \exp\left\{-\frac{\rho(\boldsymbol{x}_1, \boldsymbol{x}_k)}{2\sigma_2^2}\right\}, \quad (3)$$

where ρ stands for the topographic distance between the pixels on the image domain. Thus, the BF combines the closeness of pixels in the color space and their topographic proximity. The parameters σ_1 and σ_2 control the influence of the weights on the final filter output.

Since its introduction, various modifications of the BF, aiming mainly at increasing its ability to preserve image details, have been introduced [17, 39, 40, 56, 57]. However, although effective in the case of Gaussian noise, the bilateral filter cannot remove the impulses, which are treated as tiny details. To alleviate this problem, various designs introduce additional weighting factors which diminish the influence of outliers on the final filtering outcome.

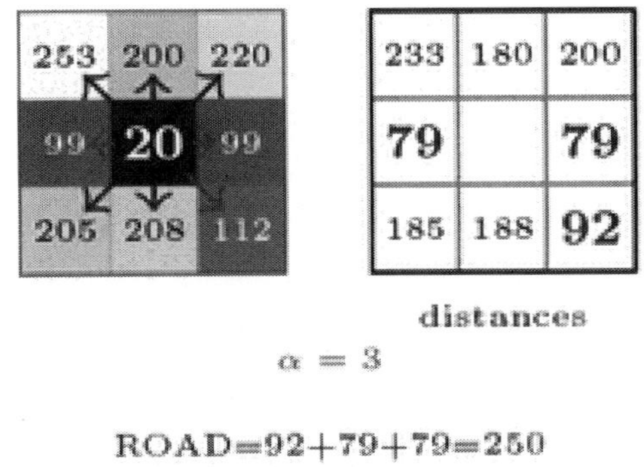

Figure. 2: Calculation of the ROAD measure

In [14], an extension of the bilateral filter based on the ROAD measure was proposed. The so-called trilateral filter assigns to each pixel its measure of impulsiveness, which is considered when building the average over the pixels belonging to the local processing region. The indicator of pixel's distortion is defined as

$$\text{ROAD} = \sum_{k=1}^{\alpha} d_{1(k)}, \quad (4)$$

where $d_{1(k)}$ is the *k*th smallest Euclidean distance in the RGB color space between the central pixel x_1 of a small window W centered at x_1 and α denotes the number of close neighbors. If the central pixel is distorted by impulsive noise, then the corresponding value of ROAD is high. This value will be high even if the central pixel is an impulse and has close, similarly damaged neighbors. For example, in a 3×3 window, setting α=5, the ROAD will fail to detect the impulses and will be small if the filtering window contains 5 similar impulses, which is unlikely even for strong impulsive noise.

Figure 2 shows the calculation of the ROAD coefficient for the gray-scale image, when α=3. As can be observed, generally the value of the αα parameter is of high importance, as in this exemplary situation, for higher αα the central pixel does not possess close neighbors and the ROAD would be substantially increased [12, 23, 24, 43–45, 54, 55].

The proposed filtering design, which will be called *Robust Local Similarity Filter* (RLSF), is based on the ROAD concept. We assign pixels x_j, j=1,...,N from a region B surrounding the central pixel x_1, the ROAD calculated as the sum of the smallest distances between x_j∈Band the pixels contained in a small window W, typically of size 3×3 centered at x_1. The construction of the modified ROAD is illustrated again for a gray-scale image in Fig. 3. The central pixel x_1 of a processing region has the intensity 154 and the ROAD of pixel with intensity 55 is calculated summing the α smallest distances to the pixel belonging to W in the processing block center (red window).

In this way, if a pixel from the processing region is corrupted, its ROAD will be high, as it is unlikely that the small window in the center will also have similar outliers. The important feature of the new design is that the central pixel x_1 may be corrupted and the ROAD assigned to a pixel x_j from B will be not affected, as this pixel will find uncorrupted close pixels in the small window W centered at x_1.

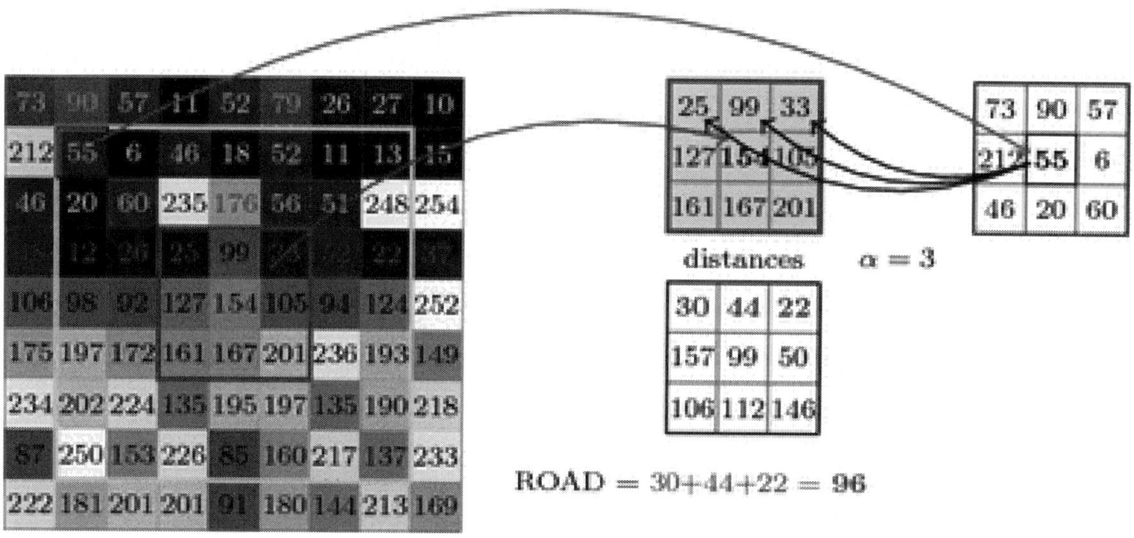

Figure. 3: Calculation of the similarity measure between a pixel with intensity 55 (*upper left corner*) and the 3×3 window centered at the processed pixel x_1 with intensity 154. The processing block B is depicted by *green* and the window W by *red square* (color figure online)

The structure of the proposed design is similar to the BF; however, to decrease the computational load, we neglect the topographic distance between pixels

$$y_1 = \frac{\sum_{j=1}^{N} w_j \cdot x_j}{\sum_{j=1}^{N} w_j}, \tag{5}$$

with

$$w_j = \mathcal{K}\left(\frac{1}{\alpha}\sum_{k=1}^{\alpha} d_{j(k)}\right), \tag{6}$$

where K denotes the kernel function (e.g., Gaussian) and $d_{j(k)}$ is the *k*th smallest Euclidean distance in the RGB color space between x_j and the pixels of W.

EXPERIMENTS

The proposed filter was evaluated on a set of color test images depicted in Fig. 4. The images were first distorted by Gaussian noise with standard deviation in the range 10–50, (with step 10), and then, 10%–50% of the pixels were replaced by random-valued impulsive noise, so that every RGB channel of a corrupted pixel was assigned a value drawn from the uniform distribution from the range [0,255]. To simplify the notation, the noise intensity *p* will denote a Gaussian noise with standard deviation *p* and with p% pixels with randomly corrupted channels.

Figure. 4: Color test images used in experiments

The restoration efficiency has been assessed using the commonly used PSNR measure defined as

$$\text{MSE} = \frac{1}{3N}\sum_{j=1}^{N}\sum_{q=1}^{3}(x_{j,q} - \hat{x}_{j,q})^2, \tag{7}$$

$$\text{PSNR} = 10\log_{10}\left(\frac{255^2}{\text{MSE}}\right), \tag{8}$$

where $x_{j,q}$, q=1,2,3 are the RGB channels of the original image pixels and $\hat{x}_{j,q}$ are the restored components.

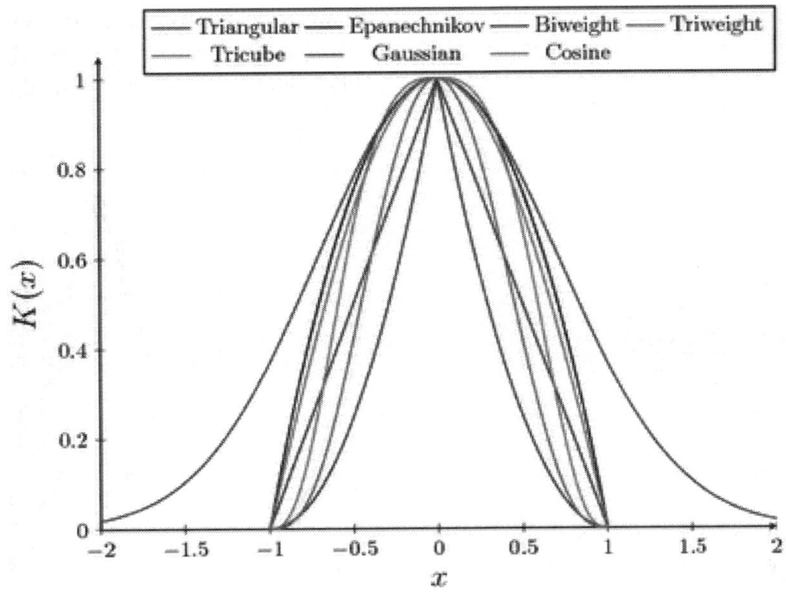

Figure. 5: Plots of the kernel functions

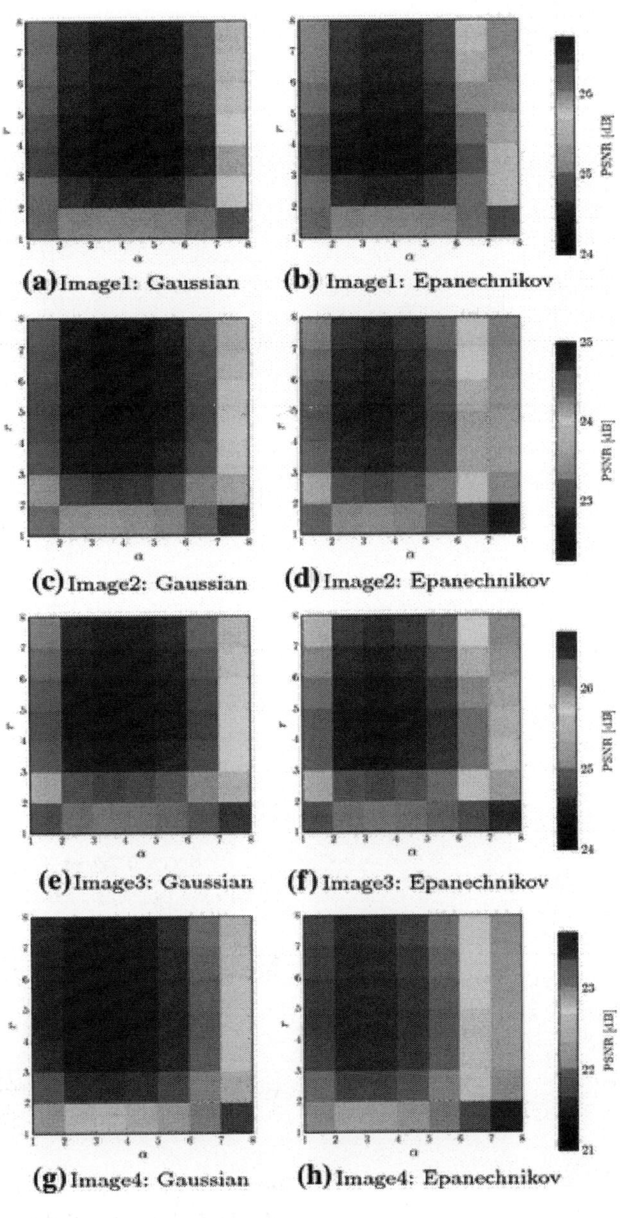

Figure. 6: Dependence of the optimal PSNR obtained using the proposed RLSF on α and *r* parameters using the Gaussian and Epanechnikov kernels. The test images shown in Fig. 4 were contaminated with mixed noise of intensity p=30

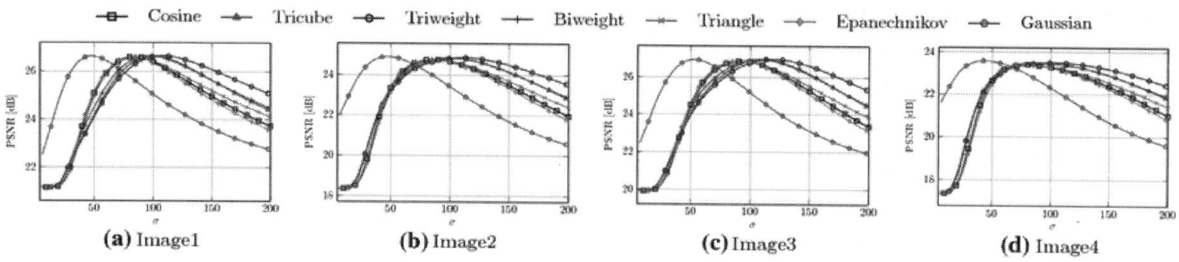

Figure. 7: Dependence of PSNR values obtained using the proposed RLSF with the size of processing block r=4 and setting α=4, on the σ parameter value for different kernel functions K. The color test images depicted in Fig. 4 were contaminated with mixed noise of intensity p=30.

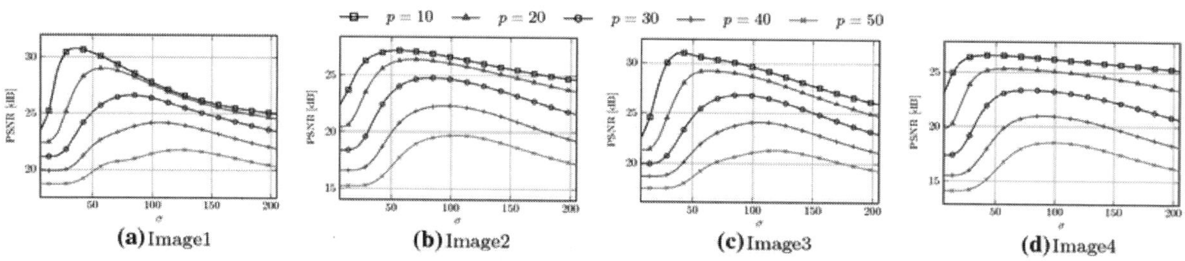

Figure. 8: Dependence of PSNR values obtained using the proposed RLSF with Epanechnikov kernel on σ value, for r=4, α=4 on test images distorted with various noise level.

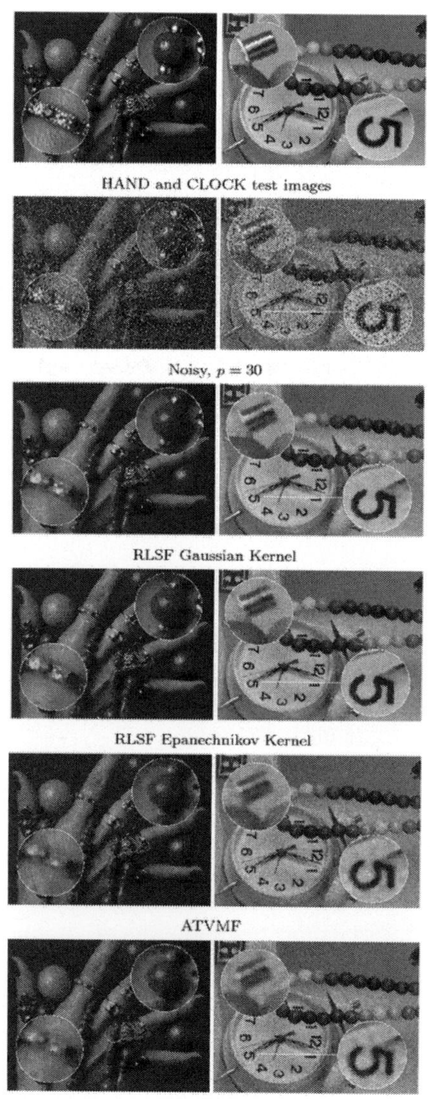

Figure. 9: Comparison of the efficiency of the new RLSF with r=4, α=4 and p=30, using the Gaussian and Epanechnikov kernels with the ATVMF and FVMF

In our experiments, various kernel functions were used to determine the weight in (6). The suitability of the following functions, whose graphs are presented in Fig. 5, was evaluated:

- Triangular: $K(x) = 1 - (x/\sigma), \quad |x/\sigma| \leq 1$,
- Epanechnikov: $K(x) = 1 - (x/\sigma)^2, \quad |x/\sigma| \leq 1$,
- Biweight: $K(x) = (1 - (x/\sigma)^2)^2, \quad |x/\sigma| \leq 1$,
- Triweight: $K(x) = (1 - (x/\sigma)^2)^3, \quad |x/\sigma| \leq 1$,
- Tricube: $K(x) = (1 - |x/\sigma|^3)^3, \quad |x/\sigma| \leq 1$,
- Gaussian: $K(x) = \exp(-(x/\sigma))^2$,
- Cosine: $K(x) = \cos(\pi x/2\sigma), \quad |x/\sigma| \leq 1$.

First, we investigated the influence of the block size r and α parameter in (6) on the filtering efficiency expressed in terms of the PSNR measure applying various kernels. The σ parameter of the kernels was adjusted so that the best PSNR value was achieved. The simulations revealed that independently of the noise contamination level and the applied kernel, for various natural color images, the optimal settings of the block BB size r and the αα parameter of the ROAD measure are in the range [3–5], and the setting r=α=4 was used for the comparison of the proposed design with competitive denoising methods. Figure 6 depicts the dependence of the best possible PSNR metric on r and α parameters for the test images shown in Fig. 4 using the Gaussian and Epanechnikov kernels. As can be observed, the structure of the visualizations is similar.

Figure 7 depicts the influence of the σ parameter on the restoration results for a block B size r=4 and α=4 in WW, when using various kernel functions. As expected, the optimal PSNR values were not dependent on the choice of the kernel shape; however, the Triweight, Tricube or Epanechnikov functions are less dependent on the σ value than the Gaussian kernel. This effect is beneficial, as the optimal value of the kernel smoothing parameter σ is slightly dependent on the image structure and also on the contamination intensity as exhibited in Fig. 8. Therefore, for further experiments, the Epanechnikov kernel was chosen, as it not only is less sensitive to the σvalue but is also computationally less demanding than the widely used Gaussian function. Due to the fact that this kernel drops to 0, it is also more robust to the outliers introduced by the impulsive noise process.

Analyzing the plots of this figure, the setting σ=100 enables to achieve results very close to the optimal ones for considerable amount of noise distortions. For images degraded by less intensive mixed noise, lower values of the smoothing parameter, e.g., σ=50, should be used.

Table 1: Dependence of PSNR in dB for various images denoised using the proposed RLSF (r=4r=4,α=4α=4) with various kernels, as compared with competitive filters

	RLSF - Kernels							Competitive denoising methods										
p	Cosine	Tricube	Triweight	Biweight	Triangle	Epanechnikov	Gaussian	FVMF [32]	FOVMF [34]	FVDF [32]	ANNF [31]	ANNMF [35]	DDF [16]	ATVMF [53]	HDF [7]	FOVDF [34]	EVMF [7]	AHDF [7]
Image1																		
10	30.74	30.73	30.78	30.78	30.70	30.73	30.89	**31.11**	30.54	29.12	28.94	30.48	28.85	30.86	29.74	28.53	28.05	29.91
20	28.99	29.05	**29.06**	**29.06**	28.93	28.99	**29.06**	26.82	26.60	24.24	24.43	25.81	25.12	26.93	25.92	23.37	22.40	26.02
30	26.61	26.64	**26.65**	**26.65**	26.58	26.61	26.63	24.40	24.17	20.79	21.91	22.95	22.66	24.50	23.41	20.21	19.48	23.51
40	24.18	24.22	**24.23**	**24.23**	24.17	24.18	24.21	22.09	21.99	18.32	19.84	20.36	20.80	22.39	21.45	17.67	17.58	21.47
50	21.77	**21.80**	**21.80**	**21.80**	21.77	21.77	21.78	20.30	20.22	16.61	18.33	18.50	19.24	20.48	19.73	15.76	16.21	19.69
Image2																		
10	27.17	27.27	27.32	27.30	27.17	27.15	27.45	**28.24**	27.89	25.61	25.59	27.58	26.81	28.07	27.31	24.92	26.74	27.43
20	26.38	26.48	26.50	26.50	26.36	26.38	**26.54**	25.13	24.64	20.94	21.36	23.52	23.42	25.26	23.88	20.01	21.63	23.92
30	24.76	24.82	24.87	24.85	24.73	24.76	**24.91**	22.39	22.18	17.62	18.50	20.30	21.08	22.60	21.48	16.68	18.75	21.51
40	22.33	22.42	22.45	22.43	22.29	22.32	**22.48**	19.86	19.72	15.32	16.37	17.21	18.88	20.34	19.16	14.40	16.65	19.14
50	19.70	19.78	19.81	19.79	19.67	19.70	**19.82**	17.65	17.56	13.74	14.82	15.15	16.88	18.15	17.10	12.86	15.05	17.03
Image3																		
10	31.04	30.53	30.59	30.67	31.07	31.02	30.82	**31.72**	31.09	29.66	29.50	30.33	29.31	31.25	30.10	28.92	28.18	30.28
20	29.24	29.19	29.27	29.22	29.24	29.23	**29.37**	26.93	26.68	25.41	25.20	25.84	25.22	26.87	25.88	24.75	22.28	25.97
30	26.85	26.94	**26.95**	26.94	26.81	26.85	26.92	24.10	23.85	21.99	22.41	22.73	22.41	24.26	23.04	21.51	19.28	23.09
40	24.10	**24.18**	**24.18**	**24.18**	24.07	24.11	24.12	21.60	21.39	19.85	20.39	20.50	20.34	21.97	20.81	19.40	17.28	20.76
50	21.30	**21.35**	21.34	**21.35**	21.28	21.30	21.30	19.52	19.42	17.90	18.62	18.58	18.51	19.82	18.93	17.48	15.83	18.83
Image4																		
10	26.57	26.57	26.61	26.59	26.57	26.56	26.64	**27.04**	26.82	24.64	24.68	26.64	26.13	26.94	26.10	23.96	25.99	26.41
20	25.36	25.39	25.43	25.42	25.35	25.35	**25.47**	24.57	24.41	20.28	20.93	23.28	23.53	24.56	23.82	19.78	21.63	23.85
30	23.39	23.45	23.49	23.48	23.37	23.39	**23.57**	21.85	21.67	17.18	18.00	19.63	20.85	22.09	21.04	16.75	18.61	21.04
40	20.99	21.05	21.11	21.08	20.95	20.99	**21.19**	19.11	18.98	14.91	15.82	16.54	18.30	19.57	18.40	14.42	16.35	18.37
50	18.55	18.63	18.67	18.64	18.52	18.55	**18.71**	16.80	16.70	13.28	14.19	14.43	16.15	17.28	16.22	12.70	14.56	16.14

Best values are indicated in bold

The proposed filtering design has been compared with a set of competitive filters whose structure has been described in detail in [7]

- Fuzzy vector median filter (FVMF),
- Fuzzy ordered vector median filter (FOVMF),
- Fuzzy vector directional filter (FVDF),
- Adaptive nearest-neighbor filter (ANNF),
- Adaptive nearest-neighbor multichannel filter (ANNMF),
- Directional distance filter (DDF),
- Alpha-trimmed vector median filter (ATVMF),
- Hybrid directional filter (HDF),
- Fuzzy ordered vector directional filter (FOVDF),

IEEE Trans. Circuits Syst. Video Technol. 4(2), 129–142, 208–210 (1994)

55. Xu, G., Tan, J., Zhong, J.: An improved trilateral filter for image denoising using an effective impulse detector. In: Proceedings of the 4th International Congress on Image and Signal Processing (CISP), 2011, vol. 1, pp. 90–94 (2011)

56. Yu, H., Zhao, L., Wang, H.: An efficient procedure for removing random-valued impulse noise in images. IEEE Signal Process. Lett. 15, 922–925 (2008)

57. Zhang, M., Gunturk, B.: Multiresolution bilateral filtering for image denoising. IEEE Trans. Image Process. 17(12), 2324–2333 (2008)

58. Zhang, M., Gunturk, B.: A new image denoising method based on the bilateral filter. In: IEEE International Conference on Acoustics, Speech and Signal Processing, ICASSP 2008, pp. 929–932 (2008)

59. Zheng, J., Valavanis, K., Gauch, J.: Noise removal from color images. J. Intell. Robot. Syst. 7(1), 257–285 (1993)

In our experiments, various kernel functions were used to determine the weight in (6). The suitability of the following functions, whose graphs are presented in Fig. 5, was evaluated:

- Triangular: $K(x) = 1 - (x/\sigma), \quad |x/\sigma| \leq 1,$
- Epanechnikov: $K(x) = 1 - (x/\sigma)^2, \quad |x/\sigma| \leq 1,$
- Biweight: $K(x) = (1 - (x/\sigma)^2)^2, \quad |x/\sigma| \leq 1,$
- Triweight: $K(x) = (1 - (x/\sigma)^2)^3, \quad |x/\sigma| \leq 1,$
- Tricube: $K(x) = (1 - |x/\sigma|^3)^3, \quad |x/\sigma| \leq 1,$
- Gaussian: $K(x) = \exp(-(x/\sigma))^2,$
- Cosine: $K(x) = \cos(\pi x/2\sigma), \quad |x/\sigma| \leq 1.$

First, we investigated the influence of the block size r and α parameter in (6) on the filtering efficiency expressed in terms of the PSNR measure applying various kernels. The σ parameter of the kernels was adjusted so that the best PSNR value was achieved. The simulations revealed that independently of the noise contamination level and the applied kernel, for various natural color images, the optimal settings of the block BB size r and the αα parameter of the ROAD measure are in the range [3–5], and the setting r=α=4 was used for the comparison of the proposed design with competitive denoising methods. Figure 6 depicts the dependence of the best possible PSNR metric on r and α parameters for the test images shown in Fig. 4 using the Gaussian and Epanechnikov kernels. As can be observed, the structure of the visualizations is similar.

Figure 7 depicts the influence of the σ parameter on the restoration results for a block B size r=4 and α=4 in WW, when using various kernel functions. As expected, the optimal PSNR values were not dependent on the choice of the kernel shape; however, the Triweight, Tricube or Epanechnikov functions are less dependent on the σ value than the Gaussian kernel. This effect is beneficial, as the optimal value of the kernel smoothing parameter σ is slightly dependent on the image structure and also on the contamination intensity as exhibited in Fig. 8. Therefore, for further experiments, the Epanechnikov kernel was chosen, as it not only is less sensitive to the σvalue but is also computationally less demanding than the widely used Gaussian function. Due to the fact that this kernel drops to 0, it is also more robust to the outliers introduced by the impulsive noise process.

Analyzing the plots of this figure, the setting σ=100 enables to achieve results very close to the optimal ones for considerable amount of noise distortions. For images degraded by less intensive mixed noise, lower values of the smoothing parameter, e.g., σ=50, should be used.

Table 1: Dependence of PSNR in dB for various images denoised using the proposed RLSF (r=4r=4,α=4α=4) with various kernels, as compared with competitive filters

	RLSF - Kernels							Competitive denoising methods											
p	Cosine	Tricube	Triweight	Biweight	Triangle	Epanechnikov	Gaussian	FVMF [32]	FOVMF [34]	FVDF [32]	ANNF [31]	ANNMF [35]	DDF [16]	ATVMF [53]	HDF [7]	FOVDF [34]	EVMF [7]	AHDF [7]
Image1																			
10	30.74	30.73	30.78	30.78	30.70	30.73	30.89	**31.11**	30.54	29.12	28.94	30.48	28.85	30.86	29.74	28.53	28.05	29.91
20	28.99	29.05	**29.06**	**29.06**	28.93	28.99	**29.06**	26.82	26.60	24.24	24.43	25.81	25.12	26.93	25.92	23.37	22.40	26.02
30	26.61	26.64	**26.65**	**26.65**	26.58	26.61	26.63	24.40	24.17	20.79	21.91	22.95	22.66	24.50	23.41	20.21	19.48	23.51
40	24.18	24.22	**24.23**	**24.23**	24.17	24.18	24.21	22.09	21.99	18.32	19.84	20.36	20.80	22.39	21.45	17.67	17.58	21.47
50	21.77	**21.80**	**21.80**	**21.80**	21.77	21.77	21.78	20.30	20.22	16.61	18.33	18.50	19.24	20.48	19.73	15.76	16.21	19.69
Image2																			
10	27.17	27.27	27.32	27.30	27.17	27.15	27.45	**28.24**	27.89	25.61	25.59	27.58	26.81	28.07	27.31	24.92	26.74	27.43
20	26.38	26.48	26.50	26.50	26.36	26.38	**26.54**	25.13	24.64	20.94	21.36	23.52	23.42	25.26	23.88	20.01	21.63	23.92
30	24.76	24.82	24.87	24.85	24.73	24.76	**24.91**	22.39	22.18	17.62	18.50	20.30	21.08	22.60	21.48	16.68	18.75	21.51
40	22.33	22.42	22.45	22.43	22.29	22.32	**22.48**	19.86	19.72	15.32	16.37	17.21	18.88	20.34	19.16	14.40	16.65	19.14
50	19.70	19.78	19.81	19.79	19.67	19.70	**19.82**	17.65	17.56	13.74	14.82	15.15	16.88	18.15	17.10	12.86	15.05	17.03
Image3																			
10	31.04	30.53	30.59	30.67	31.07	31.02	30.82	**31.72**	31.09	29.66	29.50	30.33	29.31	31.25	30.10	28.92	28.18	30.28
20	29.24	29.19	29.27	29.22	29.24	29.23	**29.37**	26.93	26.68	25.41	25.20	25.84	25.22	26.87	25.88	24.75	22.28	25.97
30	26.85	26.94	**26.95**	26.94	26.81	26.85	26.92	24.10	23.85	21.99	22.41	22.73	22.41	24.26	23.04	21.51	19.28	23.09
40	24.10	**24.18**	**24.18**	**24.18**	24.07	24.11	24.12	21.60	21.39	19.85	20.39	20.50	20.34	21.97	20.81	19.40	17.28	20.76
50	21.30	**21.35**	21.34	**21.35**	21.28	21.30	21.30	19.52	19.42	17.90	18.62	18.58	18.51	19.82	18.93	17.48	15.83	18.83
Image4																			
10	26.57	26.57	26.61	26.59	26.57	26.56	26.64	**27.04**	26.82	24.64	24.68	26.64	26.13	26.94	26.10	23.96	25.99	26.41
20	25.36	25.39	25.43	25.42	25.35	25.35	**25.47**	24.57	24.41	20.28	20.93	23.28	23.53	24.56	23.82	19.78	21.63	23.85
30	23.39	23.45	23.49	23.48	23.37	23.39	**23.57**	21.85	21.67	17.18	18.00	19.63	20.85	22.09	21.04	16.75	18.61	21.04
40	20.99	21.05	21.11	21.08	20.95	20.99	**21.19**	19.11	18.98	14.91	15.82	16.54	18.30	19.57	18.40	14.42	16.35	18.37
50	18.55	18.63	18.67	18.64	18.52	18.55	**18.71**	16.80	16.70	13.28	14.19	14.43	16.15	17.28	16.22	12.70	14.56	16.14

Best values are indicated in bold

The proposed filtering design has been compared with a set of competitive filters whose structure has been described in detail in [7]

- Fuzzy vector median filter (FVMF),
- Fuzzy ordered vector median filter (FOVMF),
- Fuzzy vector directional filter (FVDF),
- Adaptive nearest-neighbor filter (ANNF),
- Adaptive nearest-neighbor multichannel filter (ANNMF),
- Directional distance filter (DDF),
- Alpha-trimmed vector median filter (ATVMF),
- Hybrid directional filter (HDF),
- Fuzzy ordered vector directional filter (FOVDF),

- Entropy vector median filter (EVMF),
- Adaptive hybrid directional filter (AHDF).

The comparison with the state-of-the-art denoising methods presented in Table 1 and also in Fig. 9 shows that the proposed filter clearly outperforms the competitive techniques for medium and high noise contamination levels. The Epanechnikov kernel yields slightly lower PSNR values; however, for applications in which the computational complexity plays a crucial role, the use of this kernel is recommended.

The RLSF better preserves image edges and produces visually pleasing results, without noticeable unfiltered noise. It is also worth noticing that the proposed design is significantly faster than the NLM and BM3D methods and also much faster than the above-listed filters for the default 9×99×9 processing block.

CONCLUSIONS

In this paper, a new approach to the suppression of mixed Gaussian and impulsive noise has been presented. The main novelty of this work lies in the introduction of a novel efficient measure of the similarity of a pixel from the processing region and a small filtering window centered at the pixel which is being restored. The experimental results revealed that the new filter is competitive with existing denoising solutions. Future work will be focused on the introduction of additional measures which evaluate the impulsiveness of pixels taken for the weighted averaging process.

ACKNOWLEDGMENTS

This work was supported by the Polish National Science Center (NCN) under the Grant: DEC-2012/05/B/ST6/ 03428 and POIG. 02.03.01-24-099/13 Grant: GeCONiI—Upper Silesian Center for Computational Science and Engineering.

REFERENCES

1. Astola, J., Haavisto, P., Neuvo, Y.: Vector median filters. Proc. IEEE 78(4), 678–689 (1990)
2. Boncelet, C.: Image noise models. In: Bovik, A. (ed.) Handbook of Image and Video Processing, Communications, Networking and Multimedia, pp. 397–410. Academic Press, New York (2005)
3. Buades, A., Coll, B., Morel, J.M.: A non-local algorithm for image denoising. In: IEEE Computer Society Conference on Computer Vision and Pattern Recognition, CVPR 2005, vol. 2, pp. 60–65 (2005)
4. Buades, A., Coll, B., Morel, J.M.: A review of image denoising algorithms, with a new one. Multiscale Model. Simul. 4(2), 490–530 (2005)
5. Kenney, C., D, Y., Manjunath, B.S.: Peer group image enhancement. IEEE Trans. Image Process. 10(2), 326–334 (2001)
6. Camarena, J., Gregori, V., Morillas, S., Sapena, A.: A simple fuzzy method to remove mixed Gaussian-impulsive noise from color images. IEEE Trans. Fuzzy Syst. 21(5), 971–978 (2013)
7. Celebi, M.E., Kingravi, H.A., Aslandogan, Y.A.: Nonlinear vector filtering for impulsive noise removal from color images. J. Electron. Imaging 16(3), 033,008–033,008–21 (2007)
8. Criminisi, A., Sharp, T., Rother, C., Pérez, P.: Geodesic image and video editing. ACM Trans. Graph. 29(5), 134:1–134:15 (2010)
9. Dabov, K., Foi, A., Katkovnik, V., Egiazarian, K.: Image denoising with block-matching and 3D filtering. In: Electronic Imaging 2006, vol. 6064 (2006)
10. Deergha Rao, K., Plotkin, E., Swamy, M.: An hybrid filter for restoration of color images in the mixed noise environment. In: IEEE International Conference on Acoustics, Speech, and Signal Processing (ICASSP), 2002, vol. 4, pp. IV–3680–IV–3683 (2002)
11. Deng, Y., Kenney, C., Moore, M., Manjunath, B.: Peer group filtering and perceptual color image quantization, In: Proceedings of IEEE International Symposium on Circuits and Systems, vol. 4, pp. 21–24. Springer, Berlin (1999)
12. Dong, Y., Chan, R., Xu, S.: A detection statistic for random-valued impulse noise. IEEE Trans. Image Process. 16(4), 1112–1120 (2007)
13. Faraji, H., MacLean, W.: CCD noise removal in digital images. IEEE Trans. Image Process. 15(9), 2676–2685 (2006)
14. Garnett, R., Huegerich, T., Chui, C., He, W.: A universal noise removal algorithm with an impulse detector. IEEE Trans. Image Process. 14(11), 1747–1754 (2005)
15. Hwang, H., Haddad, R.: Adaptive median filters: new algorithms and results. IEEE Trans. Image Process. 4(4), 499–502 (1995)
16. Karakos, D., Trahanias, P.: Combining vector median and vector directional filters: the directional-distance filters. In: International Conference on Image Processing, 1995, vol. 1, pp. 171–174 (1995)
17. Kishan, H., Seelamantula, C.: Sure-fast bilateral filters. In: IEEE International Conference on Acoustics, Speech and Signal Processing (ICASSP), 2012, pp. 1129–1132 (2012)
18. Lin, L.C.H., Tsai, S., Ching-Te, C.: Switching bilateral filter with a texture/noise detector for universal noise removal. In: 2010 IEEE International Conference on Acoustics Speech and Signal Processing (ICASSP), pp. 1434–1437 (2010)
19. Liu, C., Szeliski, R., Kang, S., Zitnick, C., Freeman, W.: Automatic estimation and removal of noise from a single image. IEEE Trans. Pattern Anal. Mach. Intell. 30(2), 299–314 (2008)
20. López-Rubio, E.: Restoration of images corrupted by Gaussian and uniform impulsive noise. Pattern Recogn. 43(5), 1835–1846 (2010)
21. Lukac, R.: Adaptive vector median filtering. Pattern Recogn. Lett. 24(12), 1889–1899 (2003)
22. Lukac, R., Plataniotis, K.: A taxonomy of color image filtering and enhancement solutions. In: Hawkes, P. (ed.) Advances in Imaging and Electron Physics, Advances in Imaging and Electron Physics, vol. 140, pp. 187–264. Elsevier, Amsterdam (2006)

23. Lukac, R., Smolka, B., Martin, K., Plataniotis, K., Venetsanopoulos, A.: Vector filtering for color imaging. IEEE Signal Process. Mag. 22(1), 74–86 (2005)
24. Lukac, R., Smolka, B., Plataniotis, K.: Sharpening vector median filters. Signal Process. 87, 2085–2099 (2007)
25. Malik, K., Machala, B., Smolka, B.: Novel approach to noise reduction in ultrasound images based on geodesic paths. In: Computer Vision and Graphics, Lecture Notes in Computer Science, vol. 8671, pp. 409–417. Springer, Berlin (2014)
26. Malik, K., Smolka, B.: Modified bilateral filter for the restoration of noisy color images. In: Advanced Concepts for Intelligent Vision Systems, Lecture Notes in Computer Science, vol. 7517, pp. 72–83. Springer, Berlin (2012)
27. Morillas, S., Gregori, V., Hervas, A.: Fuzzy peer groups for reducing mixed gaussian-impulse noise from color images. IEEE Trans. Image Process. 18(7), 1452–1466 (2009)
28. Morillas, S., Gregori, V., Peris-Fajarnés, G.: Isolating impulsive noise pixels in color images by peer group techniques. Comput. Vis. Image Underst. 110(1), 102–116 (2008)
29. Morillas, S., Gregori, V., Sapena, A.: Adaptive marginal median filter for colour images. Sensors 11(3), 3205 (2011)
30. Pitas, I., Venetsanopoulos, A.: Nonlinear Digital Filters: Principles and Applications. Kluwer Academic Publishers, Boston (1990)
31. Plataniotis, K., Androutsos, D., Sri, V., Venetsanopoulos, A.: Nearest-neighbour multichannel filter. Electron. Lett. 31(22), 1910–1911 (1995)
32. Plataniotis, K., Androutsos, D., Venetsanopoulos, A.: Fuzzy adaptive filters for multichannel image processing. Signal Process. 55(1), 93–106 (1996)
33. Plataniotis, K., Androutsos, D., Venetsanopoulos, A.: Color image processing using adaptive vector directional filters. IEEE Trans. Circuits Syst. II Analog Digit. Signal Process. 45(10), 1414–1419 (1998)
34. Plataniotis, K., Androutsos, D., Venetsanopoulos, A.: Adaptive fuzzy systems for multichannel signal processing. Proc. IEEE 87(9), 1601–1622 (1999)
35. Plataniotis, K., Sri, V., Androutsos, D., Venetsanopoulos, A.: An adaptive nearest neighbor multichannel filter. IEEE Trans. Circuits Syst. Video Technol. 6(6), 699–703 (1996)
36. Plataniotis, K., Venetsanopoulos, A.: Color Image Processing and Applications. Springer, Berlin (2000)
37. Ponomaryov, V., Gallegos-Funes, F., Rosales-Silva, A.: Real-time color image processing using order statistics filters. J. Math. Imaging Vis. 23(3), 315–319 (2005)
38. MathSciNetCrossRef
39. Ponomaryov, V., Montengro, H., Rosales, A., Duchen, G.: Fuzzy 3D filter for color video sequences contaminated by impulsive noise. J. Real Time Image Process. 10(2), 313–328 (2012)
40. Radlak, K., Smolka, B.: Trimmed non-local means technique for mixed noise removal in color images. In: IEEE International Symposium on Multimedia (ISM), pp. 405–406 (2013)
41. Sable, A., Jondhale, K.: Modified double bilateral filter for sharpness enhancement and noise removal. In: International Conference on Advances in Computer Engineering (ACE), 2010, pp. 295–297 (2010)
42. Smolka, B.: Peer group vector median filter. In: Pattern Recognition, Lecture Notes in Computer Science, vol. 4713, pp. 314–323. Springer, Berlin (2007)
43. Smolka, B.: Peer group switching filter for impulse noise reduction in color images. Pattern Recogn. Lett. 31(6), 484–495
44. Smolka, B.: Soft switching technique for impulsive noise removal in color images. In: Proceedings of the Fifth International Conference on Computational Intelligence, Communication Systems and Networks (CICSyN), pp. 222–227 (2013)
45. Smolka, B.: Robustified vector median filter. In: Proceedings of the 9th International Conference on Computer Science Education (ICCSE), 2014, pp. 362–367 (2014)
46. Smolka, B., Andrzejczak, A., Nabialkowski, P., Nelip, A.: Thresholded median filter for the impulsive noise removal in digital images. In: Proceedings of the 5th International Conference on Information, Intelligence, Systems and Applications, IISA 2014, pp. 355–360 (2014)
47. Smolka, B., Malik, K.: Fast technique for mixed Gaussian and impulsive noise suppression in color images. In: AFRICON, 2013, pp. 1–5 (2013)
48. Smolka, B., Malik, K.: Reduced ordering technique of impulsive noise removal in color images. Lect. Notes Comput. Sci. 7786, 296–310 (2013)
49. Smolka, B., Malik, K., Malik, D.: Adaptive rank weighted switching filter for impulsive noise removal in color images. J. Real Time Image Process. 10(2), 289–311 (2015)
50. Smolka, B., Plataniotis, K., Venetsanopoulos, A.: Nonlinear signal and image processing: theory, methods, and applications. In: Nonlinear Techniques for Color Image Processing, pp. 445–505. CRC Press, Boca Raton (2004)
51. Smolka, B., Venetsanopoulos, A.: Color Image Processing: Methods and Applications, chap. In: Noise Reduction and Edge Detection in Color Images, pp. 75–100. CRC Press, Boca Raton (2006)
52. Szczepanski, M., Smolka, B., Plataniotis, K., Venetsanopoulos, A.: On the geodesic paths approach to color image filtering. Signal Process. 83(6), 1309–1342 (2003)
53. Tomasi, C., Manduchi, R.: Bilateral filtering for gray and color images. Sixth Int. Conf. Comput. Vis. 1998, 839–846 (1998)
54. Viero, T., Oistamo, K., Neuvo, Y.: Three-dimensional median-related filters for color image sequence filtering.

IEEE Trans. Circuits Syst. Video Technol. 4(2), 129–142, 208–210 (1994)

55. Xu, G., Tan, J., Zhong, J.: An improved trilateral filter for image denoising using an effective impulse detector. In: Proceedings of the 4th International Congress on Image and Signal Processing (CISP), 2011, vol. 1, pp. 90–94 (2011)
56. Yu, H., Zhao, L., Wang, H.: An efficient procedure for removing random-valued impulse noise in images. IEEE Signal Process. Lett. 15, 922–925 (2008)
57. Zhang, M., Gunturk, B.: Multiresolution bilateral filtering for image denoising. IEEE Trans. Image Process. 17(12), 2324–2333 (2008)
58. Zhang, M., Gunturk, B.: A new image denoising method based on the bilateral filter. In: IEEE International Conference on Acoustics, Speech and Signal Processing, ICASSP 2008, pp. 929–932 (2008)
59. Zheng, J., Valavanis, K., Gauch, J.: Noise removal from color images. J. Intell. Robot. Syst. 7(1), 257–285 (1993)

CHAPTER 16

A NON-STATIONARY DENSITY MODEL TO SEPARATE OVERLAPPED TEXTS IN DEGRADED DOCUMENTS

A. Tonazzini[1] P. Savino[1,2] E. Salerno[1,3]

[1]Istituto di Scienza e Tecnologie dell'Informazione-CNR, Via G. Moruzzi, 1, 56124 Pisa, Italy

[2]e-mail: pasquale.savino@isti.cnr.it

[3]e-mail: emanuele.salerno@isti.cnr.it

ABSTRACT

We address the problem of the removal of a text superimposed to a more important one, in a document image, considering the two instances of canceling back-to-front interferences from recto and verso images of archival documents and of recovering the erased text in palimpsests from multispectral images. Both problems are approached through a model where the ideal images of the two texts are considered as individual source patterns, mixed through some parametric operator. To cope with occlusions, ink saturation, and space variability of the mixing operator, a data model for this problem should be nonlinear and space variant. Here, we show that if a pointwise non-stationarity is allowed, a linear model can compensate for the lack of a suitable nonlinearity and for other modeling errors.

INTRODUCTION

One of the most common degradations affecting archival documents written or printed on both sides of the page is the presence of back-to-front interferences (or see-through, show-through/bleed-through). These are undesired patterns caused by either transparency or seeping of the ink of the text printed in the reverse side of the page. Such distortion can significantly degrade the readability of the document.

Several approaches for see-through reduction have been already investigated, the most effective being those that exploit information from scans or images of both sides (*recto* and *verso*). The earliest solutions proposed were based on thresholding [1], wavelet techniques for enhancing the foreground strokes and smearing the interferences [2], or segmentation and classification [3]. Other more recent methods use energy minimization. In [4], the Chan-Vese active contour model is modified to incorporate information from both sides of the document, and a subsequent function minimization technique is then applied to correct broken strokes where the two texts overlap. In [5], to classify the pixels in the two sides as foreground, bleed-through or background, a regularized energy function is defined that uses as data term likelihoods derived from small sets of user-labeled pixels, and as smoothness term a dual-layer Markov Random Field (MRF). A 4-class classification approach is proposed in [6], by segmenting the recto–verso joint histogram with the aid of a smoothness term built from a learning set based on the availability of ground truths. Possible pixel misclassifications are then iteratively corrected by analyzing the connected components in the segmented image, and the regions classified

Citation: Anna Tonazzini Pasquale Savino · Emanuele Salerno, A non-stationary density model to separate overlapped texts in degraded documents, DOI 10.1007/s11760-014-0735-3.

as bleed-through are inpainted with suitable patterns extracted from the background texture.

Model-based techniques have also raised much interest. The recto and verso appearances of the degraded document are modeled as two parametric superimpositions of the uncorrupted front- and rear-side images, to be recovered by blind or semi-blind source separation techniques. When a linear mixing model is assumed, fast algorithms are available, such as those based on Independent Component Analysis (ICA) and data decorrelation [7, 8]. Within this approach, compensations for the apparent nonlinearity and/or non-stationarity of the physical phenomenon have used regularization techniques using MRF [9, 10] or penalized nonnegative matrix factorization (NMF) [11]. In [12], a nonlinear convolutional mixing model is derived by approximating the physical model of the show-through in modern scanners, and adaptive linear filters are used to estimate the unknown model parameters. Based on the same model with known nonlinearity, in [13], the problem is regularized by a total variation stabilizer. In [14] and [15], we experimented this model for bleed-through as well, employing a constrained maximum likelihood technique for blindly estimating both the mixing parameters and the source images, or estimating off-line the model parameters. A quadratic mixing model accounting for a blur kernel on the interfering patterns is assumed for show-through in [16] and solved through maximum likelihood. Still for show-through, a nonlinear but invertible model is proposed in [17], whose parameters are learned by nonlinear ICA. Finally, in [18], variational approaches, based on nonlinear diffusion and wavelet transforms, have been proposed to model and remove bleed-through.

Another issue in ancient document analysis is revealing the whole contents of a document, which can aid scholars in dating or authenticating. Often, interesting document features are hidden or barely detectable in the original color document. Although these features can sometimes be highlighted by multispectral acquisitions in the non-visible range (e.g., faint traces of the erased text in palimpsested manuscripts are often enhanced by ultraviolet fluorescence), they are still overlapped to other paper contents. It is thus important to look for digital image processing strategies to make the pattern of interest more readable and possibly free from interferences. Thus, as for the case of see-through in recto–verso pairs, also multispectral images of palimpsested manuscripts can be viewed as mixtures of two individual patterns, to be separately extracted. According to this point of view, a unified approach to the two seemingly different problems of see-through removal and underwriting extraction can be taken, where the choice of a suitable mixing model is crucial.

Nevertheless, we believe that a comprehensive and feasible model for text overlapping in ancient documents is still far from being available. Whereas nonlinearity can be useful to describe text superposition in the occlusion areas, non-stationarity is important as well. Indeed, whereas in the scans of modern documents it is expected that the percentage of the rear-side text shining through the front-side (and vice versa) is almost constant throughout the page, the seeping of ink in bleed-through can be highly variable in space, due, for example, to humidity spots. Similarly, in palimpsests, usually very old manuscripts, the aging process is likely to have produced irregular erosion and/or diffusion of the materials. We argue that the use of accurately estimated pixel-dependent parameters, even within an additive or whatever (reasonable) model, could compensate not only for non-stationarity, but also for the lack or the imprecise knowledge of the nonlinearity. By using the same nonlinearity as in [15], we tested this possibility in [19] and in [20] within a regularization framework.

Here, we propose a non-stationary linear model combining the ideal optical densities of the two individual texts, to describe two of the available observations. In the see-through case, these observations will be the recto and verso images, both acquired in the same wavelength range, and the model includes blur kernels to account for the smearing of the interfering patterns, due to light transmission or ink diffusion through the medium. In the case of palimpsests, these observations will be two images of the manuscript, acquired at two different channels, and the model is assumed to be instantaneous. The challenge is to estimate the pixel-dependent coefficients of the linear mixture. We propose to estimate them from the data alone, based on simple and intuitive criteria. A few characteristics that are peculiar of each problem will entail the adoption of some minor modifications in the models and, consequently, in the related way to estimate those coefficients. The algorithms derived are very fast and only require a little user intervention to estimate, still from the data, mean background values and the blur kernels. However, as explained later on, the algorithms could be easily made fully automatic and even faster. Another interesting feature of our method is that it meets the requirement, fundamental when dealing with the restoration of ancient documents, to remove interferences while preserving the original appearance (i.e., color and texture) of the characters and the background. For the see-through case, we show experimental results that are competitive with those of more complex recent methods.

The paper is organized as follows. In Sect. 2, we describe the data model we consider when the application is see-through removal, discuss in detail our strategy to estimate the model parameters, and derive the scheme employed to estimate the two ideal texts. The adaptation of the data model and the restoration algorithm to the instance of underwriting recovery from palimpsests is summarized in Sect. 3. Section 4 analyzes some experimental results for both applications. Finally, Sect. 5 concludes the paper, by discussing some ideas that could help to overcome the present limitations of the method.

SEE-THROUGH INTERFERENCE REMOVAL: DATA MODEL AND ESTIMATION STRATEGY

Assuming preregistered recto and verso images, the recto–verso data model we consider is the following non-stationary, linear convolutional model:

$$\begin{aligned} D_r^{obs}(t) &= D_r(t) + q_v(t)\,[h_v(t) \otimes D_v(t)] \\ D_v^{obs}(t) &= D_v(t) + q_r(t)\,[h_r(t) \otimes D_r(t)] \\ t &= 1, 2, \ldots, \mathrm{T} \end{aligned} \tag{1}$$

with

$$D(t) = -log\left(\frac{s(t)}{R}\right) \tag{2}$$

where $D_r^{obs}(t)$ and $D_v^{obs}(t)$ are the observed optical densities, and $D_r(t)$ and $D_v(t)$ are the ideal optical densities, of the front and back side, respectively, at pixel tt. Each density is related to the corresponding (ideal or observed, recto or verso) reflectance ss through Eq. (2), involving a suitable constant R that represents, in the two instances R_r and R_v, the mean reflectance values of the background in the recto and verso side, respectively. The model also includes two unit volume point spread functions (PSF), hrhr and hvhv, describing the smearing of ink that penetrates or shines through the paper, thus allowing a pattern in a side to match the corresponding one in the opposite side. These PSFs are stationary throughout the image, but characterized by different gains $q_r(t)$ and $q_v(t)$, which, while representing our way to account for non-stationarity, have also the physical meaning of interference levels from the front to the back and from the back to the front, respectively, at each pixel.

In order to invert system (1) for D_r and D_v, the model parameters R_r and R_v, $q_r(t)$ and $q_v(t)$, $\forall t$, and h_r and h_v, and hrhr and hvhv, need to be estimated. We do this off-line, by exploiting a minimum intervention by the user. R_r and R_v are computed by averaging the pixel intensity values within two selected pure background areas. Note that fully automatic procedures to estimate mean background values could also be easily devised for document images, usually characterized by dark texts on light backgrounds. As per the PSFs, normally, if the two sides are perfectly aligned, it is reasonable to assume that both h_r and h_v have the form of a centered, gaussian-like function. To estimate h_v, we isolate a small area in the recto side, where pure see-through is present $(D_r = 0)$ and the interference levels can be assumed almost constant. Using also the corresponding area of pure foreground text in the opposite verso side, h_v can be estimated from the two areas by constrained least squares [21], as detailed in [15], and hrhr can be estimated in specular way. In the presence of a possible small residual displacement between the two sides, due to registration errors, the maximum of h_v will appear shifted of a corresponding value, which can be used to correct the misalignment, by just translating one of the two sides. We have experimentally verified that, in the majority of the cases, it is $h_r = h_v$, so, so that the above procedure can be executed for a side only. Furthermore, for documents belonging to the same class, e.g., same ink and paper used, even a fixed Gaussian PSF can be effectively used, thus avoiding the need for user intervention.

As per the estimation of the space-variant maps $q_r(t)$ and $q_v(t)$ at each pixel tt, we adopted the following formulas:

$$q_r(t) = \frac{D_v^{obs}(t)}{h_r(t) \otimes D_r^{obs}(t) + \epsilon}$$

$$q_v(t) = \frac{D_r^{obs}(t)}{h_v(t) \otimes D_v^{obs}(t) + \epsilon} \tag{3}$$

where ϵ is a small positive number to avoid indeterminacies or infinity.

It is apparent that Eq. (3) make both sense when the ideal density is zero, that is in the pixels corresponding to the background in both sides. For the pixels of pure see-through in one side and foreground text in the opposite side only one of the two equations makes sense. Finally, for pixels corresponding to occlusion areas none of the two equations is valid. More specifically, given a pixel t, four situations can occur: 1) t is a background pixel in both the recto and verso side, 2) t is a pure show-through pixel in the recto side and foreground text in the verso side, 3) t is a foreground text pixel in the recto side and pure show-through in the verso side, and 4) t is an occlusion pixel. Unfortunately, it is in general impossible to determine a priori which of the four situations a pixel corresponds to. Therefore, we first compute q_r and q_v at each pixel ttaccording to Eq. (3), and then, based on simple heuristic criteria, correct the possible wrongly computed interference levels, thus also discriminating the various situations.

Ideally, in the background pixels (case 1), we would like Eq. (3) to give null interference levels, so that the reconstructed images retain there the same values of the observed images, thus preserving at best the original appearance. This should occur automatically, since in those pixels both densities should be zero. Nevertheless, due to small fluctuations around the mean background values, and to the fact that the density is always close to 0 in those areas, even high values of $q_r(t)$ and $q_v(t)$ can be obtained. When inverting the system of Eq. (1), this might lead to negative densities, which however simply means reconstructed background values greater than RR. Alternatively, the negative values of the estimated density can be set to zero, or to the density of the lightest pixel in the data.

For the pixels of foreground text in one side and see-through in the opposite side (cases 2 and 3), we would like to obtain $q_v > 0, q_r = 0$ in case 2 and $q_v = 0, q_r > 0$ in case 3. In other words,q_r and qvqv should be mutually exclusive there. Nevertheless, we will obtain $q_r > q_v > 0$ in case 2, since the first of Eq. (3) does not hold true, and, symmetrically, $q_v > q_r > 0$ in case 3. However, correcting this error is trivial: it suffices, at each t, to maintain the smallest between the two computed interference levels, and set the other to zero, thus automatically discriminating cases 2 and 3.

The last possible situation occurs in the occlusion pixels (case 4), i.e., when the densities in the two sides are both high since the two texts overlap. According to the data model (1), a correct estimation of the interference levels should account for the ideal density, which is however unknown. Assuming instead the ideal density to be zero, as done in (3), would overestimate the interference level thus producing "holes" in correspondence of the occlusion areas. However, provided that the two inks in the two sides reflect similarly under the same wavelength, we can expect that the densities of the occlusion areas in the two sides are high and close to each other. If the degradation is not the strongest possible, i.e., interferences as dark as foreground text, only the pixels in the occluding areas have this property, so that they can be located. In the most obvious way, this can be accomplished with the aid of manually selected thresholds. However, in many cases, we have found that a simple procedure allows us to automatically discriminate the occlusion pixels: if the gray levels of the two foreground texts are similar, we compute the absolute difference between the reflectance maps of the recto and verso observations, and then binarize (e.g., through the Otsu algorithm [22]) the difference map. This will return a binary map where the zero pixels correspond either to the occlusion pixels or to the background pixels, where both interference levels will be set to zero.

Once the model parameters are known, D_r and D_v are computed through the following single step:

$$D_r(t) = D_r^{obs}(t) - q_v(t)\left[h_v(t) \otimes D_v^{obs}(t)\right]$$
$$D_v(t) = D_v^{obs}(t) - q_r(t)\left[h_r(t) \otimes D_r(t)\right] \quad (4)$$

RECOVERY OF UNDERWRITING IN PALIMPSESTS: DATA MODEL AND ESTIMATION STRATEGY

In the case of palimpsests, we assume the availability of multiple observation channels. Let us consider two such channels, one centered at wavelength λ_1 and the other at wavelength λ_2. Assuming preregistration of the two images, and neglecting the support (paper or parchment), i.e., considering that its density is zero at the two wavelengths, we can write:

$$D_{\lambda 1}^{obs}(t) = D_{\lambda_1}^{u}(t) + D_{\lambda_1}^{o}(t)$$
$$D_{\lambda 2}^{obs}(t) = D_{\lambda_2}^{o}(t) + D_{\lambda_2}^{u}(t) \quad (5)$$

where $D_{\lambda_1}^{obs}(t)$ and $D_{\lambda_2}^{obs}(t)$ are the observed maps of the optical density at wavelengths λ_1 and λ_2, respectively, and at pixel t, and D^o and D^u are the ideal optical densities of the overwriting and underwriting texts, observed at wavelength λ_1 in the first equation and at wavelength λ_2 in the second equation. Let us consider two positive, space-variant coefficients, $q^o(t)$ and $q^u(t)$,, such that the following relationships are satisfied:

$$D_{\lambda_1}^{o}(t) = q^o(t) D_{\lambda_2}^{o}(t)$$
$$D_{\lambda_2}^{u}(t) = q^u(t) D_{\lambda_1}^{u}(t) \quad (6)$$

Hence, Eq. (5) can be rewritten as:

$$D_{\lambda 1}^{obs}(t) = D_{\lambda_1}^{u}(t) + q^o(t) D_{\lambda_2}^{o}(t)$$
$$D_{\lambda 2}^{obs}(t) = D_{\lambda_2}^{o}(t) + q^u(t) D_{\lambda_1}^{u}(t) \quad (7)$$

As in Eq. (1), the pixel-dependent parameters $q^o(t)$ and $q^u(t)$ account for the modeling errors, even if in this case, they do not have any precise physical meaning. Note that no blur kernels are included in the above data model. In fact, the filters used for imaging the palimpsested manuscript at the two wavelengths might have different transfer functions, which means that the maps D^o and D^u do not match perfectly in the two equations above. However, since, differently from the see-through case, here the blur kernels affect the whole observations and are known, they can be removed off-line. Thus, without losing generality, we can assume D^o and D^u to be at the same resolution in both equations.

To invert system (7) and estimate the ideal overwriting and underwriting texts, the same scheme of (4) can be used, once the two maps q^o and q^u are known. In particular, we expect to obtain the ideal underwriting image as depicted at wavelength λ_1 from the first of Eq. (7), i.e., from the first data image, and the ideal overwriting image as it appears at wavelength λ_2 from the second of Eq. (7), i.e., from the second data image.

To estimate the two maps q^o and q^u, we first observe that, given a pixel t, one of four situations occurs in both the observations: 1) t is a background pixel, 2) t is an underwriting pixel, 3) t is an overwriting pixel, and 4) t is an occlusion pixel. In general, to determine a priori which of the four situations a pixel corresponds to is more difficult than in the see-through case. However, the two wavelengths λ_1 and λ_2 can be chosen to drive the estimation of the correct $q^o(t)$ and $q^u(t)$

In palimpsests, the two different writings normally feature different spectral signatures, thus tending to fade over different wavelength ranges. Often, there exists a wavelength where the underwriting is more visible than at the others, although still mixed to the overwriting. We thus choose this wavelength as λ_1, to obtain the first data image. Furthermore, there exists another wavelength where the underwriting almost disappears. We choose this second wavelength as λ_2; it is apparent that from this second data image, the overwriting can easily be located. Neglecting for the moment the fact that the overwriting includes also the occlusions, we can assume that, at the locations of the overwriting pixels, the ideal first image should have null density, so that q^o can be computed at those locations as the ratio between $D_{\lambda 1}^{obs}$, and $D_{\lambda 2}^{obs}$, and set to zero elsewhere. Now we know that the underwriting, except the occlusions, is located somewhere outside the overwriting, in both images. Consequently, since the density of the ideal second image should be zero in those pixels, we set q^u as the ratio between $D_{\lambda 2}^{obs}$ and $D_{\lambda 1}^{obs}$ where q^o is zero, and zero elsewhere, i.e., in correspondence of the overwriting. In formulas, it is:

$$q^o(t) = \begin{cases} \dfrac{D_{\lambda 1}^{obs}(t)}{D_{\lambda 2}^{obs}(t)+\epsilon} & \text{if } t \text{ is overwriting} \\ 0 & \text{elsewhere} \end{cases} \quad (8)$$

$$q^u(t) = \begin{cases} \dfrac{D_{\lambda 2}^{obs}(t)}{D_{\lambda 1}^{obs}(t)+\epsilon} & \text{if } q^o(t) = 0 \\ 0 & \text{elsewhere} \end{cases} \quad (9)$$

where ϵ is a small positive number to avoid indeterminacies or infinity.

Note that, in this way, q^u is forced to zero at the occlusion pixels, thus preserving the occlusion areas of the overwriting, but can be positive and high in the background pixels. By managing this latter situation as done for the see-through case, scheme (4) will return a map where the overwriting is extracted cleansed from the faint traces of the older text, also retaining the values of the data in the occlusions. Conversely, q^o is forced to zero in the background pixels, thus preserving the textures of the background outside the underwriting, but has wrong positive and possibly high values at the occlusion pixels. Indeed, in those areas, the density observed at wavelength λ_1 is normally higher than that observed at wavelength λ_2. Thus, scheme (4) will return a map where the underwriting is extracted cleansed from the even strong strokes of the overwriting, but with "holes" in the occlusion areas. To solve this problem, we assume herein that, at wavelength λ_1, the occlusion pixels are significantly darker than the other pixels, which leads to the highest values of q^o. Thus, in those pixels where q^u is zero and q^o exceeds the average of its positive values, we set q^o to its inverse value. In this way, the occlusion pixels of the cleansed underwriting map are attenuated without reaching the background value. More efficiently, the occlusions so located can be filled in both maps with the average value of the recovered texts outside them, or, equivalently, these average values can be used to refine q^o and q^u through inversion of the complete data equation.

DISCUSSION OF THE EXPERIMENTAL RESULTS

We experimented the method proposed above for the separation of overlapped texts in recto–verso documents and for the extraction of the underwriting from multispectral images of palimpsests. The algorithms derived are very fast since the data model can be inverted in a single step.

As per the see-through problem, we tested the method on several grayscale real recto–verso pairs, affected by either show-through, bleed-through, or both. In the following, we analyze in detail the results obtained in some of such experiments, by also providing comparisons with the results obtained by a stationary linear model, the stationary nonlinear model in [15], and other state-of-the-art methods.

Figure 1a, e shows the original degraded recto and verso images. It is not easy to decide whether the degradation is caused by show-through, bleed-through, or both. As apparent, the interfering patterns are highly non-stationary. The results of the application of ICA, improved by histogram clipping, are shown in Fig. 1b, f, whereas Fig. 1c, g shows the results obtained with the stationary nonlinear model of [15]. The results of the method proposed in this paper are shown in Fig. 1d, h. The stationary linear model produces lower ink density in the occlusion areas and leaves some interferences as well. On the other hand, the results of the stationary nonlinear model highlight that a single value of the interference level is not sufficient to remove all the interference, although "holes" are not present anymore in the occlusions, for this image. The superior performance of the method proposed here is apparent. In other words, the adoption of a stationary data model, even nonlinear and convolutional, necessarily entails a compromise between see-through removal and preservation of the occlusions. Transforming the model, either linear or nonlinear, from stationary to non-stationary allows us to meet both requirements.

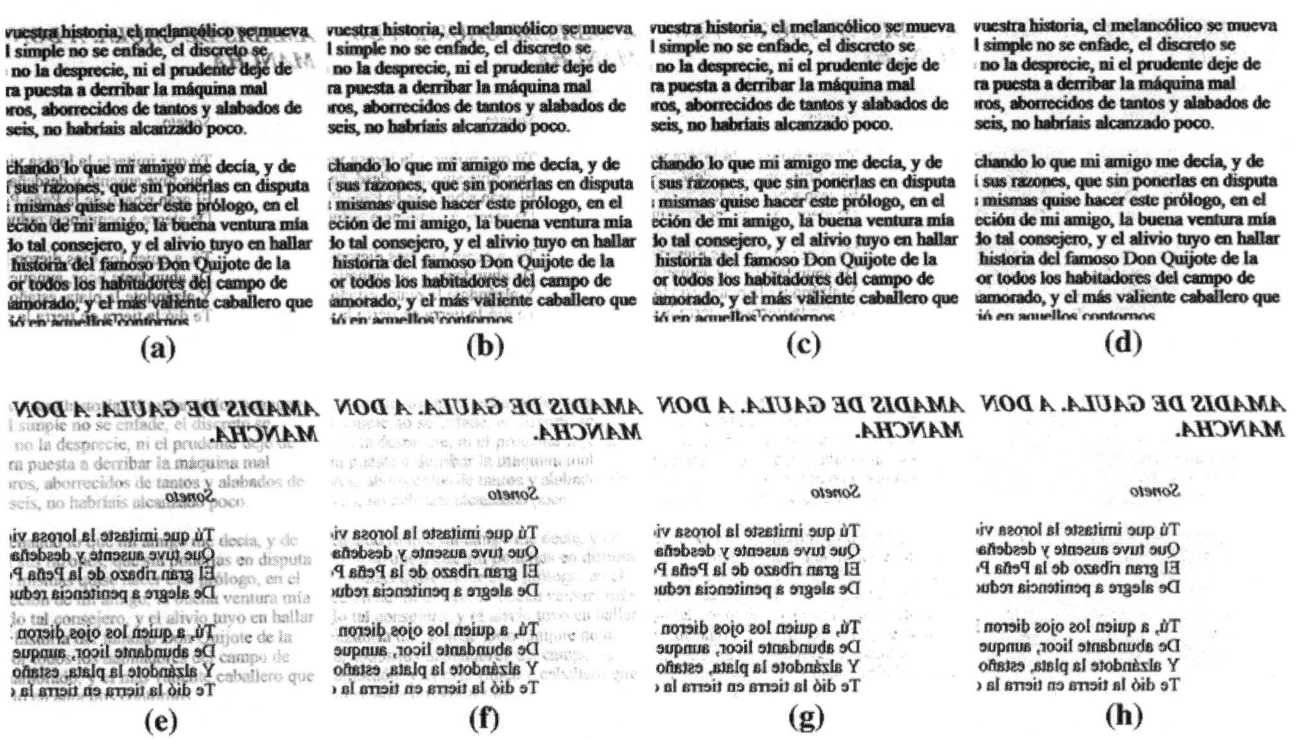

Figure. 1: Restoration of a recto–verso pair of a real document: **a** degraded recto, **b** recto restored with ICA, **c** recto restored with the method in [15] (stationary nonlinear model), **d** recto restored with the proposed method (non-stationary linear model), **e** degraded verso, **f**verso restored with ICA, **g** verso restored with the method in [15] (stationary nonlinear model), **h** verso restored with the proposed method (non-stationary linear model)

Figure 2 shows the maps of the estimated interference levels q_r and q_v and of the estimated occlusion pixels. Darker pixels in the interference level maps indicate higher values of q_r and q_v. In the third map, black pixels represent occlusion areas.

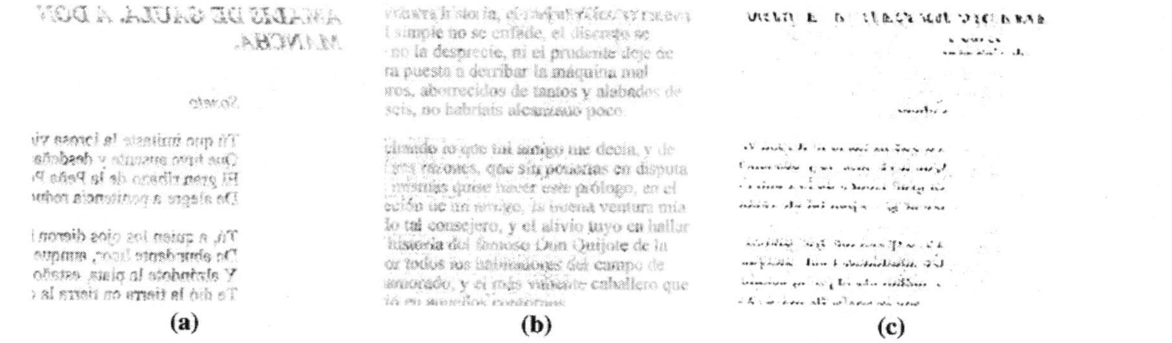

Figure. 2: Maps of the estimated interference levels qvqv (**a**) and qrqr (**b**) and of the estimated occlusion pixels (**c**), for the images in Fig. 1a, e. *Darker pixels* in the interference level maps indicate higher values; *black pixels* in the third map represent occlusion areas

Recently, a database of high-resolution grayscale images of ancient documents affected by bleed-through has been published online [23]. This database comprises 2525 registered recto–verso sample grayscale image pairs, taken from larger high-resolution manuscript images, with varied degrees of bleed-through. In addition, for each image, a binary ground truth mask of the foreground text is provided. Although these ground truth images are synthetic, i.e., created manually, they can be useful for a quantitative analysis of the results. This database is diffusely described in [24].

In Fig. 3, we present the results obtained on one of such images, a manuscript belonging to the Allan and Maria Myers Academic Centre, University of Melbourne (Fig. 3a, d). The results of using the stationary nonlinear model of [15] are shown in Fig. 3b, e, whereas Fig. 3c, f shows the much better results obtained by using the non-stationary, linear model proposed in this paper. From a visual point of view, the higher quality of the reconstructions in this case can be appreciated especially in the recto side. Indeed, in Fig. 3b, the lower ink density in the occlusion areas is well visible, whereas it does not affect the homologous image of Fig. 3c.

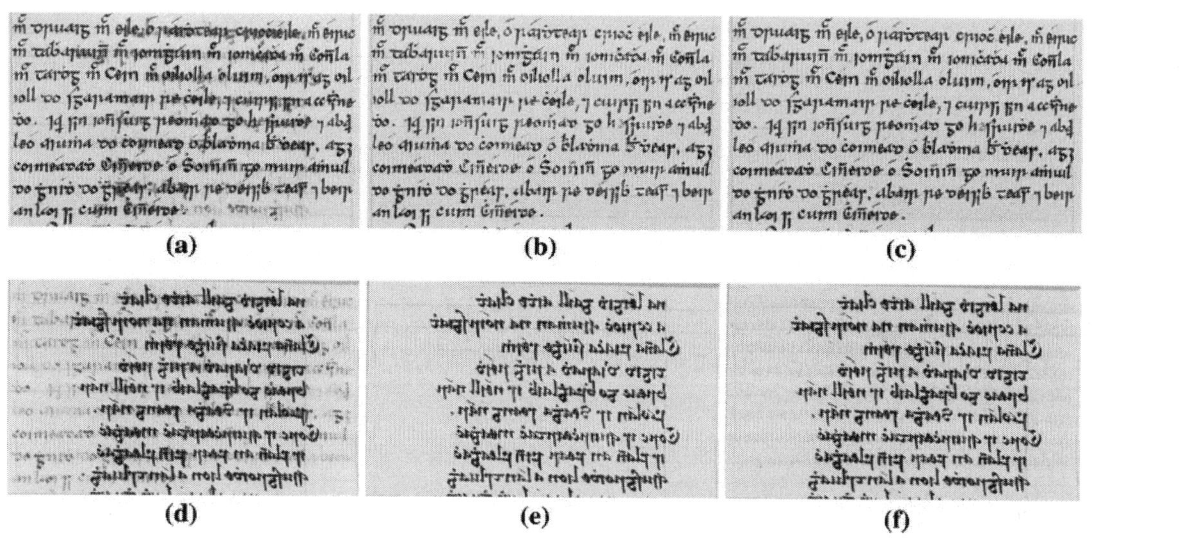

Figure. 3: Application of the proposed method to a real recto–verso pair: **a** original degraded recto,**b** recto restored with the method in [15] (stationary nonlinear model), **c** recto restored with the proposed method (non-stationary linear model), **d** original degraded verso, **e**verso restored with the method in [15] (stationary nonlinear model), **f** verso restored with the proposed method (non-stationary linear model). Original images (**a**) and (**d**): reproduction by courtesy of The Allan and Maria Myers Academic Centre, University of Melbourne, digitized by Irish Script On Screen (www.isos.dias.ie)

In Fig. 4, the maps of the estimated interference levels q_r and q_v and of the estimated occlusion pixels are shown.

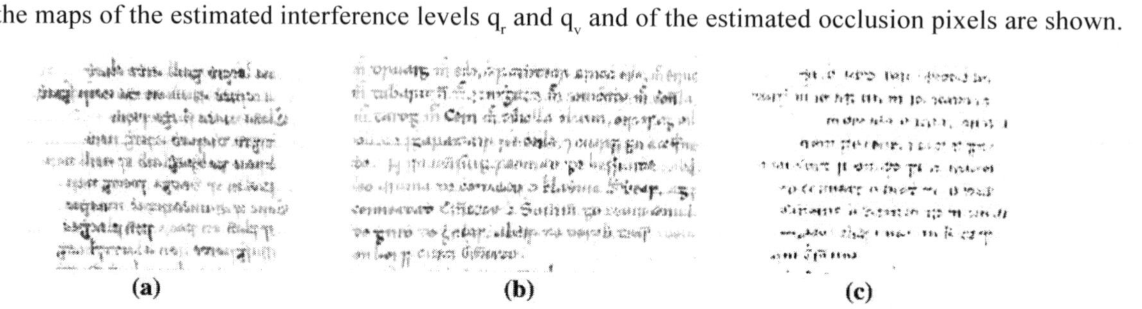

Figure. 4: Maps of the estimated interference levels qvqv (**a**) and qrqr (**b**) and of the estimated occlusion pixels (**c**), for the images in Fig. 3a, d. *Darker pixels* in the interference level maps indicate higher values; *black pixels* in the third map represent occlusion areas

Finally, Fig. 5 shows, for a detail of the images of Fig. 3a, d, the comparisons with recent state-of-the-art methods. The recto sides are depicted in the left and the verso sides in the right, respectively. From top to bottom, we see the original degraded images, the results from the methods in [5, 10, 18], and [6], respectively, and finally, the results from the method proposed herein. It is apparent that the results obtained with our method are fully qualitatively comparable to the best ones, produced by the method in [6].

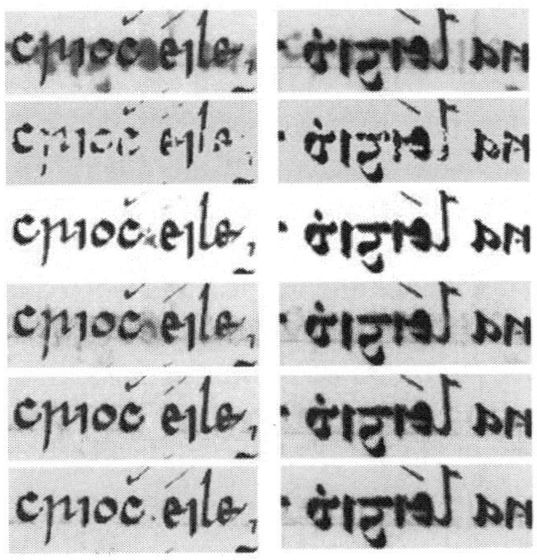

Figure. 5: Comparison among state-of-the-art methods on a detail of the images in Fig. 3a, d. From*top to bottom*: originals, results from the method in [5], results from the method in [18], results from the method in [10], results from the method in [6], and results from our method

In another experiment, we processed a second pair from the same dataset [23]. The manuscript belongs to The James Hardiman Library, National University of Ireland, Galway. The original recto–verso pair and the corresponding reconstructions obtained with the method proposed in [6], which was the best performing for the previous experiment, and with our method are shown in Fig. 6. Again, our results are of very similar quality.

For a quantitative analysis, having available binary ground truth images, we binarized the reconstructions with the adaptive Sauvola algorithm [25] and then computed as quality indices the probability FgErrorFgError that a pixel in the foreground text was classified as background, the probability BgErrorBgError that a background or bleed-through pixel was classified as foreground, and the WTotErrorWTotError, that is, the weighted mean of FgErrorFgError and BgErrorBgError, with the weights being the numbers of the foreground pixels and the background pixels as they result from the corresponding ground truth images, indicating the probability that any pixel in the image was misclassified. According to [24], these quality indices are defined as:

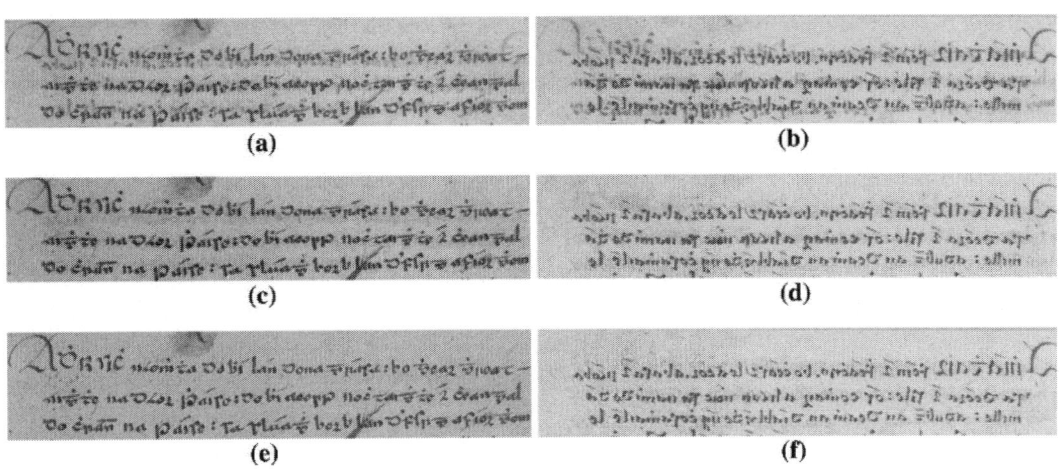

Figure. 6: Application of the proposed method to a real recto–verso pair from dataset [23, 24]: **a**original degraded recto, **b** original degraded verso, **c** recto restored with the method in [6], **d** verso restored with the method in [6], **e** recto restored with the proposed method, and **f** verso restored with the proposed method

$$FgError = \frac{1}{N} \sum_{t \in GT(Fg)} |GT(t) - B(t)|$$

$$BgError = \frac{1}{N} \sum_{t \in GT(Bg)} |GT(t) - B(t)|$$

$$WTotError = \frac{N_{Fg} FgError + N_{Bg} BgError}{N} \quad (10)$$

where GT is the ground truth, B is the binarized restoration result, GT(Fg) is the foreground region of the ground truth image constituted of N_{Fg} pixels, $GT(Bg)$ is the complementary background region of the ground truth image constituted of N_{Bg} pixels, and N is the total number of pixels in the image. The values of these indices for the two experiments above are summarized in Table 1.

20. Tonazzini, A., Salerno, E., Savino, P., Bedini, L.: Removal of non-stationary see-through interferences from recto-verso documents. In: Proceedings of International Workshop on Intelligent Pattern Recognition and Applications WIPRA 2013, pp. 151–158 (2013)

21. 20.

22. Gerace, I., Martinelli, F., Tonazzini, A.: Restoration of recto-verso archival documents through a regularized nonlinear model. In: Proceedings of Eusipco 2012, Bucharest, August 27–31 pp. 1588–1592 (2012)

23. Luenberger, D.G.: Linear and Nonlinear Programming. Addison-Wesley, Reading (1984)MATH

24. Otsu, N.: A threshold selection method from gray-level histograms. IEEE Trans. Syst. Man. Cybern. 9(1), 62–66 (1979)MathSciNetCrossRef

25. Irish Script On Screen Project: http://www.isos.dias.ie (2012)

26. Rowley-Brooke, R., Piti, F., Kokaram, A.: A ground truth bleed-through document image database. In: Adn, G., Buchanan, P.Z., Rasmussen, E., Loizides, F. (eds.) Theory and Practice of Digital Libraries. Volume 7489 of Lecture Notes in Computer Science, pp. 185–196. Springer, Berlin (2012)

27. Sauvola, J., Pietikäinen, M.: Adaptive document image binarization. Pattern Recognit. 33, 225236 (2000) CrossRef

28. Gatos, B., Pratikakis, I., Perantonis, S.J.: Adaptive degraded document image binarization. Pattern Recognit. 39(3), 317–327 (2006)

Finally, Fig. 5 shows, for a detail of the images of Fig. 3a, d, the comparisons with recent state-of-the-art methods. The recto sides are depicted in the left and the verso sides in the right, respectively. From top to bottom, we see the original degraded images, the results from the methods in [5, 10, 18], and [6], respectively, and finally, the results from the method proposed herein. It is apparent that the results obtained with our method are fully qualitatively comparable to the best ones, produced by the method in [6].

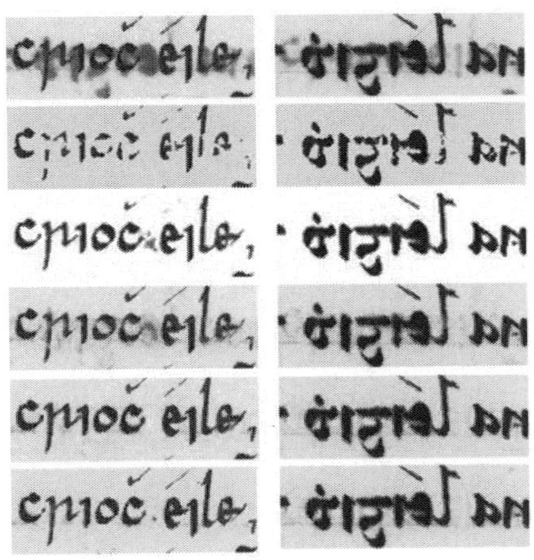

Figure. 5: Comparison among state-of-the-art methods on a detail of the images in Fig. 3a, d. From*top to bottom*: originals, results from the method in [5], results from the method in [18], results from the method in [10], results from the method in [6], and results from our method

In another experiment, we processed a second pair from the same dataset [23]. The manuscript belongs to The James Hardiman Library, National University of Ireland, Galway. The original recto–verso pair and the corresponding reconstructions obtained with the method proposed in [6], which was the best performing for the previous experiment, and with our method are shown in Fig. 6. Again, our results are of very similar quality.

For a quantitative analysis, having available binary ground truth images, we binarized the reconstructions with the adaptive Sauvola algorithm [25] and then computed as quality indices the probability FgErrorFgError that a pixel in the foreground text was classified as background, the probability BgErrorBgError that a background or bleed-through pixel was classified as foreground, and the WTotErrorWTotError, that is, the weighted mean of FgErrorFgError and BgErrorBgError, with the weights being the numbers of the foreground pixels and the background pixels as they result from the corresponding ground truth images, indicating the probability that any pixel in the image was misclassified. According to [24], these quality indices are defined as:

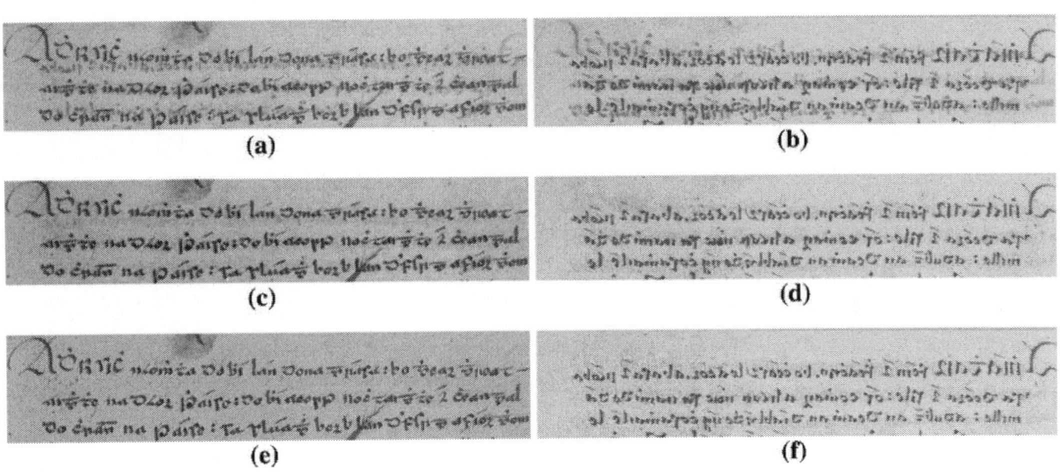

Figure. 6: Application of the proposed method to a real recto–verso pair from dataset [23, 24]: **a**original degraded recto, **b** original degraded verso, **c** recto restored with the method in [6], **d** verso restored with the method in [6], **e** recto restored with the proposed method, and **f** verso restored with the proposed method

$$FgError = \frac{1}{N} \sum_{t \in GT(Fg)} |GT(t) - B(t)|$$

$$BgError = \frac{1}{N} \sum_{t \in GT(Bg)} |GT(t) - B(t)|$$

$$WTotError = \frac{N_{Fg} FgError + N_{Bg} BgError}{N} \qquad (10)$$

where GT is the ground truth, B is the binarized restoration result, GT(Fg) is the foreground region of the ground truth image constituted of N_{Fg} pixels, $GT(Bg)$ is the complementary background region of the ground truth image constituted of N_{Bg} pixels, and N is the total number of pixels in the image. The values of these indices for the two experiments above are summarized in Table 1.

Table 1: Quality indices for the results in Figs. 3c, f, and 6e, f

Image	FgErrorFgError	BgErrorBgError	WTotErrorWTotError
Figure 3c	0.0097	0.0070	0.0074
Figure 3f	0.0071	0.0106	0.0100
Figure 6e	0.0151	0.0110	0.0035
Figure 6f	0.0143	0.0104	0.0108

The authors of [6] quantified the performance of their method and those of the methods in [5,18] and [10] by binarizing the reconstructions with Gatos binarization [26] and then computing the mean quality indices of Eq. (10) for the entire dataset. Their method was the best, with $FgError = 0.0696$, $BgError = 0.0085$ and $WTotError = 0.0196$.. Though no comparison can be done on each single image, we observe that the quality indices for the two images shown above are below those mean values. We also computed the quality indices on the whole dataset. The measured errors are summarized in the following Table 2, where the mean errors are completed with the standard deviations and the best and worst quality indices. The mean errors reported in [6] are shown in the last column.

Table 2: Quality indices for the entire dataset

	Mean	SD	Best	Worst	[6]
FgErrorFgError	0.0176	0.011255	0.0049	0.0548	0.0696
BgErrorBgError	0.0285	0.012949	0.0033	0.0669	0.0085
WTotErrorWTotError	0.0165	0.012115	0.0023	0.0542	0.0196

From Table 2, it appears that, on average, our results have lower FgError and WeightedTotError, and higher BgError, than those of [6]. From a check on some of our worst results, we observed that the wrong background pixels are usually confined at the character boundaries. This means that the binarization algorithm underestimates the threshold between foreground and background. In general, we expect different values of the quality indices on the same image when using different binarization algorithms. Thus, for a fair quantitative comparison using this peculiar kind of ground truth images, the same binarization algorithm, with the same parameters, should be used. Finally, it is worth highlighting again the simplicity of our method, which leads to a very fast algorithm. Indeed, a non-optimized Matlab code takes 0.77 s. for restoring the 1,745×1,070 images of Fig. 3a, d, on an Intel core i7 3GHz CPU.

In the case of palimpsests, we show the potentialities of our model by presenting the results obtained on a real–fake palimpsest, i.e., a document created by printing a first vertical text in a light color and then overprinting it with a horizontal darker text. Scans of the documents have then been used for processing. With this choice of the colors, in the red channel, the underwriting (the vertical text) almost disappears, whereas it is slightly visible in the green channel and well visible in the blue channel. We chose as observation at wavelength λ_1 the blue channel, and as observation at wavelength λ_2 the green channel. Note that if we had chosen as second observation the red channel, we could have set $q^u(t) = 0, \forall t$, thus making the data system to reduce to a single equation. Note also that none of the standard thresholding techniques applied to the blue channel alone could permit separation of the two texts. Figure 7a, d shows the green and blue channels of the RGB original scan. Figure 7b, e shows the results of applying ICA to the density data images, with the aim at showing that global coefficients for the linear model do not allow the two texts to be separated. The promising results of our method are shown in Fig. 7c, f.

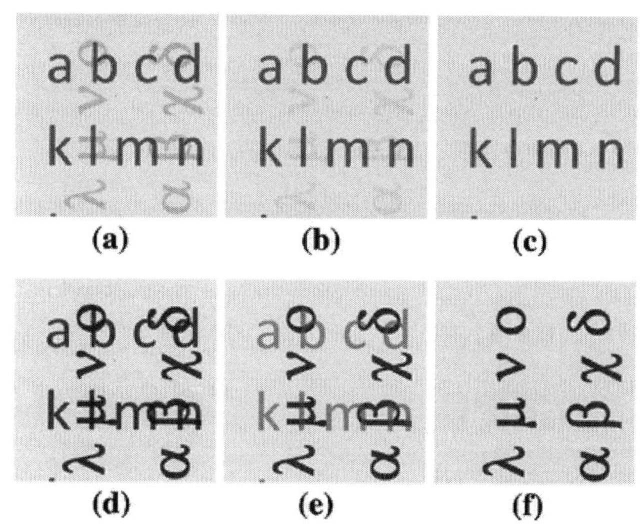

Figure. 7: Application of the proposed method to an RGB real-fake palimpsest: **a** green channel, **d**blue channel, **b** and **e** results of ICA applied to the density maps, **c** overwriting extracted with our method, and **f** underwriting extracted with our method

CONCLUSIONS

We propose a non-stationary, linear mixing model of the optical densities to describe text overlapping in recto–verso images of archival documents, affected by see-through, and in multispectral images of palimpsests. Based on this model, we derive two simple, partially supervised algorithms to separate the two texts in the two different applications. These algorithms act in two phases: In the first one, the model parameters are estimated off-line from the data; in the second, the restored images are recovered by inverting the data model in a single step. Both the algorithms we propose are very fast, apart from the trivial need of selecting one or two small areas in one of the two observations. The experimental results on real recto–verso printed documents and manuscripts show that the parameters involved in the data model can be estimated with satisfactory accuracy, so as to obtain clean images of the individual texts. The method outperforms standard blind source separation techniques such as ICA and produces results comparable to those of the best performing among recent state-of-the-art methods. Also, the preliminary results obtained for the application to the recovery of the erased text in palimpsests are promising.

The two fundamental features of the data model adopted are non-stationarity, which permits to describe both penetration and transparency of the ink through the medium, as well as the variability of the degradation strength, and the inclusion of a PSF, which allows a pattern in a side to match the corresponding one in the opposite side.

Currently, we are working at the full automatization of the method and at its extension to the RGB case. This simply entails restoring the three pairs of recto–verso channels independently and then recomposing them to recover the RGB appearance of the cleansed document. Other issues regard relaxing the present assumption of lower density of the interferences with respect to that of the foreground text and finding better strategies for a more accurate selection of the model parameters in correspondence with the occlusion areas. These could range from the iterative refining of an initial, rough estimate, to the use of the average density of the foreground text computed in the same area used to estimate the PSF, and so on. Another possibility could be to include priors on the ideal images and estimate jointly all the parameters. Even if we already pursued this way with some success in [20], more specialized priors should be sought, and the computational costs are still high.

ACKNOWLEDGMENTS

This work has been supported by European funds, through the program POR Calabria FESR 2007-2013-PIA Regione Calabria Pacchetti Integrati di Agevolazione Industria Artigianato Servizi, project ITACA (Innovative Tools for cultural heritage ArChiving and restorAtion).

REFERENCES

1. Dubois, E., Pathak, A.: Reduction of bleed-through in scanned manuscript documents. In: Proceedings of IS&T Image Processing, Image Quality, Image Capture Systems Conference, pp. 177–180 (2001)
2. Tan, C.L., Cao, R., Shen, P.: Restoration of archival documents using a wavelet technique. IEEE Trans. PAMI 24(10), 1399–1404 (2002)CrossRef
3. Wang, Q., Tan, C.L.: Matching of double-sided document images to remove interference. In: Proceedings of IEEE CVPR 2001, p. 1084 (2001)
4. Hanasusanto, G.A., Wu, Z., Brown, M.S.: Ink-bleed reduction using functional minimization. In: Proceedings of the IEEE Conference on Computer Vision and Pattern Recognition, pp. 825–832 (2010)
5. Huang, Y., Brown, M.S., Xu, D.: User assisted ink-bleed reduction. IEEE Trans. Image Process. 19(10), 2646–2658 (2010)MathSciNetCrossRef
6. Rowley-Brooke, R., Piti, F., Kokaram, A.: A non-parametric framework for document bleed-through removal. In: Proceedings of the IEEE Conference on Computer Vision and Pattern Recognition, pp. 2954–2960 (2013)
7. Tonazzini, A., Bedini, L., Salerno, E.: Independent component analysis for document restoration. Int. J. Doc. Anal. Recognit. 7, 17–27 (2004)CrossRef
8. Tonazzini, A., Salerno, E., Bedini, L.: Fast correction of bleed-through distortion in grayscale documents by a blind source separation technique. Int. J. Doc. Anal. Recognit.10, 17–25 (2007)CrossRef
9. Tonazzini, A., Gerace, I., Martinelli, F.: Multichannel blind separation and deconvolution of images for document analysis. IEEE Trans. Image Process. 19(4), 912–925 (2010)MathSciNetCrossRef
10. Rowley-Brooke, R., Kokaram, A.: Bleed-through removal in degraded documents. In: Proceedings of SPIE 8297 Document Recognition and Retrieval XIX, 82970T–10 (2012)
11. 11.
12. Merrikh-Bayat, F., Babaie-Zadeh, M., Jutten, C.: Using non-negative matrix factorization for removing show-through. In: Proceedings of LVA/ICA, pp. 482–489 (2010)
13. Sharma, G.: Show-through cancellation in scans of duplex printed documents. IEEE Tans. Image Process. 10(5), 736–754 (2001)CrossRef
14. Ophir, B., Malah, D.: Show-through cancellation in scanned images using blind source separation techniques. In: Proceedings of International Conference on Image Processing ICIP. Volume III. pp. 233–236 (2007)
15. Martinelli, F., Salerno, E., Gerace, I., Tonazzini, A.: Non-linear model and constrained ml for removing back-to-front interferences from recto-verso documents. Pattern Recognit.45, 596–605 (2012)CrossRef
16. Salerno, E., Martinelli, F., Tonazzini, A.: Nonlinear model identification and seethrough cancellation from recto-verso data. Int. J. Doc. Anal. Recognit. 16, 177–187 (2013)CrossRef
17. Merrikh-Bayat, F., Babaie-Zadeh, M., Jutten, C.: Linear-quadratic blind source separating structure for removing show-through in scanned documents. IJDAR 14, 319–333 (2011)CrossRef
18. Almeida, M.S.C., Almeida, L.B.: Nonlinear separation of show-through image mixtures using a physical model trained with ica. Signal Process. 92, 872884 (2012) CrossRef
19. Moghaddam, R.F., Cheriet, M.: A variational approach to degraded document enhancement. IEEE Trans. Pattern Anal. Mach. Intell. 32(8), 1347–1361 (2010)CrossRef

20. Tonazzini, A., Salerno, E., Savino, P., Bedini, L.: Removal of non-stationary see-through interferences from recto-verso documents. In: Proceedings of International Workshop on Intelligent Pattern Recognition and Applications WIPRA 2013, pp. 151–158 (2013)
21. 20.
22. Gerace, I., Martinelli, F., Tonazzini, A.: Restoration of recto-verso archival documents through a regularized nonlinear model. In: Proceedings of Eusipco 2012, Bucharest, August 27–31 pp. 1588–1592 (2012)
23. Luenberger, D.G.: Linear and Nonlinear Programming. Addison-Wesley, Reading (1984)MATH
24. Otsu, N.: A threshold selection method from gray-level histograms. IEEE Trans. Syst. Man. Cybern. 9(1), 62–66 (1979)MathSciNetCrossRef
25. Irish Script On Screen Project: http://www.isos.dias.ie (2012)
26. Rowley-Brooke, R., Piti, F., Kokaram, A.: A ground truth bleed-through document image database. In: Adn, G., Buchanan, P.Z., Rasmussen, E., Loizides, F. (eds.) Theory and Practice of Digital Libraries. Volume 7489 of Lecture Notes in Computer Science, pp. 185–196. Springer, Berlin (2012)
27. Sauvola, J., Pietikäinen, M.: Adaptive document image binarization. Pattern Recognit. 33, 225236 (2000) CrossRef
28. Gatos, B., Pratikakis, I., Perantonis, S.J.: Adaptive degraded document image binarization. Pattern Recognit. 39(3), 317–327 (2006)

INDEX

N

O

P

R

S

T

V

W